Equations of
Mathematical Physics

PURE AND APPLIED MATHEMATICS
A Series of Monographs

COORDINATOR OF THE EDITORIAL BOARD
S. Kobayashi
UNIVERSITY OF CALIFORNIA AT BERKELEY

1. KENTARO YANO. Integral Formulas in Riemannian Geometry (1970)
2. S. KOBAYASHI. Hyperbolic Manifolds and Holomorphic Mappings (1970)
3. V. S. VLADIMIROV. Equations of Mathematical Physics (A. Jeffrey, editor; A. Littlewood, translator) (1970)

In Preparation:

L. NARICI, E. BECKENSTEIN, and G. BACHMAN. Functional Analysis and Valuation Theory

W. BOOTHBY and G. L. WEISS (eds.). Geometry and Harmonic Analysis of Symmetric Spaces

Equations of Mathematical Physics

Contents

Preface iii

CHAPTER 1. FORMULATION OF BOUNDARY VALUE PROBLEMS IN MATHEMATICAL PHYSICS 1

1. Some Concepts and Propositions Concerning the Theory of Sets, Theory of Functions, and the Theory of Operators 1
2. Basic Equations of Mathematical Physics 27
3. Classification of Linear (Second-Order) Differential Equations 38
4. Formulation of Boundary Value Problems for Linear Second-Order Differential Equations 52

CHAPTER 2. GENERALIZED FUNCTIONS 63

5. Test and Generalized Functions 63
6. Differentiation of Generalized Functions 79
7. The Direct Product and Convolution of Generalized Functions 95
8. Generalized Functions of Slow Growth (Tempered Distributions) 113
9. Fourier Transform of Generalized Functions of Slow Growth 121

CHAPTER 3. FUNDAMENTAL SOLUTIONS AND THE CAUCHY PROBLEM 139

10. Fundamental Solutions of Linear Differential Operators 139
11. Retarded Potential 157
12. The Cauchy Problem for the Wave Equation 171
13. Wave Propagation 178
14. The Cauchy Problem for the Equation of Heat Conduction 193

Chapter 4. Integral Equations 201

15. The Method of Successive Approximations 202
16. Fredholm's Theorems 217
17. Integral Equations with an Hermitian Kernel 231
18. The Hilbert-Schmidt Theorem and its Corollaries 237

Chapter 5. Boundary Value Problems For Elliptic Equations 255

19. The Eigenvalue Problem 255
20. The Sturm-Liouville Problem 270
21. Harmonic Functions 278
22. Newtonian Potential 291
23. Boundary Value Problems for Laplace and Poisson Equations in Space 307
24. The Green's Function for the Dirichlet Problem 318
25. Spherical Functions 333
26. Boundary Value Problems for Laplace's Equation in a Plane 347
27. Helmholtz's Equation 362

Chapter 6. The Mixed Problem 373

28. Fourier's Method 373
29. The Mixed Problem for an Equation of Hyperbolic Type 388
30. The Mixed Problem for an Equation of Parabolic Type 404

Bibliography 411

Index 415

CHAPTER
1

Formulation of Boundary Value Problems in Mathematical Physics

§ 1. Some Concepts and Propositions Concerning the Theory of Sets, the Theory of Functions, and the Theory of Operators

Let A be an arbitrary set. If element a is contained (is not contained) in set A, we shall write it thus: $a \in A$ ($a \notin A$). Let B be another set. If A is contained in B we denote this by writing $A \subset B$: we denote by $A = B$ the coincidence of A with B, by $A \cup B$ the union of A and B, by $A \cap B$ the intersection of A and B, by $A \setminus B$ the complement of B relative to A (Fig. 1), by $A \times B$ the product of A and B [the set of pairs (a, b), $a \in A$, $b \in B$]; $\emptyset$ denotes an empty set.

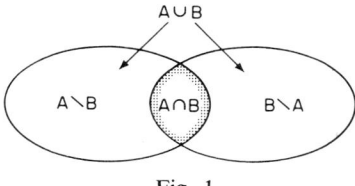

Fig. 1

1. Point Sets in R^n. We shall denote an n-dimensional Euclidean space by R^n, and its points by $x = (x_1, x_2, \ldots, x_n)$, y, ξ and so on, where x_i,

$i = 1, 2, \ldots, n$, are coordinates of the point x. We shall use the symbols (x, y) and $|x|$ to denote, respectively, the *scalar product and length* (*norm*) in R^n

$$(x, y) = x_1 y_1 + x_2 y_2 + \cdots + x_n y_n,$$
$$|x| = \sqrt{(x, x)} = \sqrt{x_1^2 + x_2^2 + \cdots + x_n^2}$$

In this way, the number $|x - y|$ is the Euclidean distance between points x and y.

The set of points x belonging to R^n satisfying the inequality $|x - x_0| < R$ is called an *open sphere* of radius R with its center at the point x_0. We shall denote this sphere by $U(x_0; R)$; $U_R = U(0; R)$ denotes the sphere of radius R centered on the origin.

The sequence of points $x_k = (x_{1k}, x_{2k}, \ldots, x_{nk})$, $k = 1, 2, \ldots$, is said to *converge* to the point x in R^n, written $x_k \to x$ as $k \to \infty$, if $|x_k - x| \to 0$ as $k \to \infty$. The sequence x_k, $k = 1, 2, \ldots$, is said to *converge in itself* in R^n if $|x_k - x_p| \to 0$ as $k \to \infty$, $p \to \infty$.

The following proposition expresses the property of completeness of the space R^n (*Cauchy's principle of convergence*): *In order that a sequence of points should converge in R^n, it is necessary and sufficient that it should converge in itself in R^n.*

A set is said to be bounded in R^n if there is a sphere containing this set.

The following proposition expresses the property of compactness of the space R^n (*Bolzano–Weierstrass theorem*): *It is possible to choose a converging sequence from every infinite bounded set in R^n.*

The point x_0 is called an *interior* point of a set if there is a sphere $U(x_0; R)$ contained in this set. A set is called *open* if all its points are interior. A set is called *connected* if any two of its points can be joined by an unbroken line lying in this set. A connected open set is called a *region*. The point x_0 is called a limit point of the set A if there is a sequence x_k, $k = 1, 2, \ldots$, such that $x_k \in A$, $x_k \to x_0$ as $k \to \infty$. If we add all its limit points to the set A, the set obtained is called the *closure* of the set A and is denoted by $\bar{A}$; it is clear that $A \subset \bar{A}$. If a set coincides with its closure; it is called *closed*. A closed bounded set is called a *compactum*. Each open set containing A is called a *neighborhood* of the set A; the union of the spheres $U(x; \varepsilon)$, when x ranges over A, written symbolically $A_\varepsilon = \bigcup_{x \in A} U(x; \varepsilon)$, is called an *$\varepsilon$-neighborhood* of the set A.

The function $\chi_A(x)$, which is equal to 1 when $x \in A$ and to 0 when $x \notin A$, is called the *characteristic function* of the set A.

§ 1. SOME CONCEPTS AND PROPOSITIONS

The following covering theorem is true (*Heine–Borel lemma*): *If the compactum K is covered by a system of open spheres, then a finite subsystem covering K can be chosen from this covering.*

Let G be a region. The points of closure $\bar{G}$ not belonging to G form a closed set S, called the *boundary* of the region G, so that $S = \bar{G} \setminus G$. For instance, the spherical surface $|x - x_0| = R$ is the boundary of the open sphere $U(x_0; R)$. We shall denote this spherical surface by $S(x_0; R)$; $S_R = S(0; R)$ denotes the spherical surface of radius R centered on the origin.

We shall say that a surface S belongs to the class C^p, $p \geq 1$, if in some neighborhood of each point $x_0 \in S$ it may be represented by the equation $\omega_{x_0}(x) = 0$, where grad $\omega_{x_0}(x) \neq 0$, and the function $\omega_{x_0}(x)$ is continuous together with all its derivatives up to order p inclusive in the neighborhood referred to. A surface S is called *piecewise*, or *sectionally smooth*, if it consists of a finite number of surfaces of class C^1.

Henceforth we shall examine only regions with piecewise smooth boundaries; we shall denote the unit vector along the external normal towards the boundary S at the point $x \in S$ by $\mathbf{n} = \mathbf{n}_x$. Let the point x_0 lie on the piecewise smooth surface S. The connected part of the set $S \cap U(x_0; R)$ which contains the point x_0 is called the *neighborhood* of the point x_0 on the surface S.

The bounded region G' is called a subregion, strictly lying in the region G, if $\bar{G}' \subset G$; in this connection it is written $G' \Subset G$. By virtue of the Heine–Borel lemma, there is a number $\varepsilon > 0$ such that $G'_\varepsilon \Subset G$.

2. Classes of Functions $C^p(G)$ and $C^p(\bar{G})$. Let $\alpha = (\alpha_1, \alpha_2, \ldots, \alpha_n)$ be a vector with integral nonnegative components α_j. We shall signify by $D^\alpha f(x)$ the derivative of the function $f(x)$ of order $|\alpha| = \alpha_1 + \alpha_2 + \cdots + \alpha_n$,

$$D^\alpha f(x) = \frac{\partial^{|\alpha|} f(x_1, x_2, \ldots, x_n)}{\partial x_1^{\alpha_1} \partial x_2^{\alpha_2} \cdots \partial x_n^{\alpha_n}}, \qquad D^0 f(x) = f(x)$$

We shall also use the following abbreviations:

$$x^\alpha = x_1^{\alpha_1} x_2^{\alpha_2} \cdots x_n^{\alpha_n}, \qquad \alpha! = \alpha_1! \alpha_2! \cdots \alpha_n!$$

A set of (complex) functions f, continuous together with the derivatives $D^\alpha f(x)$, $|\alpha| \leq p$ ($0 \leq p < \infty$), forms in the region G a *class of functions* $C^p(G)$. Functions f of class $C^p(G)$, for which all the derivatives $D^\alpha f(x)$, $|\alpha| \leq p$, allow continuous extension onto the closure $\bar{G}$, form a *class of* functions $C^p(\bar{G})$. We also write

$$C^\infty(G) = \bigcap_{p \geq 0} C^p(G), \qquad C^\infty(\bar{G}) = \bigcap_{p \geq 0} C^p(\bar{G}).$$

In this way the classes $C^0(G)$ and $C^0(\bar{G})$ are seen to be sets of continuous functions in G and on $\bar{G}$, respectively. To abbreviate the notation we shall write

$$C(G) = C^0(G), \qquad C(\bar{G}) = C^0(\bar{G})$$

Sometimes the argument G or $\bar{G}$ associated with the class C^p will be omitted.

The classes of functions introduced are *linear sets*; that is, if the functions f and g belong to any of these classes, then any linear combination $\lambda f + \mu g$, where λ and μ are arbitrary complex numbers, also belongs to this class.

Let $\varphi \in C(R^n)$. The closure of the set of those points for which $\varphi(x) \neq 0$ is called the *support* (*carrier*) of the continuous function φ; we denote this support by supp φ. If supp φ is a bounded set, the function φ is said to have *compact support*.

We shall use $\mathscr{D}(G)$ to denote the set of functions with bounded support belonging to the class $C^\infty(G)$ which has compact support in the region G.

3. Space of Continuous Functions $C(T)$. Let T be a closed set—for instance, the closure $\bar{G}$ or the boundary S of region G. We shall denote by $C(T)$ the class of continuous functions bounded on T. We shall provide the class $C(T)$ with a *norm* by setting

$$\|f\|_C = \max_{x \in T} |f(x)|, \qquad f \in C(T) \tag{1}$$

With this we convert the class $C(T)$ into a (linear) normed space.

Norm (1) has the following three properties which are characteristic of norms:

(a) $\|f\|_C > 0$; $\|f\|_C = 0$, if, and only if, $f = 0$;
(b) $\|\lambda f\|_C = |\lambda| \|f\|_C$, where λ is any complex number;
(c) $\|f + g\|_C \leq \|f\|_C + \|g\|_C$ (triangle inequality).

In general, each linear set provided with a norm which has properties (a)–(c) is called a *linear normed space*.

The sequence of functions f_k, $k = 1, 2, \ldots$, belonging to $C(T)$ is said to *converge* to the function $f \in C(T)$ in the space $C(T)$, written $f_k \to f$ as $k \to \infty$ in $C(T)$, if $\|f_k - f\|_C \to 0$ as $k \to \infty$. Obviously the convergence

§ 1. SOME CONCEPTS AND PROPOSITIONS

$f_k \to f$ as $k \to \infty$ in $C(T)$ is equivalent to the *uniform convergence* of the sequence of functions $f_k(x)$, $k = 1, 2, \ldots$, to the function $f(x)$ on the set T.

$$f_k(x) \xrightarrow{x \in T} f(x), \qquad k \to \infty$$

The sequence of functions f_k, $k = 1, 2, \ldots$, belonging to $C(T)$ is said to *converge in itself* in $C(T)$ if $\|f_k - f_p\|_C \to 0$ as $k \to \infty$, $p \to \infty$.

The following proposition expresses the property of completeness of the space $C(T)$ (*Cauchy's theorem*): *In order that a sequence of functions belonging to $C(T)$ should converge in $C(T)$, it is necessary and sufficient that it should converge in itself in $C(T)$.*

The following useful propositions are true.

WEIERSTRASS THEOREM. *If G is a bounded region, and $f \in C^p(\bar{G})$, then for any $\varepsilon > 0$ there is a polynomial P such that*

$$\| D^\alpha f - D^\alpha P \|_C < \varepsilon \qquad \text{for all} \quad |\alpha| \leq p$$

DINI'S LEMMA.* *If a monotonic sequence of continuous functions on a compactum K converges at each point to a continuous function on K, it converges uniformly on K.*

The series composed of the functions $u_k \in C(T)$ is said to be *regularly convergent* on T if the series of absolute values of $|u_k(x)|$ converges in $C(T)$, that is, converges uniformly on T.

A set $\mathcal{M} \subset C(T)$ is said to be *equicontinuous* on T if for any $\varepsilon > 0$ there is a number δ_ε such that for all $f \in \mathcal{M}$ the inequality $|f(x_1) - f(x_2)| < \varepsilon$ is satisfied whenever $|x_1 - x_2| < \delta$, $x_1, x_2 \in T$.

The function $f \in C(T)$ is said to be Hölder *continuous on T* if there are numbers $C > 0$, and α, $0 < \alpha \leq 1$, such that for all $x_1 \in T$ and $x_2 \in T$ the inequality

$$|f(x_1) - f(x_2)| \leq C |x_1 - x_2|^\alpha$$

is true; if $\alpha = 1$, the function $f(x)$ is said to be *Lipschitz continuous on T*.

4. The Lebesgue Integral. It is said that the set $A \subset R^n$ has *measure zero* if for any $\varepsilon > 0$ it can be covered with open spheres of total volume less than ε.

---

* Compare, for instance, V. I. Smirnov (2, Chap. 1).

It follows from this definition that every subset of a set of measure zero also has measure zero, and a union of a no greater than countable number of sets of measure zero also has measure zero. For example, each countable set and every piecewise smooth surface has measure zero.

It is said that a certain property is realized *almost everywhere in the region* $G \subset R^n$ if the set of points of the region G which do not possess this property have measure zero; in this case instead of "almost everywhere in R^n" we shall say simply "almost everywhere."

We shall consider that all functions are defined in the whole space R^n.

A function f is said to be *measurable* if it coincides almost everywhere with the limit of an almost everywhere converging sequence of piecewise continuous functions.

It follows from this definition: If the functions f and g are measurable, the functions $f + g$, fg, $\max(f, g)$, $\min(f, g)$, $|f|$, and f/g, if $g \neq 0$, are also measurable; each function which coincides almost everywhere with the limit of an almost everywhere converging sequence of measurable functions is measurable.

A set $A \subset R^n$ is said to be *measurable* if its characteristic function $\chi_A(x)$ (cf. Sec. 1.1) is measurable.

Nonmeasurable functions (and sets) are arranged quite irregularly, and none of them are constructed in an explicit form; their existence can only be proved theoretically, using the so-called axiom of choice. This asserts that all the functions and sets which we can possibly encounter will be measurable. Therefore we shall suppose in future, without arguing it out every time, that all sets considered are measurable and functions are measurable and almost everywhere finite.

Let $f(x)$ be a piecewise continuous bounded function with compact support. An element of volume in R^n will be denoted by $dx = dx_1 dx_2 \cdots dx_n$, so that the n-multiple Riemann integral of the function f in R^n will be abbreviated in the form

$$\int f(x)\, dx = \int \int_{R^n} \cdots \int f(x_1, x_2, \ldots, x_n)\, dx_1\, dx_2 \cdots dx_n$$

The nonnegative (measurable and almost everywhere finite) function $f(x)$ is said to be *Lebesgue integrable* (*summable*) if it coincides almost everywhere with the limit of a nondecreasing sequence of piecewise continuous functions with compact support $f_k(x)$, $k = 1, 2, \ldots$, possessing a bounded se-

§ 1. SOME CONCEPTS AND PROPOSITIONS

quence of integrals $\int f_k(x)\, dx$, $k = 1, 2, \ldots$. The limit of the nondecreasing bounded sequence of these integrals is called *the Lebesgue integral of the function f* and is denoted by the symbol $\int f(x)\, dx$, so that

$$\int f(x)\, dx = \lim_{k \to \infty} \int f_k(x)\, dx$$

First of all we shall prove that

$$\lim_{k \to \infty} \int f_k(x)\, dx \geq 0$$

For the proof we shall take an arbitrary $\varepsilon > 0$. Let $f_1(x) = 0$, $|x| > R$ and $f_1(x) \geq -M$, so that $f_k(x) > 0$, $|x| \geq R$, and $f_k(x) \geq -M$, $k = 1, 2, \ldots$. Let us construct the set A_0 which consists of the points of discontinuity of all the functions f_k and of the points at which the sequence $\{f_k\}$ does not tend to f. The set A_0 has measure zero, and hence can be covered with spheres, the sum of the volumes of which is less than ε. On the set $A_1 = R^n \setminus A_0$, $f_k(x) \to f(x) \geq 0$ as $k \to \infty$, and therefore for any point $x \in A$, a number $N = N_x$ can be found such that $f_N(x) > -\varepsilon$. However, the set A_1 consists of points of continuity of the functions f_k, and therefore the inequality $f_N(x') > -\varepsilon$ is true even in a sphere $U(x; r_x)$. In this way, the compactum $\bar{U}_R$ may be covered with a system of open spheres. According to the Heine–Borel lemma (cf. Sec. 1.1), from this covering we can extract a finite covering. Let N_0 be the largest of the corresponding numbers N_x. Then since the functions f_k do not decrease, by virtue of the choice of the number N_0,

$$f_{N_0}(x) \geq f_N(x) \geq -\varepsilon, \qquad x \in A_1 \cap \bar{U}_R$$

Therefore, for all $k > N_0$,

$$\int f_k(x)\, dx \geq \int f_{N_0}(x)\, dx \geq \int_{U_R} f_{N_0}(x)\, dx \geq -\varepsilon \left(M + \int_{U_R} dx \right)$$

from which the required inequality occurs.

The Lebesgue integral of the function $f \geq 0$ does not depend on the sequence $\{f_k\}$.

In fact, if $\{g_k\}$ is another such sequence, from the inequality

$$\lim_{k \to \infty} [f_k(x) - g_j(x)] = f(x) - g_j(x) \geq 0 \qquad \text{(almost everywhere)}$$

there follows the inequality

$$\lim_{k \to \infty} \int [f_k(x) - g_j(x)] \, dx$$
$$= \lim_{k \to \infty} \int f_k(x) \, dx - \int g_j(x) \, dx \geq 0, \qquad j = 1, 2, \ldots$$

and therefore

$$\lim_{j \to \infty} \int g_j(x) \, dx \leq \lim_{k \to \infty} \int f_k(x) \, dx.$$

Changing the roles of the sequences $\{f_k\}$ and $\{g_k\}$ reverses the inequality, and consequently the equation

$$\lim_{k \to \infty} \int g_k(x) \, dx = \lim_{k \to \infty} \int f_k(x) \, dx = \int f(x) \, dx$$

is true, as was to be shown.

In order that the function $f(x) \geq 0$ be equal to zero almost everywhere, it is necessary and sufficient that $\int f(x) \, dx = 0$.

In fact if $f(x) = 0$ almost everywhere, when $f_k = 0$ we obtain

$$\int f(x) \, dx = \lim_{k \to \infty} \int f_k(x) \, dx = 0$$

Conversely, if the function $f(x) \geq 0$ coincides almost everywhere with the limit of the nondecreasing sequence of piecewise continuous functions with compact support $f_k(x)$, $k = 1, 2, \ldots$,

$$f_k^+(x) = \max(f_k, 0) \to f(x), \qquad k \to \infty \qquad \text{(almost everywhere)}$$

and, consequently,

$$\lim_{k \to \infty} \int f_k^+(x) \, dx = \int f(x) \, dx = 0$$

From this, as $f_k^+(x) \geq 0$, we deduce that $f_k^+(x) \equiv 0$.
Therefore $f(x) = 0$ almost everywhere, as was to be shown.

Let $f(x)$ be an arbitrary real (measurable) function. We shall introduce the nonnegative functions

$$f^+(x) = \max[f(x), 0], \qquad f^-(x) = \max[-f(x), 0]$$

Evidently,

$$f(x) = f^+(x) - f^-(x), \qquad |f(x)| = f^+(x) + f^-(x)$$

§ 1. SOME CONCEPTS AND PROPOSITIONS

A real function $f(x)$ is said to be *Lebesgue integrable* (*summable*) if the functions $f^+(x)$ and $f^-(x)$ are Lebesgue integrable; the number

$$\int f^+(x)\, dx - \int f^-(x)\, dx = \int f(x)\, dx$$

is called the *Lebesgue integral of the function f*. A complex function $f(x)$ is said to be Lebesgue integrable if the functions $\operatorname{Re} f(x)$ and $\operatorname{Im} f(x)$ are Lebesgue integrable; the number

$$\int \operatorname{Re} f(x)\, dx + i \int \operatorname{Im} f(x)\, dx = \int f(x)\, dx$$

is called the *Lebesgue integral of the function f*.

We shall say that a function $f(x)$ is Lebesgue integrable on the measurable set A if the function $f(x)\chi_A(x)$ is Lebesgue integrable; the number

$$\int f(x)\chi_A(x)\, dx = \int_A f(x)\, dx$$

is called the *Lebesgue integral of the function f over the set A*.

A function $f(x)$ is said to be *locally Lebesgue integrable in the region G* if it is Lebesgue integrable over any subregion $G' \Subset G$.

In correspondence with the definition, each piecewise continuous function with compact support is Lebesgue integrable, and its Riemann and Lebesgue integrals coincide. On the other hand, there are functions which are Lebesgue integrable but which are not Riemann integrable, for instance the Dirichlet function:

$$f_0(x) = \begin{cases} 0, & x \text{ irrational} \\ 1, & x \text{ rational} \end{cases}$$

The Lebesgue integral of the function $f_0(x)$ over any finite interval (a, b) is equal to zero since it is possible to take $f_k(x) = 0$.

The following properties of the Lebesgue integral follow directly from these definitions.

(a) *The functions f and $|f|$ are simultaneously Lebesgue integrable, moreover*

$$\left| \int f(x)\, dx \right| \leq \int |f(x)|\, dx$$

(b) *The Lebesgue integral is linear with respect to f.* If functions f and g are Lebesgue integrable, and λ and μ are complex numbers, then the function $\lambda f + \mu g$ is also Lebesgue integrable, and the equation

$$\int [\lambda f(x) + \mu g(x)]\, dx = \lambda \int f(x)\, dx + \mu \int g(x)\, dx$$

is true.

(c) *Change of Variables in the Lebesgue Integral.* Let the transformation $x = x(y)$ be of the class $C^1(\bar{G})$; that is, $x_k = x_k(y_1, y_2, \ldots, y_n)$ for $k = 1, 2, \ldots, n$, $x_k \in C^1(\bar{G})$, represents a one-to-one mapping of the region G onto the region G_1, and let $D(x/y)$ be the Jacobian of this transformation.

In order that the function $f(x)$ should be Lebesgue integrable over the region G, it is necessary and sufficient that the function $f[x(y)]\,|\,D(x/y)\,|$ should be Lebesgue integrable over the region G_1. When this is so the equation

$$\int_G f(x)\, dx = \int_{G_1} f[x(y)] \left| D\!\left(\frac{x}{y}\right) \right| dy$$

is true.

This statement is true for piecewise continuous functions. For Lebesgue integrable functions, this result may be proved by employing piecewise continuous functions in conjunction with the definition of the Lebesgue integral.

We shall accept the following properties of the Lebesgue integral without proof.*

(d) *If the functions $f(x)$ and $|f(x)|$ are Riemann integrable (possibly in an improper sense), then they are also Lebesgue integrable and both integrals coincide.*

Taking this property of the Lebesgue integral into account, we shall henceforth refer to Lebesgue integrable functions simply as integrable functions.

(e) *Lebesgue's Theorem* (passage to the limit under the Lebesgue integral sign). *Let the sequence of (measurable) functions $f_k(x)$, $k = 1, 2, \ldots$, converge almost everywhere to the function $f(x)$. If there is an integrable function $g(x)$ such that for all $k = 1, 2, \ldots$,*

$$|f_k(x)| \leq g(x) \qquad \text{almost everywhere}$$

---

* Proofs of these statements can be found, for example, in the books of A. N. Kolmogorov and S. V. Fomin (*1*, Vol. II), F. Riesz and B. Sz. Nagy (*1*, Chap. 2), and G. E. Shilov (*1*).

§ 1. SOME CONCEPTS AND PROPOSITIONS

then the function $f(x)$ is also integrable and

$$\lim_{k \to \infty} \int f_k(x)\, dx = \int f(x)\, dx$$

Specifically, if the function $g(x)$ is integrable, and $|f(x)| \leq g(x)$ almost everywhere, the function $f(x)$ is also integrable and the inequality

$$\int |f(x)|\, dx \leq \int g(x)\, dx$$

is true.

Hence it follows that each bounded measurable function is integrable over any bounded (measurable) set A. Specifically, the integral

$$\int_A dx = \int \chi_A(x)\, dx$$

exists; it is called the *Lebesgue measure of the set A*. Evidently the measure of a bounded region with a piecewise smooth boundary coincides with its volume.

(f) *Fubini's Theorem* (change of the order of integration). *If a function $f(x, y)$ defined in R^{n+m}, $x \in R^n$, $y \in R^m$, is measurable, and there exists a repeated integral of the function $|f(x, y)|$*

$$\int \left[\int |f(x, y)|\, dx \right] dy < \infty$$

then $f(x, y)$ is integrable, and the integrals

$$\int f(x, y)\, dx, \quad \int f(x, y)\, dy$$

exist almost everywhere and are integrable, and the equations

$$\int f(x, y)\, dx\, dy = \int \left[\int f(x, y)\, dy \right] dx = \int \left[\int f(x, y)\, dx \right] dy$$

are true.

We note that if the function $f(x, y)$ is nonintegrable, the repeated integrals can either not exist or not be equal, e.g.,

$$\int_0^1 \left[\int_0^1 \frac{x^2 - y^2}{(x^2 + y^2)^2}\, dy \right] dx = \frac{\pi}{4}$$

$$\int_0^1 \left[\int_0^1 \frac{x^2 - y^2}{(x^2 + y^2)^2}\, dx \right] dy = -\frac{\pi}{4}$$

1. FORMULATION OF BOUNDARY VALUE PROBLEMS

4. Remark. A Lebesgue integral over a piecewise smooth surface S may be constructed analogously. In this case the corresponding Fubini theorem exists for the functions $f(x, y)$ defined on $R^n \times S$.

5. The Space of Functions $\mathscr{L}_2(G)$. $\mathscr{L}_2(G)$ will be used to denote the set of all functions f, for which the function $|f(x)|^2$ is integrable over the region G.

The set of functions $\mathscr{L}_2(G)$ is linear.

In fact, if $f \in \mathscr{L}_2(G)$ and $g \in \mathscr{L}_2(G)$, it follows from the inequality

$$|\lambda f + \mu g|^2 \leq 2|\lambda|^2|f|^2 + 2|\mu|^2|g|^2$$

that any linear combination $\lambda f + \mu g$ also belongs to $\mathscr{L}_2(G)$.

We shall establish an important inequality (*Cauchy–Buniakowski inequality*): If f and $g \in \mathscr{L}_2(G)$, then

$$\left| \int_G f(x)g(x)\,dx \right| \leq \left(\int_G |f(x)|^2\,dx \right)^{1/2} \left(\int_G |g(x)|^2\,dx \right)^{1/2} \quad (2)$$

In fact, when f and $g \in \mathscr{L}_2(G)$, for all λ, $|f| + \lambda|g| \in \mathscr{L}_2(G)$, and by virtue of this

$$0 \leq \int_G (|f(x)| + \lambda|g(x)|)^2\,dx$$

$$= \int_G |f(x)|^2\,dx + 2\lambda \int_G |f(x)g(x)|\,dx + \lambda^2 \int_G |g(x)|^2\,dx$$

Consequently the discriminant of this quadratic form is nonpositive, that is,

$$\left[\int_G |f(x)g(x)|\,dx \right]^2 - \int_G |f(x)|^2\,dx \int_G |g(x)|^2\,dx \leq 0$$

The required inequality (2) follows directly from this.

We shall note the discrete analog of the Cauchy–Buniakowski inequality: If complex numbers a_k and b_k, $k = 1, 2, \ldots$, are such that

$$\sum_{k=1}^{\infty} |a_k|^2 < \infty \quad \text{and} \quad \sum_{k=1}^{\infty} |b_k|^2 < \infty$$

then

$$\left| \sum_{k=1}^{\infty} a_k b_k \right| \leq \left(\sum_{k=1}^{\infty} |a_k|^2 \right)^{1/2} \left(\sum_{k=1}^{\infty} |b_k|^2 \right)^{1/2} \quad (3)$$

§ 1. SOME CONCEPTS AND PROPOSITIONS

If $f \in \mathscr{L}_2(G)$ and G is a bounded region, the function $f(x)$ is integrable over G.

In fact, applying the Cauchy–Buniakowski inequality with $g \equiv 1$, we obtain

$$\int_G |f(x)|\, dx \leq \left(\int_G |f(x)|^2\, dx\right)^{1/2} \left(\int_G dx\right)^{1/2} < \infty$$

On the set of functions $\mathscr{L}_2(G)$ we shall introduce a scalar product and norm according to the formulas

$$(f, g) = \int_G f(x)\bar{g}(x)\, dx$$
$$\|f\| = \sqrt{(f,f)} = \left(\int_G |f(x)|^2\, dx\right)^{1/2} \tag{4}$$

by which we convert $\mathscr{L}_2(G)$ into a (linear) normed space. Here $\bar{g}(x)$ is the complex conjugate of $g(x)$.

Evidently, the scalar product introduced has the properties:

$$(f, g) = \overline{(g, f)}, \quad (\lambda f + \mu g, h) = \lambda(f, h) + \mu(g, h) \tag{5}$$

Moreover, in terms of the norm and scalar product the Cauchy–Buniakowski inequality assumes the form

$$|(f, g)| \leq \|f\|\,\|g\|, \quad f, g \in \mathscr{L}_2(G) \tag{6}$$

From this inequality there follows the *Minkowski inequality*:

$$\|f + g\| \leq \|f\| + \|g\|, \quad f, g \in \mathscr{L}_2(G) \tag{7}$$

In fact

$$\|f + g\|^2 = (f+g, f+g) = (f,f) + (f,g) + (g,f)$$
$$+ (g,g) \leq \|f\|^2 + |(f,g)| + |(g,f)| + \|g\|^2$$
$$\leq \|f\|^2 + \|f\|\,\|g\| + \|g\|\,\|f\| + \|g\|^2 = (\|f\| + \|g\|)^2$$

In this way we see that the norm (4) satisfies conditions (a)–(c) of Sec. 1.3.

The sequence of functions f_k, $k = 1, 2, \ldots$, belonging to $\mathscr{L}_2(G)$ is said to *converge* to the function $f \in \mathscr{L}_2(G)$ in the space $\mathscr{L}_2(G)$ (or to *converge in the mean in G*) if $\|f_k - f\| \to 0$ as $k \to \infty$; we shall write

$$f_k \to f, \quad k \to \infty \quad \text{in} \quad \mathscr{L}_2(G)$$

The following proposition expresses the property of completeness of the space $\mathscr{L}_2(G)$ (Riesz–Fischer theorem*): *If the sequence of functions f_k, $k = 1, 2, \ldots$, belonging to $\mathscr{L}_2(G)$ converges in itself in $\mathscr{L}_2(G)$, that is, if $\|f_k - f_p\| \to 0$ as $k \to \infty$, $p \to \infty$, then there is a function $f \in \mathscr{L}_2(G)$ such that $\|f_k - f\| \to 0$ as $k \to \infty$.* The space $\mathscr{L}_2(G)$ belongs to the class of so-called *Hilbert spaces*.

A set of functions $\mathscr{M} \subset \mathscr{L}_2(G)$ is called dense in $\mathscr{L}_2(G)$ if for any $f \in \mathscr{L}_2(G)$ there is a sequence of functions belonging to $\mathscr{M}$ which converges to f in $\mathscr{L}_2(G)$. For instance, a set of polynomials is dense in $\mathscr{L}_2(G)$ if G is a bounded region, by virtue of Weierstrass's theorem (cf. Sec. 1.3).

LEMMA. *The set $\mathscr{D}(G)$ is dense in $\mathscr{L}_2(G)$.*

Proof. Let $f \in \mathscr{L}_2(G)$ and $\varepsilon > 0$ be any number. Then we can find a bounded region $G_1 \subset G$ such that

$$\int_{G \setminus G_1} |f(x)|^2 \, dx < \frac{\varepsilon^2}{5} \tag{8}$$

Since the set of polynomials is dense in $\mathscr{L}_2(G)$, there is a polynomial P such that

$$\int_{G_1} |f(x) - P(x)|^2 \, dx < \frac{\varepsilon^2}{5} \tag{9}$$

Now we shall choose a subregion $G' \Subset G_1$ which is sufficiently close to region G_1 (Fig. 2) that

$$\int_{G_1 \setminus G'} |P(x)|^2 \, dx < \frac{\varepsilon^2}{5} \tag{10}$$

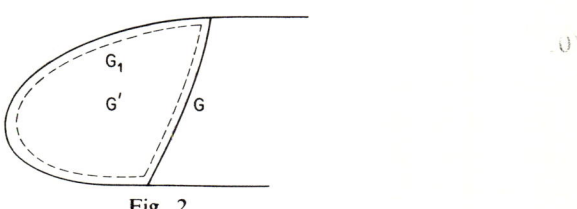

Fig. 2

We now take the function η belonging to $\mathscr{D}(G_1)$ such that $0 \leq \eta \leq 1$, $\eta(x) = 1$, $x \in G'$. (In Sec. 5.2 it will be shown that such functions exist.)

---

* Compare, F. Riesz and B. Sz. Nagy (*1*, Chap. 2).

§ 1. SOME CONCEPTS AND PROPOSITIONS

Then $P\eta \in \mathscr{D}(G)$ and, by virtue of inequalities (8)–(10),

$$\|f - P\eta\|^2 = \int_G |f - P\eta|^2 \, dx = \int_{G_1} |f - P\eta|^2 \, dx$$
$$+ \int_{G \setminus G_1} |f|^2 \, dx < \frac{\varepsilon^2}{5} + \left[\left(\int_{G_1} |f - P|^2 \, dx\right)^{1/2}\right.$$
$$+ \left.\left(\int_{G_1} |P|^2 (1 - \eta)^2 \, dx\right)^{1/2}\right]^2 < \frac{\varepsilon^2}{5}$$
$$+ \left[\frac{\varepsilon}{\sqrt{5}} + \left(\int_{G_1 \setminus G'} |P|^2 \, dx\right)^{1/2}\right]^2 < \frac{\varepsilon^2}{5} + \frac{4\varepsilon^2}{5}$$
$$= \varepsilon^2$$

Here Minkowski's inequality (7) has been used several times. The lemma is thus proved.

6. Orthonormal Systems. Functions f and g belonging to $\mathscr{L}_2(G)$ are called *orthogonal* if $(f, g) = 0$; the function f belonging to $\mathscr{L}_2(G)$ is said to be *normalized* if $\|f\| = 1$. The system of functions $\{\varphi_k\}$ belonging to $\mathscr{L}_2(G)$ is said to be *orthonormal* in $\mathscr{L}_2(G)$ if $(\varphi_k, \varphi_i) = \delta_{ki}$, where δ_{ki} is the Krönecker delta symbol $\delta_{ki} = 0$, $k \neq i$, $\delta_{ii} = 1$.

An example of an orthonormal system in $\mathscr{L}_2(-\pi, \pi)$ is the trigonometrical system

$$\varphi_k(x) = \frac{1}{\sqrt{2\pi}} e^{ikx}, \qquad k = 0, \pm 1, \ldots$$

Each orthonormal system $\{\varphi_k\}$ consists of linearly independent functions.

In fact, in the corresponding case involving a sequence of (complex) numbers $\{c_k\}$ of which only a finite number are distinct from zero, we have the equation $\sum_k c_k \varphi_k = 0$, from which, by virtue of the orthonormality of the system $\{\varphi_k\}$, we obtain

$$0 = \left(\sum_k c_k \varphi_k, \varphi_i\right) = \sum_k (c_k \varphi_k, \varphi_i) = \sum_k c_k (\varphi_k, \varphi_i) = c_i$$

Each system of linearly independent functions $\psi_1, \psi_2, \ldots$, belonging to

$\mathscr{L}_2(G)$ may be converted into an orthonormal system $\varphi_1, \varphi_2, \ldots$, by the following Gram–Schmidt orthogonalization process:

$$\varphi_1 = \frac{\psi_1}{\|\psi_1\|}, \qquad \varphi_2 = \frac{\psi_2 - (\psi_2, \varphi_1)\varphi_1}{\|\psi_2 - (\psi_2, \varphi_1)\varphi_1\|}, \ldots, \qquad (11)$$

$$\varphi_k = \frac{\psi_k - (\psi_k, \varphi_{k-1})\varphi_{k-1} - \cdots - (\psi_k, \varphi_1)\varphi_1}{\|\psi_k - (\psi_k, \varphi_{k-1})\varphi_{k-1} - \cdots - (\psi_k, \varphi_1)\varphi_1\|}, \ldots$$

Example. If in the space $\mathscr{L}_2(-1, 1)$ we orthogonalize the powers $1, x, x^2, \ldots$, by means of the Gram–Schmidt process, the system of normalized Legendre polynomials is obtained.

Let a system of functions φ_k, $k = 1, 2, \ldots$, be orthonormal in $\mathscr{L}_2(G)$ and let $f \in \mathscr{L}_2(G)$. Then the numbers (f, φ_k) are called *Fourier coefficients* and the formal series

$$\sum_{k=1}^{\infty} (f, \varphi_k)\varphi_k \qquad (12)$$

is called the *Fourier series* of the function f in terms of the system $\{\varphi_k\}$.

If a system of functions φ_k, $k = 1, 2, \ldots$, is orthonormal in $\mathscr{L}_2(G)$, then the equation

$$\left\| f - \sum_{k=1}^{N} a_k \varphi_k \right\|^2 = \left\| f - \sum_{k=1}^{N} (f, \varphi_k)\varphi_k \right\|^2 + \sum_{k=1}^{N} |(f, \varphi_k) - a_k|^2 \qquad (13)$$

is true for each $f \in \mathscr{L}_2(G)$ and any (complex) numbers $\alpha_1, \alpha_2, \ldots, \alpha_N$, $N = 1, 2, \ldots$.

In fact, writing

$$f_N = f - \sum_{k=1}^{N} (f, \varphi_k)\varphi_k, \qquad c_k = (f, \varphi_k) - a_k \qquad (14)$$

we obtain when $i = 1, 2, \ldots, N$

$$(f_N, \varphi_i) = \left(f - \sum_{k=1}^{N} (f, \varphi_k)\varphi_k, \varphi_i \right) = (f, \varphi_i) - \sum_{k=1}^{N} (f, \varphi_k)(\varphi_k, \varphi_i) = 0$$

§ 1. SOME CONCEPTS AND PROPOSITIONS

Consequently,

$$\left\| f - \sum_{k=1}^{N} a_k \varphi_k \right\|^2 = \left\| f_N + \sum_{k=1}^{N} c_k \varphi_k \right\|^2$$

$$= \left(f_N + \sum_{k=1}^{N} c_k \varphi_k, f_N + \sum_{k=1}^{N} c_k \varphi_k \right) = (f_N, f_N)$$

$$+ \sum_{k=1}^{N} (f_N, c_k \varphi_k) + \sum_{k=1}^{N} (c_k \varphi_k, f_N) + \sum_{k,i=1}^{N} (c_k \varphi_k, c_i \varphi_i)$$

$$= \| f_N \|^2 + \sum_{k,i=1}^{N} c_k \bar{c}_i (\varphi_k, \varphi_i) = \| f_N \|^2 + \sum_{k=1}^{N} |c_k|^2$$

from which, by virtue of (14), Eq. (13) follows.

From Eq. (13) follows the inequality

$$\left\| f - \sum_{k=1}^{N} (f, \varphi_k) \varphi_k \right\|^2 \leq \left\| f - \sum_{k=1}^{N} a_k \varphi_k \right\|^2 \tag{15}$$

Further, supposing in (13) that $a_k = 0$, $k = 1, 2, \ldots, N$, we obtain the equation

$$\left\| f - \sum_{k=1}^{N} (f, \varphi_k) \varphi_k \right\|^2 = \| f \|^2 - \sum_{k=1}^{N} |(f, \varphi_k)|^2 \tag{16}$$

From Eq. (16) there then follows the inequality

$$\sum_{k=1}^{\infty} |(f, \varphi_k)|^2 \leq \| f \|^2 \tag{17}$$

which is known as *Bessel's inequality*.

Moreover, from Eq. (16) and from the Riesz–Fischer theorem (cf. Sec. 1.5) we obtain the following proposition: *In order that the Fourier series* (12) *converge to the function f in* $\mathscr{L}_2(G)$, *it is necessary and sufficient that Parseval's equation* (*the equation of closure*) *be satisfied*

$$\sum_{k=1}^{\infty} |(f, \varphi_k)|^2 = \| f \|^2 \tag{18}$$

7. Complete Orthonormal Systems. Let the system of functions $\varphi_1, \varphi_2, \ldots,$ be orthonormal in $\mathscr{L}_2(G)$. If for any $f \in \mathscr{L}_2(G)$ its Fourier series in terms of the system $\{\varphi_k\}$ converges to f in $\mathscr{L}_2(G)$, then this system is said to be complete (closed) in $\mathscr{L}_2(G)$.

The trigonometrical system serves as an example of a complete orthonormal system in $\mathscr{L}_2(0, 2\pi)$.

THEOREM. *In order that the orthonormal system $\{\varphi_k\}$ be complete in $\mathscr{L}_2(G)$, it is necessary and sufficient that each function f belonging to the set $\mathscr{M}$, dense in $\mathscr{L}_2(G)$, be capable of arbitrarily close approximation in $\mathscr{L}_2(G)$ by linear combinations of functions of this system.*

The necessity of the condition is evident; we shall prove its sufficiency. Let $f \in \mathscr{L}_2(G)$ and $\varepsilon > 0$ be any number. Then since $\mathscr{M}$ is dense in $\mathscr{L}_2(G)$, there is a function $f_0 \in \mathscr{M}$ such that

$$\|f - f_0\| < \varepsilon/2 \tag{19}$$

According to the condition of the theorem, the function f may be approximated arbitrarily closely in $\mathscr{L}_2(G)$ by a linear combination of the functions of the system $\{\varphi_k\}$. Therefore numbers $m, c_1, c_2, \ldots, c_m$, can be found such that

$$\left\| f_0 - \sum_{k=1}^{m} c_k \varphi_k \right\| < \varepsilon/2$$

As a result, from (19), and by virtue of Minkowski's inequality, we obtain

$$\left\| f - \sum_{k=1}^{m} c_k \varphi_k \right\| \leq \|f - f_0\| + \left\| f_0 - \sum_{k=1}^{m} c_k \varphi_k \right\| < \varepsilon/2 + \varepsilon/2 = \varepsilon$$

However, by virtue of inequality (15) we have

$$\left\| f - \sum_{k=1}^{N} (f, \varphi_k) \varphi_k \right\| < \varepsilon, \qquad N \geq m$$

as was to be shown.

COROLLARY. *If G is a bounded region, there exists in $\mathscr{L}_2(G)$ a countable complete orthonormal system of polynomials.*

In fact, the set of polynomials is dense in $\mathscr{L}_2(G)$ (cf. Sec. 1.5), countable, and can be made orthonormal, using the Gram–Schmidt orthogonalization process (cf. Sec. 1.6).

LEMMA. *Let the regions $G \subset R^n$ and $D \subset R^m$ be bounded, the system of functions $\psi_j(y)$, $j = 1, 2, \ldots$, be orthonormal and complete in $\mathscr{L}_2(D)$, and for each $j = 1, 2, \ldots$, the system of functions $\varphi_{kj}(x)$, $k = 1, 2, \ldots$, be ortho-*

§ 1. SOME CONCEPTS AND PROPOSITIONS

normal and complete in $\mathscr{L}_2(G)$. Then the system of functions

$$\chi_{kj}(x, y) = \varphi_{kj}(x)\psi_j(y), \qquad k, j = 1, 2, \ldots \tag{20}$$

is orthonormal and complete in $\mathscr{L}_2(G \times D)$.

Proof. The orthonormality of the system $\{\chi_{kj}\}$ in $\mathscr{L}_2(G \times D)$ is easily established as follows:

$$\begin{aligned}(\chi_{kj}, \chi_{k'j'}) &= \int_{G \times D} \chi_{kj} \bar{\chi}_{k'j'} \, dx \, dy \\ &= \int_G \varphi_{kj} \bar{\varphi}_{k'j'} \, dx \int_D \psi_j \bar{\psi}_{j'} \, dy = (\varphi_{kj}, \varphi_{k'j'})(\psi_j, \psi_{j'}) \\ &= (\varphi_{kj}, \varphi_{k'j'})\delta_{jj'} = (\varphi_{kj}, \varphi_{k'j})\delta_{jj'} = \delta_{kk'}\delta_{jj'}\end{aligned}$$

We shall prove the completeness of this system in $\mathscr{L}_2(G \times D)$. Since $C(\bar{G} \times \bar{D})$ is dense in $\mathscr{L}_2(G \times D)$ (cf. Sec. 1.5), then according to the theorem of Sec. 1.7, it is sufficient to establish the truth of Parseval's equation for all $f \in C(\bar{G} \times \bar{D})$. Let $f \in C(\bar{G} \times \bar{D})$. Since the system $\{\psi_j\}$ is complete in $\mathscr{L}_2(D)$, for each $x \in \bar{G}$ the Parseval equation (cf. Sec. 1.6) is true and

$$\sum_{j=1}^{\infty} |a_j(x)|^2 = \int_D |f(x, y)|^2 \, dy \tag{21}$$

where

$$a_j(x) = \int_D f(x, y) \bar{\psi}_j(y) \, dy \tag{22}$$

By virtue of the boundedness of the region D, the functions ψ_j are integrable over D (cf. Sec. 1.5) and therefore, since $f \in C(\bar{G} \times \bar{D})$, $a_j \in C(\bar{G})$. Since for each $j = 1, 2, \ldots$, the system $\{\varphi_{kj}\}$ is complete in $\mathscr{L}_2(G)$, the Perseval equation (cf. Sec. 1.6) is true:

$$\sum_{k=1}^{\infty} |a_{kj}|^2 = \int_G |a_j(x)|^2 \, dx \tag{23}$$

where, by virtue of (22), (20), and Fubini's theorem (cf. Sec. 1.4),

$$\begin{aligned}a_{kj} = (a_j, \varphi_{kj}) &= \int_G a_j \bar{\varphi}_{kj} \, dx \\ &= \int_G \left[\int_D f(x, y) \bar{\psi}_j(y) \, dy \right] \bar{\varphi}_{kj}(x) \, dx \\ &= \int_{G \times D} f(x, y) \bar{\varphi}_{kj}(x) \bar{\psi}_j(y) \, dx \, dy \\ &= \int_{G \times D} f \bar{\chi}_{kj} \, dx \, dy = (f, \chi_{kj})\end{aligned} \tag{24}$$

According to Dini's lemma (cf. Sec. 1.3), the series (21) converges uniformly on $\bar{G}$. Integrating this series term by term over the region G and using Eqs. (23) and (24), we obtain the desired Parseval equation for the function f

$$\sum_{j=1}^{\infty}\sum_{k=1}^{\infty} |a_{kj}|^2 = \sum_{j,k=1}^{\infty} |(f, \chi_{kj})|^2 = \int_G \int_D |f(x,y)|^2 \, dx \, dy = \|f\|^2$$

The lemma is proved.

Remark. All that has been said of the space $\mathscr{L}_2(G)$ is true also of the spaces $\mathscr{L}_2(G; \varrho)$ and $\mathscr{L}_2(S)$ with the scalar products

$$(f, g)_\varrho = \int_G \varrho(x) f(x) \bar{g}(x) \, dx, \qquad f, g \in \mathscr{L}_2(G; \varrho)$$

$$(f, g) = \int_S f(x) \bar{g}(x) \, dS, \qquad f, g \in \mathscr{L}_2(S)$$

where the weight function $\varrho \in C(\bar{G})$, $\varrho(x) > 0$, $x \in \bar{G}$, and S is a piecewise smooth surface.

8. Linear Operators and Functionals. Let $\mathscr{M}$ and $\mathscr{N}$ be linear sets. The operator L, transforming elements of set $\mathscr{M}$ into elements of set $\mathscr{N}$, is said to be linear if for any elements f and g belonging to $\mathscr{M}$, and complex numbers λ and μ the equation

$$L(\lambda f + \mu g) = \lambda L f + \mu L g$$

s true. In this case the set $\mathscr{M} = \mathscr{M}_L$ is called the *domain of definition* of the operator L. If $Lf = f$ for all $f \in \mathscr{M}$, the operator L is called an *identity* (*unit*) operator. We shall denote a unit operator by I.

Let the convergence of the elements be defined on the linear sets $\mathscr{M}$ and $\mathscr{N}$. The linear operator L, mapping $\mathscr{M}$ into $\mathscr{N}$, is said to be continuous from $\mathscr{M}$ to $\mathscr{N}$ if the convergence $Lf_k \to Lf$ as $k \to \infty$ in $\mathscr{N}$ follows from the convergence $f_k \to f$ as $k \to \infty$ in $\mathscr{M}$.

Let $\mathscr{M}$ and $\mathscr{N}$ be linear normed spaces with the norms $\|\ \|_{\mathscr{M}}$ and $\|\ \|_{\mathscr{N}}$, respectively [e.g., $\mathscr{N} = C(T)$, $\mathscr{M} = \mathscr{L}_2(G)$]. The linear operator L, mapping $\mathscr{M}$ into $\mathscr{N}$, is said to be bounded from $\mathscr{M}$ to $\mathscr{N}$ if there is a number $C > 0$ such that the inequality

$$\|Lf\|_{\mathscr{N}} \leq C \|f\|_{\mathscr{M}} \tag{25}$$

is true for any $f \in \mathscr{M}$.

§1. SOME CONCEPTS AND PROPOSITIONS

From these definitions it follows that if the linear operator L is bounded from $\mathscr{M}$ to $\mathscr{N}$, then it is continuous from $\mathscr{M}$ to $\mathscr{N}$.

In fact, if $f_k \to f$ as $k \to \infty$ in $\mathscr{M}$, that is, if $\|f - f_k\|_{\mathscr{M}} \to 0$ as $k \to \infty$, by virtue of the linearity and boundedness of the operator L, then

$$\|Lf_k - Lf\|_{\mathscr{N}} = \|L(f_k - f)\|_{\mathscr{N}} \leq C\|f_k - f\|_{\mathscr{M}}$$

and therefore $Lf_k \to Lf$ as $k \to \infty$ in $\mathscr{N}$. This also means that the operator L is continuous from $\mathscr{M}$ to $\mathscr{N}$.

A set $\mathscr{B}$ of the linear normed space $\mathscr{M}$ is said to be *bounded* in $\mathscr{M}$ if there is a number A such that $\|f\|_{\mathscr{M}} < A$ whenever $f \in \mathscr{B}$.

Let the linear operator L map $\mathscr{M}$ into $\mathscr{N}_1$ and the linear operator K map $\mathscr{N}_1$ into $\mathscr{N}$. The linear operator

$$KLf = K(Lf)$$

mapping $\mathscr{M}$ into $\mathscr{N}$, is called the product KL of the operators K and L; specifically, $K^p f = K(K^{p-1} f) = K^{p-1}(Kf)$, $K^1 = K$, $K^0 = I$.

Linear functionals are a particular case of linear operators. If the linear operator l transforms the set of elements $\mathscr{M}$ into the set of complex numbers lf, $f \in \mathscr{M}$, then l is said to be a linear functional on the set $\mathscr{M}$. We shall denote by (l, f) the effect of the functional l on element f—the complex number lf. In this way, by the continuity of the linear functional l we mean the following: If $f_k \to f$ as $k \to \infty$ in $\mathscr{M}$, then the sequence of complex numbers (l, f_k) as $k \to \infty$ tends to (l, f).

The linear functional $\tilde{l}$ on the set $\tilde{\mathscr{M}} \supset \mathscr{M}$ is said to be a continuation of the linear functional l on $\mathscr{M}$ if

$$(\tilde{l}, f) = (l, f), \qquad f \in \mathscr{M}$$

EXAMPLES OF LINEAR OPERATORS AND FUNCTIONALS

(a) The linear operator of the form

$$Kf = \int_G \mathscr{K}(x, y) f(y)\, dy, \qquad x \in G \tag{26}$$

is called the linear *integral operator*, and the function $\mathscr{K}(x, y)$ is its *kernel*. If the kernel $\mathscr{K} \in \mathscr{L}_2(G \times G)$,

$$\int_{G \times G} |\mathscr{K}(x, y)|^2\, dx\, dy = C^2 < \infty \tag{27}$$

then the operator K is bounded (and, consequently, continuous) from $\mathscr{L}_2(G) = \mathscr{M}$ to $\mathscr{L}_2(G) = \mathscr{N}$.

In fact, applying the Cauchy–Buniakowski inequality and the Fubini theorem (cf. Sec. 1.4) and using (27), for all $f \in \mathscr{L}_2(G)$ we obtain the inequality

$$\| Kf \|^2 = \int_G \left| \int_G \mathscr{K}(x, y) f(y)\, dy \right|^2 dx$$
$$\leq \int_G \left(\int_G |\mathscr{K}(x, y)|^2\, dy \int_G |f(y)|^2\, dy \right) dx = C^2 \| f \|^2$$

that is,

$$\| Kf \| \leq C \| f \|, \qquad f \in \mathscr{L}_2(G) \tag{28}$$

which also shows that the operator K is bounded in $\mathscr{L}_2(G)$.

(b) The linear operator of the form

$$Lf = \sum_{|\alpha| \leq m} a_\alpha(x) D^\alpha f(x), \qquad \sum_{|\alpha| = m} |a_\alpha(x)| \not\equiv 0, \qquad m > 0 \tag{29}$$

is called the *linear differential operator* of order m, and the functions $a_\alpha(x)$ are its *coefficients*. If the coefficients $a_\alpha(x)$ are continuous functions in the region $\bar{G} \subset R^n$, the operator L maps $C^m(\bar{G}) = \mathscr{M}$ into $C(\bar{G}) = \mathscr{N}$. However, the operator L is not continuous from $C^m(\bar{G})$ to $C(\bar{G})$. Indeed, the sequence

$$f_k(x) = \frac{1}{k} e^{ik(x,\alpha)} \to 0, \qquad k \to \infty \quad \text{in} \quad C(\bar{G})$$

as well as the sequence

$$Lf_k = \sum_{|\alpha| \leq m} a_\alpha(x) D^\alpha f_k(x) = \sum_{|\alpha| \leq m} a_\alpha(x)(i\alpha)^\alpha k^{|\alpha|-1} e^{ik(x,\alpha)}$$

has no limit in $C(\bar{G})$. We remark in passing that the operator L is not defined on the whole space $C(\bar{G})$, but only on a part of it—namely, on the set of functions $C^m(\bar{G})$.

(c) The linear operator

$$Lf = \sum_{|\alpha| \leq m} \left[\int_G \mathscr{K}_\alpha(x, y) D^\alpha f(y)\, dy + a_\alpha(x) D^\alpha f(x) \right] \tag{30}$$

is called the *linear integrodifferential operator*.

§1. SOME CONCEPTS AND PROPOSITIONS

(d) An example of a linear continuous functional l on $\mathscr{L}_2(G)$ is the scalar product $(l, f) = (f, g)$, where g is a fixed function belonging to $\mathscr{L}_2(G)$. The linearity of this functional is expressed by property (5), and, by virtue of the Cauchy–Buniakowski inequality (6), it is bounded,

$$|(l, f)| = |(f, g)| \leq \|g\| \|f\|$$

and therefore it is continuous.

9. Linear Equations. Let L be a linear operator with domain of definition $\mathscr{M}_L$. The equation

$$Lu = F \tag{31}$$

is called a *linear (inhomogeneous)* equation. In Eq. (31) the element F is called the *inhomogeneous term* (*free term*, or *right-hand side*), and the unknown element u belonging to $\mathscr{M}_L$ is called the *solution* of this equation. If in Eq. (31) we assume that the inhomogeneous term F is equal to zero, then the equation obtained,

$$Lu = 0 \tag{32}$$

is called the *linear homogeneous equation* corresponding to Eq. (31).

By virtue of the linearity of the operator L, the set of solutions of the homogeneous equation (32) forms a linear set; specifically, $u = 0$ is always a solution of this equation.

Each solution u of the linear inhomogeneous equation (31) (if it exists) appears in the form of a sum of a particular solution u_0 of this equation and of the general solution $\tilde{u}$ of the corresponding linear homogeneous equation (32),

$$u = u_0 + \tilde{u} \tag{33}$$

In fact, if u is an arbitrary solution of Eq. (31), $Lu = F$, $u \in \mathscr{M}_L$, while u_0 is a particular solution of this equation, $Lu_0 = F$, $u_0 \in \mathscr{M}_L$, then, by virtue of the linearity of operator L, their difference $u - u_0 = \tilde{u} \in \mathscr{M}_L$ also satisfies the homogeneous equation (32):

$$L\tilde{u} = L(u - u_0) = Lu - Lu_0 = F - F = 0$$

This proves the representation equation (33) for the solution u.

It follows directly from this that: *in order that the solution of Eq. (31) should be unique in $\mathscr{M}_L$, it is necessary and sufficient that the corresponding homogeneous equation (32) have only a zero solution in $\mathscr{M}_L$.*

Let the homogeneous equation (32) have only a zero solution in $\mathscr{M}_L$. We shall denote by $\mathscr{R}_L$ the range of values of the operator L, that is, the (linear) set of elements of the form $\{Lf\}$ where f belongs to $\mathscr{M}_L$. Then for any $F \in \mathscr{R}_L$, Eq. (31) has a unique solution $u \in \mathscr{M}_L$; in this way there appears a certain operator associating with each element F belonging to $\mathscr{R}_L$ an element u belonging to $\mathscr{M}_L$, the solution of Eq. (31). This operator is known as the *inverse operator* to the operator L and is denoted L^{-1} so that

$$u = L^{-1}F \tag{34}$$

The operator L^{-1}, evidently, is linear and transforms $\mathscr{R}_L$ onto $\mathscr{M}_L$. It follows immediately from the definition of the operator L^{-1}, and also from (31) and (34), that

$$LL^{-1}F = F, \quad F \in \mathscr{R}_L; \qquad L^{-1}Lu = u, \quad u \in \mathscr{M}_L$$

that is,

$$LL^{-1} = I \quad \text{and} \quad L^{-1}L = I$$

Let us consider the linear homogeneous equation

$$Lu = \lambda u \tag{35}$$

where λ is a complex parameter. This equation has a zero solution for all λ. It can happen that for some λ it has nonzero solutions belonging to $\mathscr{M}_L$. Those complex values λ for which Eq. (35) has nonzero solutions belonging to $\mathscr{M}_L$ are called the *eigenvalues* or *characteristic values* of the operator L and the corresponding solutions are the *eigenfunctions* or the *characteristic functions* corresponding to this eigenvalue. The integral number r ($1 \leq r \leq \infty$) of linearly independent eigenfunctions, corresponding to the given eigenvalue λ, is called the *multiplicity* of this eigenvalue; if the multiplicity $r = 1$, then λ is said to be a *simple* eigenvalue.

If the multiplicity r of the eigenvalue λ of the operator L is finite and $u_1, u_2, \ldots, u_r$ are the corresponding linearly independent eigenfunctions, any linear combination of them

$$u_0 = c_1 u_1 + c_2 u_2 + \cdots + c_r u_r \tag{36}$$

§1. SOME CONCEPTS AND PROPOSITIONS

is also an eigenfunction corresponding to this eigenvalue, and this combination represents a general solution of Eq. (35). From this, and from formula (33), it follows: If a solution of the equation

$$Lu = \lambda u + f \tag{37}$$

exists, its general solution is given by the formula

$$u = u^* + \sum_{k=1}^{r} c_k u_k \tag{38}$$

where u^* is a particular solution of this equation and c_k, $k = 1, 2, \ldots, r$, are arbitrary constants.

10. Hermitian Operators. The linear operator L mapping $\mathcal{M}_L \subset \mathcal{L}_2$ into $\mathcal{L}_2$ is said to be *Hermitian* (or *self-adjoint in the Lagrangian sense*) if for any f and g belonging to $\mathcal{M}_L$ it is true that

$$(Lf, g) = (f, Lg) \tag{39}$$

The expressions (Lf, g) and (Lf, f) are called *bilinear* and *quadratic forms*, respectively, generated by the operator L.

In order that the operator L should be Hermitian, it is necessary and sufficient that the quadratic form (Lf, f), $f \in \mathcal{M}_L$, generated by it should assume only real values.

In fact, if the operator L is Hermitian, then by virtue of (5) and (39)

$$(Lf, f) = \overline{(f, Lf)} = \overline{(Lf, f)}, \quad f \in \mathcal{M}_L$$

so that the quadratic form (Lf, f) can assume only real values.

Conversely, if the quadratic form (Lf, f) assumes only real values, for all f and g belonging to $\mathcal{M}_L$ we have

$$\operatorname{Re}[(Lg, f) - (Lf, g)] = \operatorname{Re} \frac{1}{i}[(L(f+ig), f+ig) - (Lf, f) - (Lg, g)] = 0$$

$$\operatorname{Im}[(Lg, f) + (Lf, g)] = \operatorname{Im}[(L(f+g), f+g) - (Lf, f) - (Lg, g)] = 0$$

and therefore,

$$(Lf, g) = \operatorname{Re}(Lf, g) + i \operatorname{Im}(Lf, g)$$
$$= \operatorname{Re}(Lg, f) - i \operatorname{Im}(Lg, f) = \overline{(Lg, f)} = (f, Lg)$$

so that the operator L is Hermitian.

A linear operator is said to be *positive* if the quadratic form (Lf, f), $f \in \mathcal{M}_L$, generated by it assumes only nonnegative values.

From the assertion which has been proved it follows that each positive operator is Hermitian.

THEOREM. *If the operator L is Hermitian (positive), all its eigenvalues are real (nonnegative) and its eigenfunctions, corresponding to different eigenvalues, are orthogonal.*

Proof. Let λ_0 be an eigenvalue and u_0 a corresponding normalized eigenfunction of the Hermitian operator L, $Lu_0 = \lambda_0 u_0$. Scalar multiplication of this equation by u_0 will give

$$(Lu_0, u_0) = (\lambda_0 u_0, u_0) = \lambda_0 (u_0, u_0) = \lambda_0 \| u_0 \|^2 = \lambda_0 \quad (40)$$

But for an Hermitian (positive) operator the quadratic form (Lf, f) assumes only real (nonnegative) values, and consequently, by virtue of (40), λ_0 is a real (nonnegative) number.

We shall prove that any eigenfunctions u_1 and u_2, corresponding to different eigenvalues λ_1 and λ_2, are orthogonal. In fact, from the results

$$Lu_1 = \lambda_1 u_1, \qquad Lu_2 = \lambda_2 u_2$$

and from the Hermitian property of the operator L we obtain the sequence of equations

$$\lambda_1 (u_1, u_2) = (\lambda_1 u_1, u_2) = (Lu_1, u_2) = (u_1, Lu_2) = (u_1, \lambda_2 u_2) = \lambda_2 (u_1, u_2)$$

that is,

$$\lambda_1 (u_1, u_2) = \lambda_2 (u_1, u_2)$$

From this, as $\lambda_1 \neq \lambda_2$, it follows that $(u_1, u_2) = 0$. The theorem is proved.

We shall assume that a set of eigenfunctions of the Hermitian operator L is at most countable. We shall enumerate all its eigenvalues: $\lambda_1, \lambda_2, \ldots$, repeating λ_k as many times as its multiplicity. The corresponding eigenfunctions will be denoted by $u_1, u_2, \ldots$, so that only one eigenfunction u_k corresponds to each eigenvalue,

$$Lu_k = \lambda_k u_k, \qquad k = 1, 2, \ldots \quad (41)$$

The eigenfunctions corresponding to the same eigenvalue can be chosen to be orthonormal by using the Gram–Schmidt orthogonalization process (cf. Sec. 1.6). As a consequence of this, eigenfunctions corresponding to the same eigenvalue are again obtained. According to the theorem of Sec. 1.10, eigenfunctions corresponding to different eigenvalues are orthogonal.

In this way, *if a system of eigenfunctions $\{u_k\}$ of the Hermitian operator L is at most countable, it can be chosen so as to be orthonormal*

$$(Lu_k, u_i) = \lambda_k(u_k, u_i) = \lambda_k \delta_{ki} \tag{42}$$

§ 2. Basic Equations of Mathematical Physics

The mathematical description of many physical processes leads to linear differential and integral equations, or even to integrodifferential second-order equations. A fairly wide class of physical problems may be reduced to linear second order differential equations (cf. Sec. 1.8)

$$\sum_{i,j=1}^{n} a_{ij}(x) \frac{\partial^2 u}{\partial x_i \, \partial x_j} + \sum_{i=1}^{n} b_i(x) \frac{\partial u}{\partial x_i} + c(x)u = F(x) \tag{1}$$

In this section we shall consider typical physical problems leading to various equations of mathematical physics.

1. Vibration Equations. Many problems of mechanics (vibration of strings, rods, membranes, and three-dimensional volumes) and of physics (electromagnetic waves) lead to a wave equation of the form

$$\varrho \frac{\partial^2 u}{\partial t^2} = \text{div}(p \text{ grad } u) - qu + F(x, t) \tag{2}$$

where the unknown function $u(x, t)$ depends on n ($n = 1, 2, 3$), the spatial coordinates $x = (x_1, x_2, \ldots, x_n)$, and the time t; coefficients ϱ, p, and q are defined as properties of the surrounding medium in which the vibration takes place; the inhomogeneous term $F(x, t)$ expresses the intensity of the exterior perturbation or disturbance. In Eq. (2), in agreement with the definition of the operators div (divergence) and grad (gradient),

$$\text{div}(p \text{ grad } u) = \sum_{i=1}^{n} \frac{\partial}{\partial x_i} \left(p \frac{\partial u}{\partial x_i} \right)$$

Let us demonstrate the meaning of Eq. (2) by using the example of small transverse vibrations of a string. An elastic thread which does not resist flexion is called a string.

Suppose that in the plane (x, u) the string performs small transverse vibrations around its state of equilibrium which coincides with the x axis. The magnitude of the string's displacement from a state of equilibrium at point x at an instant of time t is denoted by $u(x, t)$, so that $u = u(x, t)$ is the equation of the string at the instant of time t. Limiting ourselves to the consideration of small vibrations of the string, we shall ignore magnitudes of order greater than $\tan \alpha = \partial u/\partial x$.

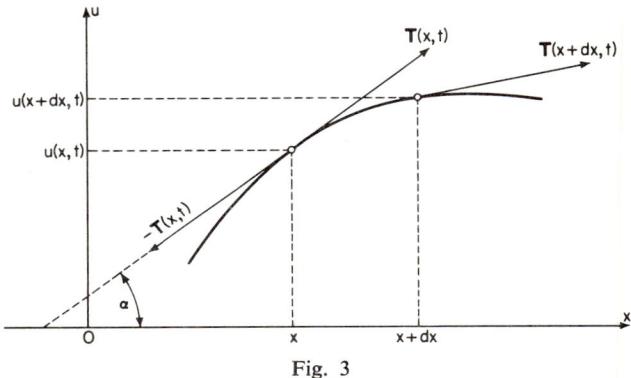

Fig. 3

Since the string does not resist flexion, its tension $\mathbf{T}(x, t)$ at the point x at an instant of time t is directed along the tangent to the string at the point x (Fig. 3). Any section of the string (a, b), after displacement from a state of equilibrium within the range of our approximation, does not change its length

$$l = \int_a^b \sqrt{1 + \left(\frac{\partial u}{\partial x}\right)^2} \, dx \simeq b - a$$

and therefore, in agreement with Hooke's law, the magnitude of the tension $|\mathbf{T}(x, t)|$ will remain constant, depending neither on x nor on t, $|\mathbf{T}(x, t)| = T_0$. We shall use $F(x, t)$ to denote the magnitude of the external forces acting on the string at the point x at an instant of time t and directed perpendicular to the x axis. Finally, let $\varrho(x)$ be the linear density of the string at point x, so that $\varrho(x) \, dx$ is the mass of the element of the string $(x, x+dx)$.

We now construct an equation of motion for the element $(x, x + dx)$ of the string. The tension force $\mathbf{T}(x + dx, t)$, $-\mathbf{T}(x, t)$ (Fig. 3) and the external force, the sum of which, according to Newton's laws, must be equal to the product of the mass of the element and its acceleration, are acting upon the element$(x, x + dx)$. Projecting this vector equation onto the u axis, on account of all that has been said, we obtain the equation

$$T_0 \sin \alpha \,|_{x+dx} - T_0 \sin \alpha \,|_x + F(x, t) \, dx = \varrho(x) \, dx \, \frac{\partial^2 u(x, t)}{\partial t^2} \qquad (3)$$

However, within the range of our approximation

$$\sin \alpha = \frac{\tan \alpha}{\sqrt{1 + \tan^2 \alpha}} \approx \tan \alpha = \frac{\partial u}{\partial x}$$

and therefore from (3) we have

$$\varrho \, \frac{\partial^2 u(x, t)}{\partial t^2} = T_0 \, \frac{1}{dx} \left[\frac{\partial u(x + dx, t)}{\partial x} - \frac{\partial u(x, t)}{\partial x} \right] + F(x, t)$$

that is,

$$\varrho \, \frac{\partial^2 u}{\partial t^2} = T_0 \, \frac{\partial^2 u}{\partial x^2} + F \qquad (4)$$

This is indeed the *equation of small transverse vibrations of a string*.

If the density ϱ is constant, $\varrho(x) = \varrho$, the equation of vibration of the string assumes the form

$$\frac{\partial^2 u}{\partial t^2} = a^2 \, \frac{\partial^2 u}{\partial x^2} + f \qquad (5)$$

where we have set $a^2 = T_0/\varrho$ (constant), $f = F/\varrho$. We shall also call Eq. (5) the *one-dimensional wave equation*.

From physical considerations it follows that to give a unique description of the process of vibration of a string it is necessary, in addition, to specify the magnitude of the displacement u and the velocity $\partial u/\partial t$ of the string at the initial instant of time (*initial conditions*), and also the behavior at the ends of the string (*boundary conditions*). Examples of boundary conditions are: $u\,|_{x=x_0} = 0$, if the end of the string x_0 is fixed; $(\partial u/\partial x)\,|_{x=x_0} = 0$, if the end x_0 is free, by virtue of the equation $\alpha(x_0, t) = 0$.

An equation of the form (2) also describes *small longitudinal vibrations of an elastic rod*

$$\varrho S \, \frac{\partial^2 u}{\partial t^2} = \frac{\partial}{\partial x} \left(ES \, \frac{\partial u}{\partial x} \right) + F(x, t) \qquad (6)$$

where $S(x)$ is the area of cross section of the rod and $E(x)$ is Young's modulus at the point x.

The equation of *small transverse vibrations of a membrane* may be derived analogously

$$\varrho \frac{\partial^2 u}{\partial t^2} = T_0 \left(\frac{\partial^2 u}{\partial x_1^2} + \frac{\partial^2 u}{\partial x_2^2} \right) + F \tag{7}$$

If the density ϱ is constant, the equation of the vibrations of a membrane assumes the form

$$\frac{\partial^2 u}{\partial t^2} = a^2 \left(\frac{\partial^2 u}{\partial x_1^2} + \frac{\partial^2 u}{\partial x_2^2} \right) + f, \qquad a^2 = \frac{T_0}{\varrho}, \qquad f = \frac{F}{\varrho} \tag{8}$$

We shall call Eq. (8) the *two-dimensional wave equation*.

The *three-dimensional wave equation*

$$\frac{\partial^2 u}{\partial t^2} = a^2 \left(\frac{\partial^2 u}{\partial x_1^2} + \frac{\partial^2 u}{\partial x_2^2} + \frac{\partial^2 u}{\partial x_3^2} \right) + f \tag{9}$$

describes the processes of propagation of sound in homogeneous media and also of electromagnetic waves in homogeneous nonconducting media. The density of a gas, its pressure and velocity potential, and also the components of electric and magnetic field strength and corresponding potentials satisfy this equation (cf. Sec. 2.6).

We shall express the wave equations (5), (8), and (9) by the single formula:

$$\square_a u = f \tag{10}$$

where $\square_a$ is the *wave operator* (*D'Alembert's operator*)

$$\square_a = \frac{\partial^2}{\partial t^2} - a^2 \triangle \qquad (\square = \square_1)$$

and $\triangle$ is *Laplace's operator*

$$\triangle = \frac{\partial^2}{\partial x_1^2} + \frac{\partial^2}{\partial x_2^2} + \cdots + \frac{\partial^2}{\partial x_n^2}$$

2. Diffusion Equation. The process of heat diffusion or the diffusion of particles in a medium are described by the following general *diffusion equation*:

$$\varrho \frac{\partial u}{\partial t} = \operatorname{div}(p \operatorname{grad} u) - qu + F(x, t) \tag{11}$$

§ 2. BASIC EQUATIONS OF MATHEMATICAL PHYSICS

We shall deduce the equation of *heat diffusion*. We shall use $u(x, t)$ to denote the temperature of the medium at the point $x = (x_1, x_2, x_3)$ at an instant of time t. We shall consider the medium to be isotropic and shall denote its density by $\varrho(x)$, its specific heat by $c(x)$, and its coefficient of thermal conductivity at point x by $k(x)$. $F(x, t)$ denotes the intensity of heat sources at the point x at the instant of time t. We shall calculate the balance of heat in an arbitrary volume V during the time interval $(t, t + dt)$. The boundary of V is denoted by S and $\mathbf{n}$ is the external normal to it. In agreement with Fourier's law, an amount of heat

$$Q_1 = \int_S k \frac{\partial u}{\partial \mathbf{n}} \, dS \, dt = \int_S (k \operatorname{grad} u, \mathbf{n}) \, dS \, dt$$

equal, by virtue of the Gauss–Ostrogradski formula, to

$$Q_1 = \int_V \operatorname{div}(k \operatorname{grad} u) \, dx \, dt$$

enters volume V through surface S.

An amount of heat

$$Q_2 = \int_V F(x, t) \, dx \, dt$$

arises in volume V as a result of heat sources. Since the temperature in the volume V during the interval of time $(t, t + dt)$ has increased in magnitude by

$$u(x, t + dt) - u(x, t) \simeq \frac{\partial u}{\partial t} \, dt$$

it follows that for this it is necessary to expend an amount of heat

$$Q_3 = \int_V c\varrho \frac{\partial u}{\partial t} \, dx \, dt$$

On the other hand, $Q_3 = Q_1 + Q_2$, and thus

$$\int_V \left[\operatorname{div}(k \operatorname{grad} u) + F - c\varrho \frac{\partial u}{\partial t} \right] dx \, dt = 0$$

from which, since volume V is arbitrary, we obtain the equation of heat diffusion

$$c\varrho \frac{\partial u}{\partial t} = \operatorname{div}(k \operatorname{grad} u) + F(x, t) \qquad (12)$$

If the medium is homogeneous, that is, if c, ϱ, and k are constants, Eq. (12) assumes the form

$$\frac{\partial u}{\partial t} = a^2 \Delta u + f \tag{13}$$

where

$$a^2 = \frac{k}{c\varrho}, \quad f = \frac{F}{c\varrho}$$

Equation (13) is called the *heat conduction equation*. The number n of spatial variables $x_1, x_2, \ldots, x_n$ in this equation can be arbitrary.

As in the case of the wave equation, it is necessary to give the initial distribution of the temperature u in the medium (initial condition) and the behavior at the boundary of this medium (boundary condition) so as to obtain a complete description of the process of heat diffusion.

EXAMPLES OF BOUNDARY CONDITIONS

(a) If a given temperature distribution u_0 is maintained over the boundary S, then

$$u\,|_S = u_0 \tag{14}$$

(b) If a given flux of heat u_1 is maintained over S,

$$-k\,\frac{\partial u}{\partial \mathbf{n}}\bigg|_S = u_1 \tag{15}$$

(c) If heat exchange in agreement with Newton's law takes place over S,

$$k\,\frac{\partial u}{\partial \mathbf{n}} + h(u - u_0)\bigg|_S = 0 \tag{16}$$

where h is the heat transfer coefficient and u_0 is the temperature of the surrounding medium.

A *diffusion equation for particles* may be deduced analogously. In this case, instead of Fourier's law it is necessary to use Nernst's law for the flux of particles through an element of surface dS during a unit time: $dQ = -D(\partial u/\partial n)\,dS$, where $D(x)$ is the diffusion coefficient and $u(x, t)$ is the density of the particles at the point x at an instant of time t. The equation for the density u will have the form shown in (11), where ϱ now denotes the porosity coefficient, $p = D$, and q characterizes the absorption of the medium.

§ 2. BASIC EQUATIONS OF MATHEMATICAL PHYSICS

3. Steady State Equation. For steady state processes $F(x, t) = F(x)$, $u(x, t) = u(x)$, and both the wave equation (2) and the diffusion equation (11) assume the form

$$-\text{div}(p \text{ grad } u) + qu = F(x) \tag{17}$$

When $p = \text{const}$ and $q = 0$, Eq. (17) is called *Poisson's equation*

$$\Delta u = -f, \quad f = \frac{F}{p} \tag{18}$$

when $f = 0$, Eq. (18) is called *Laplace's equation*

$$\Delta u = 0 \tag{19}$$

In the wave equation (10), let the external disturbance $f(x, t)$ be periodic with frequency ω and amplitude $f_0(x)$,

$$f(x, t) = f_0(x)e^{i\omega t}$$

If we seek periodic solutions $u(x, t)$ with the same frequency and unknown amplitude $u_0(x)$,

$$u(x, t) = u_0(x)e^{i\omega t}$$

then for the function $u_0(x)$ we obtain the equation

$$\Delta u_0 + k^2 u_0 = -\frac{f_0(x)}{a^2}, \quad k^2 = \frac{\omega^2}{a^2} \tag{20}$$

called the *Helmholtz equation*.

4. Transport Equation. If the mean free path of the particles is significantly greater than their dimensions, the process of particle propagation may be described by a more accurate equation than the diffusion equation: namely, by the so-called transport equation (kinetic equation). We shall cite the transport equation subject to the following assumptions: (1) The speed of all particles is the same, and is equal to v. (2) Collisions among the particles themselves may be ignored. (3) Particles collide with motionless nuclei of the medium; $l(x)$ is the mean free path at the point x. (4) When the particles collide with a motionless nucleus at point x, one of the three following accidental events occurs: (a) with probability $p_1(x)$

the particle is scattered by the nucleus, bouncing away from it like an elastic ball; (b) with probability $p_2(x)$ the particle is captured by the nucleus; (c) with probability $p_3 = 1 - p_1 - p_2$ the particle divides the nucleus, as a result of which $v(x) \geq 1$ such particles appear (in this case it is considered that the particle which has divided the nucleus disappears). (5) The distribution of particles with respect to direction is isotropic after scattering as well as after division.

We shall denote by $n(x, \mathbf{s}, t)$ the density of particles at the point x, moving in the direction $\mathbf{s} = (s_1, s_2, s_3)$, $|\mathbf{s}| = 1$, at an instant of time t, and by $F(x, \mathbf{s}, t)$ the density of the sources. Then the function $\psi = vn$—the flux of particles—satisfies the following integrodifferential equation:

$$\frac{1}{v}\frac{\partial \psi}{\partial t} + (\mathbf{s}, \operatorname{grad} \psi) + \alpha\psi = \frac{\alpha h}{4\pi} \int_{S_1} \psi(x, \mathbf{s}', t)\, d\mathbf{s}' + F \qquad (21)$$

where $\alpha = 1/l$, $h = p_1 + vp_3$. This is the one-velocity transport equation for processes with isotropic scattering. More general transport equations and related research can be found in G. I. Marchuk (1) and V. S. Vladimirov (1).

If the transport process is stationary, then

$$F(x, \mathbf{s}, t) = F(x, \mathbf{s}), \qquad \psi(x, \mathbf{s}, t) = \psi(x, \mathbf{s})$$

when the transport equation (21) assumes the form

$$(\mathbf{s}, \operatorname{grad} \psi) + \alpha\psi = \frac{\alpha h}{4\pi} \int_{S_i} \psi(x, \mathbf{s}')\, d\mathbf{s}' + F \qquad (22)$$

For a complete description of the process of particle transport it is necessary to give the initial distribution of the flux of particles ψ in the medium (initial conditions) and the behavior on the boundary of this medium (boundary conditions). For example, if the region G, where the transport process is taking place, is convex, a boundary condition of the form

$$\psi(x, \mathbf{s}, t) = 0, \quad x \in S, \qquad (\mathbf{s}, \mathbf{n}_x) < 0 \qquad (23)$$

expresses the absence of a flux of particles incident on the region G from the outside (Fig. 4).

Finally we remark that the transport equation describes the processes of the transport of neutrons in a nuclear reactor, the transfer of radiant

§ 2. BASIC EQUATIONS OF MATHEMATICAL PHYSICS

energy, the passage of γ quanta through matter, the movement of gases, and other processes.

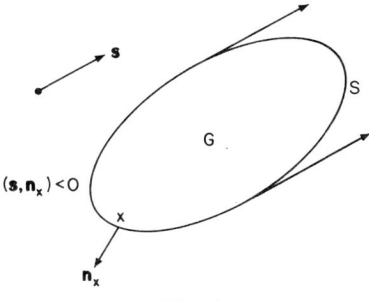

Fig. 4

5. Hydrodynamic Equations. We shall consider the movement of an ideal liquid (gas); that is, a liquid in which viscosity forces are absent. Let $\mathbf{V}(x, t) = (v_1, v_2, v_3)$ be the velocity vector of the liquid, $\varrho(x, t)$ be its density, $p(x, t)$ its pressure, $f(x, t)$ the strength of sources, and $\mathbf{F}(x, t) = (F_1, F_2, F_3)$ the strength of the body forces. Then these magnitudes satisfy the following (nonlinear) system of equations, which are called the *hydrodynamic equations*:

$$\frac{\partial \varrho}{\partial t} + \operatorname{div}(\varrho \mathbf{V}) = f \qquad (24)$$

$$\frac{\partial \mathbf{V}}{\partial t} + (\mathbf{V}, \operatorname{grad}) \mathbf{V} + \frac{1}{\varrho} \operatorname{grad} p = \mathbf{F} \qquad (25)$$

Equations (24) and (25) are called the continuity equation and Euler's equation of motion, respectively. In order to close this system of equations, it is also necessary to give the connection between the pressure and density:

$$\Phi(p, \varrho) = 0 \qquad (26)$$

the so-called equation of state. For instance, for an incompressible liquid the equation of state has the form $\varrho = \text{const}$, while for adiabatic gas motion

$$p\varrho^{-\varkappa} = \text{const}, \qquad \varkappa = \frac{c_p}{c_v}$$

where c_p and c_v are the specific heats of the gas at constant pressure and constant volume, respectively.

Specifically, if a liquid is incompressible ($\varrho = $ const) and has a velocity potential ($V = -\operatorname{grad} u$), it follows from the continuity equation (24) that the velocity potential u satisfies Poisson's equation (18).

6. Maxwell's Equations. Let there be a variable electromagnetic field in a certain medium. Let $\mathbf{E}(x, t) = (E_1, E_2, E_3)$ be the electric field strength; $\mathbf{H}(x, t) = (H_1, H_2, H_3)$ be the magnetic field strength; $\varrho(x)$ be the density of charges; ε be the dielectric constant of the medium; μ be the coefficient of magnetic permeability of the medium; and $\mathbf{I}(x, t) = (I_1, I_2, I_3)$ be the conduction current. Then these magnitudes satisfy the following (linear) system of differential equations, called *Maxwell's equations*:

$$\operatorname{div}(\varepsilon \mathbf{E}) = 4\pi\varrho, \quad \operatorname{div}(\mu \mathbf{H}) = 0 \tag{27}$$

$$\operatorname{curl} \mathbf{E} = -\frac{1}{c} \frac{\partial(\mu \mathbf{H})}{\partial t} \tag{28}$$

$$\operatorname{curl} \mathbf{H} = \frac{1}{c} \frac{\partial(\varepsilon \mathbf{E})}{\partial t} + \frac{4\pi}{c} \mathbf{I} \tag{29}$$

where $c = 3 \cdot 10^{10}$ cm/sec is the speed of light in a vacuum.

Equation (28) expresses Faraday's law, and Eq. (29) expresses Ampere's law (Bio–Savart law).

We shall note particular cases of Maxwell's equations.

(a) $\varrho = 0$; $\varepsilon = $ const; $\mu = $ const; and $\mathbf{I} = \lambda \mathbf{E}$ (Ohm's law), $\lambda = $ const. Applying the operator curl to Eqs. (28) and (29) and using Eqs. (27), we obtain for the components of vectors $\mathbf{E}$ and $\mathbf{H}$ the so-called *telegrapher's equation*.

$$\Box_a u + \frac{4\pi\lambda}{\varepsilon} \frac{\partial u}{\partial t} = 0, \quad a = \frac{c}{\sqrt{\varepsilon\mu}} \tag{30}$$

(b) $\mathbf{I} = 0$; $\varepsilon = $ const; and $\mu = $ const. Introducing a four-component *electromagnetic potential* $(\varphi_0, \boldsymbol{\varphi})$, $\boldsymbol{\varphi} = (\varphi_1, \varphi_2, \varphi_3)$, we may present the solution of Maxwell's equations in the form

$$\mathbf{E} = \operatorname{grad} \varphi_0 - \frac{1}{c} \frac{\partial \boldsymbol{\varphi}}{\partial t}, \quad \mathbf{H} = \frac{1}{\mu} \operatorname{curl} \boldsymbol{\varphi} \tag{31}$$

From this it follows that the components of the electromagnetic potential

§ 2. BASIC EQUATIONS OF MATHEMATICAL PHYSICS

must satisfy the wave equations

$$\Box_a \varphi_0 = -\frac{4\pi c^2}{\varepsilon^2 \mu} \varrho, \qquad \Box_a \boldsymbol{\varphi} = 0 \tag{32}$$

and the Lorentz condition

$$\frac{\mu\varepsilon}{c} \frac{\partial \varphi_0}{\partial t} - \operatorname{div} \boldsymbol{\varphi} = 0 \tag{33}$$

(c) If the process is stationary, Maxwell's equations are transformed into the *equations of electrostatics*

$$\operatorname{div}(\varepsilon \mathbf{E}) = 4\pi\varrho, \qquad \operatorname{curl} \mathbf{E} = 0 \tag{34}$$

and into the *equations of magnetostatics*

$$\operatorname{div}(\mu \mathbf{H}) = 0, \qquad \operatorname{curl} \mathbf{H} = \frac{4\pi}{c} \mathbf{I} \tag{35}$$

When $\varepsilon = \text{const}$, the electrostatic potential φ_0 satisfies, by virtue of (32), Poisson's equation (18) with $f = -(4\pi/\varepsilon)\varrho$.

When transforming Maxwell's equations we made use of the following formulas of vector analysis:

$$\begin{aligned} \operatorname{div} \operatorname{grad} &= \varDelta, & \operatorname{curl} \operatorname{curl} &= \operatorname{grad} \operatorname{div} - \varDelta \\ \operatorname{curl} \operatorname{grad} &= 0, & \operatorname{div} \operatorname{curl} &= 0 \end{aligned} \tag{36}$$

7. Schrödinger's Equation. Let a quantum particle of mass m move in an external force field with potential $V(x)$. We shall denote by $\psi(x, t)$ the wave function of this particle, so that $|\psi(x, t)|^2 \, dx$ is the probability that the particle will be in a neighborhood $d(x)$ of the point x at an instant of time t; here dx is the volume of $d(x)$. Then the function ψ satisfies Schrödinger's equation

$$i\hbar \frac{\partial \psi}{\partial t} = -\frac{\hbar^2}{2m} \varDelta \psi + V\psi \tag{37}$$

where $\hbar = 1.054 \cdot 10^{-27}$ erg sec is Planck's constant [cf. L. D. Landau and E. M. Lifschitz (1, Chap. III)].

8. Klein–Gordon Equation and Dirac's Equation.

The wave function $\varphi(x_0, x)$, $x_0 = ct$, $x = (x_1, x_2, x_3)$ (where $c = 3 \cdot 10^{10}$ cm/sec is the speed of light), describing a free relativistic (pseudo) scalar particle of mass m, satisfies the *Klein–Gordon equation*

$$(\Box + m^2)\varphi = 0 \tag{38}$$

To describe a free relativistic particle of mass m with spin $\frac{1}{2}$ (electron, proton, neutron, neutrino, etc.) a four-component wave function (spinor)

$$\Psi(x_0, x) = (\psi_1, \psi_2, \psi_3, \psi_4)$$

is used. It satisfies *Dirac's equation*, a system of four linear differential equations of first order:

$$\left(i \sum_{k=0}^{3} \gamma^k \frac{\partial}{\partial x_k} - m\mathbf{I}\right)\Psi(x) = 0 \tag{39}$$

where I is a unit matrix and γ^k are Dirac's matrices,

$$\gamma^0 = \begin{pmatrix} 1 & 0 & 0 & 0 \\ 0 & 1 & 0 & 0 \\ 0 & 0 & -1 & 0 \\ 0 & 0 & 0 & -1 \end{pmatrix}, \quad \gamma^1 = \begin{pmatrix} 0 & 0 & 0 & 1 \\ 0 & 0 & 1 & 0 \\ 0 & -1 & 0 & 0 \\ -1 & 0 & 0 & 0 \end{pmatrix}$$

$$\gamma^2 = \begin{pmatrix} 0 & 0 & 0 & -i \\ 0 & 0 & i & 0 \\ 0 & i & 0 & 0 \\ -i & 0 & 0 & 0 \end{pmatrix}, \quad \gamma^3 = \begin{pmatrix} 0 & 0 & 1 & 0 \\ 0 & 0 & 0 & -1 \\ -1 & 0 & 0 & 0 \\ 0 & 1 & 0 & 0 \end{pmatrix}$$

Dirac's matrices satisfy the conditions

$$\left(\sum_{k=0}^{3} \gamma^k \frac{\partial}{\partial x_k}\right)^2 = \Box \mathbf{I} \tag{40}$$

[cf. N. N. Bogolyubov and D. V. Shirkov (1, Sec. 6)].

§ 3. Classification of Linear (Second-Order) Differential Equations

Before formulating mathematically the solution of various physical problems leading to linear second-order differential equations, it is necessary to classify these equations.

§ 3. LINEAR (SECOND-ORDER) DIFFERENTIAL EQUATIONS

1. Classification of Equations at a Point. Let us consider the second-order differential equation

$$\sum_{i,j=1}^{n} a_{ij}(x) \frac{\partial^2 u}{\partial x_i \, \partial x_j} + \Phi(x, u, \text{grad } u) = 0 \tag{1}$$

with continuous coefficients $a_{ij}(x)$. We shall first of all make clear by what laws the coefficients a_{ij} are transformed as a result of an arbitrary change of independent variables $y = y(x)$, that is,

$$y_l = y_l(x_1, x_2, \ldots, x_n), \quad l = 1, 2, \ldots, n \tag{2}$$

$$y_l \in C^2, \quad D\left(\frac{y_1, y_2, \ldots, y_n}{x_1, x_2, \ldots, x_n}\right) \neq 0$$

Since $D \neq 0$, in a certain neighborhood it is possible to express the variables x in terms of the variables y, writing $x = x(y)$. We shall set $u(x(y)) = \tilde{u}(y)$; then $\tilde{u}(y(x)) = u(x)$. We have

$$\frac{\partial u}{\partial x_i} = \sum_{l=1}^{n} \frac{\partial \tilde{u}}{\partial y_l} \frac{\partial y_l}{\partial x_i}$$

$$\frac{\partial^2 u}{\partial x_i \, \partial x_j} = \frac{\partial}{\partial x_j}\left(\frac{\partial u}{\partial x_i}\right) = \sum_{k,l=1}^{n} \frac{\partial^2 \tilde{u}}{\partial y_l \, \partial y_k} \frac{\partial y_l}{\partial x_i} \frac{\partial y_k}{\partial x_j} + \sum_{l=1}^{n} \frac{\partial \tilde{u}}{\partial y_l} \frac{\partial^2 y_l}{\partial x_i \, \partial x_j} \tag{3}$$

Substituting expressions (3) into Eq. (1), we obtain

$$\sum_{k,l=1}^{n} \frac{\partial^2 \tilde{u}}{\partial y_l \, \partial y_k} \sum_{i,j=1}^{n} a_{ij} \frac{\partial y_l}{\partial x_i} \frac{\partial y_k}{\partial x_j} + \sum_{l=1}^{n} \frac{\partial \tilde{u}}{\partial y_l} \sum_{i,j=1}^{n} a_{ij} \frac{\partial^2 y_l}{\partial x_i \, \partial x_j}$$
$$+ \Phi^*(y, \tilde{u}, \text{grad } \tilde{u}) = 0 \tag{4}$$

Denoting now by $\tilde{a}_{lk}$ the new coefficients of the second derivatives

$$\tilde{a}_{lk}(y) = \sum_{i,j=1}^{n} a_{ij}(x) \frac{\partial y_l}{\partial x_i} \frac{\partial y_k}{\partial x_j} \tag{5}$$

we rewrite Eq. (4) in the form (1):

$$\sum_{k,l=1}^{n} \tilde{a}_{lk}(y) \frac{\partial^2 \tilde{u}}{\partial y_l \, \partial y_k} + \tilde{\Phi}(y, \tilde{u}, \text{grad } \tilde{u}) = 0 \tag{6}$$

We shall fix the point x_0 and write $y_0 = y(x_0)$, $\alpha_{li} = (\partial y_l(x_0))/\partial x_i$.

Then formula (5) at the point x_0 can be written in the form

$$\tilde{a}_{lk}(y_0) = \sum_{i,j=1}^{n} a_{ij}(x_0)\alpha_{li}\alpha_{kj} \tag{7}$$

This transformation formula for the coefficients a_{ij} at the point x_0 coincides with the transformation formula for the coefficients of the quadratic form

$$\sum_{i,j=1}^{n} a_{ij}(x_0)p_i p_j \tag{8}$$

subject to the nonsingular linear transformation

$$p_i = \sum_{l=1}^{n} \alpha_{li} q_l, \quad \det(\alpha_{li}) \neq 0 \tag{9}$$

transforming form (8) into form

$$\sum_{k,l=1}^{n} \tilde{a}_{lk}(y_0) q_l q_k \tag{10}$$

So, in order to simplify Eq. (1) at the point x_0 by means of a change of variables (2), it is sufficient to simplify the quadratic form (8) by means of a nonsingular linear transformation (9). But it is proved in courses on linear algebra that there is always a nonsingular transformation (9) which can make the quadratic form (8) assume the following canonical form:

$$\sum_{l=1}^{r} q_l^2 - \sum_{l=r+1}^{m} q_l^2, \quad m \leq n \tag{11}$$

Moreover, by virtue of the law of inertia of quadratic forms, the integers r and m do not depend on the transformation (9).* This allows us to classify differential equations (1) according to the significance of the coefficients a_{ij} at the point x_0.

If in the quadratic form (11) $m = n$ and all the terms are of one sign (i.e., either $r = m$ or 0), Eq. (1) is called an equation of *elliptic type*; if $m = n$ but there are terms of various signs (i.e., $1 \leq r \leq n - 1$), Eq. (1) is of the *hyperbolic type* (when $r = 1$ or $n - 1$ it is of the *normally hyperbolic type*); lastly, if $m < n$, Eq. (1) is of the *parabolic type* (when $r = 1$ or $n - 1$ it is of the *normally parabolic type*).

* Compare, for instance, A. I. Maltsev (*1*, Chap. 6).

§ 3. LINEAR (SECOND-ORDER) DIFFERENTIAL EQUATIONS

We must stress that this classification depends on the point x_0, since the numbers r and m depend on x_0. For instance, Tricomi's equation

$$y \frac{\partial^2 u}{\partial x^2} + \frac{\partial^2 u}{\partial y^2} = 0 \tag{12}$$

is of mixed type: when $y < 0$ it is of hyperbolic type; when $y > 0$ it is of elliptic type; and when $y = 0$ it is of parabolic type.

Let the coefficients a_{ij} in Eq. (1) be constant, that is, independent of x, and let the transformation (9) reduce the quadratic form (8) to a canonical form (11). Then the linear change of independent variables

$$y_l = \sum_{i=1}^{n} \alpha_{li} x_i$$

transforms Eq. (1) into the following canonical form:

$$\sum_{l=1}^{r} \frac{\partial^2 \tilde{u}}{\partial y_l^2} - \sum_{l=r+1}^{m} \frac{\partial^2 \tilde{u}}{\partial y_l^2} + \tilde{\Phi}(y, \tilde{u}, \operatorname{grad} \tilde{u}) = 0 \tag{13}$$

Examples. Laplace's equation is of elliptic type; the wave equation is of hyperbolic type; and the equation of heat conduction is of parabolic type.

2. Expression of Laplace's Operator in Spherical and Cylindrical Coordinates. To illustrate the transformations of Sec. 3.1 we shall find the expression of the three-dimensional Laplace operator ($n = 3$, $\alpha_{ij} = \delta_{ij}$, $\Phi = 0$) in spherical and cylindrical coordinates.

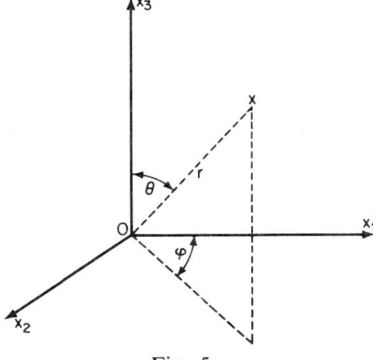

Fig. 5

(a) Spherical coordinates (Fig. 5)

$$x_1 = r \sin \theta \cos \varphi, \quad x_2 = r \sin \theta \sin \varphi, \quad x_3 = r \cos \theta$$

We have

$$\frac{\partial r}{\partial x_i} = \frac{x_i}{r}, \quad i = 1, 2, 3, \quad \Delta r = \frac{2}{r}$$

$$\frac{\partial \theta}{\partial x_1} = \frac{\cos \theta \cos \varphi}{r}, \quad \frac{\partial \theta}{\partial x_2} = \frac{\cos \theta \sin \varphi}{r}$$

$$\frac{\partial \theta}{\partial x_3} = -\frac{\sin \theta}{r}, \quad \Delta \theta = \frac{\cos \theta}{r^2 \sin \theta}$$

$$\frac{\partial \varphi}{\partial x_1} = -\frac{\sin \varphi}{r \sin \theta}, \quad \frac{\partial \varphi}{\partial x_2} = \frac{\cos \varphi}{r \sin \theta}, \quad \frac{\partial \varphi}{\partial x_3} = 0, \quad \Delta \varphi = 0$$

Substituting these expressions into formula (4) with $n = 3$, $a_{ij} = \delta_{ij}$, and $\Phi = 0$, and collecting similar terms, we obtain

$$\Delta = \frac{1}{r^2} \frac{\partial}{\partial r} \left(r^2 \frac{\partial}{\partial r} \right) + \frac{1}{r^2 \sin \theta} \frac{\partial}{\partial \theta} \left(\sin \theta \frac{\partial}{\partial \theta} \right) + \frac{1}{r^2 \sin^2 \theta} \frac{\partial^2}{\partial \varphi^2} \quad (14)$$

(b) Cylindrical (polar) coordinates (Fig. 6)

$$x_1 = r \cos \varphi, \quad x_2 = r \sin \varphi, \quad x_3 = z$$

Deriving an analogous result we obtain

$$\Delta = \frac{1}{r} \frac{\partial}{\partial r} \left(r \frac{\partial}{\partial r} \right) + \frac{1}{r^2} \frac{\partial^2}{\partial \varphi^2} + \frac{\partial^2}{\partial z^2} \quad (15)$$

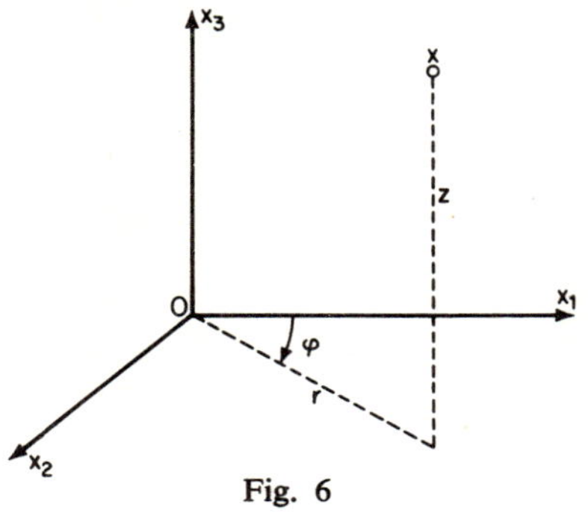

Fig. 6

§3. LINEAR (SECOND-ORDER) DIFFERENTIAL EQUATIONS

3. Characteristic Surfaces (Characteristics). Let the function $\omega(x)$, $x = (x_1, x_2, \ldots, x_n)$, $n \geq 2$, of the class C^1 be such that on the surface $\omega(x) = 0$, grad $\omega(x) \neq 0$ and

$$\sum_{i,j=1}^{n} a_{ij}(x) \frac{\partial \omega(x)}{\partial x_i} \frac{\partial \omega(x)}{\partial x_j} = 0 \qquad (16)$$

Then the surface $\omega(x) = 0$ is called a *characteristic surface* of the differential equation (1), while Eq. (16) is called a *characteristic equation*. When $n = 2$ the characteristic surface is called a *characteristic curve*.

Let us suppose that each surface of the family $\omega(x) - C = 0$, $a < C < b$, is a characteristic of Eq. (1). Insofar as on each characteristic grad $\omega \neq 0$, this family fills a certain, sufficiently small, region G, through each point of which passes one and only one characteristic. Let $\omega \in C^2(G)$. Then if in transformation (2) we take $y_1 = \omega(x)$, by virtue of (5) and (16), the coefficient $\tilde{a}_{11}$ becomes zero in the corresponding region. Therefore the knowledge of one or several families of characteristics of a differential equation makes it possible to reduce this equation to a simpler form.

EXAMPLES OF CHARACTERISTICS

(a) *Wave Equation* [cf. Eq. (10), Sec. 2.1]. Its characteristic equation has the form

$$\left(\frac{\partial \omega}{\partial t}\right)^2 - a^2 \sum_{i=1}^{n} \left(\frac{\partial \omega}{\partial x_i}\right)^2 = 0$$

The surface

$$a^2(t - t_0)^2 - |x - x_0|^2 = 0 \qquad (17)$$

called a *characteristic cone*, with its vertex at the point (x_0, t_0), is a characteristic of a wave equation.

The characteristic cone (17) is the boundary of the cones

$$\Gamma^+(x_0, t_0) = [a(t - t_0) > |x - x_0|]$$

and

$$\Gamma^-(x_0, t_0) = [-a(t - t_0) > |x - x_0|]$$

which are called, respectively, the *future* and *past light cones* with vertex at the point (x_0, t_0) (Fig. 7). We shall write $\Gamma^\pm = \Gamma^\pm(0, 0)$.

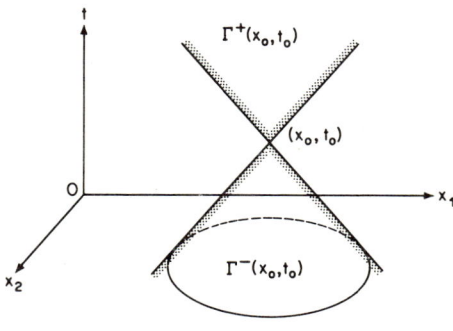

Fig. 7

A wave equation has another family of characteristic surfaces—a family of tangent planes to the characteristic cones

$$at + (x, b) = c \tag{18}$$

where $b = (b_1, b_2, \ldots, b_n)$, b_k and c are any real numbers, and $|b| = 1$.

(b) *Equation of Heat Conduction* [cf. Eq. (13), Sec. 2.2]. Its characteristics, evidently, are the family of planes $t = C$.

(c) *Poisson's Equation* [cf. Eq. (18), Sec. 2.3]. It has no (real) characteristics, since from the characteristic equation

$$\sum_{i=1}^{n} \left(\frac{\partial \omega}{\partial x_i} \right)^2 = 0 \quad \text{on} \quad \omega = 0$$

it follows that grad $\omega = 0$ on $\omega = 0$, which is impossible.

4. Canonical Form of Equations with Two Independent Variables. In Sec. 3.1 we considered the means of reducing a second-order differential equation to canonical form at each separate point at which the equation is specified. The question then arises: Is it possible to use a single transformation (2) to reduce Eq. (1) to canonical form (13) in a sufficiently small neighborhood of each point? To make this reduction for any equation, it is necessary that the number of conditions

$$\tilde{a}_{lk} = 0, \quad l \neq k, \quad l, k = 1, 2, \ldots, n$$
$$\tilde{a}_{ll} = \varepsilon_l \tilde{a}_{11}, \quad l = 2, 3, \ldots, n; \quad \tilde{a}_{11} \neq 0$$

§ 3. LINEAR (SECOND-ORDER) DIFFERENTIAL EQUATIONS

where $\varepsilon_l = 0, \pm 1$, should not be greater than the number of unknown functions y_l, $l = 1, 2, \ldots, n$:

$$\frac{n(n-1)}{2} + n - 1 \leq n, \quad \text{that is} \quad n \leq 2$$

We shall show that for $n = 2$ (and, obviously, for $n = 1$) it is always possible to make this reduction.

Let us consider a second-order differential equation with two independent variables

$$a \frac{\partial^2 u}{\partial x^2} + 2b \frac{\partial^2 u}{\partial x \, \partial y} + c \frac{\partial^2 u}{\partial y^2} + \Phi\left(x, y, u, \frac{\partial u}{\partial x}, \frac{\partial u}{\partial y}\right) = 0 \quad (19)$$

where we shall suppose that the coefficients a, b, and c belong to class C^2 in a certain neighborhood and do not become zero simultaneously anywhere in this neighborhood. For definiteness we consider that $a \neq 0$ in this neighborhood. In fact, in other circumstances it may happen that $c \neq 0$. But then, by reversing the roles of x and y, we obtain an equation in which $a \neq 0$. If then a and c become zero simultaneously at a point, then $b \neq 0$ in the neighborhood of this point. In this case the change of variables $x' = x + y$, $y' = x - y$ leads to an equation in which $a \neq 0$.

By using the new variables

$$\xi = \xi(x, y), \quad \eta = \eta(x, y), \quad \xi \in C^2, \quad \eta \in C^2, \quad D\left(\frac{\xi, \eta}{x, y}\right) \neq 0 \quad (20)$$

we reduce Eq. (19) to the form

$$\tilde{a} \frac{\partial^2 \tilde{u}}{\partial \xi^2} + 2\tilde{b} \frac{\partial^2 \tilde{u}}{\partial \xi \, \partial \eta} + \tilde{c} \frac{\partial^2 \tilde{u}}{\partial \eta^2} + \tilde{\Phi}\left(\xi, \eta, \tilde{u}, \frac{\partial \tilde{u}}{\partial \xi}, \frac{\partial \tilde{u}}{\partial \eta}\right) = 0 \quad (21)$$

where, by virtue of (5),

$$\tilde{a} = a\left(\frac{\partial \xi}{\partial x}\right)^2 + 2b \frac{\partial \xi}{\partial x} \frac{\partial \xi}{\partial y} + c\left(\frac{\partial \xi}{\partial y}\right)^2$$

$$\tilde{b} = a \frac{\partial \xi}{\partial x} \frac{\partial \eta}{\partial x} + b\left(\frac{\partial \xi}{\partial x} \frac{\partial \eta}{\partial y} + \frac{\partial \xi}{\partial y} \frac{\partial \eta}{\partial x}\right) + c \frac{\partial \xi}{\partial y} \frac{\partial \eta}{\partial y} \quad (22)$$

$$\tilde{c} = a\left(\frac{\partial \eta}{\partial x}\right)^2 + 2b \frac{\partial \eta}{\partial x} \frac{\partial \eta}{\partial y} + c\left(\frac{\partial \eta}{\partial y}\right)^2$$

We shall require that the functions $\xi(x, y)$ and $\eta(x, y)$ make the coefficients $\tilde{a}$ and $\tilde{c}$ vanish, that is, by virtue of (22), they should satisfy the equations

$$a\left(\frac{\partial \xi}{\partial x}\right)^2 + 2b \frac{\partial \xi}{\partial x} \frac{\partial \xi}{\partial y} + c\left(\frac{\partial \xi}{\partial y}\right)^2 = 0$$
$$a\left(\frac{\partial \eta}{\partial x}\right)^2 + 2b \frac{\partial \eta}{\partial x} \frac{\partial \eta}{\partial y} + c\left(\frac{\partial \eta}{\partial y}\right)^2 = 0 \qquad (23)$$

Since $a \neq 0$, Eqs. (23) are equivalent to the linear equations

$$\frac{\partial \xi}{\partial x} + \lambda_1(x, y) \frac{\partial \xi}{\partial y} = 0, \qquad \frac{\partial \eta}{\partial x} + \lambda_2(x, y) \frac{\partial \eta}{\partial y} = 0 \qquad (24)$$

where

$$\lambda_1 = \frac{b - \sqrt{d}}{a}, \qquad \lambda_2 = \frac{b + \sqrt{d}}{a}$$
$$\lambda_2 - \lambda_1 = \frac{2\sqrt{d}}{a}, \qquad d = b^2 - ac \qquad (25)$$

According to the classification set out in Sec. 3.1, the following three types of Eq. (19) are possible:

I. Hyperbolic type, if $d > 0$.
II. Parabolic type, if $d = 0$.
III. Elliptic type, if $d < 0$.

We shall consider the three separately.

I. HYPERBOLIC TYPE, $d > 0$. In this case Eq. (19) reduces to the canonical form

$$\frac{\partial^2 \tilde{u}}{\partial \xi \, \partial \eta} + \tilde{\Phi} = 0 \qquad (26)$$

We remark that the change of variables $\varrho = \xi + \eta$, $\sigma = \xi - \eta$ reduces Eq. (19) to another, equivalent, canonical form

$$\frac{\partial^2 u_1}{\partial \varrho^2} - \frac{\partial^2 u_1}{\partial \sigma^2} + \Phi_1 = 0 \qquad (27)$$

To prove form (26) we establish the existence of at least one pair of solutions ξ, η of Eqs. (24) which satisfy the conditions (20). We first establish the connection of these solutions with the characteristics of Eq. (19).

§ 3. LINEAR (SECOND-ORDER) DIFFERENTIAL EQUATIONS

We shall suppose there to be solutions of Eqs. (24) such that grad $\xi \neq 0$ and grad $\eta \neq 0$ in the neighborhood we are considering. Then, according to the definition (cf. Sec. 3.3), the curves

$$\xi(x, y) = C_1, \qquad \eta(x, y) = C_2 \tag{28}$$

define two families of characteristics of Eq. (19).

For further progress we shall need the following lemma.

LEMMA. *Let the function $\omega(x, y)$ of class C^1 be such that $\partial \omega / \partial y \neq 0$. In order that the family of curves $\omega(x, y) = C$ be the characteristics of Eq. (19), it is necessary and sufficient that the expression $\omega(x, y) = C$ be the general integral of one of the ordinary differential equations*

$$\frac{dy}{dx} = \lambda_1(x, y), \qquad \frac{dy}{dx} = \lambda_2(x, y) \tag{29}$$

Equations (29) are called the *differential equations of the characteristics* of Eq. (19).

Proof. Let $\omega(x, y) = C$ be a family of characteristics of Eq. (19). From the condition $\partial \omega / \partial y \neq 0$ it follows that the curves $\omega(x, y) = C$ fill the whole neighborhood under consideration. Therefore in this neighborhood the function ω satisfies one of the equations (24), for instance the equation

$$\frac{\partial \omega}{\partial x} + \lambda_1(x, y) \frac{\partial \omega}{\partial y} = 0 \tag{24'}$$

Further, on each characteristic $\omega(x, y) = C$ the result

$$\frac{\partial \omega}{\partial x} + \frac{\partial \omega}{\partial y} \frac{dy}{dx} = 0 \tag{30}$$

is true. From this, and from (24'), we conclude, by virtue of condition $\partial \omega / \partial y \neq 0$, that $\omega(x, y) = C$ is the general integral of the first of the equations (29).

Conversely, if $\omega(x, y) = C$ is the general integral of one of the equations (29), for instance the equation $y' = \lambda_1(x, y)$, then, by virtue of (30), on each line $\omega(x, y) = C$ result (24') is satisfied. However, according to the theorem of existence and uniqueness for ordinary differential equations, one integral curve $\omega(x, y) = C$ of the equation $y' = \lambda_1$ passes through each point of the neighborhood we are considering. Therefore Eq. (24′) is satisfied at all points of this neighborhood. We conclude, since $\omega \in C^1$, $\partial \omega/\partial y \neq 0$, that the curves $\omega(x, y) = C$ are characteristics of Eq. (19). The lemma is proved.

On the basis of the lemma we have just proved, the general integrals of Eqs. (29), $\xi(x, y) = C_1$ and $\eta(x, y) = C_2$, that are such that ξ and $\eta \in C^1$, $\partial \xi/\partial y \neq 0$ and $\partial \eta/\partial y \neq 0$ define two families of characteristics of Eq. (19). As follows from the general theory of ordinary differential equations*, such integrals exist in, possibly, a smaller neighborhood. Using this result, insofar as $\lambda_i \in C^2$, then ξ and $\eta \in C^2$ and, by virtue of (29) and (25),

$$D\left(\frac{\xi, \eta}{x, y}\right) = \frac{\partial \xi}{\partial x}\frac{\partial \eta}{\partial y} - \frac{\partial \xi}{\partial y}\frac{\partial \eta}{\partial x} = \frac{\partial \xi}{\partial y}\frac{\partial \eta}{\partial y}(\lambda_2 - \lambda_1)$$

$$= 2\frac{\sqrt{d}}{a}\frac{\partial \xi}{\partial y}\frac{\partial \eta}{\partial y} \neq 0 \qquad (31)$$

In this way, the family of characteristics (28) forms a family of coordinate lines (Fig. 8) and the functions $\xi(x, y)$ and $\eta(x, y)$ can be taken as new

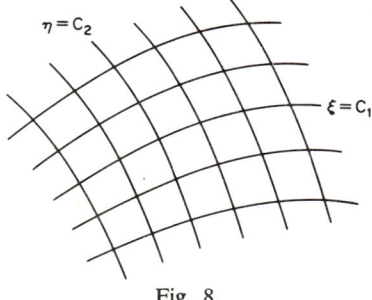

Fig. 8

variables. Using this in Eq. (21) gives $\tilde{a} \equiv \tilde{c} \equiv 0$ and, by virtue of (22) and (29),

$$\tilde{b} = [a\lambda_1\lambda_2 - b(\lambda_1 + \lambda_2) + c]\frac{\partial \xi}{\partial y}\frac{\partial \eta}{\partial y} = -\frac{2d}{a}\frac{\partial \xi}{\partial y}\frac{\partial \eta}{\partial y} \neq 0$$

Dividing Eq. (21) by the coefficient $2\tilde{b} \neq 0$, we obtain an equation in the canonical form (26).

* See, for instance, L. S. Pontryagin (*1*, Chap. 4).

§ 3. LINEAR (SECOND-ORDER) DIFFERENTIAL EQUATIONS

II. PARABOLIC TYPE, $d = 0$. Let $d \equiv 0$ in a certain neighborhood. Then Eq. (19) leads to the canonical form

$$\frac{\partial^2 \tilde{u}}{\partial \eta^2} + \tilde{\Phi} = 0 \tag{32}$$

In this case, by virtue of (25), $\lambda_1 = \lambda_2 = b/a$, so that the two differential equations (24) coincide and yield the single equation

$$\frac{\partial \xi}{\partial x} + \frac{b}{a} \frac{\partial \xi}{\partial y} = 0 \tag{33}$$

Therefore, there is one family $\xi(x, y) = C_1$ of characteristics of Eq. (19), definable, by virtue of the lemma, in terms of a general integral of the equation $y' = b/a$ such that $d\xi/dy \neq 0$; and $\xi \in C^2$. We shall select straight lines $x = C_2$ as the second family of coordinate lines. As a result of the change of variables

$$\xi = \xi(x, y), \qquad \eta = x, \qquad D\!\left(\frac{\xi, \eta}{x, y}\right) = -\frac{\partial \xi}{\partial y} \neq 0$$

we find, by virtue of (22) and (33), that

$$\tilde{a} = 0, \qquad \tilde{b} = a\frac{\partial \xi}{\partial x} + b\frac{\partial \xi}{\partial y} = 0, \qquad \tilde{c} = a$$

Dividing Eq. (21) by the coefficient $\tilde{c} = a \neq 0$, we obtain an equation in the canonical form (32).

III. ELLIPTIC TYPE, $d < 0$. Let the coefficients a, b, and c of Eq. (19) be analytic functions of the variables (x, y) in a neighborhood of a certain point (cf. Sec. 4.7). Then this equation is reducible to the canonical form.

$$\frac{\partial^2 \tilde{u}}{\partial \xi^2} + \frac{\partial^2 \tilde{u}}{\partial \eta^2} + \tilde{\Phi} = 0 \tag{34}$$

In this case, by virtue of (25), the coefficients λ_1 and λ_2 of Eqs. (24) are analytic functions, and with real (x, y), $\lambda_1 = \bar{\lambda}_2$. It follows from Kowalewski's theorem (cf. Sec. 4.7) that in a sufficiently small neighborhood there is an analytic solution $\omega(x, y)$ of the equation*

$$\frac{\partial \omega}{\partial x} + \lambda_1(x, y) \frac{\partial \omega}{\partial y} = 0 \qquad (24')$$

satisfying the condition $\partial \omega/\partial y \neq 0$. Let us suppose

$$\xi = \frac{\omega(x, y) + \bar{\omega}(x, y)}{2}, \qquad \eta = \frac{\omega(x, y) - \bar{\omega}(x, y)}{2i} \qquad (35)$$

where $\bar{\omega} = \xi - i\eta$ is the complex function conjugate to $\omega = \xi + i\eta$; it satisfies the second of Eqs. (24):

$$\frac{\partial \bar{\omega}}{\partial x} + \lambda_2(x, y) \frac{\partial \bar{\omega}}{\partial y} = 0$$

Functions ξ and $\eta \in C^\infty$ have, by virtue of (35) and (31), a nonzero Jacobian:

$$D\left(\frac{\xi, \eta}{x, y}\right) = D\left(\frac{\xi, \eta}{\omega, \bar{\omega}}\right) D\left(\frac{\omega, \bar{\omega}}{x, y}\right)$$

$$= -\frac{1}{2i} 2 \frac{\sqrt{d}}{a} \frac{\partial \omega}{\partial y} \frac{\partial \bar{\omega}}{\partial y} = -\frac{\sqrt{-d}}{a} \left|\frac{\partial \omega}{\partial y}\right|^2 \neq 0$$

Therefore the functions ξ and η may be taken as new variables. Let us see what form Eq. (19) will take in these variables. According to this the function ω satisfies the equation

$$a\left(\frac{\partial \omega}{\partial x}\right)^2 + 2b \frac{\partial \omega}{\partial x} \frac{\partial \omega}{\partial y} + c\left(\frac{\partial \omega}{\partial y}\right)^2 = 0$$

Separating the real and the imaginary parts and using (35), we obtain

$$a\left(\frac{\partial \xi}{\partial x}\right)^2 + 2b \frac{\partial \xi}{\partial x} \frac{\partial \xi}{\partial y} + c\left(\frac{\partial \xi}{\partial y}\right)^2 = a\left(\frac{\partial \eta}{\partial x}\right)^2 + 2b \frac{\partial \eta}{\partial x} \frac{\partial \eta}{\partial y} + c\left(\frac{\partial \eta}{\partial y}\right)^2$$

$$a \frac{\partial \xi}{\partial x} \frac{\partial \eta}{\partial x} + b\left(\frac{\partial \xi}{\partial x} \frac{\partial \eta}{\partial y} + \frac{\partial \xi}{\partial y} \frac{\partial \eta}{\partial x}\right) + c \frac{\partial \xi}{\partial y} \frac{\partial \eta}{\partial y} = 0$$

Taking formulas (22) into account, we conclude from this that in terms of

---

* This assertion is true even without postulating the analyticity of the coefficients a, b, and c [cf. I. N. Vekua (*1*)]. The assumption of the analyticity of the coefficients allows us to use Kowalewski's theorem about the solubility of Eq. (24') with complex coefficients (when $d < 0$ this equation is called Beltrami's equation).

§ 3. LINEAR (SECOND-ORDER) DIFFERENTIAL EQUATIONS

the variables ξ, η we have $\tilde{a} = \tilde{c}$ and $\tilde{b} = 0$. Further, since $d < 0$ and $\partial \xi / \partial y \neq 0$, then $\tilde{a} = \tilde{c} \neq 0$. Dividing Eq. (21) by $\tilde{a} = \tilde{c} \neq 0$, we reduce it to the canonical form (34).

5. Example. Tricomi's Equation. As was mentioned in Sec. 3.1, Tricomi's equation (12)

$$y \frac{\partial^2 u}{\partial x^2} + \frac{\partial^2 u}{\partial y^2} = 0 \tag{12}$$

belongs to the mixed type: when $y < 0$ it is of hyperbolic type, but when $y > 0$ it is of elliptic type, since $d = -y$. Tricomi's equation is of interest in gas dynamics.

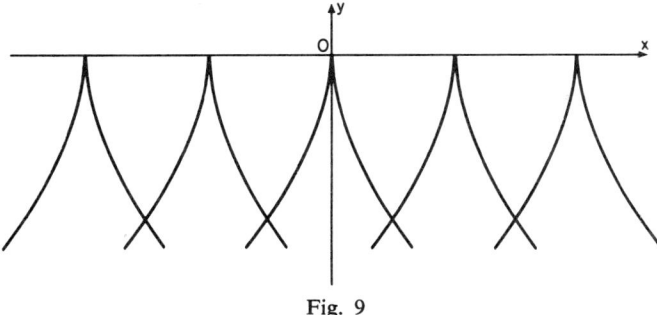

Fig. 9

With $y < 0$ the equations of the characteristics (29) assume the form $y' = \pm 1/\sqrt{-y}$. Therefore the curves (Fig. 9)

$$\tfrac{2}{3} x + \sqrt{-y^3} = C_1, \qquad \tfrac{2}{3} x - \sqrt{-y^3} = C_2$$

are characteristics of Tricomi's equation. The transformation

$$\xi = \tfrac{2}{3} x + \sqrt{-y^3}, \qquad \eta = \tfrac{2}{3} x - \sqrt{-y^3}$$

reduces Tricomi's equation to the canonical form

$$\frac{\partial^2 \tilde{u}}{\partial \xi \partial \eta} - \frac{1}{6(\xi - \eta)} \left(\frac{\partial \tilde{u}}{\partial \xi} - \frac{\partial \tilde{u}}{\partial \eta} \right) = 0, \qquad \xi > \eta$$

If $y > 0$, then in agreement with the theory of Sec. 3.4, $\omega = \tfrac{2}{3} x - i\sqrt{y^3}$ and insertion into (35) to give

$$\xi = \tfrac{2}{3} x, \qquad \eta = -\sqrt{y^3}$$

reduces Tricomi's equation to the canonical form

$$\frac{\partial^2 \tilde{u}}{\partial \xi^2} + \frac{\partial^2 \tilde{u}}{\partial \eta^2} + \frac{1}{3\eta} \frac{\partial \tilde{u}}{\partial \eta} = 0, \quad \eta < 0$$

§ 4. Formulation of Boundary Value Problems for Linear Second-Order Differential Equations

1. Classification of Boundary Value Problems. As was shown in Sec. 2.2, the second-order linear differential equation

$$\varrho \frac{\partial^2 u}{\partial t^2} = \text{div}(p \text{ grad } u) - qu + F(x, t) \tag{1}$$

describes the processes of vibration, equation

$$\varrho \frac{\partial u}{\partial t} = \text{div}(p \text{ grad } u) - qu + F(x, t) \tag{2}$$

describes the processes of diffusion, and lastly, equation

$$-\text{div}(p \text{ grad } u) + qu = F(x) \tag{3}$$

describes the corresponding steady state processes.

Let $G \subset R^n$ be the region in which the process is taking place, and S be its boundary, which we shall consider to be a piecewise smooth surface. In this way, G is the region over which the argument x in Eq. (3) varies—*the*

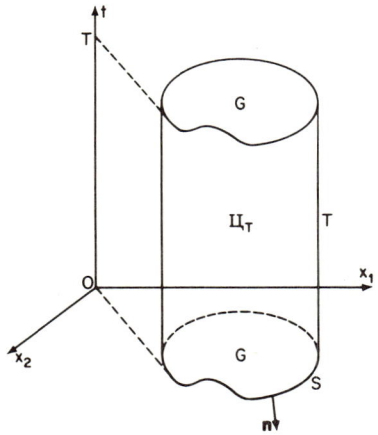

Fig. 10

§ 4. FORMULATION OF BOUNDARY VALUE PROBLEMS

region of definition of Eq. (3). The region of definition of Eqs. (1) and (2) will be a cylinder $\Pi_T = G \times (0, T)$ of height T and with base G. Its boundary consists of a side surface $S \times [0, T]$ and two bases, a lower $\bar{G} \times \{0\}$ and an upper $\bar{G} \times \{T\}$ (Fig. 10).

We shall suppose that the coefficients ϱ, p, and q of Eqs. (1)–(3) do not depend on the time t; further, in agreement with their physical sense, we shall suppose that $\varrho(x) > 0$, $p(x) > 0$, $q(x) \geq 0$, $x \in \bar{G}$. Finally, in agreement with the mathematical meaning of Eqs. (1)–(3), it is necessary to consider that $\varrho \in C(\bar{G})$, $p \in C^1(\bar{G})$, and $q \in C(\bar{G})$.

With these suppositions, according to the classification of Sec. 3, the equation of vibration (1) is of hyperbolic type, the equation of diffusion (2) is of parabolic type, and the steady state form of Eq. (3) is of elliptic type. In this way the difference in type of the equations we are considering is closely linked with the difference in the physical processes which they describe.

As mentioned in Sec. 2, it is necessary, for a full description of a physical process, to give not only the equation which describes this process, but also to give the initial state of this process (initial conditions) and the behavior on the boundary of the region in which the process is taking place (boundary conditions). Mathematically this is linked with the nonuniqueness of the solution of a differential equation. In fact, even for ordinary differential equations of nth order the general solution depends on n arbitrary constants. For equations involving partial derivatives the solution, generally speaking, depends on arbitrary functions; for instance, the general solution of the equation $(\partial^2 u/\partial x\, \partial y) = 0$ has the form $u(x, y) = f(x) + g(y)$, where f and g are arbitrary functions belonging to the class C^2. Therefore, to isolate a particular solution describing a real physical process, it is necessary to give additional conditions. These additional conditions are *boundary value conditions: or initial and boundary conditions*. The corresponding problem is called a boundary value problem. In this way, the following three basic types of boundary value problems for differential equations may be distinguished.

(a) *Cauchy's problem* for equations of hyperbolic and parabolic type: Initial conditions are given, the region G coincides with the whole space R^n and boundary conditions are absent.

(b) *A boundary value problem* for equations of elliptic type: Boundary conditions on the boundary S are given, the initial conditions, naturally, are absent.

(c) *A mixed problem* for equations of hyperbolic and parabolic type: Initial and boundary conditions are given, $G \neq R^n$.

We now describe in more detail the formulation of each of the boundary value problems enumerated here for Eqs. (1)–(3).

2. Cauchy's Problem. For the vibration equation (1) (hyperbolic type) Cauchy's problem is formulated in the following way: to find the function $u(x, t)$ of class $C^2(t > 0) \cap C^1(t \geq 0)$ satisfying Eq. (1) in the half-space $t > 0$ and the initial conditions when $t = 0$:

$$u|_{t=+0} = u_0(x), \quad \frac{\partial u}{\partial t}\bigg|_{t=+0} = u_1(x) \tag{4}$$

With this it is necessary that

$$F \in C(t > 0), \quad u_0 \in C^1(R^n), \quad u_1 \in C(R^n)$$

For the diffusion equation (2) (parabolic type) Cauchy's problem is formulated thus: to find the function $u(x, t)$ of class $C^2(t > 0) \cap C(t \geq 0)$ satisfying Eq. (2) in the half-space $t > 0$ and the initial condition with $t = +0$:

$$u|_{t=+0} = u_0(x) \tag{5}$$

With this, it is necessary that $F \in C(t > 0)$, $u_0 \in C(R^n)$.

This formulation of Cauchy's problem permits the following generalization. Let the second-order differential equation

$$\frac{\partial^2 u}{\partial t^2} = \sum_{i,j=1}^{n} a_{ij} \frac{\partial^2 u}{\partial x_i \partial x_j} + \sum_{i=1}^{n} a_{i0} \frac{\partial^2 u}{\partial x_i \partial t}$$
$$+ \Phi\left(x, t, u, \frac{\partial u}{\partial x_i}, \ldots, \frac{\partial u}{\partial x_n}, \frac{\partial u}{\partial t}\right) \tag{6}$$

the piecewise smooth surface $\Sigma = [t = \sigma(x)]$, and the functions u and u_1 on Σ (Cauchy data) be given. Cauchy's problem for Eq. (6) consists of finding, in a certain part of the region $t > \sigma(x)$, adjoining the surface Σ, a solution $u(x, t)$ satisfying on Σ the boundary value conditions

$$u|_{\Sigma} = u_0, \quad \frac{\partial u}{\partial \mathbf{n}}\bigg|_{\Sigma} = u_1 \tag{7}$$

where **n** is the normal to Σ in the direction of increasing t (Fig. 11).

§ 4. FORMULATION OF BOUNDARY VALUE PROBLEMS

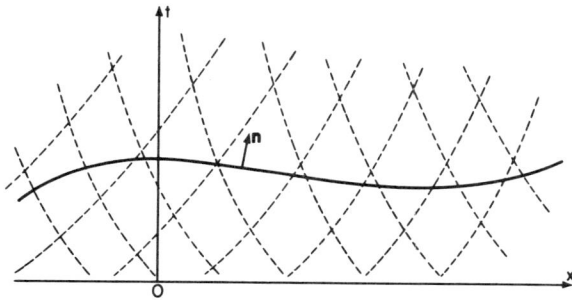

Fig. 11

3. The Role of Characteristics in Formulating Cauchy's Problem. Let us suppose that the surface Σ belongs to the class C^2 (cf. Sec. 1.1) and that at no point does it touch the characteristic surface (cf. Sec. 3.3) of Eq. (6); that is, on Σ the inequality

$$\tilde{a}_{00} \equiv 1 - \sum_{i,j=1}^{n} a_{ij} \frac{\partial \sigma}{\partial x_i} \frac{\partial \sigma}{\partial x_j} - \sum_{i=1}^{n} a_{i0} \frac{\partial \sigma}{\partial x_i} \neq 0 \qquad (8)$$

is fulfilled.

Let us transform the Cauchy problem (6)–(7) to the Cauchy problem in which the Cauchy data are given on the plane $\tau = 0$. For this, instead of the variable t, we shall introduce a new variable $\tau = t - \sigma(x)$. With this change of variable, Eq. (6) for the function

$$\tilde{u}(x, \tau) = u(x, \tau + \sigma(x)) \qquad (9)$$

in the neighborhood of the surface Σ assumes the form (cf. Sec. 3.1)

$$\frac{\partial^2 \tilde{u}}{\partial \tau^2} = \frac{1}{\tilde{a}_{00}} \sum_{i,j=1}^{n} \tilde{a}_{ij} \frac{\partial^2 \tilde{u}}{\partial x_i \partial x_j} + \frac{1}{\tilde{a}_{00}} \sum_{i=1}^{n} \tilde{a}_{i0} \frac{\partial^2 \tilde{u}}{\partial x_i \partial \tau} + \tilde{\Phi} \qquad (10)$$

insofar as, by virtue of (8), $\tilde{a}_{00} \neq 0$ on Σ. Because of this the surface Σ maps into the plane $\tau = 0$, and the boundary value conditions (7), by virtue of (9), assume the form

$$\tilde{u}\big|_{\tau=0} = u\big|_{\Sigma} = u_0(x), \qquad \frac{\partial \tilde{u}}{\partial \tau}\bigg|_{\tau=0} = \frac{\partial u}{\partial t}\bigg|_{\Sigma} \qquad (11)$$

It remains to find $\partial u/\partial t$ on Σ. Differentiating the first of the boundary value conditions (7), $u_0(x) = u(x, \sigma(x))$, along x_i, we obtain the n results on Σ

$$\frac{\partial u_0}{\partial x_i} = \frac{\partial u}{\partial t} \frac{\partial \sigma}{\partial x_i} + \frac{\partial u}{\partial x_i}, \qquad i = 1, 2, \ldots, n \qquad (12)$$

Differentiating the function $u(x, t)$ along the normal

$$\mathbf{n} = \left(\frac{1}{\Delta}, -\frac{1}{\Delta} \operatorname{grad} \sigma\right), \quad \Delta = \sqrt{1 + |\operatorname{grad} \sigma|^2}$$

and taking into account the second of the boundary value conditions (7), we obtain one further result on Σ:

$$u_1 = \frac{\partial u}{\partial t} \frac{1}{\Delta} - \frac{1}{\Delta} \sum_{i=1}^{n} \frac{\partial u}{\partial x_i} \frac{\partial \sigma}{\partial x_i} \tag{13}$$

The system of linear algebraic equations (12)–(13) has a unique solution as regards the magnitudes of $\partial u/\partial x_i$, $i = 1, 2, \ldots, n$, and $\partial u/\partial t$, at each point of the surface Σ, since the determinant

$$\begin{vmatrix} \dfrac{\partial \sigma}{\partial x_1} & 1 & \cdots & 0 \\ \vdots & \vdots & & \vdots \\ \dfrac{\partial \sigma}{\partial x_n} & 0 & \cdots & 1 \\ \dfrac{1}{\Delta} & -\dfrac{1}{\Delta}\dfrac{\partial \sigma}{\partial x_1} & \cdots & -\dfrac{1}{\Delta}\dfrac{\partial \sigma}{\partial x_n} \end{vmatrix} = (-1)^n \Delta \neq 0$$

4. Boundary Value Problem for Equations of Elliptic Type. A boundary value problem for Eq. (3) (elliptic type) consists of finding the function $u(x)$ of class $C^2(G) \cap C^1(\bar{G})$, satisfying, in the region G, Eq. (3) together with a boundary condition on S of the form

$$\left. \alpha u + \beta \frac{\partial u}{\partial \mathbf{n}} \right|_S = \nu \tag{14}$$

where α, β, and ν are given continuous functions on S, with $\alpha \geq 0$, $\beta \geq 0$, $\alpha + \beta > 0$.

The following types of boundary conditions (14) are distinguished:

Boundary condition of the first kind ($\alpha = 1$, $\beta = 0$)

$$u|_S = u_0 \tag{15}$$

Boundary condition of the second kind ($\alpha = 0$, $\beta = 1$)

$$\left. \frac{\partial u}{\partial \mathbf{n}} \right|_S = u_1 \tag{16}$$

§ 4. FORMULATION OF BOUNDARY VALUE PROBLEMS

Boundary condition of the third kind ($\beta = 1$, $\alpha \geq 0$)

$$\frac{\partial u}{\partial \mathbf{n}} + \alpha u \bigg|_S = u_2 \tag{17}$$

The corresponding boundary value problems are called *boundary value problems of the first, second, and third kinds*.

For the equations of Laplace and Poisson (cf. Sec. 2.3) the boundary value problem of the first kind

$$\Delta u = -f, \qquad u|_S = u_0 \tag{18}$$

is called *Dirichlet's problem*; a boundary value problem of the second kind

$$\Delta u = -f, \qquad \frac{\partial u}{\partial \mathbf{n}} \bigg|_S = u_1 \tag{19}$$

is called *Neumann's problem*.

5. Mixed Problem. For the vibration equation (1) (hyperbolic type) a mixed problem is posed in the following way: to find the function $u(x, t)$ of class $C^2(\Pi_T) \cap C^1(\bar{\Pi}_T)$ satisfying Eq. (1) in the cylinder Π_T, the initial conditions (4) when $t = 0$, $x \in \bar{G}$ (on the lower base of the cylinder Π_T), and the boundary conditions (14) where $x \in S$, $0 \leq t \leq T$ (on the side surface of the cylinder Π_T). In addition to this it is necessary that the smoothness condition

$$F \in C(\Pi_T), \qquad u_0 \in C^1(\bar{G}), \qquad u_1 \in C(\bar{G}), \qquad v \in C(S \times [0, T))$$

and the consistency condition

$$\alpha u_0 + \beta \frac{\partial u_0}{\partial \mathbf{n}} \bigg|_S = v|_{t=0}$$

be satisfied.

Analogously, for the diffusion equation (2) (parabolic type) a mixed problem is posed thus: to find the function $u(x, t)$ of class $C^2(\Pi_T) \cap C(\bar{\Pi}_T)$, $\mathrm{grad}_x u \in C(\bar{\Pi}_T)$, satisfying Eq. (2) in Π_T, the initial condition (5), and the boundary condition (14).

Note. The solutions of the boundary value problems just formulated with smoothness C^1 up to the boundary of the region of definition of the problem do not always exist. Therefore we sometimes cannot insist on such smoothness, but we can require, for example, that the solution only be continuous up to the boundary of the region. This formulation is natural in problems which do not contain first derivatives in the boundary value conditions, e.g., for Eqs. (2) and (3) with boundary conditions of the first kind. If first derivatives enter into boundary value conditions, in each particular situation it is necessary to show the sense in which these boundary value conditions must be fulfilled. For example, for a mixed problem for Eq. (1), the fulfillment of the second of the initial conditions (4) may be required in the sense $\mathscr{L}_2(G)$:

$$\left\| \frac{\partial u}{\partial t} - u_1 \right\| \to 0, \qquad t \to +0 \qquad (20)$$

For Neumann's problem for the Laplace equation the fulfillment of boundary condition (16) may be required in the following sense

$$\frac{\partial u(x')}{\partial \mathbf{n}_x} \xrightarrow{x \in S} u_1(x), \qquad x' \to x, \qquad x' \in G, \qquad x' \in -\mathbf{n}_x \qquad (21)$$

6. The Correctness of Formulation of Problems of Mathematical Physics. Insofar as problems of mathematical physics describe real physical processes, the mathematical formulation of these problems must satisfy the following natural requirements:

(a) The solution must *exist* within a certain class of functions $\mathscr{M}_1$.

(b) The solution must be *unique* within a certain class of functions $\mathscr{M}_2$.

(c) The solution must depend *continuously* on the data of the problem (initial and boundary data, inhomogeneous term, coefficients of the equation, etc.). The continuous dependence of the solution u on the data of the problem $\tilde{u}$ signifies the following:

Let the sequence of data $\tilde{u}_k$, $k = 1, 2, \ldots$, in some sense tend toward $\tilde{u}$ and let u_k, $k = 1, 2, \ldots$, and u be the corresponding solutions of the problem; then $u_k \to u$, $k \to \infty$ in a manner of convergence properly chosen. For example, let the problem lead to the equation $Lu = F$, where L is the linear operator mapping $\mathscr{M}$ into $\mathscr{N}$, where $\mathscr{M}$ and $\mathscr{N}$ are linear normed spaces. In this way the continuous dependence of the solution u on the inhomogeneous term F will be guaranteed if the operator L^{-1} exists and is bounded from $\mathscr{N}$ into $\mathscr{M}$ (cf. Secs. 1.8 and 1.9). The continuous dependence of the solution is necessary, because as a rule the data of a physical problem

§ 4. FORMULATION OF BOUNDARY VALUE PROBLEMS

are defined from experiment, approximately, and so one must be sure that the solution of the problem is not essentially dependent on errors in measurement.

A problem which satisfies the requirements (a)–(c) listed above is called *correctly posed*, and the corresponding set of functions $\mathcal{M}_1 \cap \mathcal{M}_2$ is called the *correctness class*.

Finding the correct formulations for problems of mathematical physics and methods of constructing their solutions (exact or approximate) is the main content of the study of the equations of mathematical physics.

7. Kowalewski's Theorem. At this point we shall select a fairly general class of Cauchy problems for which a solution exists and is unique. First of all we shall introduce two definitions.*

(1) The system of N differential equations with N unknown functions $u_1, u_2, \ldots, u_N$

$$\frac{\partial^{k_i} u_i}{\partial t^{k_i}} = \Phi_i\left(x, t, u_1, u_2, \ldots, u_N, \ldots, \frac{\partial^{\alpha_0}}{\partial t^{\alpha_0}} D_x^\alpha u_j, \ldots\right) \qquad (22)$$

$$i = 1, 2, \ldots, N$$

is called *normal with respect to the variable* t if the right-hand sides Φ_i do not contain derivatives of an order higher than k_i or derivatives in t of an order higher than $k_i - 1$; that is,

$$\alpha_0 + \alpha_1 + \cdots + \alpha_n \leq k_i, \qquad \alpha_0 \leq k_i - 1$$

For example, the wave equation, Laplace's equation, and the equation of heat conduction are normal with respect to each variable x; the wave equation, moreover, is normal with respect to t.

(2) A function $f(x)$, $x = (x_1, x_2, \ldots, x_n)$, is called *analytic at the point* x_0 if in a certain neighborhood of this point it can be represented in the form of a uniformly converging power series

$$f(x) = \sum_{|\alpha| \geq 0} c_\alpha (x - x_0)^\alpha = \sum_{|\alpha| \geq 0} \frac{D^\alpha f(x_0)}{\alpha!} (x - x_0)^\alpha$$

(the point x_0 can also be complex). If a function $f(x)$ is analytic at each point of some region G, then it is said to be *analytic in the region* G.

---

* The notation used below was introduced in Sec. 1.2.

For a system of equations (22), normal with respect to t, we shall formulate the following Cauchy problem: to find the solution $u_1, u_2, \ldots, u_N$ of this system, satisfying the initial conditions when $t = t_0$:

$$\left.\frac{\partial^k u_i}{\partial t^k}\right|_{t=t_0} = \varphi_{ik}(x), \quad k = 0, 1, \ldots, k_i - 1, \quad i = 1, 2, \ldots, N \qquad (23)$$

where $\varphi_{ik}(x)$ are the given functions in a certain region $G \subset R^n$.

THEOREM OF KOWALEWSKI. *If all the functions $\varphi_{ik}(x)$ are analytic in a certain neighborhood of the points x_0 and all the functions $\Phi_i(x, t, \ldots, u_{j\alpha_0\alpha_1\ldots\alpha_n}, \ldots)$ are analytic in a certain neighborhood of the point*

$$(x_0, t_0, \ldots, D^\alpha \varphi_{j\alpha_0}(x_0), \ldots)$$

then Cauchy's problem (22)–(23) has an analytic solution in a certain neighborhood of the point (x_0, t_0) and is, moreover, unique in the class of analytic functions.

To prove this theorem, a solution $u_1, u_2, \ldots, u_N$ in the neighborhood of a point (x_0, t_0) is sought in the form of a power series

$$u_i(x, t) = \sum_{\alpha_0 \geq 0, |\alpha| \geq 0} \frac{(\partial^{\alpha_0}/\partial t^{\alpha_0}) D_x^\alpha u_i(x_0, t_0)}{\alpha_0! \alpha!} (t - t_0)^{\alpha_0}(x - x_0)^\alpha \qquad (24)$$

From the initial conditions (23) and from Eqs. (22), all the derivatives $(\partial^{\alpha_0}/\partial t^{\alpha_0}) D_x^\alpha u_i$ at the point (x_0, t_0) may be successively determined. The uniform convergence of the series (24) in a certain neighborhood of the point (x_0, t_0) may be proved by the method of majorants. The uniqueness of this solution constructed in the class of analytic functions follows from the theory of uniqueness for analytic functions.

Detailed proofs of Kowalewski's theorem are contained, for instance, in the books of I. G. Petrovsky (*1*), R. Courant (*1*), and G. N. Polozhy (*1*).

8. Hadamard's Example. Kowalewski's theorem, in spite of its general character, does not fully answer the question of the correctness of formulation of Cauchy's problem for a normal system of differential equations. In fact, this theorem guarantees the existence and uniqueness of the solution only in a sufficiently small neighborhood or, as it is said, *in the small*; usually these facts need to be established in preassigned (and not at all small) regions or, as it is said, *in the large*. Further, the initial data and the

§4. FORMULATION OF BOUNDARY VALUE PROBLEMS

inhomogeneous term of the equation, as a rule, are nonanalytic functions. Finally, there may not be continuous dependence of the solution on the initial data. This is shown by an example first constructed by Hadamard.

The solution of the Cauchy problem

$$u|_{t=0} = 0, \quad \frac{\partial u}{\partial t}\bigg|_{t=0} = \frac{1}{k}\sin kx$$

for Laplace's equation

$$\frac{\partial^2 u}{\partial t^2} = -\frac{\partial^2 u}{\partial x^2}$$

is

$$u_k(x, t) = \frac{\sinh kt}{k^2}\sin kx$$

If $k \to +\infty$, then $(1/k)\sin kx \to 0$ for all x; nevertheless, when $x \neq j\pi$, $j = 0 \pm 1, \ldots$,

$$u_k(x, t) = \frac{\sinh kt}{k^2}\sin kx \not\to 0, \quad k \to \infty$$

In this way, the Cauchy problem for Laplace's equation is incorrectly posed (in the sense of the definition of Sec. 4.6). Nevertheless, correct formulations of this problem are possible. For instance, in a class of functions bounded by a fixed constant, this problem is correctly posed under the condition that its solution exists (the latter requirement leads to fully defined restrictions on a set of permissible initial data u_0 and u_1).

For correct formulation of Cauchy's problem for Laplace's equation and methods of solution, see M. M. Lavrentiev (1).

9. Classical and Generalized Solutions. The formulations of boundary value problems which have been set out in the previous pages are characterized by the fact that their solutions are assumed to be sufficiently smooth and to satisfy the equation at each point inside the region of definition of these problems. We shall call such solutions *classical* and the formulation of the corresponding boundary value problem will be a *classical formulation*. In this way, the classical formulation of a problem assumes, for instance, the smoothness of the right-hand side of an equation within its region of definition. However, in the most interesting problems these right-hand sides (let us note that they characterize the intensity of the external perturbations) have fairly marked singularities. Therefore, for problems such as these, classical formulations are not sufficient. To formulate these problems, one must refrain (partially or completely) from insisting on the smoothness of the solution inside the region, and introduce so-called *generalized solutions*. But then the question arises: Which functions can be called the solutions of the equation? In order to answer this, it is necessary, in essence, to generalize the idea of a derivative and the idea of a function; that is, to introduce so-called *generalized functions*. We devote the following chapter to the study of this question.

CHAPTER

2

Generalized Functions

Generalized functions were first introduced into science as a result of Dirac's research into quantum mechanics, where he systematically uses the δ function. The foundations of the mathematical theory of generalized functions were laid by S. L. Sobolev (2, 1936) and L. Schwartz (2, 1950–1951). Later the theory of generalized functions was developed intensively by many mathematicians; its rapid development was stimulated mainly by the requirements of mathematical physics and especially by the theory of differential equations and the quantum field theory. At the present time the theory of generalized functions is far advanced and has numerous applications in physics and mathematics; it is becoming an essential tool for the physicist, the mathematician, and the engineer.

§ 5. Test and Generalized Functions

1. Introduction. A generalized function is a generalization of the classical concept of a function. This generalization, on one hand, allows us to express in mathematical form idealized concepts such as, for instance, the density of a material point, the strength of a point charge or dipole, the density of a simple or double layer, the strength of an instantaneous point source, the intensity of a force applied at a point, and so on. On the other hand, we find in the concept of a generalized function a reflection of the fact that it is really impossible, for instance, to measure the density of a material at a point; we can only measure its average density in a sufficiently small neighborhood of this point and call this the density at a given point. Roughly speaking, a generalized function is defined by its "average values" in the neighborhood of each point.

To clarify what has just been said, we shall examine in more detail the question of the density created by a material point of mass 1. We shall consider that this point coincides with the origin of the coordinates. We shall denote this density by $\delta(x)$.

To define the density we shall distribute (or, as it is said, spread) the unit mass uniformly inside the sphere U_ε. As a result we shall obtain the average density

$$f_\varepsilon(x) = \begin{cases} \dfrac{3}{4\pi\varepsilon^3}, & |x| < \varepsilon \\ 0, & |x| > \varepsilon \end{cases}$$

We shall first take as the density $\delta(x)$ the point limit of the sequence of average densities $f_\varepsilon(x)$, that is,

$$\delta(x) = \lim_{\varepsilon \to 0} f_\varepsilon(x) = \begin{cases} +\infty, & \text{if } x = 0 \\ 0, & \text{if } x \neq 0 \end{cases} \tag{1}$$

It is naturally required that the integral of the density δ over any volume G should give the mass of this volume, that is,

$$\int_G \delta(x)\,dx = \begin{cases} 1, & \text{if } 0 \in G \\ 0, & \text{if } 0 \bar{\in} G \end{cases}$$

But, by virtue of (1), the left-hand side of this equation is always equal to zero if the integral is taken to be improper. The contradiction here shows that the point limit of the sequence $f_\varepsilon(x)$ as $\varepsilon \to 0$ cannot be taken as the density $\delta(x)$.

We shall now calculate the *weak limit* of the sequence of functions $f_\varepsilon(x)$ as $\varepsilon \to 0$, that is, for any continuous function φ we shall find the limit of the numerical sequence $\int f_\varepsilon \varphi \, dx$ when $\varepsilon \to 0$.

We show that

$$\lim_{\varepsilon \to 0} \int f_\varepsilon(x)\varphi(x)\,dx = \varphi(0)$$

In fact, on account of the continuity of function $\varphi(x)$ for any $\eta > 0$

§ 5. TEST AND GENERALIZED FUNCTIONS

there is a $\varepsilon_0 > 0$ such that $|\varphi(x) - \varphi(0)| < \eta$ whenever $|x| < \varepsilon_0$. From this, for all $\varepsilon \leq \varepsilon_0$, we obtain

$$\left| \int f_\varepsilon(x) \varphi(x)\, dx - \varphi(0) \right| = \frac{3}{4\pi\varepsilon^3} \left| \int_{|x|<\varepsilon} [\varphi(x) - \varphi(0)]\, dx \right|$$

$$\leq \frac{3}{4\pi\varepsilon^3} \int_{|x|<\varepsilon} |\varphi(x) - \varphi(0)|\, dx$$

$$< \eta \frac{3}{4\pi\varepsilon^3} \int_{|x|<\varepsilon} dx = \eta$$

as was to be shown.

In this way, the weak limit of the sequence of functions $f_\varepsilon(x)$ as $\varepsilon \to 0$ is the functional $\varphi(0)$, assigning to each continuous function $\varphi(x)$ the number $\varphi(0)$—its value at the point $x = 0$. It is this functional which is taken as the definition of the density $\delta(x)$, and this is the well-known Dirac δ function.

So $f_\varepsilon(x) \to \delta(x)$ as $\varepsilon \to 0$, in the sense that for any continuous function $\varphi(x)$ the limiting result

$$\int f_\varepsilon(x) \varphi(x)\, dx \to (\delta, \varphi), \qquad \varepsilon \to 0$$

is valid, where the symbol (δ, φ) denotes the number $\varphi(0)$—the value of the functional δ acting on function φ.

To recover the complete mass, it is necessary to act with the functional (density) $\delta(x)$ on the function $\varphi(x) = 1$, $(\delta, 1) = 1$.

If the mass m is concentrated at the point $x = 0$, the corresponding density must be considered equal to $m\delta(x)$. If mass m is concentrated at the point x_0, its density is naturally considered equal to $m\delta(x - x_0)$, where $(m\delta(x - x_0), \varphi) = m\varphi(x_0)$. In general, if masses m_k are concentrated at different points x_k, $k = 1, 2, \ldots, N$, the corresponding density is equal to

$$\sum_{k=1}^{N} m_k \delta(x - x_k)$$

2. The Space of Test Functions $\mathscr{D}$. This example of the δ function shows that it is defined by means of continuous functions as a linear continuous functional over these functions (cf. Sec. 1.8). The continuous functions are said to be *test functions* for the δ function. This point of view is also taken as the basis of definition of an arbitrary generalized function as a linear continuous functional over a space of sufficiently "good" (test) functions. It is clear that the narrower the space of test functions, the more linear continuous functionals there are over it. On the other hand, the supply of test functions must be sufficiently large. In this subsection we introduce an important space of test functions $\mathscr{D}$. We first give some definitions.

Let us relate to the set of *test functions* $\mathscr{D} = \mathscr{D}(R^n)$ *all the infinitely differentiable functions in R^n with compact support*. We shall define convergence in $\mathscr{D}$ as follows: The sequence of functions $\varphi_1, \varphi_2, \ldots,$ from $\mathscr{D}$ converges to the function φ (belonging to $\mathscr{D}$) if: (a) there is a number $R > 0$ such that supp $\varphi_k \subset U_R$; (b) for each $\alpha = (\alpha_1, \alpha_2, \ldots, \alpha_n)$

$$D^\alpha \varphi_k(x) \xrightarrow{x \in R^n} D^\alpha \varphi(x), \quad k \to \infty$$

In this case we shall write: $\varphi_k \to \varphi$ as $k \to \infty$ in $\mathscr{D}$.

Evidently $\mathscr{D}$ is a linear space (cf. Sec. 1.2).

The operation of differentiation $D^\beta \varphi(x)$ is continuous from $\mathscr{D}$ into $\mathscr{D}$ (cf. Sec. 1.8). In fact, from the definition of convergence in $\mathscr{D}$ it follows that if $\varphi_k \to \varphi$ as $k \to \infty$ in $\mathscr{D}$, then $D^\beta \varphi_k \to D^\beta \varphi$ as $k \to \infty$ in $\mathscr{D}$.

Analogously, *the operations of nonsingular linear change of variable $\varphi(Ay + b)$ and multiplication by a function $a \in C^\infty(R^n)$, $a(x)\varphi(x)$, are continuous from $\mathscr{D}$ into $\mathscr{D}$*.

The set of test functions, the supports of which are contained in the given region G, is denoted by $\mathscr{D}(G)$ (cf. Sec. 1.2); in this way

$$\mathscr{D}(G) \subset \mathscr{D}(R^n) = \mathscr{D}$$

The question then arises: Are there test functions distinct from being identically zero? It is clear that such functions cannot be analytic in R^n (cf. Sec. 4.7). As an example of a test function distinct from a zero function, there is the "cap-shaped function" (Fig. 12)

$$\omega_\varepsilon(x) = \begin{cases} C_\varepsilon \exp\left(-\dfrac{\varepsilon^2}{\varepsilon^2 - |x|^2}\right), & |x| \leq \varepsilon \\ 0, & |x| > \varepsilon \end{cases}$$

We shall choose a constant C_ε so that

$$\int \omega_\varepsilon(x)\, dx = 1$$

§ 5. TEST AND GENERALIZED FUNCTIONS   67

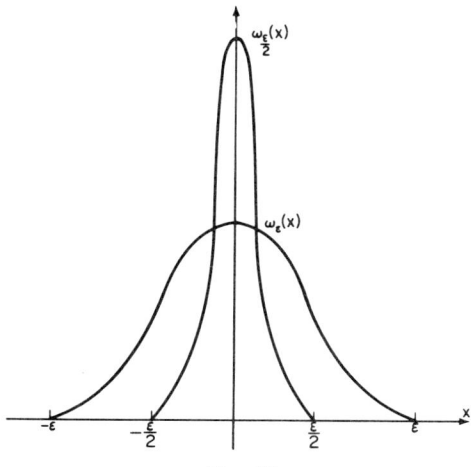

Fig. 12

that is,

$$C_\varepsilon \varepsilon^n \int_{U_1} \exp\left(-\frac{1}{1-|\xi|^2}\right) d\xi = 1$$

The following lemma gives numerous other examples of test functions.

LEMMA 1. *For any region G and any number $\varepsilon > 0$ there is a function $\eta \in C^\infty(R^n)$ such that*

$$0 \le \eta(x) \le 1; \quad \eta(x) = 1, \ x \in G_\varepsilon; \quad \eta(x) = 0, \ x \bar{\in} G_{3\varepsilon}$$

Proof. Let $\chi(x)$ be the characteristic function of the set $G_{2\varepsilon}$: $\chi(x) = 1$ for $x \in G_{2\varepsilon}$; $\chi(x) = 0$ for $x \notin G_{2\varepsilon}$. Then the function

$$\eta(x) = \int \chi(y) \omega_\varepsilon(x - y) \, dy$$

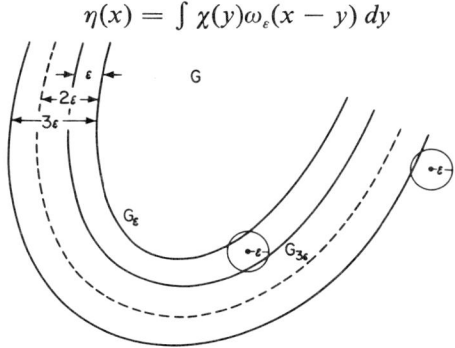

Fig. 13

has the required properties. In fact, since $\omega_\varepsilon \in \mathscr{D}$, $0 \leq \omega_\varepsilon(x)$, supp $\omega_\varepsilon = \bar{U}_\varepsilon$, $\int \omega_\varepsilon(x)\,dx = 1$, then (Fig. 13)

$$\eta(x) = \int_{G_{2\varepsilon}} \omega_\varepsilon(x-y)\,dy \in C^\infty(R^n)$$

$$0 \leq \eta(x) \leq \int \omega_\varepsilon(x-y)\,dy = \int \omega_\varepsilon(\xi)\,d\xi = 1$$

$$\eta(x) = \int_{U(x;\varepsilon)} \chi(y)\omega_\varepsilon(x-y)\,dy$$

$$= \begin{cases} \int_{U(x;\varepsilon)} \omega_\varepsilon(x-y)\,dy = \int \omega_\varepsilon(\xi)\,d\xi = 1, & x \in G_\varepsilon \\ 0, & x \bar{\in} G_{3\varepsilon} \end{cases}$$

The lemma is proved.

It follows from the lemma just proved: (a) If the region G is bounded, there is a function $\eta \in \mathscr{D}$ such that $\eta(x) = 1$, for $x \in G_\varepsilon$. (b) If $G' \Subset G$, there is a function $\eta \in \mathscr{D}(G)$ such that $\eta(x) = 1$ for $x \in G'$.

LEMMA 2 (expansion of the unit function). *Let the support of the test function φ be covered with a finite number of neighborhoods $U(x_k; r_k)$, $k = 1, 2, \ldots, N$. Then there will be functions $\varphi h_k \in \mathscr{D}$ such that*

$$\varphi(x) = \sum_{k=1}^{N} \varphi(x) h_k(x) \tag{2}$$

$$\operatorname{supp} \varphi h_k \subset U(x_k; r_k)$$

Proof. Let us take the diminished neighborhoods $U(x_k; r'_k)$, $r'_k < r_k$, $k = 1, 2, \ldots, N$, covering the compactum supp φ (Fig. 14). According to

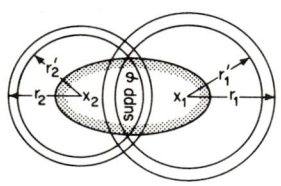

Fig. 14

Lemma 1 of Sec. 5.2 there are test functions $e_k(x)$ such that

$$e_k(x) = 1, \quad x \in U(x_k; r_k'), \quad \operatorname{supp} e_k \subset U(x_k; r_k)$$

We shall set

$$h(x) = \sum_{k=1}^{N} e_k(x), \quad h_k(x) = \frac{e_k(x)}{h(x)}$$

According to the construction $h(x) \geq 1$, $\sum_{k=1}^{N} h_k(x) = 1$ in the neighborhood of $\operatorname{supp} \varphi$, and therefore $\varphi h_k \in \mathscr{D}$, $\operatorname{supp} \varphi h_k \subset U(x_k; r_k)$, and Eq. (2) is true. The lemma is proved.

3. The Space of Generalized Functions $\mathscr{D}'$. *Each linear continuous functional over the space of test functions $\mathscr{D}$ is known as a generalized function in the Sobolev–Schwartz sense.* In correspondence with the notation of Sec. 1.8, we shall denote by (f, φ) the effect of the functional (generalized function) f over the test function φ. We shall also formally denote the generalized function f by $f(x)$, with the understanding that x is the argument of the test functions on which the functional f acts.

Let us interpret the definition of the generalized function f.

(1) The generalized function f is a *functional* over $\mathscr{D}$; that is, a (complex) number (f, φ) is associated with each $\varphi \in \mathscr{D}$.

(2) The generalized function f is a *linear* function over $\mathscr{D}$; that is, if $\varphi \in \mathscr{D}$, $\psi \in \mathscr{D}$, and λ, μ are complex numbers, then

$$(f, \lambda\varphi + \mu\psi) = \lambda(f, \varphi) + \mu(f, \psi)$$

(3) The generalized function f is a *continuous* functional over $\mathscr{D}$; that is, if $\varphi_k \to \varphi$ as $k \to \infty$ in $\mathscr{D}$, then

$$(f, \varphi_k) \to (f, \varphi), \quad k \to \infty$$

We shall denote the set of all generalized functions by $\mathscr{D}' = \mathscr{D}'(R^n)$.

The set $\mathscr{D}'$ is linear if the linear combination $\lambda f + \mu g$ of generalized functions f and g is defined as a functional acting in accordance with the formula

$$(\lambda f + \mu g, \varphi) = \lambda(f, \varphi) + \mu(g, \varphi), \quad \varphi \in \mathscr{D}$$

We shall verify that the functional $\lambda f + \mu g$ is linear and continuous over $\mathscr{D}$; that is, that it belongs to $\mathscr{D}'$. In fact, if $\varphi \in \mathscr{D}$, $\psi \in \mathscr{D}$, and α, β are any complex numbers, then, according to the definition,

$$(\lambda f + \mu g, \alpha \varphi + \beta \psi) = \lambda(f, \alpha \varphi + \beta \psi) + \mu(g, \alpha \varphi + \beta \psi)$$
$$= \alpha[\lambda(f, \varphi) + \mu(g, \varphi)] + \beta[\lambda(f, \psi) + \mu(g, \psi)]$$
$$= \alpha(\lambda f + \mu g, \varphi) + \beta(\lambda f + \mu g, \psi)$$

and so this functional is linear. Its continuity follows from the continuity of the functionals f and g: If $\varphi_k \to \varphi$ as $k \to \infty$ in $\mathscr{D}$, then

$$(\lambda f + \mu g, \varphi_k) = \lambda(f, \varphi_k) + \mu(g, \varphi_k) \to \lambda(f, \varphi) + \mu(g, \varphi) = (\lambda f + \mu g, \varphi)$$

We shall define convergence in $\mathscr{D}'$ in the following way: The sequence of generalized functions $f_1, f_2, \ldots$, belonging to $\mathscr{D}'$ converges to the generalized function $f \in \mathscr{D}'$ if, for any $\varphi \in \mathscr{D}$, $(f_k, \varphi) \to (f, \varphi)$ as $k \to \infty$. In this case we shall write $f_k \to f$ as $k \to \infty$ in $\mathscr{D}'$. This convergence is called *weak convergence*. The linear set $\mathscr{D}'$ with the convergence which has been introduced into it is known as the *space of generalized functions* $\mathscr{D}'$.

Note. The space $\mathscr{D}'$ is complete; that is, if the sequence $f_1, f_2, \ldots$, belonging to $\mathscr{D}'$ is such that for any test function $\varphi \in \mathscr{D}$ there is a limit of the numerical sequence $(f_1, \varphi), (f_2, \varphi), \ldots$, then there will be a generalized function (obviously unique) $f \in \mathscr{D}'$, such that

$$\lim_{k \to \infty} (f_k, \varphi) = (f, \varphi), \qquad \varphi \in \mathscr{D}$$

Proof of this theorem may be found, for instance, in the book of G. E. Shilov (2, Sec. 9). In future, the completeness of the space $\mathscr{D}'$ will not be used.

4. The Support of a Generalized Function. It is clear from the definition that generalized functions, roughly speaking, have no values at separate points. Nonetheless it is possible to say that a generalized function becomes zero in a region.

The generalized function f *becomes zero in the region* G if $(f, \varphi) = 0$ for all $\varphi \in \mathscr{D}(G)$. We shall write this fact thus: $f = 0$ for $x \in G$. In correspondence with this definition, generalized functions f and g are said to be *equal in the region* G if $f - g = 0$ for $x \in G$; in this case we write $f = g$ for $x \in G$. Specifically, the generalized functions f and g are said to be equal, $f = g$, if for all $\varphi \in \mathscr{D}$, $(f, \varphi) = (g, \varphi)$.

§ 5. TEST AND GENERALIZED FUNCTIONS

We shall say that the generalized function f *belongs to the class* $C^p(G)$ if in the region G it coincides with the function $f_0(x)$ of the class $C^p(G)$; that is, for any $\varphi \in \mathscr{D}(G)$

$$(f, \varphi) = \int f_0(x)\varphi(x)\, dx \tag{3}$$

LEMMA. *Let the space R^n be covered with a countable system of neighborhoods $U(x_k; r_k)$, $k = 1, 2, \ldots$, and in each neighborhood $U(x_k; r_k)$ let the generalized function f coincide with the generalized function f_k. Then f is defined in a one-to-one manner by its local elements f_k.*

Proof. Let $\varphi \in \mathscr{D}$. According to the Heine–Borel lemma (cf. Sec. 1.1), the compactum $\operatorname{supp} \varphi$ is covered by a finite number of neighborhoods $U(x_k; r_k)$, $k = 1, 2, \ldots, N$, $N = N(\varphi)$. According to Lemma 2 of Sec. 5.2, there are functions $\varphi_k \in \mathscr{D}$ such that

$$\varphi(x) = \sum_{k=1}^{N} \varphi_k(x), \qquad \operatorname{supp} \varphi_k \subset U(x_k; r_k)$$

Consequently,

$$(f, \varphi) = \sum_{k=1}^{N} (f, \varphi_k) = \sum_{k=1}^{N} (f_k, \varphi_k) \tag{4}$$

that is, the generalized function f is defined by its local elements f_k. We shall prove the uniqueness of f. Let there be another generalized function g with the same local elements f_k. Then, by virtue of (4), for any $\varphi \in \mathscr{D}$

$$(f, \varphi) = \sum_{k=1}^{N} (f_k, \varphi_k) = (g, \varphi)$$

and therefore $f = g$. The lemma is proved.

It follows from the lemma which has just been proved: *In order that the generalized function f should become zero in the region G, it is necessary and sufficient that it become zero in the neighborhood of each point of this region.*

The set of all those points such that in no neighborhood of each point of this set does $f \neq 0$ is known as the *support* of the generalized function f. We shall denote the support of f by $\operatorname{supp} f$. It is evident that $\operatorname{supp} f$ is a closed set. If $\operatorname{supp} f$ is a bounded set, the generalized function f is said to have *compact support*.

It follows that in any region lying outside supp f, the generalized function f becomes zero, that is,

$$(f, \varphi) = 0, \qquad \varphi \in \mathscr{D}, \qquad \text{supp} f \cap \text{supp} \varphi = \varnothing \tag{5}$$

In fact, if the point $x_0 \notin \text{supp} f$, there will be a neighborhood of this point such that $f = 0$ in this neighborhood. As a corollary to the lemma of Sec. 5.4, $f = 0$ also in any region consisting of such neighborhoods.

5. Regular Generalized Functions. The simplest example of a generalized function is the functional generated by the function $f(x)$ locally integrable in R^n:

$$(f, \varphi) = \int f(x)\varphi(x) \, dx, \qquad \varphi \in \mathscr{D} \tag{6}$$

From the property of linearity of the integral follows the linearity of this functional:

$$(f, \lambda\varphi + \mu\psi) = \int f(x)[\lambda\varphi(x) + \mu\psi(x)] \, dx$$
$$= \lambda \int f(x)\varphi(x) \, dx + \mu \int f(x)\psi(x) \, dx = \lambda(f, \varphi) + \mu(f, \psi)$$

and from the theorem which concerns proceeding to the limit under the integral sign follows the continuity of this functional over $\mathscr{D}$:

$$(f, \varphi_k) = \int_{U_R} f(x)\varphi_k(x) \, dx \to \int f(x)\varphi(x) \, dx = (f, \varphi), \qquad k \to \infty$$

if $\varphi_k \to \varphi$ as $k \to \infty$ in $\mathscr{D}$. In this way the functional (6) defines a generalized function belonging to $\mathscr{D}'$.

Generalized functions which are definable in terms of functions locally integrable in R^n according to formula (6) are said to be *regular generalized functions*. The remaining generalized functions are said to be *singular generalized functions*.

LEMMA (Du Bois Reymond). *In order that the function $f(x)$, locally integrable in G, should become zero in the region G in the sense of generalized functions, it is necessary and sufficient that $f(x) = 0$ almost everywhere in G.*

Proof. The sufficiency of the condition is evident. Let us prove its necessity. Let a be an arbitrary point of the region G. There will be a closed sphere $\bar{U}(a; \varepsilon)$ which is wholly contained in the region G and in which,

§ 5. TEST AND GENERALIZED FUNCTIONS

consequently, $f = 0$ in the sense of Sec. 5.4. Since for each $k = (k_1, k_2, \ldots, k_n)$ the function

$$\psi_k(x) = \exp\left[\frac{i}{\varepsilon}(k, x)\right]\omega_\varepsilon(x - a)$$

where ω_ε is a "cap-shaped function" (cf. Sec. 5.2), belongs to $\mathscr{D}(G)$, then

$$(f, \psi_k) = \int f(x)\omega_\varepsilon(x - a) \exp\left[\frac{i}{\varepsilon}(k, x)\right] dx = 0$$

In this way, all the Fourier coefficients corresponding to the trigonometric system $\{\exp[(i/\varepsilon)(k, x)]\}$ of the function $f(x)\omega_\varepsilon(x - a)$, which is integrable over the sphere $U(a; \varepsilon)$, are equal to zero. It follows* that this function is equal to zero almost everywhere and, consequently, $f(x) = 0$ almost everywhere in this sphere. Since a is an arbitrary point of the region G, then $f(x) = 0$ almost everywhere in G. The lemma is proved.

Each function locally integrable in R^n defines, according to formula (6), a regular generalized function. It follows from Du Bois Reymond's lemma that each regular generalized function is defined by a unique[†] function locally integrable in R^n. Consequently, there is a mutual one-to-one correspondence between the functions locally integrable in R^n and regular generalized functions. Therefore we shall identify a locally integrable function $f(x)$ and the generalized function generated by it according to formula (6) — the functional (f, φ). In this sense the "usual," that is, locally integrable functions in R^n, are (regular) generalized functions.

It follows also from Du Bois Reymond's lemma that both definitions of the support of a continuous function given in Secs. 1.2 and 5.4 coincide.

Finally, we remark that if the sequence $f_k(x)$ for $k = 1, 2, \ldots$, of functions locally integrable in R^n converges uniformly over each compactum to a function $f(x)$, then it converges to $f(x)$ in $\mathscr{D}'(R^n)$.

In fact, for any $\varphi \in \mathscr{D}$ we have

$$(f_k, \varphi) = \int f_k(x)\varphi(x) \, dx \to \int f(x)\varphi(x) \, dx = (f, \varphi), \qquad k \to \infty$$

---

* Compare, for instance, G. M. Fichtengolz (*1*, Vol. III, Chap. XX), and G. E. Shilov (*1*, Chap. VII).
[†] With an accuracy as far as the values on a set of measure zero.

6. Singular Generalized Functions. By the definition we have just given, it is impossible to identify a singular generalized function with any locally integrable function. The simplest example of a singular generalized function is the Dirac δ function (cf. Sec. 5.1)

$$(\delta, \varphi) = \varphi(0), \qquad \varphi \in \mathscr{D}$$

Evidently, $\delta \in \mathscr{D}'$, $\delta(x) = 0$ for $x \neq 0$, so that supp $\delta = \{0\}$. We shall prove that $\delta(x)$ is a *singular generalized function*. Let there be, on the other hand, a function $f(x)$ locally integrable in R^n such that for any function $\varphi \in \mathscr{D}$

$$\int f(x)\varphi(x)\,dx = \varphi(0) \tag{7}$$

Since $x_1\varphi(x) \in \mathscr{D}$ if $\varphi \in \mathscr{D}$, then it follows from (7) that

$$\int f(x)x_1\varphi(x)\,dx = x_1\varphi(x)\big|_{x=0} = 0 = (x_1 f, \varphi)$$

for all $\varphi \in \mathscr{D}$; here x_1 is the first coordinate of x. In this way, the function $x_1 f(x)$, locally integrable in R^n, is equal to zero in the sense of generalized functions. According to Du Bois Reymond's lemma, $x_1 f(x) = 0$ almost everywhere and therefore $f(x) = 0$ almost everywhere. But this contradicts Eq. (7). The contradiction obtained proves the singularity of the δ function.

Let $\omega_\varepsilon(x)$ be a "cap-shaped function" (cf. Sec. 5.2). We shall prove that

$$\omega_\varepsilon(x) \to \delta(x), \qquad \varepsilon \to +0 \quad \text{in } \mathscr{D}' \tag{8}$$

In fact, according to the definition of convergence in $\mathscr{D}'$, Eq. (8) is equivalent to the equality

$$\lim_{\varepsilon \to +0} \int \omega_\varepsilon(x)\varphi(x)\,dx = \varphi(0), \qquad \varphi \in \mathscr{D}$$

On account of the continuity of the function $\varphi(x)$ for any $\eta > 0$ there is an $\varepsilon_0 > 0$ such that $|\varphi(x) - \varphi(0)| < \eta$ if and only if $|x| < \varepsilon_0$. From this, using the properties of the "cap-shaped function" $\omega_\varepsilon(x)$, for all $\varepsilon \leq \varepsilon_0$, we obtain

$$\left|\int \omega_\varepsilon(x)\varphi(x)\,dx - \varphi(0)\right| \leq \int \omega_\varepsilon(x)|\varphi(x) - \varphi(0)|\,dx < \eta \int \omega_\varepsilon(x)\,dx = \eta$$

as was to be shown.

§ 5. TEST AND GENERALIZED FUNCTIONS

A simple layer over a surface is a generalization of the δ function. Let S be a piecewise smooth surface and $\mu(x)$ be a continuous function defined over S. We shall introduce the generalized function $\mu\delta_S$, acting according to the rule:

$$(\mu\delta_S, \varphi) = \int_S \mu(x)\varphi(x)\, dS, \qquad \varphi \in \mathscr{D}$$

It is evident that $\mu\delta_S \in \mathscr{D}'$; $\mu\delta_S(x) = 0$ for $x \notin S$ so that $\operatorname{supp} \mu\delta_S \subset S$. The generalized function $\mu\delta_S$ is said to be a *simple layer* over the surface S, with density μ.

Note. Locally integrable functions and δ functions describe the distributions (densities) of masses, charges, forces, and so on (cf. Sec. 5.1). Therefore, generalized functions are also known as distributions [cf. L. Schwartz (1, 2)]. If, for example, the generalized function f is the density of masses or charges, then the expression $(f, 1)$ is a complete mass or a charge, respectively (if f acts on a function identically equal to 1; this function does not belong to $\mathscr{D}$!). Specifically, $(\delta, 1) = 1$; $(f, 1) = \int f(x)\, dx$ if f is an absolutely integrable function over R^n.

7. Sokhotsky's Formulas. We shall introduce a linear functional $\mathscr{P}(1/x)$, acting according to the formula

$$\left(\mathscr{P}\frac{1}{x}, \varphi\right) = Pv \int \frac{\varphi(x)}{x}\, dx = \lim_{\varepsilon \to +0} \left(\int_{-\infty}^{-\varepsilon} + \int_{\varepsilon}^{\infty}\right) \frac{\varphi(x)}{x}\, dx, \qquad \varphi \in \mathscr{D}$$

Since for any $\varphi \in \mathscr{D}$, $\varphi(x) = 0$ for $|x| > R$, the inequality

$$\left|\left(\mathscr{P}\frac{1}{x}, \varphi\right)\right| = \left|Pv \int \frac{\varphi(x)}{x}\, dx\right| = \left|Pv \int_{-R}^{R} \frac{\varphi(0) + x\varphi'(x')}{x}\, dx\right|$$

$$\leq \int_{-R}^{R} |\varphi'(x')|\, dx \leq 2R \max_{x \in R^1} |\varphi'(x)|, \qquad x' \in (-R, R)$$

is true, then $\mathscr{P}(1/x) \in \mathscr{D}'$. The generalized function $\mathscr{P}(1/x)$ coincides (in the sense of Sec. 5.4) with the function $1/x$ when $x \neq 0$. It is said to be the *finite part* (*partie finie*) or the *principal value of the integral* of $1/x$.

We shall now establish the equation

$$\lim_{\varepsilon \to +0} \int \frac{\varphi(x)}{x + i\varepsilon}\, dx = -i\pi\varphi(0) + Pv \int \frac{\varphi(x)}{x}\, dx, \qquad \varphi \in \mathscr{D} \qquad (9)$$

In fact, if $\varphi(x) = 0$ when $|x| > R$, then

$$\lim_{\varepsilon \to +0} \int \frac{\varphi(x)}{x + i\varepsilon} \, dx = \lim_{\varepsilon \to +0} \int_{-R}^{R} \frac{x - i\varepsilon}{x^2 + \varepsilon^2} \varphi(x) \, dx$$

$$= \varphi(0) \lim_{\varepsilon \to +0} \int_{-R}^{R} \frac{x - i\varepsilon}{x^2 + \varepsilon^2} \, dx$$

$$+ \lim_{\varepsilon \to +0} \int_{-R}^{R} \frac{x - i\varepsilon}{x^2 + \varepsilon^2} \, [\varphi(x) - \varphi(0)] \, dx$$

$$= -2i\varphi(0) \lim_{\varepsilon \to +0} \arctan \frac{R}{\varepsilon} + \int_{-R}^{R} \frac{\varphi(x) - \varphi(0)}{x} \, dx$$

$$= -i\pi\varphi(0) + Pv \int \frac{\varphi(x)}{x} \, dx$$

Equation (9) shows that there is a continuous limit of the sequence $1/(x + i\varepsilon)$, $\varepsilon \to +0$ in $\mathscr{D}'$ which we denote by $1/(x + i0)$ and this limit is equal to $-i\pi\delta(x) + \mathscr{P}(1/x)$. So,

$$\frac{1}{x + i0} = -i\pi\delta(x) + \mathscr{P}\frac{1}{x} \tag{10}$$

Analogously,

$$\frac{1}{x - i0} = i\pi\delta(x) + \mathscr{P}\frac{1}{x} \tag{10'}$$

Formulas (10) and (10') are known as *Sokhotsky's formulas* and are used in quantum physics.

8. Linear Change of Variables in Generalized Functions. Let $f(x)$ be a function locally integrable in R^n and let $x = Ay + b$ with $\det A \neq 0$ be a nonsingular linear transformation of the space R^n onto itself. Then for any $\varphi \in \mathscr{D}$ we obtain

$$(f(Ay + b), \varphi) = \int f(Ay + b)\varphi(y) \, dy$$

$$= \frac{1}{|\det A|} \int f(x)\varphi[A^{-1}(x - b)] \, dx$$

$$= \frac{1}{|\det A|} (f, \varphi[A^{-1}(x - b)])$$

We shall take this equation as the definition of the generalized function

§ 5. TEST AND GENERALIZED FUNCTIONS

$f(Ay + b)$ for any $f(x) \in \mathscr{D}'$:

$$(f(Ay + b), \varphi) = \left(f, \frac{\varphi[A^{-1}(x - b)]}{|\det A|}\right), \qquad \varphi \in \mathscr{D} \qquad (11)$$

Since the operation $\varphi[A^{-1}(x - b)]$ is linear and continuous from $\mathscr{D}$ into $\mathscr{D}$ (cf. Sec. 5.2), the functional $f(Ay + b)$, defined by the right-hand side of Eq. (11), belongs to $\mathscr{D}'$.

Specifically, if A is a rotation, that is, if $A' = A^{-1}$, and $b = 0$, then

$$(f(Ay), \varphi) = (f, \varphi(A'x))$$

if A is a dilatation (or reflection), $A = cI$, and $b = 0$, then

$$(f(cy), \varphi) = \frac{1}{|c^n|}\left(f, \varphi\left(\frac{x}{c}\right)\right)$$

if $A = I$, then

$$(f(y + b), \varphi) = (f, \varphi(x - b))$$

The generalized function $f(x + b)$ is called the *translation* of the generalized function $f(x)$ by the vector b. For instance, $\delta(x - x_0)$ is the translation of $\delta(x)$ by the vector $-x_0$, and it acts according to the formula

$$(\delta(x - x_0), \varphi) = (\delta, \varphi(x + x_0)) = \varphi(x_0)$$

These equations allow us to define spherically symmetrical, centrally symmetrical, homogeneous, translation invariants with respect to displacements, Lorentz invariants, and so on, and other generalized functions.

9. Multiplication of Generalized Functions. Let $f(x)$ be a function locally integrable in R^n and $a(x) \in C^\infty(R^n)$. Then for any $\varphi \in \mathscr{D}$ the equation

$$(af, \varphi) = (f, a\varphi), \qquad \varphi \in \mathscr{D} \qquad (12)$$

is true. We take Eq. (12) as the definition of the product af for any $f \in \mathscr{D}'$. Since the operation of multiplication by the function $a \in C^\infty(R^n)$ is linear and continuous from $\mathscr{D}$ into $\mathscr{D}$ (cf. Sec. 5.2), then the functional af, definable by the right-hand side of Eq. (12), belongs to $\mathscr{D}'$.

If $f \in \mathscr{D}'$, then the equation

$$f = \eta f \qquad (13)$$

is true, where η is any function of the class $C^\infty(R^n)$ equal to 1 in the neighborhood of the support of f.

In fact, for any $\varphi \in \mathscr{D}$ the supports of f and $(1 - \eta)\varphi$ do not have common points, and therefore, by virtue of (5),

$$(f - \eta f, \varphi) = (f, (1 - \eta)\varphi) = 0$$

Examples.

(a) $a(x)\delta(x) = a(0)\delta(x)$, since for all $\varphi \in \mathscr{D}$

$$(a\delta, \varphi) = (\delta, a\varphi) = a(0)\varphi(0) = (a(0)\delta, \varphi)$$

(b) $x\mathscr{P}(1/x) = 1$, since for all $\varphi \in \mathscr{D}(R^1)$

$$\left(x\mathscr{P}\frac{1}{x}, \varphi\right) = \left(\mathscr{P}\frac{1}{x}, x\varphi\right) = Pv \int \frac{x\varphi(x)}{x}\, dx = \int \varphi(x)\, dx = (1, \varphi)$$

We may then ask: can we not define the product of generalized functions as another generalized function? The product of locally integrable functions need not be locally integrable [for instance $(|x|^{-1/2})^2 = |x|^{-1}$ in R^1], and this holds true for generalized functions: L. Schwartz has shown that it is impossible to define the kind of product that would be associative and commutative. In fact, if it were to exist, then, using Examples (a) and (b), we should obtain the following contradictory sequence of equations:

$$0 = 0\mathscr{P}\frac{1}{x} = (x\delta(x))\mathscr{P}\frac{1}{x} = (\delta(x)x)\mathscr{P}\frac{1}{x} = \delta(x)\left(x\mathscr{P}\frac{1}{x}\right) = \delta(x)$$

To define the product of generalized functions f and g, they must have the following properties: insofar as f is "nonregular" in the neighborhood of an arbitrary point, so must g be "regular" in this neighborhood. For example, $\delta(x - a)\delta(x - b) = 0$, if $a \neq b$.

10. Exercises. (a) Prove that the functions

$$\frac{1}{2\sqrt{\pi\varepsilon}}\exp\left(-\frac{x^2}{4\varepsilon}\right), \quad \frac{1}{\pi x}\sin\frac{x}{\varepsilon}, \quad \frac{1}{\pi}\frac{\varepsilon}{x^2 + \varepsilon^2}$$

tend to $\delta(x)$ when $\varepsilon \to +0$.

(b) Using Sokhotsky's formula, prove the following limits when $t \to +\infty$:

$$\frac{e^{ixt}}{x-i0} \to 2\pi i \delta(x), \qquad \frac{e^{-ixt}}{x-i0} \to 0$$

$$\frac{e^{ixt}}{x+i0} \to 0, \qquad \frac{e^{-ixt}}{x+i0} \to -2\pi i \delta(x)$$

§ 6. Differentiation of Generalized Functions

Generalized functions have a number of convenient properties. For instance, with the proper generalization of the concept of a derivative, any generalized function is infinitely differentiable, and converging series of generalized functions can be differentiated term by term an infinite number of times.

1. Derivatives of Generalized Functions. Let $f \in C^p(R^n)$. Then whenever α, $|\alpha| \leq p$, and $\varphi \in \mathscr{D}$ the formula for integration by parts,

$$(D^\alpha f, \varphi) = \int D^\alpha f(x)\varphi(x)\,dx = (-1)^{|\alpha|} \int f(x) D^\alpha \varphi(x)\,dx = (-1)^{|\alpha|}(f, D^\alpha \varphi)$$

is valid.

We shall also take this equation as the definition of the (generalized) derivative $D^\alpha f$ of the generalized function $f \in \mathscr{D}'$:

$$(D^\alpha f, \varphi) = (-1)^{|\alpha|}(f, D^\alpha \varphi), \qquad \varphi \in \mathscr{D} \tag{1}$$

We shall check that $D^\alpha f \in \mathscr{D}'$. In fact, since $f \in \mathscr{D}'$, the functional $D^\alpha f$, definable by the right-hand side of Eq. (1), is linear:

$$(D^\alpha f, \lambda\varphi + \mu\psi) = (-1)^{|\alpha|}(f, D^\alpha(\lambda\varphi + \mu\psi))$$
$$= (-1)^{|\alpha|}(f, \lambda D^\alpha \varphi + \mu D^\alpha \psi)$$
$$= \lambda(-1)^{|\alpha|}(f, D^\alpha \varphi) + \mu(-1)^{|\alpha|}(f, D^\alpha \psi)$$
$$= \lambda(D^\alpha f, \varphi) + \mu(D^\alpha f, \psi)$$

and continuous:

$$(D^\alpha f, \varphi_k) = (-1)^{|\alpha|}(f, D^\alpha \varphi_k) \to (-1)^{|\alpha|}(f, D^\alpha \varphi) = (D^\alpha f, \varphi)$$

for, if $\varphi_k \to \varphi$ as $k \to \infty$ in $\mathscr{D}$, then also $D^\alpha \varphi_k \to D^\alpha \varphi$ as $k \to \infty$ in $\mathscr{D}$ (cf. Sec. 5.2).

We shall denote by $\{D^\alpha f(x)\}$ the classical derivative (where it exists). It follows from the definition of the generalized derivative that if the gener-

alized function $f \in C^p(G)$, then
$$D^\alpha f = \{D^\alpha f(x)\}, \quad x \in G, \quad |\alpha| \le p$$
according to the definition of Sec. 5.4.

2. Properties of Generalized Derivatives. The following properties of the operation of differentiation of generalized functions are true.

(a) *Any generalized function is infinitely differentiable.*
In fact, if $f \in \mathscr{D}'$, then $\partial f/\partial x_i \in \mathscr{D}'$; in its turn $(\partial/\partial x_j)(\partial f/\partial x_i) \in \mathscr{D}'$; and so on.

(b) *The result of differentiation does not depend on the order of differentiation*; for example,
$$\frac{\partial}{\partial x_1}\left(\frac{\partial f}{\partial x_2}\right) = \frac{\partial}{\partial x_2}\left(\frac{\partial f}{\partial x_1}\right) = D^{(1,1)}f \qquad (2)$$

In fact, for any $\varphi \in \mathscr{D}$ we obtain
$$(D^{(1,1)}f, \varphi) = \left(f, \frac{\partial^2 \varphi}{\partial x_1 \partial x_2}\right) = \left(\frac{\partial}{\partial x_1}\left(\frac{\partial f}{\partial x_2}\right), \varphi\right) = \left(\frac{\partial}{\partial x_2}\left(\frac{\partial f}{\partial x_1}\right), \varphi\right)$$
from which Eq. (2) (cf. Sec. 5.4) follows.

In general,
$$D^{\alpha+\beta}f = D^\alpha(D^\beta f) = D^\beta(D^\alpha f) \qquad (3)$$

(c) *If $f \in \mathscr{D}'$ and $a \in C^\infty(R^n)$, then Leibnitz' formula for differentiation of the product af* (cf. Sec. 5.9) *is valid.* For example:
$$\frac{\partial(af)}{\partial x_1} = \frac{\partial a}{\partial x_1}f + a\frac{\partial f}{\partial x_1} \qquad (4)$$

In fact, if φ is any basic function, then
$$\left(\frac{\partial(af)}{\partial x_1}, \varphi\right) = -\left(af, \frac{\partial \varphi}{\partial x_1}\right) = -\left(f, a\frac{\partial \varphi}{\partial x_1}\right)$$
$$= -\left(f, \frac{\partial(a\varphi)}{\partial x_1} - \frac{\partial a}{\partial x_1}\varphi\right) = -\left(f, \frac{\partial(a\varphi)}{\partial x_1}\right) + \left(f, \frac{\partial a}{\partial x_1}\varphi\right)$$
$$= \left(\frac{\partial f}{\partial x_1}, a\varphi\right) + \left(\frac{\partial a}{\partial x_1}f, \varphi\right) = \left(a\frac{\partial f}{\partial x_1}, \varphi\right) + \left(\frac{\partial a}{\partial x_1}f, \varphi\right)$$
$$= \left(a\frac{\partial f}{\partial x_1} + \frac{\partial a}{\partial x_1}f, \varphi\right)$$
from which Eq. (4) (cf. Sec. 5.4) follows.

§ 6. DIFFERENTIATION OF GENERALIZED FUNCTIONS

(d) *If the generalized function $f = 0$ for $x \in G$, then also $D^\alpha f = 0$ for $x \in G$, so that* supp $D^\alpha f \subset$ supp f.

In fact, if $\varphi \in \mathscr{D}(G)$ then $D^\alpha \varphi \in \mathscr{D}(G)$ and so

$$(D^\alpha f, \varphi) = (-1)^{|\alpha|}(f, D^\alpha \varphi) = 0, \qquad \varphi \in \mathscr{D}(G)$$

which shows that $D^\alpha f = 0$ for $x \in G$ (cf. Sec. 5.4).

(e) *The operation of differentiation is continuous from $\mathscr{D}'$ into $\mathscr{D}'$, that is, if $f_k \to f$ as $k \to \infty$ in $\mathscr{D}'$, then $D^\alpha f_k \to D^\alpha f$ as $k \to \infty$ in $\mathscr{D}'$.*

Indeed, according to the definition of convergence in the space $\mathscr{D}'$ (cf. Sec. 5.3), for all $\varphi \in \mathscr{D}$ we have

$$(D^\alpha f_k, \varphi) = (-1)^{|\alpha|}(f_k, D^\alpha \varphi) \to (-1)^{|\alpha|}(f, D^\alpha \varphi) = (D^\alpha f, \varphi),$$
$$k \to \infty$$

which shows that $D^\alpha f_k \to D^\alpha f$ as $k \to \infty$ in $\mathscr{D}'$.

(f) *If the series*

$$\sum_{k=1}^{\infty} u_k(x) = S(x)$$

composed of locally integrable functions $u_k(x)$ converges uniformly over each compactum, then it can be differentiated term by term any number of times and the series obtained will converge in $\mathscr{D}'$.

In fact, since for any $R > 0$,

$$S_p(x) = \sum_{k=1}^{p} u_k(x) \overset{|x| \leq R}{\Longrightarrow} S(x), \qquad p \to \infty$$

then $S_p \to S$ as $p \to \infty$ in $\mathscr{D}'$ (cf. Sec. 5.5). But then, by virtue of (e),

$$D^\alpha S_p = \sum_{k=1}^{p} D^\alpha u_k \to D^\alpha S, \qquad p \to \infty \quad \text{in} \quad \mathscr{D}'$$

as was to be shown.

It follows specifically from this: *If*

$$|a_k| \leq A |k|^m + B \tag{5}$$

then the trigonometric series

$$\sum_{k=-\infty}^{\infty} a_k e^{ikx} \tag{6}$$

converges in $\mathscr{D}'(R^1)$.

Indeed, by virtue of (5), the series

$$\frac{a_0 x^{m+2}}{(m+2)!} + \sum_{\substack{k=-\infty \\ k \neq 0}}^{\infty} \frac{a_k}{(ik)^{m+2}} e^{ikx}$$

converges uniformly in R^1; consequently, its derivative of order $m+2$ converges in $\mathscr{D}'(R^1)$ and coincides with series (6).

3. Examples, $n = 1$. (a) Let the function $f(x)$ be such that $f \in C^1(x \leq x_0)$ and $f \in C^1(x \geq x_0)$. We shall show that

$$f' = \{f'(x)\} + [f]_{x_0} \delta(x - x_0) \tag{7}$$

where we write $[f]_{x_0} = f(x_0 + 0) - f(x_0 - 0)$.

In fact, if $\varphi \in \mathscr{D}$, then

$$(f', \varphi) = -(f, \varphi') = -\int f(x) \varphi'(x) \, dx$$
$$= [f(x_0 + 0) - f(x_0 - 0)] \varphi(x_0) + \int \{f'(x)\} \varphi(x) \, dx$$
$$= ([f]_{x_0} \delta(x - x_0) + \{f'(x)\}, \varphi)$$

Specifically, $\theta'(x) = \delta(x)$, where θ is Heaviside's function:

$$\theta(x) = 1, \quad x \geq 0; \quad \theta(x) = 0, \quad x < 0$$

If then the function $f(x)$ has isolated discontinuities of the first kind at the points $\{x_k\}$, then formula (7) may be naturally generalized to give

$$f' = \{f'(x)\} + \sum_k [f]_{x_k} \delta(x - x_k) \tag{8}$$

Specifically, if $f_0(x) = \frac{1}{2} - (x/2\pi)$ for $x \in [0, 2\pi]$ is a 2π-periodic function (Fig. 15), then

$$f_0' = -\frac{1}{2\pi} + \sum_{k=-\infty}^{\infty} \delta(x - 2k\pi) \tag{9}$$

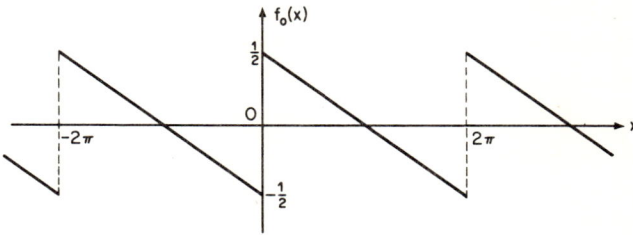

Fig. 15

§ 6. DIFFERENTIATION OF GENERALIZED FUNCTIONS

It is convenient to obtain formula (8) locally, in the neighborhood of each point x_k, by using formula (7).

In this way, roughly speaking, generalized and classical derivatives do not coincide.

(b) We shall calculate the density of charges corresponding to a dipole with electric moment $+1$ at the point $x = 0$ on a straight line.

The density of charges

$$\frac{1}{\varepsilon} \delta(x - \varepsilon) - \frac{1}{\varepsilon} \delta(x), \qquad \varepsilon > 0$$

approximately corresponds to this dipole. Proceeding to the limit when $\varepsilon \to +0$

$$\left(\frac{1}{\varepsilon} \delta(x - \varepsilon) - \frac{1}{\varepsilon} \delta(x), \varphi\right) = \frac{1}{\varepsilon} [\varphi(\varepsilon) - \varphi(0)] \to \varphi'(0)$$

$$= (\delta, \varphi') = -(\delta', \varphi)$$

we conclude that the density sought is equal to $-\delta'(x)$.

We shall make certain that the complete charge of the dipole is equal to 0:

$$(-\delta', 1) = (\delta, 1') = (\delta, 0) = 0$$

and that its moment is equal to 1:

$$(-\delta', x) = (\delta, x') = (\delta, 1) = 1$$

(c) We shall prove the formula

$$\frac{1}{2\pi} \sum_{k=-\infty}^{\infty} e^{ikx} = \sum_{k=-\infty}^{\infty} \delta(x - 2k\pi) \tag{10}$$

For this we expand the 2π-periodic function $f_0(x) = \frac{1}{2} - (x/2\pi)$ [cf. Sec. 6.3, (a)] in the Fourier series

$$f_0(x) = -\frac{i}{2\pi} \sum_{\substack{k=-\infty \\ k \neq 0}}^{\infty} \frac{1}{k} e^{ikx}$$

By virtue of the results of Sec. 6.2, (f), this series can be differentiated term by term in $\mathscr{D}'$ any number of times. Taking (9) into consideration we obtain

$$f_0' = -\frac{1}{2\pi} + \sum_{k=-\infty}^{\infty} \delta(x - 2k\pi) = \frac{1}{2\pi} \sum_{\substack{k=-\infty \\ k \neq 0}}^{\infty} e^{ikx}$$

as was to be shown.

(d) We shall show that the general solution of the equation

$$x^m u = 0 \qquad (11)$$

in $\mathscr{D}'(R^1)$ is given by the formula

$$u = \sum_{k=0}^{m-1} c_k \delta^{(k)}(x) \qquad (12)$$

where c_k are arbitrary constants.

Since for all $\varphi \in \mathscr{D}$ and $k = 0, 1, \ldots, m-1$

$$(x^m \delta^{(k)}, \varphi) = (\delta^{(k)}, x^m \varphi) = (-1)^k (\delta, (x^m \varphi)^{(k)}) = (-1)^k (x^m \varphi)^{(k)}|_{x=0} = 0$$

then

$$x^m \delta^{(k)}(x) = 0, \qquad k = 0, 1, \ldots, m-1$$

and, consequently, the generalized function (12) satisfies Eq. (11).

We shall prove that formula (12) gives a general solution of this equation in $\mathscr{D}'$. Let $\eta(x)$ be a test function, equal to 1 in the neighborhood of the point $x = 0$. (According to Lemma 1 of Sec. 5.2 such a function exists.) Then for any $\varphi \in \mathscr{D}$ the function

$$\psi(x) = \frac{1}{x^m} \left[\varphi(x) - \eta(x) \sum_{k=0}^{m-1} \frac{\varphi^{(k)}(0)}{k!} x^k \right] \in \mathscr{D} \qquad (13)$$

Consequently, if $u \in \mathscr{D}'$ is the solution of Eq. (11), then

$$(u, \varphi) = \left(u, \eta(x) \sum_{k=0}^{m-1} \frac{\varphi^{(k)}(0)}{k!} x^k \right) + (u, x^m \psi(x))$$

$$= \sum_{k=0}^{m-1} \frac{\varphi^{(k)}(0)}{k!} (u, \eta(x) x^k) + (x^m u, \psi)$$

$$= \sum_{k=0}^{m-1} (-1)^k c_k \varphi^{(k)}(0) = \sum_{k=0}^{m-1} c_k (\delta^{(k)}, \varphi)$$

§ 6. DIFFERENTIATION OF GENERALIZED FUNCTIONS

as was to be shown. Here
$$c_k = \frac{(-1)^k}{k!}(u, \eta x^k)$$
are arbitrary constants.

Note. The result obtained follows directly from the more general assertion that each generalized function whose support is a point appears in the form of a linear combination of the δ function and its derivatives at this point (cf. Sec. 8.4).

Let us remark that, in the class of locally integrable functions, Eq. (11) has the unique solution $u = 0$.

(e) The general solution of the differential equation $u^{(m)} = 0$ in $\mathscr{D}'(R^1)$ is an arbitrary polynomial of degree $m - 1$.

It is sufficient to prove this assertion for $m = 1$: If the generalized function u satisfies the condition

$$(u, \varphi') = 0, \quad \varphi \in \mathscr{D} \tag{14}$$

then $u = $ const.

In fact, let $\omega_1(x)$ be a "cap-shaped function." Then for any $\varphi \in \mathscr{D}$ the function

$$\psi(x) = \int_{-\infty}^{x} \left[\varphi(t) - \omega_1(t) \int \varphi(\xi)\, d\xi \right] dt \in \mathscr{D}$$

and, consequently, by virtue of (14),

$$0 = (u, \psi') = (u, \varphi - \omega_1 \int \varphi(\xi)\, d\xi) = (u, \varphi) - (u, \omega_1) \int \varphi(\xi)\, d\xi$$

From this, writing $(u, \omega_1) = C$, we obtain

$$(u, \varphi) = C \int \varphi(\xi)\, d\xi = (C, \varphi)$$

as was to be shown.

(f) We shall check that the function $\mathscr{E}(t) = \theta(t)Z(t)$, where $Z(t)$ is the solution of the homogeneous differential equation

$$LZ \equiv Z^{(m)} + a_1(t)Z^{(m-1)} + \cdots + a_m(t)Z = 0$$

satisfying the conditions

$$Z(0) = Z'(0) = \cdots = Z^{(m-2)}(0) = 0, \quad Z^{(m-1)}(0) = 1$$

satisfies the equation $L\mathscr{E} = \delta(t)$.

In fact, using formula (7), we obtain

$$\mathscr{E}'(t) = \theta(t)Z'(t), \quad \ldots, \quad \mathscr{E}^{(m-1)}(t) = \theta(t)Z^{(m-1)}(t)$$
$$\mathscr{E}^{(m)}(t) = \delta(t) + \theta(t)Z^{(m)}(t)$$

from which

$$L\mathscr{E} = \theta(t)LZ + \delta(t) = \delta(t)$$

4. Exercises. (a) Prove that

$$\frac{d}{dx}\ln|x| = \mathscr{P}\frac{1}{x}, \qquad \frac{d}{dx}\mathscr{P}\frac{1}{x} = -\mathscr{P}\frac{1}{x^2}$$

$$\frac{d}{dx}\frac{1}{x \pm i0} = \mp i\pi\delta'(x) - \mathscr{P}\frac{1}{x^2}$$

where

$$\left(\mathscr{P}\frac{1}{x^2}, \varphi\right) = Pv\int \frac{\varphi(x) - \varphi(0)}{x^2}\,dx$$

(b) Show that the generalized functions displayed on the right are generalized solutions in $\mathscr{D}'(R^1)$ of the equations:

$$xu' = 1, \qquad u = c_1 + c_2\theta(x) + \ln|x|$$

$$xu' = \mathscr{P}\frac{1}{x}, \qquad u = c_1 + c_2\theta(x) - \mathscr{P}\frac{1}{x}$$

$$x^2u' = 1, \qquad u = c_1 + c_2\theta(x) + c_3\delta(x) - \mathscr{P}\frac{1}{x}$$

$$xu = \operatorname{sgn} x, \qquad u = c\delta(x) + \mathscr{P}\frac{1}{|x|}$$

where

$$\left(\mathscr{P}\frac{1}{|x|}, \varphi\right) = \int_{|x|<1}\frac{\varphi(x) - \varphi(0)}{|x|}\,dx + \int_{|x|>1}\frac{\varphi(x)}{|x|}\,dx$$

Let us note that the classical solutions of first-order differential equations contain only one arbitrary constant!

5. Examples, $n \geq 2$. (a) Let the region G be bounded by the piecewise smooth surface S and let the function $f \in C^1(\bar{G}) \cap C^1(\bar{G}_1)$, where $G_1 = R^n \setminus \bar{G}$. Then

$$\frac{\partial f}{\partial x_i} = \left\{\frac{\partial f}{\partial x_i}\right\} + [f]_S \cos(\mathbf{n}x_i)\delta_S, \qquad i = 1, 2, \ldots, n \qquad (15)$$

§ 6. DIFFERENTIATION OF GENERALIZED FUNCTIONS

where $\mathbf{n} = \mathbf{n}_x$ is the external normal to S at the point $x \in S$ and $[f]_S$ is the jump of the function f across the surface S:

$$\lim_{x' \to x, x' \in G_1} f(x') - \lim_{x' \to x, x' \in G} f(x') = [f]_S(x), \qquad x \in S$$

To obtain formula (15) we shall use Green's formula and the definition of a simple layer (cf. Sec. 5.6):

$$\left(\frac{\partial f}{\partial x_i}, \varphi\right) = -\left(f, \frac{\partial \varphi}{\partial x_i}\right) = -\int f(x) \frac{\partial \varphi(x)}{\partial x_i} dx$$

$$= \int \left\{\frac{\partial f(x)}{\partial x_i}\right\} \varphi(x) \, dx + \int_S [f]_S(x) \cos(\mathbf{n} x_i) \varphi(x) \, dS$$

$$= \left(\left\{\frac{\partial f}{\partial x_i}\right\} + [f]_S \cos(\mathbf{n} x_i) \delta_S, \varphi\right), \qquad \varphi \in \mathscr{D}$$

(b) *The double layer over a surface* is the generalization $-\delta'(x)$. Let S be a piecewise smooth double-sided surface, let $\mathbf{n}$ be the normal to S, and $\nu(x)$ a continuous function defined over S. We shall introduce the generalized function $-(\partial/\partial \mathbf{n})(\nu \delta_S)$, acting according to the rule

$$\left(-\frac{\partial}{\partial \mathbf{n}}(\nu \delta_S), \varphi\right) = \int_S \nu(x) \frac{\partial \varphi(x)}{\partial \mathbf{n}} dS, \qquad \varphi \in \mathscr{D}$$

Evidently,

$$-\frac{\partial}{\partial \mathbf{n}}(\nu \delta_S) \in \mathscr{D}', \qquad \mathrm{supp}\left[-\frac{\partial}{\partial \mathbf{n}}(\nu \delta_S)\right] \subset S$$

The generalized function $-(\partial/\partial \mathbf{n})(\nu \delta_S)$ is said to be the *double layer* over the surface S with the density $\nu(x)$ directed along the normal $\mathbf{n}$. This generalized function describes the density of the charges corresponding to a distribution of dipoles over the surface S with a surface density of moment $\nu(x)$ and directed along a given direction of the normal $\mathbf{n}$ to S [cf. Secs. 6.3, (b) and 5.6].

(c) Let the function $f \in C^2(\bar{G}) \cap C^2(\bar{G}_1)$ be that defined in Example (a). Then

$$\frac{\partial^2 f}{\partial x_i \partial x_j} = \left\{\frac{\partial^2 f}{\partial x_i \partial x_j}\right\} + \frac{\partial}{\partial x_j}([f]_S \cos(\mathbf{n} x_i) \delta_S) + \left[\left\{\frac{\partial f}{\partial x_i}\right\}\right]_S \cos(\mathbf{n} x_j) \delta_S \quad (16)$$

To obtain formula (16) we shall differentiate Eq. (15) with respect to

x_j and, in differentiating the function $\{\partial f(x)/\partial x_i\}$, we shall use formula (15):

$$\frac{\partial}{\partial x_j}\left\{\frac{\partial f}{\partial x_i}\right\} = \left\{\frac{\partial^2 f}{\partial x_j \partial x_i}\right\} + \left[\left\{\frac{\partial f}{\partial x_i}\right\}\right]_S \cos(\mathbf{n}x_j)\delta_S$$

Supposing in (16) that $i = j$, and summing with respect to $i = 1, 2, \ldots, n$, we obtain

$$\Delta f = \{\Delta f\} + \sum_{i=1}^{n} \frac{\partial}{\partial x_i} ([f]_S \cos(\mathbf{n}x_i)\delta_S) + \sum_{i=1}^{n} \left[\left\{\frac{\partial f}{\partial x_i}\right\}\right]_S \cos(\mathbf{n}x_i)\delta_S \quad (17)$$

Taking into account the equations

$$\sum_{i=1}^{n} \frac{\partial}{\partial x_i} ([f]_S \cos(\mathbf{n}x_i)\delta_S) = \frac{\partial}{\partial \mathbf{n}} ([f]_S \delta_S) \quad (18)$$

$$\sum_{i=1}^{n} \left[\left\{\frac{\partial f}{\partial x_i}\right\}\right]_S \cos(\mathbf{n}x_i)\delta_S = \left[\frac{\partial f}{\partial \mathbf{n}}\right]_S \delta_S \quad (19)$$

we rewrite Eq. (17) in the form

$$\Delta f = \{\Delta f\} + \left[\frac{\partial f}{\partial \mathbf{n}}\right]_S \delta_S + \frac{\partial}{\partial \mathbf{n}} ([f]_S \delta_S) \quad (20)$$

We now prove Eq. (18). For all $\varphi \in \mathscr{D}$ we have

$$\left(\sum_{i=1}^{n} \frac{\partial}{\partial x_i} ([f]_S \cos(\mathbf{n}x_i)\delta_S), \varphi\right) = -\sum_{i=1}^{n} \left([f]_S \cos(\mathbf{n}x_i)\delta_S, \frac{\partial \varphi}{\partial x_i}\right)$$

$$= -\sum_{i=1}^{n} \int_S [f]_S \cos(\mathbf{n}x_i) \frac{\partial \varphi}{\partial x_i} dS$$

$$= -\int_S [f]_S \sum_{i=1}^{n} \frac{\partial \varphi}{\partial x_i} \cos(\mathbf{n}x_i) dS$$

$$= -\int_S [f]_S \frac{\partial \varphi}{\partial \mathbf{n}} dS = \left(\frac{\partial}{\partial \mathbf{n}} ([f]_S \delta_S), \varphi\right)$$

Formula (19) is established analogously.

Supposing in formula (20) that $f = 0$ for $x \in G_1$, we obtain

$$\Delta f = \{\Delta f\} - \frac{\partial f}{\partial \mathbf{n}} \delta_S - \frac{\partial}{\partial \mathbf{n}} (f\delta_S) \quad (21)$$

This is Green's second formula, written in terms of generalized functions.

§ 6. DIFFERENTIATION OF GENERALIZED FUNCTIONS

Applying both sides of Eq. (21) to the test function φ, we obtain this formula in the usual notation:

$$\int_G (f \Delta\varphi - \Delta f \varphi) \, dx = \int_S \left(f \frac{\partial \varphi}{\partial \mathbf{n}} - \varphi \frac{\partial f}{\partial \mathbf{n}} \right) dS \qquad (22)$$

If G is a bounded region, then formula (22) is valid whenever $\varphi \in C^2(\bar{G})$.

(d) Let $n = 2$. We shall calculate $\Delta \ln |x|$. The function $\ln |x|$ is locally integrable in R^2. If $x \neq 0$, then $\ln |x| \in C^\infty$, and therefore $D^\alpha \ln |x| = \{D^\alpha \ln |x|\}$ (cf. Sec. 6.1). Consequently, changing to polar coordinates [cf. Eq. (15), Sec. 3.2], we obtain

$$\Delta \ln |x| = \frac{1}{r} \frac{\partial}{\partial r} \left(r \frac{\partial \ln r}{\partial r} \right) = 0, \qquad x \neq 0 \qquad (23)$$

Let $\varphi \in \mathscr{D}$, supp $\varphi \subset U_R$. Then

$$(\Delta \ln |x|, \varphi) = (\ln |x|, \Delta\varphi) = \int_{U_R} \ln |x| \, \Delta\varphi(x) \, dx$$

$$= \lim_{\varepsilon \to +0} \int_{\varepsilon < |x| < R} \ln |x| \, \Delta\varphi(x) \, dx$$

Applying formula (22) when $f = \ln |x|$ and $G = [\varepsilon < |x| < R]$ (Fig. 16) and taking (23) into account, we obtain, further,

$$(\Delta \ln |x|, \varphi) = \lim_{\varepsilon \to +0} \left[\int_{\varepsilon < |x| < R} \Delta \ln |x| \, \varphi \, dx \right.$$

$$\left. + \left(\int_{S_\varepsilon} + \int_{S_R} \right) \left(\ln |x| \frac{\partial \varphi}{\partial \mathbf{n}} - \varphi \frac{\partial \ln |x|}{\partial \mathbf{n}} \right) dS \right]$$

$$= \lim_{\varepsilon \to +0} \int_{S_\varepsilon} \left(-\ln |x| \frac{\partial \varphi}{\partial |x|} + \varphi \frac{1}{|x|} \right) dS$$

$$= \lim_{\varepsilon \to +0} \frac{1}{\varepsilon} \int_{S_\varepsilon} \varphi \, dS$$

$$= \lim_{\varepsilon \to +0} \left\{ \frac{1}{\varepsilon} \int_{S_\varepsilon} [\varphi(x) - \varphi(0)] \, dS + 2\pi \varphi(0) \right\} = 2\pi \varphi(0)$$

$$= (2\pi \delta, \varphi)$$

In this way,

$$\Delta \ln |x| = 2\pi \delta(x) \qquad (24)$$

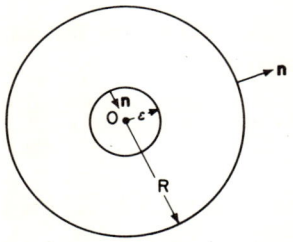

Fig. 16

Analogously, when $n \geq 3$, we obtain

$$\Delta \frac{1}{|x|^{n-2}} = -(n-2)\sigma_n \delta(x) \tag{25}$$

where σ_n is the area of the surface of a unit sphere in R^n,

$$\sigma_n = \int_{S_1} dS = \frac{2\pi^{n/2}}{\Gamma(n/2)}$$

here Γ is a Eulerian integral of the first kind.

$$\Gamma(z) = \int_0^\infty e^{-t} t^{z-1} \, dt$$

(e) We shall check that when $n = 3$ the functions

$$\mathscr{E}(x) = -\frac{e^{ik|x|}}{4\pi |x|}, \quad \bar{\mathscr{E}}(x) = -\frac{e^{-ik|x|}}{4\pi |x|} \tag{26}$$

satisfy the equation

$$\Delta \mathscr{E} + k^2 \mathscr{E} = \delta(x) \tag{27}$$

In fact, since the functions $\cos k |x|$ and $|x|^{-1} \sin k |x|$ are infinitely differentiable, when differentiating the function $|x|^{-1} e^{ik|x|}$ we may use Leibnitz' formula [cf. Sec. 6.2, (c)]. Taking into account the equation

$$\frac{\partial}{\partial x_j} \frac{1}{|x|} = -\frac{x_j}{|x|^3}$$

$$\frac{\partial}{\partial x_j} e^{ik|x|} = \frac{ikx_j}{|x|} e^{ik|x|}$$

$$\Delta e^{ik|x|} = \left(\frac{2ik}{|x|} - k^2 \right) e^{ik|x|}$$

§ 6. DIFFERENTIATION OF GENERALIZED FUNCTIONS

and using formula (25) when $n = 3$, we obtain the equation

$$(\Delta + k^2) \frac{1}{|x|} e^{ik|x|} = e^{ik|x|} \Delta \frac{1}{|x|} + 2\left(\operatorname{grad} e^{ik|x|}, \operatorname{grad} \frac{1}{|x|}\right)$$

$$+ \frac{1}{|x|} \Delta e^{ik|x|} + \frac{k^2}{|x|} e^{ik|x|} = -4\pi e^{ik|x|} \delta(x)$$

$$+ \left(-\frac{2ik}{|x|^2} + \frac{2ik}{|x|^2} - \frac{k^2}{|x|} + \frac{k^2}{|x|}\right) e^{ik|x|}$$

$$= -4\pi \delta(x)$$

as was to be shown.

(f) Let

$$\mathscr{E}(x, t) = \frac{\theta(t)}{(2a\sqrt{\pi t})^n} \exp\left(-\frac{|x|^2}{4a^2 t}\right)$$

We shall prove that

$$\frac{\partial \mathscr{E}}{\partial t} - a^2 \Delta \mathscr{E} = \delta(x, t) \tag{28}$$

The function $\mathscr{E}(x, t)$ is locally integrable in R^{n+1}, since $\mathscr{E} = 0$ for $t < 0$; $\mathscr{E} \geq 0$ for $t \geq 0$ and when $t > 0$

$$\int \mathscr{E}(x, t) \, dx = \frac{1}{(2a\sqrt{\pi t})^n} \int \exp\left(-\frac{|x|^2}{4a^2 t}\right) dx$$

$$= \prod_{i=1}^{n} \frac{1}{\sqrt{\pi}} \int_{-\infty}^{\infty} \exp(-\xi_i^2) \, d\xi_i^2 = 1 \tag{29}$$

If $t > 0$, then $\mathscr{E} \in C^\infty$, and therefore

$$\frac{\partial \mathscr{E}}{\partial t} = \left(\frac{|x|^2}{4a^2 t^2} - \frac{n}{2t}\right) \mathscr{E}$$

$$\frac{\partial \mathscr{E}}{\partial x_i} = -\frac{x_i}{2a^2 t} \mathscr{E}, \quad \frac{\partial^2 \mathscr{E}}{\partial x_i^2} = \left(\frac{x^2}{4a^4 t^2} - \frac{1}{2a^2 t}\right) \mathscr{E}$$

$$\frac{\partial \mathscr{E}}{\partial t} - a^2 \Delta \mathscr{E} = \left(\frac{|x|^2}{4a^2 t^2} - \frac{n}{2t}\right) \mathscr{E} - \left(\frac{|x|^2}{4a^2 t^2} - \frac{n}{2t}\right) \mathscr{E} = 0 \tag{30}$$

Let $\varphi \in \mathscr{D}(R^{n+1})$. Taking (30) into account, we obtain

$$\left(\frac{\partial \mathscr{E}}{\partial t} - a^2 \Delta \mathscr{E}, \varphi\right) = -\left(\mathscr{E}, \frac{\partial \varphi}{\partial t} + a^2 \Delta \varphi\right)$$

$$= -\int_0^\infty \int \mathscr{E}(x, t)\left(\frac{\partial \varphi}{\partial t} + a^2 \Delta \varphi\right) dx\, dt$$

$$= -\lim_{\varepsilon \to +0} \int_\varepsilon^\infty \int \mathscr{E}(x, t)\left(\frac{\partial \varphi}{\partial t} + a^2 \Delta \varphi\right) dx\, dt$$

$$= \lim_{\varepsilon \to +0} \left[\int \mathscr{E}(x, \varepsilon)\varphi(x, \varepsilon)\, dx \right.$$

$$\left. + \int_\varepsilon^\infty \int \left(\frac{\partial \mathscr{E}}{\partial t} - a^2 \Delta \mathscr{E}\right)\varphi\, dx\, dt\right]$$

$$= \lim_{\varepsilon \to +0} \int \mathscr{E}(x, \varepsilon)\varphi(x, 0)\, dx$$

$$+ \lim_{\varepsilon \to +0} \int \mathscr{E}(x, \varepsilon)[\varphi(x, \varepsilon) - \varphi(x, 0)]\, dx$$

$$= \lim_{\varepsilon \to +0} \int \mathscr{E}(x, \varepsilon)\varphi(x, 0)\, dx \qquad (31)$$

since, by virtue of (29),

$$\left|\int \mathscr{E}(x, \varepsilon)[\varphi(x, \varepsilon) - \varphi(x, 0)]\, dx\right| \le K\varepsilon \int \mathscr{E}(x, \varepsilon)\, dx = K\varepsilon$$

We shall now prove the result

$$\mathscr{E}(x, t) = \frac{1}{(4\pi a^2 t)^{n/2}} \exp\left(-\frac{|x|^2}{4a^2 t}\right) \to \delta(x), \quad t \to +0 \text{ in } \mathscr{D}'(R^n) \qquad (32)$$

In fact, let $\varphi(x) \in \mathscr{D}$. Then, taking into consideration that

$$\left|\int \mathscr{E}(x, t)[\varphi(x) - \varphi(0)]\, dx\right| \le \frac{K}{(4\pi a^2 t)^{n/2}} \int \exp\left(-\frac{|x|^2}{4a^2 t}\right) |x|\, dx$$

$$= \frac{K\sigma_n}{(4\pi a^2 t)^{n/2}} \int_0^\infty \exp\left(-\frac{r^2}{4a^2 t}\right) r^n\, dr$$

$$= K'\sqrt{t} \int_0^\infty e^{-u^2} u^n\, du = C\sqrt{t}$$

due to (29) we obtain as $t \to +0$ the result (32):

$$(\mathscr{E}(x, t)\varphi) = \int \mathscr{E}(x, t)\varphi(x)\, dx = \varphi(0) \int \mathscr{E}(x, t)\, dx$$
$$+ \int \mathscr{E}(x, t)[\varphi(x) - \varphi(0)]\, dx \to \varphi(0) = (\delta, \varphi)$$

§ 6. DIFFERENTIATION OF GENERALIZED FUNCTIONS

Formula (28) follows from results (31) and (32).

(g) Let
$$\mathscr{E}_1(x, t) = \frac{1}{2a} \theta(at - |x|), \qquad x = x_1$$

We shall prove that
$$\Box_a \mathscr{E}_1 = \delta(x, t) \qquad (33)$$

The function $\mathscr{E}_1$ is locally integrable in R^2 and becomes zero outside the closure of the future light cone $\bar{\Gamma}^+$ (Fig. 17). Let $\varphi \in \mathscr{D}$. Then

$$(\Box_a \mathscr{E}_1, \varphi) = (\mathscr{E}_1, \Box_a \varphi) = \int \mathscr{E}_1(x, t) \Box_a \varphi(x, t) \, dx \, dt$$

$$= \frac{1}{2a} \int_{-\infty}^{\infty} \int_{|x|/a}^{\infty} \frac{\partial^2 \varphi}{\partial t^2} \, dt \, dx - \frac{a}{2} \int_0^{\infty} \int_{-at}^{at} \frac{\partial^2 \varphi}{\partial x^2} \, dx \, dt$$

$$= -\frac{1}{2a} \int_{-\infty}^{\infty} \frac{\partial \varphi(x, (|x|/a))}{\partial t} \, dx$$

$$\quad - \frac{a}{2} \int_0^{\infty} \left[\frac{\partial \varphi(at, t)}{\partial x} - \frac{\partial \varphi(-at, t)}{\partial x} \right] dt$$

$$= -\frac{1}{2} \int_0^{\infty} \frac{d\varphi(at, t)}{dt} \, dt - \frac{1}{2} \int_0^{\infty} \frac{d\varphi(-at, t)}{dt}$$

$$= \tfrac{1}{2} \varphi(0, 0) + \tfrac{1}{2} \varphi(0, 0) = (\delta, \varphi)$$

which proves Eq. (33).

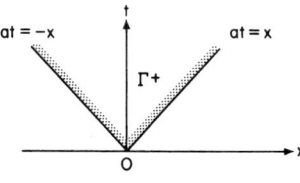

Fig. 17

(h) Let δ_{S_r} be a simple layer on the spherical surface $|x| = r$ (cf. Sec. 5.6). We shall establish the validity of the result (Pizetti's formula)

$$\frac{1}{r^2} \left(\frac{1}{\sigma_n r^{n-1}} \delta_{S_r} - \delta \right) \to \frac{1}{2n} \Delta \delta, \qquad r \to 0 \quad \text{in} \quad \mathscr{D}' \qquad (34)$$

In fact, for all $\varphi \in \mathcal{D}$ when $r \to 0$ we have

$$\left(\frac{1}{\sigma_n r^{n+1}} \delta_{S_r} - \frac{1}{r^2} \delta, \varphi\right)$$

$$= \frac{1}{\sigma_n r^{n+1}} \int_{S_r} \varphi(x)\, dS - \frac{\varphi(0)}{r^2} = \frac{1}{\sigma_n r^2} \int_{S_1} [\varphi(rs) - \varphi(0)]\, ds$$

$$= \frac{1}{\sigma_n r^2} \int_{S_1} \left[r \sum_{k=1}^{n} \frac{\partial \varphi(0)}{\partial x_k} s_k + \frac{r^2}{2} \sum_{k,i=1}^{n} \frac{\partial^2 \varphi(0)}{\partial x_k\, \partial x_i} s_k s_i + O(r^3) \right] ds$$

$$\to \frac{1}{2n} \Delta \varphi(0) = \frac{1}{2n} (\Delta \delta, \varphi)$$

since

$$\int_{S_1} s_k\, ds = 0, \qquad \int_{S_1} s_k s_i\, ds = \delta_{ki} \int_{S_1} s_k^2\, ds = \frac{\sigma_n}{n} \delta_{ki}$$

$$\int_{S_1} s_k^2\, ds = \sigma_{n-1} \int_0^\pi \sin^{n-2}\theta \cos^2\theta\, d\theta = \sigma_{n-1} \int_0^1 (1-\mu)^{(n-3)/2} \sqrt{\mu}\, d\mu$$

$$= \sigma_{n-1} B\!\left(\frac{n-1}{2}, \frac{3}{2}\right) = \frac{2\pi^{(n-1)/2} \Gamma[(n-1)/2]\Gamma(3/2)}{\Gamma[(n-1)/2]\Gamma(n/2+1)} = \frac{\sigma_n}{n}$$

Here B is the Eulerian integral (beta function) of the second kind,

$$B(p,q) = \int_0^1 (1-t)^{p-1} t^{q-1}\, dt = \frac{\Gamma(p)\Gamma(q)}{\Gamma(p+q)}$$

(i) Let $n = 2$, $z = x + iy$, $\bar{z} = x - iy$, $dz = dx + i\, dy$. The differential operator

$$\frac{\partial}{\partial \bar{z}} = \frac{1}{2}\left(\frac{\partial}{\partial x} + i \frac{\partial}{\partial y}\right)$$

is known as the *Cauchy–Riemann* operator. Let $f \in C^1(\bar{G})$ and $f(x, y) = 0$ for $z \in G_1$. Using Eq. (15), we shall deduce

$$\frac{\partial f}{\partial \bar{z}} = \left\{ \frac{\partial f}{\partial \bar{z}} \right\} - \frac{f}{2}[\cos(\mathbf{n}x) + i\cos(\mathbf{n}y)] \delta_S \qquad (35)$$

Applying both sides of Eq. (35) to the test function φ, we obtain a formula analogous to formula (22):

$$\int_G \left(f \frac{\partial \varphi}{\partial \bar{z}} + \frac{\partial f}{\partial \bar{z}} \varphi\right) dx\, dy = \frac{1}{2} \int_S f\varphi[\cos(\mathbf{n}x) + i\cos(\mathbf{n}y)]\, dS$$

$$= \frac{1}{2} \int_S f\varphi(dy - i\, dx) = -\frac{i}{2} \int_S f\varphi\, dz$$

§ 7. THE DIRECT PRODUCT AND CONVOLUTION

that is,

$$\int_G \frac{\partial}{\partial \bar{z}} (f\varphi) \, dx \, dy = -\frac{i}{2} \int_S f\varphi \, dz \tag{36}$$

(j) We shall prove that

$$\frac{\partial}{\partial \bar{z}} \frac{1}{z} = \pi \delta(x, y) \tag{37}$$

The function $1/z$ is locally integrable in R^2. Therefore, using formula (36) when $f = 1/z$ and $G = [\varepsilon < |z| < R]$ (Fig. 16), for all $\varphi \in \mathcal{D}$, supp $\varphi \subset U_R$, we obtain

$$\left(\frac{\partial}{\partial \bar{z}} \frac{1}{z}, \varphi\right) = -\left(\frac{1}{z}, \frac{\partial}{\partial \bar{z}} \varphi\right) = -\int_{U_R} \frac{1}{z} \frac{\partial \varphi}{\partial \bar{z}} \, dx \, dy$$

$$= -\lim_{\varepsilon \to +0} \int_{\varepsilon < |z| < R} \frac{1}{z} \frac{\partial \varphi}{\partial \bar{z}} \, dx \, dy$$

$$= \lim_{\varepsilon \to 0} \left[\int_{\varepsilon < |z| < R} \varphi \frac{\partial}{\partial \bar{z}} \frac{1}{z} \, dx \, dy + \frac{i}{2} \left(\int_{S_R} - \int_{S_\varepsilon}\right) \frac{\varphi}{z} \, dz\right]$$

$$= -\frac{i}{2} \lim_{\varepsilon \to +0} \int_{|z|=\varepsilon} \varphi(z) \frac{dz}{z} = -\frac{i}{2} \lim_{\varepsilon \to +0} i \int_0^{2\pi} \varphi(\varepsilon e^{i\theta}) \, d\theta$$

$$= \pi \varphi(0) = (\pi \delta, \varphi)$$

as was to be shown.

§ 7. The Direct Product and Convolution of Generalized Functions

1. Definition of the Direct Product. Let $f(x)$ and $g(y)$ be locally integrable functions in the spaces R^n and R^m, respectively. The function $f(x)g(y)$ also will be locally integrable in R^{n+m}. It defines the (regular) generalized function, acting on the test functions $\varphi(x, y) \in \mathcal{D}$, according to the formulas

$$(f(x)g(y), \varphi) = \int f(x) \int g(y) \varphi(x, y) \, dy \, dx$$
$$= (f(x), (g(y), \varphi(x, y))) \tag{1}$$

$$(g(y)f(x), \varphi) = \int g(y) \int f(x) \varphi(x, y) \, dx \, dy$$
$$= (g(y), (f(x), \varphi(x, y))) \tag{1'}$$

These equations express Fubini's theorem (cf. Sec. 1.4) concerning the equality of a repeated and a multiple integral. We shall take Eq. (1) as the definition of the *direct product* $f(x) \cdot g(y)$ of the generalized functions $f(x) \in \mathscr{D}'(R^n)$ and $g(y) \in \mathscr{D}'(R^m)$:

$$(f(x) \cdot g(y), \varphi) = (f(x), (g(y), \varphi(x, y))), \qquad \varphi \in \mathscr{D}(R^{n+m}) \qquad (2)$$

We shall check that this definition is correct, that is, that the right-hand side of Eq. (2) defines a linear continuous functional over $\mathscr{D}(R^{n+m})$.
We shall first prove the following lemma.

LEMMA. *For any $g \in \mathscr{D}'(R^m)$ and $\varphi \in \mathscr{D}(R^{n+m})$ the function*

$$\psi(x) = (g(y), \varphi(x, y)) \in \mathscr{D}(R^n)$$

moreover, for all α,

$$D^\alpha \psi(x) = (g(y), D_x^\alpha \varphi(x, y)) \qquad (3)$$

Further, if $\varphi_k \to \varphi$ as $k \to \infty$ in $\mathscr{D}(R^{n+m})$, then

$$\psi_k(x) = (g(y), \varphi_k(x, y)) \to \psi(x), \qquad k \to \infty \quad \text{in} \quad \mathscr{D}(R^n)$$

Proof. Since for each $x \in R^n$, $\varphi(x, y) \in \mathscr{D}(R^m)$, the function $\psi(x)$ is defined in R^n. We shall prove that it is continuous in R^n. Let us fix a point x,

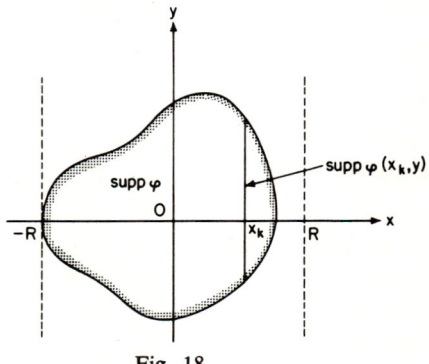

Fig. 18

and let $x_k \to x$ as $k \to \infty$. Then

$$\varphi(x_k, y) \to \varphi(x, y), \qquad k \to \infty \quad \text{in} \quad \mathscr{D}(R^m) \qquad (4)$$

since, by virtue of $\varphi \in \mathscr{D}(R^{n+m})$, the supports of $\varphi(x_k, y)$ are bounded in R^m independently of k (Fig. 18) and for all β

$$D_y^\beta \varphi(x_k, y) \xrightarrow{y \in R^m} D_y^\beta \varphi(x, y), \qquad k \to \infty$$

§ 7. THE DIRECT PRODUCT AND CONVOLUTION

As the functional $g(y)$ is continuous over $\mathscr{D}(R^m)$, the continuity of the function $\psi(x)$ at the point x follows from (4):

$$\psi(x_k) = (g(y), \varphi(x_k, y)) \to (g(y), \varphi(x, y)) = \psi(x), \qquad x_k \to x$$

We shall now prove formula (3). Let us fix a point x and write

$$h_i = (\underbrace{0, \ldots, h}_{i}, \ldots, 0)$$

Then

$$\chi_h^{(i)}(y) = \frac{1}{h} [\varphi(x + h_i, y) - \varphi(x, y)] \to \frac{\partial \varphi(x, y)}{\partial x_i} \qquad (5)$$
$$h \to 0 \quad \text{in} \quad \mathscr{D}(R^m)$$

Since, by virtue of $\varphi \in \mathscr{D}(R^{n+m})$, the supports of $\chi_h^{(i)}$ are bounded in R^m independently of h and for all β

$$D^\beta \chi_h^{(i)}(y) = \frac{1}{h} [D_y^\beta \varphi(x + h_i, y) - D_y^\beta \varphi(x, y)] \xrightarrow{y \in R^m} D_y^\beta \frac{\partial \varphi(x, y)}{\partial x_i}$$
$$h \to 0$$

Since $g \in \mathscr{D}'(R^m)$, then, using (5), we obtain

$$\frac{\psi(x + h_i) - \psi(x)}{h} = \frac{1}{h} [(g(y), \varphi(x + h_i, y)) - (g(y), \varphi(x, y))]$$
$$= \left(g(y), \frac{\varphi(x + h_i, y) - \varphi(x, y)}{h} \right)$$
$$= (g, \chi_h^{(i)}) \to \left(g(y), \frac{\partial \varphi(x, y)}{\partial x_i} \right), \qquad h \to 0$$

from which it follows that Eq. (3) is valid when $\alpha = (0, \ldots, \underbrace{1}_{i}, \ldots, 0)$:

$$\frac{\partial \psi(x)}{\partial x_i} = \left(g(y), \frac{\partial \varphi(x, y)}{\partial x_i} \right), \qquad i = 1, 2, \ldots, n$$

When we apply these arguments once more to the formula we have obtained, we establish the validity of formula (3) for all α. Since, together with φ, $D^\alpha \varphi \in \mathscr{D}(R^{n+m})$, we conclude from formula (3), by virtue of what has been proved, that $D^\alpha \psi(x)$ is a continuous function in R^n for all α. In this way, $\psi \in C^{\infty}(R^n)$. Further, the function $\psi(x)$ has compact support in R^n since $\varphi(x, y) = 0$ for $|x| > R$ (Fig. 18), and thus $\psi(x) = (g, 0) = 0$ for these values of x. Consequently, $\psi \in \mathscr{D}(R^n)$.

Let $\varphi_k(x, y) \to \varphi(x, y)$ as $k \to \infty$ in $\mathscr{D}(R^{n+m})$. We shall prove that $\psi_k(x) \to \psi(x)$ as $k \to \infty$ in $\mathscr{D}(R^n)$. Since the supports of $\varphi_k(x, y)$ are bounded in R^{n+m} independently of k, then, as we have seen earlier, the supports of $\psi_k(x)$ are also bounded in R^n independently of k. Therefore it remains to prove that for all α,

$$D^{\alpha}[\psi(x) - \psi_k(x)] \xrightarrow{x \in R^n} 0, \qquad k \to \infty$$

Suppose that this is false. Then there will be numbers $\varepsilon_0 > 0$, an index α_0, and a sequence of points x_k such that

$$|D^{\alpha_0}[\psi(x_k) - \psi_k(x_k)]| \geq \varepsilon_0, \qquad k = 1, 2, \ldots \tag{6}$$

Since the supports of $\psi - \psi_k$ are bounded in R^n independently of k, it then follows from (6) that the sequence x_k, $k = 1, 2, \ldots$, is also bounded in R^n. Therefore, according to the Bolzano–Weierstrass theorem, it is possible to choose from it a converging subsequence $x_{k_i} \to x_0$ as $i \to \infty$. But then $D_x^{\alpha_0}\varphi_{k_i}(x_{k_i}, y) \to D_x^{\alpha_0}\varphi(x_0, y)$ as $i \to \infty$ in $\mathscr{D}(R^m)$. As a result, since the functional g is continuous in $\mathscr{D}(R^m)$, from formula (3) we obtain

$$D^{\alpha_0}\psi_{k_i}(x_{k_i}) = (g(y), D_x^{\alpha_0}\varphi_{k_i}(x_{k_i}, y))$$
$$\to (g(y), D_x^{\alpha_0}\varphi(x_0, y)) = D^{\alpha_0}\psi(x_0), \qquad i \to \infty$$

which contradicts the inequalities (6). This contradiction proves the lemma.

Let us return to the definition of the direct product. According to the lemma just proved, $\psi(x) = (g(y), \varphi(x, y)) \in \mathscr{D}(R^n)$ for all $\varphi \in \mathscr{D}(R^{n+m})$. Consequently, the right-hand side of Eq. (2), equal to (f, ψ), is defined for any generalized functions f and g, and so defines the functional over $\mathscr{D}(R^{n+m})$. Moreover, the linearity of this functional follows from the linearity of the functionals f and g.

We shall prove that the functional we have constructed is continuous over $\mathscr{D}(R^{n+m})$. Let $\varphi_k \to \varphi$ as $k \to \infty$ in $\mathscr{D}(R^{n+m})$. Then, according to the lemma,

$$(g(y), \varphi_k(x, y)) \to (g(y), \varphi(x, y)), \qquad k \to \infty \quad \text{in } \mathscr{D}(R^n)$$

and therefore, by virtue of the continuity of the functional f in $\mathscr{D}(R^n)$, $(f(x), (g(y), \varphi_k(x, y))) \to (f(x), (g(y), \varphi(x, y)))$ as $k \to \infty$, which shows the

§ 7. THE DIRECT PRODUCT AND CONVOLUTION

continuity in $\mathscr{D}(R^{n+m})$ of the linear functional displayed on the right-hand side of Eq. (2).

In this way, the functional $f(x) \cdot g(y) \in \mathscr{D}'(R^{n+m})$, that is, it is a generalized function.

2. The Commutativity of the Direct Product. Let the generalized functions $f \in \mathscr{D}'(R^n)$ and $g \in \mathscr{D}'(R^m)$ be given. Side by side with the direct product $f(x) \cdot g(y)$, in agreement with formula (2), the direct product $g(y) \cdot f(x)$ is defined as:

$$(g(y) \cdot f(x), \varphi) = (g(y), (f(x), \varphi(x, y))), \qquad \varphi \in \mathscr{D}(R^{n+m}) \qquad (2')$$

It follows that

$$f(x) \cdot g(y) = g(y) \cdot f(x) \qquad (7)$$

that is, the direct product operation is *commutative*.

In fact, over the test functions $\varphi \in \mathscr{D}(R^{n+m})$ of the form

$$\varphi(x, y) = \sum_{l=1}^{N} u_l(x) v_l(y), \qquad u_l \in \mathscr{D}(R^n), \qquad v_l \in \mathscr{D}(R^m) \qquad (8)$$

Eq. (7) follows from definitions (2) and (2'):

$$(f(x) \cdot g(y), \varphi) = \left(f, \sum_{l=1}^{N} u_l(g, v_l)\right) = \sum_{l=1}^{N} (f, u_l)(g, v_l)$$

$$= \left(g, \sum_{l=1}^{N} v_l(f, u_l)\right) = (g(y) \cdot f(x), \varphi)$$

In order to extend Eq. (7) over any test functions, we shall prove a lemma to the effect that the set of test functions of the form (8) is dense in $\mathscr{D}(R^{n+m})$ (cf. Sec. 1.5).

LEMMA. *For any function $\varphi \in \mathscr{D}(R^{n+m})$ there is a sequence of test functions $\varphi_k(x, y)$, $k = 1, 2, \ldots$, of the form (8), converging to φ in $\mathscr{D}(R^{n+m})$.*

Proof. Let the support of $\varphi(x, y)$ be contained in the sphere $\bar{U}_R$ (Fig. 19). According to Weierstrass's theorem (cf. Sec. 1.3), there are polynomials $P_k(x, y)$, $k = 1, 2, \ldots$, such that

$$|D^\alpha \varphi - D^\alpha P_k| < \frac{1}{k} \quad \text{for all} \quad |\alpha| \leq k \quad \text{and} \quad |x|^2 + |y|^2 \leq 8R^2 \qquad (9)$$

Let $e(x)$ and $h(y)$ be test functions, equal to 1 in a sphere of radius R and

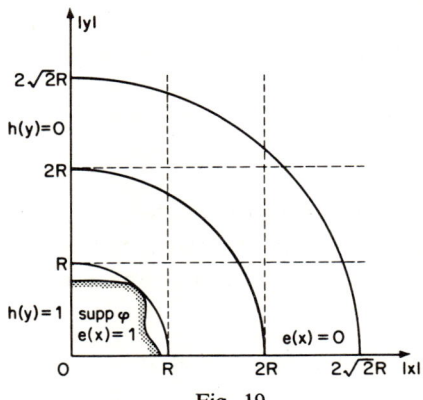

Fig. 19

to 0 outside a sphere of radius $2R$ centered at the origin. Then the sequence of test functions

$$\varphi_k(x, y) = e(x)h(y)P_k(x, y), \qquad k = 1, 2, \ldots$$

possesses the required properties. In fact, φ_k has form (8), their supports are contained in the sphere $|x|^2 + |y|^2 \leq 8R^2$, and by virtue of (9), for any α and $k \geq |\alpha|$

$$|D^\alpha \varphi - D^\alpha \varphi_k| = |D^\alpha \varphi - D^\alpha(ehP_k)| \leq \frac{C_\alpha}{k}$$
$$|x|^2 + |y|^2 \leq 8R^2$$

where C_α are numbers not depending on k. This means that $\varphi_k \to \varphi$ as $k \to \infty$ in $\mathscr{D}(R^{n+m})$. The lemma is proved.

Let φ be an arbitrary test function belonging to $\mathscr{D}(R^{n+m})$. By virtue of the lemma we have just proved there is a sequence $\varphi_1, \varphi_2, \ldots$, of test functions of the form (8) converging to φ in $\mathscr{D}(R^{n+m})$. From this, using the continuity over $\mathscr{D}(R^{n+m})$ of the functionals $f(x) \cdot g(y)$ and $g(y) \cdot f(x)$ (cf. Sec. 7.1), and using Eq. (7) over the functions of the form (8), we shall obtain Eq. (7) in the general case:

$$(f(x) \cdot g(y), \varphi) = \lim_{k \to \infty} (f(x) \cdot g(y), \varphi_k)$$
$$= \lim_{k \to \infty} (g(y) \cdot f(x), \varphi_k) = (g(y) \cdot f(x), \varphi)$$

3. Further Properties of the Direct Product

(a) THE CONTINUITY OF THE DIRECT PRODUCT. If $f_k \to f$ as $k \to \infty$ in $\mathscr{D}'(R^n)$, then $f_k(x) \cdot g(y) \to f(x) \cdot g(y)$ as $k \to \infty$ in $\mathscr{D}'(R^{n+m})$.

§ 7. THE DIRECT PRODUCT AND CONVOLUTION

In fact, if $\varphi \in \mathscr{D}(R^{n+m})$, then, according to the lemma of Sec. 7.1, $\psi(x) = (g(y), \varphi(x, y)) \in \mathscr{D}(R^n)$, and therefore

$$(f_k(x) \cdot g(y), \varphi) = (f_k(x), (g(y), \varphi(x, y)))$$
$$= (f_k, \psi) \to (f, \psi) = (f(x), (g(y), \varphi(x, y)))$$
$$= (f(x) \cdot g(y), \varphi), \qquad k \to \infty$$

(b) ASSOCIATIVITY OF THE DIRECT PRODUCT. If $f \in \mathscr{D}'(R^n), g \in \mathscr{D}'(R^m)$, and $h \in \mathscr{D}'(R^k)$, then

$$f(x) \cdot (g(y) \cdot h(z)) = (f(x) \cdot g(y)) \cdot h(z) \qquad (10)$$

In fact, if $\varphi \in \mathscr{D}(R^{n+m+k})$, then

$$(f(x) \cdot (g(y) \cdot h(z)), \varphi) = (f(x), (g(y) \cdot h(z), \varphi))$$
$$= (f(x), (g(y), (h(z), \varphi)))$$
$$= (f(x) \cdot g(y), (h(z), \varphi)) = ((f(x) \cdot g(y)) \cdot h(z), \varphi)$$

(c) DIFFERENTIATION OF THE DIRECT PRODUCT

$$D_x^\alpha(f(x) \cdot g(y)) = D^\alpha f(x) \cdot g(y) \qquad (11)$$

In fact, if $\varphi \in \mathscr{D}(R^{n+m})$, then

$$(D_x^\alpha(f(x) \cdot g(y)), \varphi) = (-1)^{|\alpha|}(f(x) \cdot g(y), D_x^\alpha \varphi)$$
$$= (-1)^{|\alpha|}(g(y), (f(x), D_x^\alpha \varphi(x, y)))$$
$$= (g(y), (D^\alpha f(x), \varphi)) = (D^\alpha f(x) \cdot g(y), \varphi)$$

(d) MULTIPLICATION OF THE DIRECT PRODUCT. If $a \in C^\infty(R^n)$, then

$$a(x)(f(x) \cdot g(y)) = (a(x)f(x)) \cdot g(y) \qquad (12)$$

In fact, if $\varphi \in \mathscr{D}(R^{n+m})$, then

$$(a(x)(f(x) \cdot g(y)), \varphi) = (f(x) \cdot g(y), a\varphi)$$
$$= (f(x), (g(y), a(x)\varphi(x, y)))$$
$$= (f(x), a(x)(g(y), \varphi(x, y)))$$
$$= (a(x)f(x), (g(y), \varphi(x, y)))$$
$$= ((a(x)f(x)) \cdot g(y), \varphi)$$

102   2. GENERALIZED FUNCTIONS

(e) TRANSLATION OF THE DIRECT PRODUCT

$$(f \cdot g)(x + h, y) = f(x + h) \cdot g(y) \tag{13}$$

In fact, if $\varphi \in \mathscr{D}(R^{n+m})$, then

$$\begin{aligned}
((f \cdot g)(x + h, y), \varphi) &= (f(x) \cdot g(y), \varphi(x - h, y)) \\
&= (g(y), (f(x), \varphi(x - h, y))) \\
&= (g(y), (f(x + h), \varphi(x, y))) \\
&= (f(x + h) \cdot g(y), \varphi)
\end{aligned}$$

(f) It is said that the generalized function of the form $f(x) \cdot 1(y)$ *does not depend on* y. It acts according to the rule: If $\varphi \in \mathscr{D}(R^{n+m})$, then

$$\begin{aligned}
(f(x) \cdot 1(y), \varphi) &= (f(x), \smallint \varphi(x, y)\, dy) = (1(y) \cdot f(x), \varphi) = (1(y), (f(x), \varphi)) \\
&= \smallint (f(x), \varphi(x, y))\, dy
\end{aligned}$$

Specifically, we obtain the equation

$$(f(x), \smallint \varphi(x, y)\, dy) = \smallint (f(x), \varphi(x, y))\, dy \tag{14}$$

valid for all $f \in \mathscr{D}'(R^n)$ and $\varphi \in \mathscr{D}(R^{n+m})$.

4. Convolution of Generalized Functions. Let $f(x)$ and $g(x)$ be functions locally integrable in R^n, where the function

$$h(x) = \smallint |\, g(y)f(x - y)\,|\, dy$$

is also locally integrable in R^n. The function

$$\begin{aligned}
(f * g)(x) &= \smallint f(y)g(x - y)\, dy \\
&= \smallint g(y)f(x - y)\, dy = (g * f)(x)
\end{aligned} \tag{15}$$

is known as the *convolution* $f * g$ of these functions. Function (15) is locally integrable in R^n and therefore defines a (regular) generalized function, acting on the test functions $\varphi \in \mathscr{D}(R^n)$ according to the rule:

$$\begin{aligned}
(f * g, \varphi) &= \smallint (f * g)(\xi)\varphi(\xi)\, d\xi \\
&= \smallint [\smallint g(y)f(\xi - y)\, dy]\varphi(\xi)\, d\xi \\
&= \smallint g(y)\, [\smallint f(\xi - y)\varphi(\xi)\, d\xi]\, dy \\
&= \smallint g(y)\, [\smallint f(x)\varphi(x + y)\, dx]\, dy
\end{aligned}$$

§ 7. THE DIRECT PRODUCT AND CONVOLUTION

(by virtue of Fubini's theorem), that is,

$$(f * g, \varphi) = \int f(x)g(y)\varphi(x+y) \, dx \, dy, \qquad \varphi \in \mathscr{D}(R^n) \quad (16)$$

We shall note three cases in which the condition of local integrability of the function $h(x)$ is satisfied and so the convolution $f * g$ exists and is defined by formula (15).

(1) One of the functions f or g has compact support, for instance supp $g \subset U_{R_1}$:

$$\int_{U_R} h(x) \, dx = \int_{U_{R_1}} |g(y)| \int_{U_R} |f(x-y)| \, dx \, dy$$

$$\leq \int_{U_{R_1}} |g(y)| \, dy \int_{U_{R+R_1}} |f(\xi)| \, d\xi < \infty$$

(2) The functions f and g become zero when $x < 0$ $(n = 1)$:

$$\int_{-R}^{R} h(x) \, dx = \int_{0}^{R} \int_{0}^{x} |g(y)| \, |f(x-y)| \, dy \, dx$$

$$= \int_{0}^{R} |g(y)| \int_{y}^{R} |f(x-y)| \, dx \, dy$$

$$\leq \int_{0}^{R} |g(y)| \, dy \int_{0}^{R} |f(\xi)| \, d\xi < \infty$$

(3) The functions f and g are integrable over R^n:

$$\int h(x) \, dx = \int |g(y)| \int |f(x-y)| \, dx \, dy$$

$$= \int |g(y)| \, dy \int |f(\xi)| \, d\xi < \infty$$

In this case the convolution $f * g$ is integrable over R^n.

We shall say that the sequence $\{\eta_k\}$ of test functions belonging to $\mathscr{D}(R^n)$ *converges to* 1 *in* R^n if for any compactum K there is a number N such that $\eta_k(x) = 1$ when $x \in K$, $k \geq N$, and the η_k are uniformly bounded in R^n together with all their derivatives $|D^\alpha \eta_k(x)| < c_\alpha$ for $x \in R^n$, $k = 1, 2, \ldots$.

We note that such sequences always exist: for example, $\eta_k(x) = \eta(x/k)$, where $\eta \in \mathscr{D}$, $\eta(x) = 1$ in U_1.

We shall prove that Eq. (16) may be written in the form

$$(f * g, \varphi) = \lim_{k \to \infty} (f(x)g(y), \eta_k(x; y)\varphi(x+y)) \qquad \varphi \in \mathscr{D}(R^n) \quad (16')$$

where $\eta_k(x; y)$ for $k = 1, 2, \ldots$, is any sequence converging to 1 in R^{2n}. In fact, according to what has been proved, the function

$$c_0 | f(x) g(y) \varphi(x + y) |$$

is integrable over R^{2n} and

$$| f(x) g(y) \eta_k(x; y) \varphi(x + y) | \leq c_0 | f(x) g(y) \varphi(x + y) |$$
$$k = 1, 2, \ldots$$

Further,

$$f(x) g(y) \eta_k(x; y) \varphi(x + y) \to f(x) g(y) \varphi(x + y)$$
$$k \to \infty \quad \text{almost everywhere in} \quad R^{2n}$$

Applying Lebesgue's theorem (cf. Sec. 1.4), we obtain the equation

$$\int f(x) g(y) \varphi(x + y) \, dx \, dy = \lim_{k \to \infty} \int f(x) g(y) \eta_k(x; y) \varphi(x + y) \, dx \, dy$$

which, by virtue of (16), is equivalent to Eq. (16').

With Eqs. (16) and (16') as points of departure, we shall define a convolution in the following way. Let a pair of generalized functions f and g belonging to $\mathscr{D}'(R^n)$ be such that their direct product $f(x) \cdot g(y)$ allows the extension (cf. Sec. 1.8) $(f(x) \cdot g(y), \varphi(x + y))$ over functions of the form $\varphi(x + y)$, where φ is any function belonging to $\mathscr{D}(R^n)$, in the following sense: Whatever might be the sequence $\{\eta_k\}$ of functions belonging to $\mathscr{D}(R^{2n})$ converging to 1 in R^{2n}, there is a limit of the numerical sequence

$$\lim_{k \to \infty} (f(x) \cdot g(y), \eta_k(x; y) \varphi(x + y)) = (f(x) \cdot g(y), \varphi(x + y))$$

and this limit does not depend on the sequence $\{\eta_k\}$. We note that for each k, $\eta_k(x; y) \varphi(x + y) \in \mathscr{D}(R^{2n})$, so that the numerical sequence is defined.

The functional

$$(f * g, \varphi) = (f(x) \cdot g(y), \varphi(x + y))$$
$$= \lim_{k \to \infty} (f(x) \cdot g(y), \eta_k(x; y) \varphi(x + y)), \quad \varphi \in \mathscr{D}(R^n) \quad (17)$$

is known as the *convolution* $f * g$ if it is continuous in $\mathscr{D}(R^n)$. Obviously this functional is linear over $\mathscr{D}(R^n)$ and in this way the convolution $f * g \in \mathscr{D}'(R^n)$.

§ 7. THE DIRECT PRODUCT AND CONVOLUTION

We note that since $\varphi(x+y)$ does not belong to $\mathscr{D}(R^{2n})$ (it is not of compact support in R^{2n}!), the right-hand side of Eq. (17) does not exist for all pairs of generalized functions f and g, and in this way the convolution does not always exist.

If the convolution $f * g$ exists, then there is also a convolution $g * f$, and they are equal,

$$f * g = g * f \qquad (18)$$

that is, the operation of convolution is commutative.

This assertion follows from the commutativity of the direct product (cf. Sec. 7.2) and from the definition of convolution:

$$\begin{aligned}
(f * g, \varphi) &= (f(x) \cdot g(y), \varphi(x+y)) \\
&= \lim_{k \to \infty} (f(x) \cdot g(y), \eta_k(x;y)\varphi(x+y)) \\
&= \lim_{k \to \infty} (g(y) \cdot f(x), \eta_k(x;y)\varphi(x+y)) \\
&= (g(y) \cdot f(x), \varphi(x+y)) = (g * f, \varphi), \qquad \varphi \in \mathscr{D}
\end{aligned}$$

Note. The condition of continuity in $\mathscr{D}$ of the functional $f * g$ in the definition of convolution follows from its existence, by virtue of the completeness of the space $\mathscr{D}'$ (cf. Sec. 5.3, note). Since the completeness of $\mathscr{D}'$ is not proved here, the continuity of this functional should be checked directly each time.

5. Condition for the Existence of a Convolution. We shall establish the condition under which the convolution $f * g$ is known to exist.

THEOREM. *Let f be an arbitrary and g a generalized function with compact support. Then the convolution $f * g$ exists in $\mathscr{D}'$ and appears in the form*

$$(f * g, \varphi) = (f(x) \cdot g(y), \eta(y)\varphi(x+y)), \qquad \varphi \in \mathscr{D} \qquad (19)$$

where η is any test function equal to 1 in the neighborhood of the support of g. For this the convolution is continuous with respect to f and g separately: (1) *If $f_k \to f$ as $k \to \infty$ in $\mathscr{D}'$, then $f_k * g \to f * g$ as $k \to \infty$ in $\mathscr{D}'$.* (2) *If $g_k \to g$ as $k \to \infty$ in $\mathscr{D}'$ and for a certain R, $\operatorname{supp} g_k \subset U_R$, then $f * g_k \to f * g$ as $k \to \infty$ in $\mathscr{D}'$.*

Proof. Let supp $g \subset U_R$ and let η be a function belonging to $\mathscr{D}(R^n)$ equal to 1 in the neighborhood of supp g and supp $\eta \subset U_R$. (According to Lemma 1 of Sec. 5.2 such functions exist.) Further, let φ be an arbitrary function belonging to $\mathscr{D}(R^n)$, let supp $\varphi \subset U_A$, and let $\eta_k(x; y)$, $k = 1$, 2, ..., be a sequence of functions belonging to $\mathscr{D}(R^{2n})$ and converging to 1 in R^{2n} (cf. Sec. 7.4). Then for all sufficiently large k

$$\eta(y)\eta_k(x; y)\varphi(x + y) = \eta(y)\varphi(x + y) \tag{20}$$

To prove Eq. (20) it is sufficient to establish that the function $\eta(y)\varphi(x+y) \in \mathscr{D}(R^{2n})$. But this follows from the fact that it is infinitely differentiable

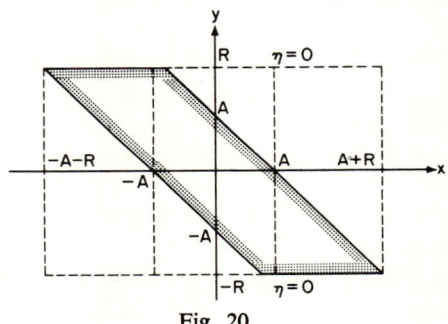

Fig. 20

and that its support is contained in the bounded set (Fig. 20):

$$[(x, y) : |x + y| \leq A, |y| \leq R] \subset U_{A+R} \times U_R$$

Taking into account result (20) and the equation $g = \eta g$ [cf. Eq. (13) of Sec. 5.9], we establish the validity of formula (19):

$$\begin{aligned}
(f * g, \varphi) &= \lim_{k \to \infty} (f(x) \cdot g(y), \eta_k(x; y)\varphi(x + y)) \\
&= \lim_{k \to \infty} (f(x) \cdot \eta(y)g(y), \eta_k(x; y)\varphi(x + y)) \\
&= \lim_{k \to \infty} (f(x) \cdot g(y), \eta(y)\eta_k(x; y)\varphi(x + y)) \\
&= (f(x) \cdot g(y), \eta(y)\varphi(x + y)), \qquad \varphi \in \mathscr{D}
\end{aligned}$$

We shall prove that the right-hand side of Eq. (19) defines a continuous functional over $\mathscr{D}$, that is, the convolution $f * g \in \mathscr{D}'$. Let $\varphi_k \to \varphi$ as

§ 7. THE DIRECT PRODUCT AND CONVOLUTION

$k \to \infty$ in $\mathscr{D}(R^n)$. Then, since

$$\eta(y)\varphi(x+y) \in \mathscr{D}(R^{2n})$$
$$\eta(y)\varphi_k(x+y) \to \eta(y)\varphi(x+y), \qquad k \to \infty \quad \text{in} \quad \mathscr{D}(R^{2n})$$

we therefore have

$$\begin{aligned}(f * g, \varphi_k) &= (f(x) \cdot g(y), \eta(y)\varphi_k(x+y)) \\ &\to (f(x) \cdot g(y), \eta(y)\varphi(x+y)) = (f*g, \varphi), \qquad k \to \infty\end{aligned}$$

as was to be shown.

The continuity of the convolution $f * g$ with respect to f and g follows from Eq. (19) and from the continuity of the direct product $f(x) \cdot g(y)$ with respect to f and g [cf. Sec. 7.3, (a)]. Moreover, in case (2) the condition supp $g_k \subset U_R$ makes it possible to choose an auxiliary function η which does not depend on k. The theorem is proved.

COROLLARY. *The convolution of any generalized function f with the δ function exists and is equal to f*

$$f * \delta = \delta * f = f \tag{21}$$

In fact, by virtue of (19), for all $\varphi \in \mathscr{D}$ we have

$$\begin{aligned}(f * \delta, \varphi) &= (f(x) \cdot \delta(y), \eta(y)\varphi(x+y)) \\ &= (f(x), (\delta(y), \eta(y)\varphi(x+y))) = (f, \varphi)\end{aligned}$$

as was to be shown.

6. Differentiation of a Convolution. *If the convolution $f * g$ exists, then the convolutions $D^\alpha f * g$ and $f * D^\alpha g$ exist, and moreover*

$$D^\alpha f * g = D^\alpha(f * g) = f * D^\alpha g \tag{22}$$

It is sufficient to prove this assertion for each first derivative $\partial/\partial x_j$, $j = 1, 2, \ldots, n$. Let $\varphi \in \mathscr{D}(R^n)$ and $\eta_k(x; y)$, $k = 1, 2, \ldots$, be a sequence of functions belonging to $\mathscr{D}(R^{2n})$ and converging to 1 in R^{2n}. Then the sequence $\eta_k + (\partial \eta_k/\partial x_j)$, $k = 1, 2, \ldots$, of functions belonging to $\mathscr{D}(R^{2n})$ also converges to 1 in R^{2n}. From this, and using the existence of the convolution $f * g$, we obtain the following sequence of equations:

$$\left(\frac{\partial}{\partial x_j}(f * g), \varphi\right) = -\left(f * g, \frac{\partial \varphi}{\partial x_j}\right) = -\left(f(x) \cdot g(y), \frac{\partial \varphi(x+y)}{\partial x_j}\right)$$

$$= -\lim_{k \to \infty}\left(f(x) \cdot g(y), \eta_k(x;y)\frac{\partial \varphi(x+y)}{\partial x_j}\right)$$

$$= -\lim_{k \to \infty}\left(f(x) \cdot g(y), \frac{\partial[\eta_k \varphi(x+y)]}{\partial x_j} - \frac{\partial \eta_k}{\partial x_j}\varphi(x+y)\right)$$

$$= \lim_{k \to \infty}\left(\frac{\partial f(x)}{\partial x_j} \cdot g(y), \eta_k \varphi(x+y)\right)$$

$$+ \lim_{k \to \infty}\left(f(x) \cdot g(y), \left(\eta_k + \frac{\partial \eta_k}{\partial x_j}\right)\varphi(x+y)\right)$$

$$- \lim_{k \to \infty}(f(x) \cdot g(y), \eta_k \varphi(x+y))$$

$$= \left(\frac{\partial f}{\partial x_j} * g, \varphi\right) + (f * g, \varphi) - (f * g, \varphi) = \left(\frac{\partial f}{\partial x_j} * g, \varphi\right)$$

from which the first Eq. (22) follows for $\partial/\partial x_j$. The second Eq. (22) follows from the first and from the commutativity of the convolution (cf. Sec. 7.4):

$$\frac{\partial}{\partial x_j}(f * g) = \frac{\partial}{\partial x_j}(g * f) = \frac{\partial g}{\partial x_j} * f = f * \frac{\partial g}{\partial x_j}$$

Equations

$$D^\alpha f = D^\alpha \delta * f = \delta * D^\alpha f, \qquad f \in \mathscr{D}' \qquad (23)$$

follow from Eqs. (21) and (22).

We note that the existence of the convolutions $D^\alpha f * g$ and $f * D^\alpha g$, $|\alpha| \geq 1$, is not sufficient for the existence of the convolution $f * g$; specifically, these convolutions need not be equal: for instance,

$$\theta' * 1 = \delta * 1 = 1, \quad \text{but} \quad \theta * 1' = \theta * 0 = 0$$

7. Regularization of Generalized Functions. Let f be a generalized function and ψ a test function. Since ψ has compact support, the convolution $f * \psi$ exists.

We shall prove that

$$f * \psi = (f(y), \psi(x-y)) \in C^\infty(R^n) \qquad (24)$$

§ 7. THE DIRECT PRODUCT AND CONVOLUTION

In fact, by virtue of (19), for all $\varphi \in \mathscr{D}$ we have

$$\begin{aligned}(f * \psi, \varphi) &= (f(y) \cdot \psi(\xi), \eta(\xi)\varphi(y + \xi)) \\ &= (f(y), \int \psi(\xi)\eta(\xi)\varphi(y + \xi)\, d\xi)) \\ &= (f(y), \int \psi(\xi)\varphi(y + \xi)\, d\xi) \\ &= (f(y), \int \varphi(x)\psi(x - y)\, dx)\end{aligned}$$

where the auxiliary function $\eta \in \mathscr{D}$ and is equal to 1 in the neighborhood of the support of ψ. Noting now that the function $\varphi(x)\psi(x - y)$ belongs to $\mathscr{D}(R^{2n})$, and using Eq. (14) (cf. Sec. 7.3), we obtain Eq. (24):

$$\begin{aligned}(f * \psi, \varphi) &= \int \varphi(x)(f(y), \psi(x - y))\, dx \\ &= ((f(y), \psi(x - y)), \varphi), \qquad \varphi \in \mathscr{D}\end{aligned}$$

The infinite differentiability of the right-hand side of Eq. (24) is established as in the proof of the lemma of Sec. 7.1.

Let $\omega_\varepsilon(x)$ be a "cap-shaped function" (cf. Sec. 5.2). Then the infinitely differentiable function

$$f_\varepsilon(x) = f * \omega_\varepsilon = (f(y), \omega_\varepsilon(x - y))$$

is known as the *regularization* of the generalized function f.

In Sec. 5.6 it was proved that $\omega_\varepsilon(x) \to \delta(x)$ as $\varepsilon \to +0$ in $\mathscr{D}'$. From this, and using the continuity of the convolution $f * \omega_\varepsilon$ with respect to ω_ε (cf. the theorem of Sec. 7.5), we obtain

$$f_\varepsilon(x) \to f(x), \qquad \varepsilon \to +0 \quad \text{in} \quad \mathscr{D}' \tag{25}$$

So, each generalized function is a weak limit of its own regularization. Using these assertions, we shall establish a more powerful result.

THEOREM. *Each generalized function f is a weak limit of test functions, that is, the set $\mathscr{D}$ is dense in $\mathscr{D}'$.*

Proof. Let $f_\varepsilon(x)$ be the regularization of f and let $\eta_\varepsilon(x)$ as $\varepsilon \to +0$ be a sequence of test functions equal to 1 in the sphere $U_{1/\varepsilon}$. Then the sequence of basic functions $\eta_\varepsilon(x)f_\varepsilon(x)$ as $\varepsilon \to +0$ tends to f in $\mathscr{D}'$, since for any $\varphi \in \mathscr{D}$, by virtue of (25), we have

$$\lim_{\varepsilon \to +0} (\eta_\varepsilon f_\varepsilon, \varphi) = \lim_{\varepsilon \to +0} (f_\varepsilon, \eta_\varepsilon \varphi) = \lim_{\varepsilon \to +0} (f_\varepsilon, \varphi) = (f, \varphi)$$

as was to be shown.

8. Examples of Convolutions. Newtonian Potential. (a) Let $f(x)$ be a continuous function in $R^n \setminus \{0\}$ and locally integrable in R^n and let $\mu \delta_S(x)$ be a simple layer over the bounded piecewise smooth surface S with a continuous density μ (cf. Sec. 5.6).

Their convolution $f * \mu \delta_S$, a locally integrable function in R^n, is expressed by the integral

$$f * \mu \delta_S = \int_S \mu(y) f(x - y) \, dS_y \qquad (26)$$

This assertion follows from Eq. (19)

$$\begin{aligned}(f * \mu \delta_S, \varphi) &= (\mu \delta_S(y) \cdot f(\xi), \eta(y)\varphi(y + \xi)) \\ &= (\mu \delta_S(y), \eta(y)(f(\xi), \varphi(y + \xi))) \\ &= \int_S \mu(y) \eta(y) \int f(\xi) \varphi(y + \xi) \, d\xi \, dS_y \\ &= \int_S \mu(y) \int f(x - y) \varphi(x) \, dx \, dS_y \\ &= \int \varphi(x) \int_S \mu(y) f(x - y) \, dS_y \, dx, \qquad \varphi \in \mathscr{D}\end{aligned}$$

(b) Let ϱ be a generalized function. The convolution

$$V_n = \frac{1}{|x|^{n-2}} * \varrho, \quad n \geq 3; \qquad V_2 = \ln \frac{1}{|x|} * \varrho, \quad n = 2 \qquad (27)$$

is known as a *Newtonian* (when $n = 2$, *logarithmic*) *potential* with a density ϱ.

If ϱ is a generalized function with compact support, then the potential V_n exists in $\mathscr{D}'$ and satisfies Poisson's equation

$$\Delta V_n = -(n - 2)\sigma_n \varrho, \quad n \geq 3; \qquad \Delta V_2 = -2\pi \varrho \qquad (28)$$

The existence of the potential V_n follows from the theorem of Sec. 7.5. Using Eqs. (22) of Sec. 7.6 and (25) of Sec. 6.5, we conclude that for $n \geq 3$ the potential V_n satisfies Poisson's equation (28):

$$\begin{aligned}\Delta V_n = \Delta\left(\frac{1}{|x|^{n-2}} * \varrho\right) &= \Delta \frac{1}{|x|^{n-2}} * \varrho \\ &= -(n - 2)\sigma_n \delta * \varrho = -(n - 2)\sigma_n \varrho\end{aligned}$$

We reason analogously when $n = 2$.

§ 7. THE DIRECT PRODUCT AND CONVOLUTION

(c) If ϱ is an (absolutely) integrable function over R^n with compact support, then the corresponding Newtonian (logarithmic) potential V_n is known as the *volume potential (area potential)*.

The volume potential V_n, a locally integrable function in R^n, is expressed by the integrals

$$V_n(x) = \int \frac{\varrho(y)}{|x-y|^{n-2}} \, dy, \qquad n \geq 3$$

$$V_2(x) = \int \varrho(y) \ln \frac{1}{|x-y|} \, dy \qquad (29)$$

This assertion follows from formula (15) for the convolution of a finite integrable function ϱ with compact support with a locally integrable function $|x|^{2-n}$ for $n \geq 3$ and $-\ln|x|$ for $n = 2$.

(d) Let S be a bounded piecewise smooth two-sided surface with a chosen direction for its normal **n** and let μ and ν be continuous functions over S. Let

$$\mu \delta_S \quad \text{and} \quad -\frac{\partial}{\partial \mathbf{n}} (\nu \delta_S)$$

be the simple and double layers over S with surface densities μ and ν [cf. Secs. 5.6 and 6.5, (b)]. The Newtonian (logarithmic) potentials generated by them

$$V_n^{(0)} = \frac{1}{|x|^{n-2}} * \mu \delta_S, \quad n \geq 3; \qquad V_2^{(0)} = \ln \frac{1}{|x|} * \mu \delta_S \qquad (30)$$

$$V_n^{(1)} = -\frac{1}{|x|^{n-2}} * \frac{\partial}{\partial \mathbf{n}} (\nu \delta_S), \qquad n \geq 3$$

$$V_2^{(1)} = -\ln \frac{1}{|x|} * \frac{\partial}{\partial \mathbf{n}} (\nu \delta_S) \qquad (31)$$

are said to be the *surface potentials of the simple* and *double layer*, respectively, with densities μ and ν.

The surface potentials $V_n^{(0)}$ and $V_n^{(1)}$ are locally integrable functions in R^n and are expressed by the formulas

$$V_n^{(0)}(x) = \int_S \frac{\mu(y)}{|x-y|^{n-2}} \, dS_y, \qquad n \geq 3$$

$$V_2^{(0)}(x) = \int_S \mu(y) \ln \frac{1}{|x-y|} \, dS_y \qquad (32)$$

$$V_n^{(1)}(x) = \int_S v(y) \frac{\partial}{\partial n_y} \frac{1}{|x-y|^{n-2}} dS_y, \quad n \geq 3$$

$$V_2^{(1)}(x) = \int_S v(y) \frac{\partial}{\partial n_y} \ln \frac{1}{|x-y|} dS_y$$

(33)

Formulas (32) are specific cases of formula (26). For clarity we shall prove formula (33) for the potential $V_n^{(1)}$, $n \geq 3$. Using Eq. (19) for convolution and using the definition of the double layer, for all $\varphi \in \mathscr{D}$ we obtain

$$(V_n^{(1)}, \varphi) = -\left(\frac{1}{|x|^{n-2}} * \frac{\partial}{\partial \mathbf{n}} (v \delta_S), \varphi \right)$$

$$= -\left(\frac{\partial}{\partial \mathbf{n}} (v \delta_S(y)) \cdot \frac{1}{|\xi|^{n-2}}, \eta(y) \varphi(y+\xi) \right)$$

$$= -\left(\frac{\partial}{\partial \mathbf{n}} (v \delta_S(y)), \eta(y) \left(\frac{1}{|\xi|^{n-2}}, \varphi(y+\xi) \right) \right)$$

$$= \int_S v(y) \frac{\partial}{\partial \mathbf{n}} \left[\eta(y) \int \frac{\varphi(y+\xi)}{|\xi|^{n-2}} d\xi \right] dS_y$$

$$= \int_S v(y) \frac{\partial}{\partial \mathbf{n}} \int \frac{\varphi(x)}{|x-y|^{n-2}} dx \, dS_y$$

$$= \int_S v(y) \int \varphi(x) \frac{\partial}{\partial n_y} \frac{1}{|x-y|^{n-2}} dx \, dS_y$$

$$= \int \varphi(x) \int_S v(y) \frac{\partial}{\partial n_y} \frac{1}{|x-y|^{n-2}} dS_y \, dx$$

from which, the required formula (33) follows. Differentiating under the integral sign and changing the order of the integration is permissible here, since the integral

$$\int |\varphi(x)| \left| \frac{\partial}{\partial n_y} \frac{1}{|x-y|^{n-2}} \right| dx$$

converges uniformly for $y \in S$.

9. Exercises. Prove the equations:

(a) $\dfrac{\partial^n [\theta(x_1)\theta(x_2) \cdots \theta(x_n)]}{\partial x_1 \partial x_2 \cdots \partial x_n} = \delta(x_1)\delta(x_2) \cdots \delta(x_n) = \delta(x);$

(b) $D_y^\alpha(f(x) \cdot 1(y)) = 0, |\alpha| \neq 0;$

§ 8. GENERALIZED FUNCTIONS OF SLOW GROWTH

(c) $f_\alpha * f_\beta = f_{\alpha+\beta}, f_\alpha(x) = \theta(x) \dfrac{x^{\alpha-1}}{\Gamma(\alpha)} e^{\alpha x}, \alpha > 0;$

(d) $f_\alpha * f_\beta = f_{\sqrt{\alpha^2+\beta^2}}, f_\alpha(x) = \dfrac{1}{\alpha\sqrt{2\pi}} \exp\left(-\dfrac{x^2}{2\alpha^2}\right), \alpha > 0;$

(e) $f_\alpha * f_\beta = f_{\alpha+\beta}, f_\alpha(x) = \dfrac{1}{\pi} \dfrac{\alpha}{\alpha^2 + x^2}, \alpha > 0$

§ 8. Generalized Functions of Slow Growth (Tempered Distributions)

One of the most effective means of solving problems in mathematical physics is by means of the transform method. Here we shall set out the Fourier transform theory for so-called generalized functions of slow growth (tempered distributions). The remarkable feature of the class of generalized functions of slow growth is that the Fourier transform does not cause us to go beyond the limits of this class.

1. The Space of Test Functions $\mathscr{S}$. *Let us relate to the set of test functions $\mathscr{S} = \mathscr{S}(R^n)$ all functions of the class $C^\infty(R^n)$ which decreases as $|x| \to \infty$, together with all their derivatives, faster than any power of $|x|^{-1}$.* We shall define convergence in $\mathscr{S}$ in the following way. The sequence of functions $\varphi_1, \varphi_2, \ldots$, belonging to $\mathscr{S}$ converges to the function $\varphi \in \mathscr{S}$, $\varphi_k \to \varphi$ as $k \to \infty$ in $\mathscr{S}$, if for all α and β

$$x^\beta D^\alpha \varphi_k(x) \xrightarrow{x \in R^n} x^\beta D^\alpha \varphi(x), \quad k \to \infty \tag{1}$$

Evidently $\mathscr{S}$ is a linear space. Moreover $\mathscr{D} \subset \mathscr{S}$ and *from the convergence in $\mathscr{D}$ follows the convergence in $\mathscr{S}$.*

In fact, if $\varphi_k \to \varphi$ as $k \to \infty$ in $\mathscr{D}$, then, since the supports of φ_k are bounded independently of k, the limiting result (1) is valid for all α and β, which means that $\varphi_k \to \varphi$ as $k \to \infty$ in $\mathscr{S}$.

However, $\mathscr{S}$ does not coincide with $\mathscr{D}$; for instance, the function $\exp(-|x|^2)$ belongs to $\mathscr{S}$ but does not belong to $\mathscr{D}$.

Nevertheless $\mathscr{D}$ is *dense* in $\mathscr{S}$; that is, for any $\varphi \in \mathscr{S}$ there is a sequence $\varphi_k \in \mathscr{D}$, $k = 1, 2, \ldots$, such that $\varphi_k \to \varphi$ as $k \to \infty$ in $\mathscr{S}$.

In fact, the sequence of functions belonging to $\mathscr{D}$,

$$\varphi_k(x) = \varphi(x)\eta\left(\frac{x}{k}\right), \qquad k = 1, 2, \ldots$$

where $\eta \in \mathscr{D}$ and $\eta(x) = 1$ for $|x| < 1$, converges to φ in $\mathscr{D}$.

The operations of differentiation $D^\beta \varphi(x)$ and of the nonsingular linear change of variable $\varphi(Ay + b)$ are continuous from $\mathscr{S}$ into $\mathscr{S}$. This follows directly from the definition of convergence in the space $\mathscr{S}$.

On the other hand, multiplication by an infinitely differentiable function can cause us to go beyond the limits of the set $\mathscr{S}$, for instance $\exp(-|x|^2) \times \exp(|x|^2) = 1 \notin \mathscr{S}$.

Let the function $a \in C^\infty(R^n)$ grow at infinity, together with all its derivatives, no faster than the polynomial

$$|D^\alpha a(x)| \le C_\alpha (1 + |x|)^{m_\alpha} \tag{2}$$

We shall denote the set of such functions by θ_M.

The operation of multiplication by the function $a \in \theta_M$ is continuous from $\mathscr{S}$ into $\mathscr{S}$.

In fact, it follows from inequality (2): If $\varphi \in \mathscr{S}$, then $a\varphi \in \mathscr{S}$, and if $\varphi_k \to \varphi$ as $k \to \infty$ in $\mathscr{S}$, then for all α and β

$$x^\beta D^\alpha(a\varphi_k) \xrightarrow{x \in R^n} x^\beta D^\alpha(a\varphi)$$

that is, $a\varphi_k \to a\varphi$ as $k \to \infty$ in $\mathscr{S}$.

2. The Space of Generalized Functions of Slow Growth (Tempered Distributions) $\mathscr{S}'$. *Each linear functional over the space of test functions $\mathscr{S}$ is known as a generalized function of slow growth (tempered distribution).* We shall denote by $\mathscr{S}' = \mathscr{S}'(R^n)$ the set of all generalized functions of slow growth. Evidently, $\mathscr{S}'$ is a linear set (cf. Sec. 5.3). We shall define convergence in $\mathscr{S}'$ as a *weak* convergence of the sequence of functionals: The sequence of generalized functions $f_1, f_2, \ldots$, belonging to $\mathscr{S}'$ converges to the generalized function $f \in \mathscr{S}'$, $f_k \to f$ as $k \to \infty$ in $\mathscr{S}'$, if for any $\varphi \in \mathscr{S}$, $(f_k, \varphi) \to (f, \varphi)$, as $k \to \infty$. The linear set $\mathscr{S}'$ with the convergence which has been introduced into it is known as the *space of generalized functions of slow growth $\mathscr{S}'$*.

It follows directly from these definitions that $\mathscr{S}' \subset \mathscr{D}'$ and from the convergence in $\mathscr{S}'$ follows the convergence in $\mathscr{D}'$.

§ 8. GENERALIZED FUNCTIONS OF SLOW GROWTH

In fact, if $f \in \mathscr{S}'$, then $f \in \mathscr{D}'$, since $\mathscr{D} \subset \mathscr{S}$ and the convergence in $\mathscr{S}$ follows from the convergence in $\mathscr{D}$ (cf. Sec. 8.1). Further, if $f_k \to f$ as $k \to \infty$ in $\mathscr{S}'$, then $(f_k, \varphi) \to (f, \varphi)$ as $k \to \infty$ for all φ belonging to $\mathscr{D} \subset \mathscr{S}$, and consequently, $f_k \to f$ as $k \to \infty$ in $\mathscr{D}'$.

THEOREM (L. Schwartz). *In order that the linear functional f over $\mathscr{S}$ should belong to $\mathscr{S}'$, (that is, be continuous over $\mathscr{S}$), it is necessary and sufficient that there be numbers $C > 0$ and $p \geq 0$, p being an integer, such that for any $\varphi \in \mathscr{S}$ the inequality*

$$|(f, \varphi)| \leq C \| \varphi \|_p \tag{3}$$

is valid, where

$$\| \varphi \|_p = \sup_{|\alpha| \leq p, x \in R^n} (1 + |x|)^p | D^\alpha \varphi(x) |$$

Proof. Sufficiency: Let the linear functional f over $\mathscr{S}$ satisfy inequality (3) for certain $C > 0$ and $p \geq 0$. We shall prove that $f \in \mathscr{S}'$. Let $\varphi_k \to \varphi$ as $k \to \infty$ in $\mathscr{S}$. Then $\| \varphi_k - \varphi \|_p \to 0$ as $k \to \infty$, and so $(f, \varphi_k) \to (f, \varphi)$ as $k \to \infty$. This means that f is a continuous function over $\mathscr{S}$.

Necessity: Let $f \in \mathscr{S}'$. We shall prove that there are numbers $C > 0$ and $p \geq 0$ such that for any $\varphi \in \mathscr{S}$, inequality (3) is valid. Suppose, on the other hand, that the numbers indicated, C and p, do not exist. Then there will be a sequence of functions φ_k, $k = 1, 2, \ldots$, belonging to $\mathscr{S}$ such that

$$|(f, \varphi_k)| \geq k \| \varphi_k \|_k \tag{4}$$

The sequence of functions

$$\psi_k(x) = \frac{\varphi_k(x)}{\sqrt{k} \| \varphi_k \|_k}, \quad k = 1, 2, \ldots$$

tends to zero in $\mathscr{S}$, for when $k \geq |\alpha|$ and $k \geq |\beta|$

$$|x^\beta D^\alpha \psi_k(x)| = \frac{|x^\beta D^\alpha \varphi_k(x)|}{\sqrt{k} \| \varphi_k \|_k} \leq \frac{1}{\sqrt{k}}$$

From this, and from the continuity of the functional f over $\mathscr{S}$, it follows that $(f, \psi_k) \to 0$ as $k \to \infty$. On the other hand, inequality (4) gives

$$|(f, \psi_k)| = \frac{1}{\sqrt{k} \|\varphi_k\|_k} |(f, \varphi_k)| \geq \sqrt{k}$$

The contradiction obtained proves the theorem.

The sense of the theorem which has just been proved is that each generalized function of slow growth is a continuous functional with respect to a certain norm $\| \ \|_p$ (in other words, it has a finite order).

3. Examples of Generalized Functions of Slow Growth. (a) *If $f(x)$ is a locally integrable function of polynomial (slow) growth at infinity, that is, for a certain $m \geq 0$*

$$\int |f(x)| (1 + |x|)^{-m} dx < \infty$$

then it defines a regular functional f belonging to $\mathscr{S}'$ according to formula (6) *of Sec. 5.5,*

$$(f, \varphi) = \int f(x) \varphi(x) \, dx, \qquad \varphi \in \mathscr{S} \qquad (5)$$

However, not every locally integrable function defines a generalized function of slow growth, for instance, $e^x \notin \mathscr{S}'(R^1)$.

On the other hand, not every locally integrable function belonging to $\mathscr{S}'$ has polynomial growth. For instance, the function $(\cos e^x)' = -e^x \sin e^x$ is not of polynomial growth, but nevertheless defines a generalized function belonging to $\mathscr{S}'$ according to the formula

$$((\cos e^x)', \varphi) = - \int \cos e^x \varphi'(x) \, dx, \qquad \varphi \in \mathscr{S}$$

Note. Using L. Schwartz's theorem (cf. Sec. 8.2), it is possible to prove* that every generalized function belonging to $\mathscr{S}'$ is a derivative of a continuous function of polynomial (slow) growth. This explains why $\mathscr{S}'$ is known as the space of generalized functions of slow growth.

(b) *If f is a finite generalized function belonging to $\mathscr{D}'$ with compact support, then it is continued over $\mathscr{S}$ in a unique way as an element belonging to $\mathscr{S}'$ according to the formula*

$$(f, \varphi) = (f, \eta \varphi), \qquad \varphi \in \mathscr{S} \qquad (6)$$

where $\eta \in \mathscr{D}$ and $\eta = 1$ in the neighborhood of the support of f.

---

*Compare L. Schwartz (2, Chap. VII) and I. M. Gelfand and G. E. Shilov (*1*, Vol. 2, Chap. II).

§ 8. GENERALIZED FUNCTIONS OF SLOW GROWTH

In fact, the linear functional $(f, \eta\varphi)$, displayed in the right-hand side of Eq. (6), is continuous in $\mathscr{S}$: If $\varphi_k \to \varphi$ as $k \to \infty$ in $\mathscr{S}$, then $\eta\varphi_k \to \eta\varphi$ as $k \to \infty$ in $\mathscr{D}$ and so

$$(f, \eta\varphi_k) \to (f, \eta\varphi), \quad k \to \infty$$

The uniqueness of the continuation of the functional f over $\mathscr{S}$ follows from the density of $\mathscr{D}$ in $\mathscr{S}$ (cf. Sec. 8.1). Specifically, continuation (6) does not depend on the auxiliary function η.

(c) *If $f \in \mathscr{S}'$, then each derivative $D^\alpha f \in \mathscr{S}'$.*

In fact, since the operation of differentiation $D^\alpha\varphi$ is continuous from $\mathscr{S}$ into $\mathscr{S}$ (cf. Sec. 8.1), the right-hand side of the equation

$$(D^\alpha f, \varphi) = (-1)^{|\alpha|}(f, D^\alpha\varphi)$$

is a linear continuous functional over $\mathscr{S}$ (cf. Sec. 6.1).

(d) *If $f \in \mathscr{S}'$ and $\det A \neq 0$, then $f(Ay + b) \in \mathscr{S}'$.*

In fact, since the operation of the transformation $\varphi[A^{-1}(x - b)]$ is continuous from $\mathscr{S}$ into $\mathscr{S}$ (cf. Sec. 8.1), the right-hand side of the equation

$$(f(Ay + b), \varphi) = \left(f, \frac{\varphi[A^{-1}(x - b)]}{|\det A|}\right)$$

is a linear continuous functional over $\mathscr{S}$ (cf. Sec. 5.8).

(e) *If $f \in \mathscr{S}'$ and $a \in \theta_M$, then $af \in \mathscr{S}'$.*

In fact, since the operation of multiplication by the function a belonging to θ_M is continuous from $\mathscr{S}$ into $\mathscr{S}$ (cf. Sec. 8.1), the right-hand side of the equation

$$(af, \varphi) = (f, a\varphi)$$

is a linear continuous functional over $\mathscr{S}$ (cf. Sec. 5.9).

4. Structure of Generalized Functions with Point Support

THEOREM. *If the support of the generalized function f is the point $\{0\}$, then in a unique manner it appears in the form*

$$f(x) = \sum_{|\alpha|=0}^{m} C_\alpha D^\alpha \delta(x) \tag{7}$$

Proof. Since the generalized function f has the support $\{0\}$, then $f \in \mathscr{S}'$

(cf. Sec. 8.3) and, by virtue of (13) of Sec. 5.9, when $k > 0$

$$f = \eta(kx)f \qquad (8)$$

where $\eta(x)$ is a test function equal to 1 in the neighborhood of the point 0 and equal to 0 when $|x| > 1$. Further, according to L. Schwartz's theorem (cf. Sec. 8.2), the inequality

$$|(f, \varphi)| \leq C \|\varphi\|_m, \qquad \varphi \in \mathcal{D} \qquad (9)$$

is true for certain $m \geq 0$ and $C > 0$, not depending on φ.

Let φ be an arbitrary function belonging to $\mathcal{D}$. Let us suppose

$$\psi_k(x) = \left[\varphi(x) - \sum_{|\alpha|=0}^{m} \frac{D^\alpha \varphi(0)}{\alpha!} x^\alpha\right] \eta(kx) \qquad (10)$$

Applying inequality (9) to the function ψ_k and using the fact that $D^\beta \eta(kx) = O(k^{|\beta|})$ as $k \to \infty$, we obtain

$$|(f, \psi_k)| \leq C \|\psi_k\|_m$$

$$= C \sup_{|\beta| \leq m, |x| \leq 1/k} (1 + |x|)^m \left| D^\beta \left[\varphi(x) - \sum_{|\alpha|=0}^{m} \frac{D^\alpha \varphi(0)}{\alpha!} x^\alpha\right] \eta(kx) \right|$$

$$= O\left(\frac{1}{k}\right), \qquad k \to \infty$$

But, by virtue of (8), (f, ψ_k) does not depend on k. Therefore, $(f, \psi_1) = (f, \psi_k) = 0$. From this, using (8) and (10) with $k = 1$, we obtain Eq. (7):

$$(f, \varphi) = (\eta f, \varphi) = (f, \eta \varphi) = \left(f, \psi_1 + \sum_{|\alpha|=0}^{m} \frac{D^\alpha \varphi(0)}{\alpha!} x^\alpha \eta(x)\right)$$

$$= (f, \psi_1) + \sum_{|\alpha|=0}^{m} \frac{D^\alpha \varphi(0)}{\alpha!} (f, x^\alpha \eta(x)) = \sum_{|\alpha|=0}^{m} C_\alpha (D^\alpha \delta, \varphi)$$

where we have written

$$C_\alpha = \frac{(-1)^{|\alpha|}}{\alpha!} (f, x^\alpha \eta)$$

We shall prove the uniqueness of Eq. (7). Let

$$f(x) = \sum_{|\alpha|=0}^{m} C'_\alpha D^\alpha \delta(x)$$

§ 8. GENERALIZED FUNCTIONS OF SLOW GROWTH

be another representation of f, so that

$$\sum_{|\alpha|=0}^{m} (C_\alpha - C'_\alpha) D^\alpha \delta(x) = 0$$

Applying this equation to the monomial x^β $|\beta| \leq m$, we obtain

$$0 = \sum_{|\alpha|=0}^{m} (C_\alpha - C'_\alpha)(D^\alpha \delta, x^\beta)$$

$$= \sum_{|\alpha|=0}^{m} (-1)^{|\alpha|}(C_\alpha - C'_\alpha)(\delta, D^\alpha x^\beta) = (-1)^{|\beta|}\beta!(C_\beta - C'_\beta)$$

that is, $C_\beta = C'_\beta$. The theorem is proved.

5. The Direct Product of Generalized Functions of Slow Growth. Let $f(x) \in \mathscr{S}'(R^n)$ and $g(y) \in \mathscr{S}'(R^m)$. Since $\mathscr{S}' \subset \mathscr{D}'$, the direct product $f(x) \cdot g(y) \in \mathscr{D}'(R^{n+m})$ (cf. Sec. 7.1). We shall prove that $f(x) \cdot g(y) \in \mathscr{S}'(R^{n+m})$.

According to the definition of the functional $f(x) \cdot g(y)$ (cf. Sec. 7.1)

$$(f(x) \cdot g(y), \varphi) = (f(x), (g(y), \varphi(x, y))) \qquad (11)$$

We shall prove that the right-hand side of Eq. (11) *is a linear continuous functional over* $\mathscr{S}(R^{n+m})$.

For this we shall establish the following lemma, analogous to the lemma of Sec. 7.1.

LEMMA. *For any* $g \in \mathscr{S}'(R^m)$ *and* $\varphi \in \mathscr{S}(R^{n+m})$ *the function*

$$\psi(x) = (g(y), \varphi(x, y)) \in \mathscr{S}(R^n)$$

and the equation

$$D^\alpha \psi(x) = (g(y), D_x^\alpha \varphi(x, y)) \qquad (12)$$

is true. Moreover, if $\varphi_k \to \varphi$ *as* $k \to \infty$ *in* $\mathscr{S}(R^{n+m})$, *then*

$$\psi_k(x) = (g(y), \varphi_k(x, y)) \to \psi(x), \qquad k \to \infty \quad \text{in } \mathscr{S}(R^n) \qquad (13)$$

Proof. As in proving the lemma of Sec. 7.1, the validity of Eq. (12) for all α and the continuity of its right-hand side are established. Consequently, $\psi \in C^\infty(R^n)$. We shall prove that $\psi \in \mathscr{S}(R^n)$. Since $g(y) \in \mathscr{S}'(R^m)$

and for each $x \in R^n$, $\varphi(x, y) \in \mathscr{S}(R^m)$, then, according to L. Schwartz's theorem (cf. Sec. 8.2), there are numbers $C > 0$ and $p \geq 0$, such that for any $\varphi \in \mathscr{S}(R^{n+m})$, α, and $x \in R^n$, the inequality

$$|(g(y), D_x^\alpha \varphi(x, y))| \leq C \sup_{y \in R^m, |\gamma| \leq p} (1 + |y|)^p |D_y^\gamma D_x^\alpha \varphi(x, y)|$$

is true. From this, by virtue of (12), we obtain the inequality

$$|x^\beta D^\alpha \psi(x)| \leq C \sup_{y \in R^m, |\gamma| \leq p} (1 + |y|)^p |x^\beta D_y^\gamma D_x^\alpha \varphi(x, y)| \qquad x \in R^n \qquad (14)$$

Since $\varphi \in \mathscr{S}(R^{n+m})$, it follows from inequality (14) that $\psi \in \mathscr{S}(R^n)$.

We shall now prove the limiting result (13). Let $\varphi_k \to \varphi$ as $k \to \infty$ in $\mathscr{S}(R^{n+m})$. From this, applying inequality (14) to the sequence $\varphi - \varphi_k$ as $k \to \infty$, we obtain

$$|x^\beta D^\alpha (\psi - \psi_k)| \leq C \sup_{y \in R^m, |\gamma| \leq p} (1 + |y|)^p |x^\beta D_y^\gamma D_x^\alpha (\varphi - \varphi_k)| \xRightarrow{x \in R^n} 0,$$
$$k \to \infty$$

that is, $\psi_k \to \psi$ as $k \to \infty$ in $\mathscr{S}(R^n)$. The lemma is proved.

It follows from the lemma which has just been proved that the right-hand side of Eq. (11), equal to (f, ψ), where $\psi(x) = (g(y), \varphi(x, y))$, is a linear and continuous functional over $\mathscr{S}(R^{n+m})$ so that $f(x) \cdot g(y) \in \mathscr{S}'(R^{n+m})$ (cf. Sec. 7.1).

The direct product of generalized functions of slow growth is commutative and associative in $\mathscr{S}'$:

$$f(x) \cdot g(y) = g(y) \cdot f(x), f(x) \cdot (g(y) \cdot h(z))$$
$$= (f(x) \cdot g(y)) \cdot h(z)$$

These assertions follow from the corresponding properties of the direct product in $\mathscr{D}'$ [cf. Secs. 7.2 and 7.3, (b)] and from the fact that $\mathscr{D}$ is dense in $\mathscr{S}$ (cf. Sec. 8.2).

Specifically, the equation $f(x) \cdot 1(y) = 1(y) \cdot f(x)$, where $f \in \mathscr{S}'(R^n)$ shows that

$$(f, \int \varphi(x, y) \, dy) = \int (f, \varphi(x, y)) \, dy, \qquad \varphi \in \mathscr{S}(R^{n+m}) \qquad (15)$$

Finally, *the direct product $f(x) \cdot g(y)$ of the generalized functions $f \in \mathscr{S}'(R^n)$ and $g \in \mathscr{S}'(R^m)$ is continuous in $\mathscr{S}'(R^{n+m})$ with respect to f or g*:

If $f_k \to f$ as $k \to \infty$ in $\mathscr{S}'(R^n)$, then $f_k(x) \cdot g(y) \to f(x) \cdot g(y)$ as $k \to \infty$ in $\mathscr{S}'(R^{n+m})$.

The proof is analogous to the corresponding proof for the space $\mathscr{D}'$ [cf. Sec. 7.3, (a)].

6. Convolution of Generalized Functions of Slow Growth. Let $f \in \mathscr{S}'$ and let g be a generalized function with compact support. Then the convolution $f * g$ exists in $\mathscr{D}'$ (cf. Sec. 7.5). We shall prove that $f * g$ belongs to $\mathscr{S}'$ and appears in the form

$$(f * g, \varphi) = (f(x) \cdot g(y), \eta(y)\varphi(x+y)), \quad \varphi \in \mathscr{S} \qquad (16)$$

when η is any function belonging to $\mathscr{D}$, equal to 1 in the neighborhood of the support of g.

In fact, according to the theorem of Sec. 7.5, formula (16) is valid over the test functions φ belonging to $\mathscr{D}$. We shall prove that the right-hand side of Eq. (16) defines a linear continuous functional over $\mathscr{S}$. Let $\varphi \in \mathscr{S}(R^n)$. Then, since the function η has compact support, $\eta(y)\varphi(x+y) \in \mathscr{S}(R^{2n})$, and since $f(x) \cdot g(y)$ is a linear functional over $\mathscr{S}(R^{2n})$, then the right-hand side of Eq. (16) is a linear functional over $\mathscr{S}(R^n)$. Now let $\varphi_k \to \varphi$ as $k \to \infty$ in $\mathscr{S}$. Then for any α, β, γ

$$x^\alpha y^\beta D^\gamma [\eta(y)\varphi_k(x+y)] \xrightarrow[x \in R^n, \, y \in R^n]{} x^\alpha y^\beta D^\gamma [\eta(y)\varphi(x+y)] \qquad k \to \infty$$

and so

$$\eta(y)\varphi_k(x+y) \to \eta(y)\varphi(x+y), \quad k \to \infty \quad \text{in } \mathscr{S}(R^{2n})$$

Since $f(x) \cdot g(y) \in \mathscr{S}'(R^{2n})$ (cf. Sec. 8.5), the continuity of the right-hand side of Eq. (16) over $\mathscr{S}(R^n)$ follows:

$$(f(x) \cdot g(y), \eta(y)\varphi_k(x+y)) \to (f(x) \cdot g(y), \eta(y)\varphi(x+y)), \quad k \to \infty$$

So, $f * g \in \mathscr{S}'$.

§9. Fourier Transform of Generalized Functions of Slow Growth

1. Fourier Transform of Test Functions Belonging to $\mathscr{S}$. Since test functions belonging to $\mathscr{S}$ are absolutely integrable over R^n, then the operation of the Fourier transform F is defined over them:

$$F[\varphi](\xi) = \int \varphi(x) e^{i(\xi, x)} \, dx, \qquad \varphi \in \mathscr{S}$$

For this the function $F[\varphi](\xi)$, *the Fourier transform of the function* $\varphi(x)$, is bounded and continuous in R^n. The test function $\varphi(x)$ decreases at infinity faster than any power of $|x|^{-1}$. Therefore, its Fourier transform may be differentiated under the integral sign any number of times,

$$D^\alpha F[\varphi](\xi) = \int (ix)^\alpha \varphi(x) e^{i(\xi, x)} \, dx = F[(ix)^\alpha \varphi](\xi) \tag{1}$$

from which it follows that $F[\varphi] \in C^\infty(R^n)$. Further, each derivative $D^\alpha \varphi$ has the same properties, and therefore

$$F[D^\alpha \varphi](\xi) = \int D^\alpha \varphi(\xi) e^{i(\xi, x)} \, dx = (-i\xi)^\alpha F[\varphi](\xi) \tag{2}$$

Finally, from Eqs. (1) and (2) we obtain

$$\xi^\beta D^\alpha F[\varphi](\xi) = \xi^\beta F[(ix)^\alpha \varphi](\xi) = i^{|\alpha|+|\beta|} F[D^\beta(x^\alpha \varphi)](\xi) \tag{3}$$

It follows from Eq. (3) that for all α and β the magnitudes $\xi^\beta D^\alpha F[\varphi](\xi)$ are uniformly bounded with respect to $\xi \in R^n$

$$|\xi^\beta D^\alpha F[\varphi](\xi)| \leq \int |D^\beta(x^\alpha \varphi)| \, dx \tag{4}$$

This means that $F[\varphi] \in \mathscr{S}$ (cf. Sec. 8.1). So the Fourier transform maps the space $\mathscr{S}$ into itself.

We note that the space of test functions $\mathscr{D}$ is not mapped into itself by the Fourier transform (since the Fourier transform of a function with compact support is an analytic function, and consequently is either not of compact support or zero).

Since the Fourier transform $F[\varphi]$ of the function φ belonging to $\mathscr{S}$ is an integrable and continuously differentiable function over R_n, then, as follows from the general theory of Fourier transforms,* the function $\varphi(x)$ may be expressed in terms of its Fourier transform $F[\varphi](\xi)$ by means of the operation of the inverse Fourier transform F^{-1}:

$$\varphi(x) = F^{-1}[F[\varphi]] = F[F^{-1}[\varphi]] \tag{5}$$

---

* Compare, for instance, L. D. Kudriavtsev (*1*, Chap. VII).

§9. FOURIER TRANSFORM OF GENERALIZED FUNCTIONS OF SLOW GROWTH

where

$$F^{-1}[\psi](x) = \frac{1}{(2\pi)^n} \int \psi(\xi) e^{-i(\xi,x)} \, d\xi = \frac{1}{(2\pi)^n} F[\psi](-x)$$

$$= \frac{1}{(2\pi)^n} \int \psi(-\xi) e^{i(\xi,x)} \, d\xi = \frac{1}{(2\pi)^n} F[\psi(-\xi)] \qquad (6)$$

It follows from Eqs. (5) and (6) that each function φ belonging to $\mathscr{S}$ is a Fourier transform of the function $\psi = F^{-1}[\varphi]$ belonging to $\mathscr{S}$, $\varphi = F[\psi]$, and if $F[\varphi] = 0$, then φ also equals 0. This means that the Fourier transform F maps $\mathscr{S}$ onto $\mathscr{S}$ and, moreover, is mutually one-to-one.

LEMMA. *The Fourier transform operation F is continuous from $\mathscr{S}$ onto $\mathscr{S}$.*

Proof. Let $\varphi_k \to \varphi$ as $k \to \infty$ in $\mathscr{S}$. Then, applying inequality (4) to the functions $\varphi_k \to \varphi$, for all α and β we obtain

$$|\xi^\beta D^\alpha F[\varphi_k - \varphi](\xi)| \leq \int |D^\beta[x^\alpha(\varphi_k - \varphi)]| \, dx$$

$$\leq \sup_{x \in R^n} |D^\beta[x^\alpha(\varphi_k - \varphi)]| (1 + |x|)^{n+1} \int \frac{dx}{(1+|x|)^{n+1}}$$

from which it follows that

$$|\xi^\beta D^\alpha F[\varphi_k] - \xi^\beta D^\alpha F[\varphi]| \xrightarrow{\xi \in R^n} 0, \qquad k \to \infty$$

that is, $F[\varphi_k] \to F[\varphi]$ as $k \to \infty$ in $\mathscr{S}$ (cf. Sec. 8.1). The lemma is proved.

The operation of the inverse Fourier transform F^{-1} has analogous properties.

2. The Fourier Transform of Generalized Functions Belonging to $\mathscr{S}'$. First let $f(x)$ be an (absolutely) integrable function over R^n. Then its Fourier transform

$$F[f](\xi) = \int f(x) e^{i(\xi,x)} \, dx, \qquad |F[f](\xi)| \leq \int |f(x)| \, dx < \infty$$

is a continuous function bounded in R^n and, consequently, defines a generalized function belonging to $\mathscr{S}'$,

$$(F[f], \varphi) = \int F[f](\xi) \varphi(\xi) \, d\xi, \qquad \varphi \in \mathscr{S}$$

Using Fubini's theorem (cf. Sec. 1.4) concerning the change of order of integration, we transform the last integral:

$$\int F[f](\xi)\varphi(\xi)\,d\xi = \int [\int f(x)e^{i(\xi,x)}\,dx]\varphi(\xi)\,d\xi$$
$$= \int f(x) \int \varphi(\xi)e^{i(\xi,x)}\,d\xi\,dx = \int f(x)F[\varphi](x)\,dx$$

that is,

$$(F[f], \varphi) = (f, F[\varphi]), \qquad \varphi \in \mathscr{S}$$

We shall take this equation as the definition of the Fourier transform $F[f]$ of any generalized function of slow growth f:

$$(F[f], \varphi) = (f, F[\varphi]), \qquad f \in \mathscr{S}', \qquad \varphi \in \mathscr{S}, \tag{7}$$

We shall check that the right-hand side of this equation defines a linear continuous functional over $\mathscr{S}$, that is, that $F[f] \in \mathscr{S}'$. In fact, since $F[\varphi] \in \mathscr{S}$ for all $\varphi \in \mathscr{S}$ (cf. Sec. 9.1), then $(f, F[\varphi])$ is a functional (obviously, linear) over $\mathscr{S}$. Let $\varphi_k \to \varphi$ as $k \to \infty$ in $\mathscr{S}$. According to the lemma of Sec. 9.1, $F[\varphi_k] \to F[\varphi]$ as $k \to \infty$ in $\mathscr{S}$ and so, by virtue of $f \in \mathscr{S}'$,

$$(f, F[\varphi_k]) \to (f, F[\varphi]), \qquad k \to \infty$$

so that the functional $(f, F[\varphi])$ is continuous over $\mathscr{S}$.

In this way, the operation of the Fourier transform F maps the space $\mathscr{S}'$ into $\mathscr{S}'$.

We shall prove that F is a *continuous operation from* $\mathscr{S}'$ *into* $\mathscr{S}'$.

In fact, let $f_k \to f$ as $k \to \infty$ in $\mathscr{S}'$. Then, by virtue of (7), for all $\varphi \in \mathscr{S}$ we obtain

$$(F[f_k], \varphi) = (f_k, F[\varphi]) \to (f, F[\varphi]) = (F[f], \varphi), \qquad k \to \infty$$

This shows that $F[f_k] \to F[f]$ as $k \to \infty$ in $\mathscr{S}'$, that is, the operation F is continuous from $\mathscr{S}'$ into $\mathscr{S}'$.

We shall introduce into $\mathscr{S}'$ one more Fourier transform operation which we shall denote by F^{-1}:

$$F^{-1}[f] = \frac{1}{(2\pi)^n} F[f(-x)], \qquad f \in \mathscr{S}' \tag{8}$$

We shall prove that the operation F^{-1} is the inverse operation to the Fourier transform F, that is,

$$F^{-1}[F[f]] = f, \qquad F[F^{-1}[f]] = f, \qquad f \in \mathscr{S}' \tag{9}$$

§ 9. FOURIER TRANSFORM OF GENERALIZED FUNCTIONS OF SLOW GROWTH

In fact, by virtue of Eqs. (5)–(8), for all $\varphi \in \mathscr{S}$ we obtain the equations

$$(F^{-1}[F[f]], \varphi) = \frac{1}{(2\pi)^n} (F[F[f]](-\xi)], \varphi)$$
$$= \frac{1}{(2\pi)^n} (F[f](-\xi), F[\varphi]) = \frac{1}{(2\pi)^n} (F[f], F[\varphi](-\xi))$$
$$= (F[f], F^{-1}[\varphi]) = (f, F[F^{-1}[\varphi]]) = (f, \varphi)$$
$$= (f, F^{-1}[F[\varphi]]) = (F^{-1}[f], F[\varphi]) = (F[F^{-1}[f]], \varphi)$$

from which Eqs. (9) follow.

It follows from Eqs. (9) that each generalized function f belonging to $\mathscr{S}'$ is a Fourier transform of the generalized function $g = F^{-1}[f]$ belonging to $\mathscr{S}'$, $f = F[g]$, and if $F[f] = 0$, then also $f = 0$. *This means that the Fourier transforms F and F^{-1} map $\mathscr{S}'$ onto $\mathscr{S}'$ mutually one-to-one (and also mutually continuously).*

Let $f(x, y) \in \mathscr{S}'(R^{n+m})$ where $x \in R^n$, $y \in R^m$. We shall introduce the Fourier transform $F_x[f]$ with respect to the variables $x = (x_1, x_2, \ldots, x_n)$, supposing that for any $\varphi(\xi, y) \in \mathscr{S}(R^{n+m})$

$$(F_x[f], \varphi) = (f, F_\xi[\varphi]) \tag{10}$$

As in the lemma of Sec. 9.1, it can be established that

$$F_\xi[\varphi](x, y) = \int \varphi(\xi, y) e^{i(\xi, x)} d\xi \in \mathscr{S}(R^{n+m})$$

and the operation $F_\xi[\varphi]$ is continuous from $\mathscr{S}(R^{n+m})$ into $\mathscr{S}(R^{n+m})$, so that Eq. (10) in fact defines a generalized function $F_x[f](\xi, y)$ belonging to $\mathscr{S}'(R^{n+m})$.

3. Properties of the Fourier Transform

(a) DIFFERENTIABILITY OF THE FOURIER TRANSFORM. If $f \in \mathscr{S}'$, then

$$D^\alpha F[f] = F[(ix)^\alpha f] \tag{11}$$

In fact, using Eq. (2), for all $\varphi \in \mathscr{S}$ we obtain

$$(D^\alpha F[f], \varphi) = (-1)^{|\alpha|} (F[f], D^\alpha \varphi) = (-1)^{|\alpha|} (f, F[D^\alpha \varphi])$$
$$= (-1)^{|\alpha|} (f, (-ix)^\alpha F[\varphi]) = ((ix)^\alpha f, F[\varphi])$$
$$= (F[(ix)^\alpha f], \varphi)$$

from which Eq. (11) follows.

(b) FOURIER TRANSFORM OF A DERIVATIVE. If $f \in \mathscr{S}'$, then

$$F[D^{\alpha}f] = (-i\xi)^{\alpha}F[f] \qquad (12)$$

In fact, using Eq. (1), for all $\varphi \in \mathscr{S}$ we obtain

$$\begin{aligned}(F[D^{\alpha}f], \varphi) &= (D^{\alpha}f, F[\varphi]) = (-1)^{|\alpha|}(f, D^{\alpha}F[\varphi]) \\ &= (-1)^{|\alpha|}(f, F[(i\xi)^{\alpha}\varphi]) = (-1)^{|\alpha|}(F[f], (i\xi)^{\alpha}\varphi) \\ &= ((-i\xi)^{\alpha}F[f], \varphi)\end{aligned}$$

from which Eq. (12) follows.

(c) FOURIER TRANSFORM OF A TRANSLATION. If $f \in \mathscr{S}'$, then,

$$F[f(x - x_0)] = \exp(i(x_0, \xi))F[f] \qquad (13)$$

In fact, for all $\varphi \in \mathscr{S}$ we have

$$\begin{aligned}(F[f(x - x_0)], \varphi) &= (f(x - x_0), F[\varphi]) = (f, F[\varphi](x + x_0)) \\ &= (f, F[\varphi \exp(i(x_0, \xi))]) = (F[f], \exp(i(x_0, \xi))\varphi) \\ &= (\exp(i(x_0, \xi))F[f], \varphi)\end{aligned}$$

from which Eq. (13) follows.

(d) TRANSLATION OF A FOURIER TRANSFORM. If $f \in \mathscr{S}'$, then

$$F[f](\xi + \xi_0) = F[e^{i(\xi_0, x)}f](\xi) \qquad (14)$$

In fact, using Eq. (13), for all $\varphi \in \mathscr{S}$ we obtain

$$\begin{aligned}(F[f](\xi + \xi_0), \varphi) &= (F[f], \varphi(\xi - \xi_0)) = (f, F[\varphi(\xi - \xi_0)]) \\ &= (f, \exp(i(\xi_0, x))F[\varphi]) = (\exp(i(\xi_0, x))f, F[\varphi]) \\ &= (F[\exp(i(\xi_0, x))f], \varphi)\end{aligned}$$

from which Eq. (14) follows.

(e) FOURIER TRANSFORM OF A DIRECT PRODUCT. If $f \in \mathscr{S}'(R^n)$ and $g \in \mathscr{S}'(R^m)$, then

$$\begin{aligned}F[f(x) \cdot g(y)] &= F_x[f(x) \cdot F[g](\eta)] \\ &= F_y[F[f](\xi) \cdot g(y)] = F[f](\xi) \cdot F[g](\eta)\end{aligned} \qquad (15)$$

§ 9. FOURIER TRANSFORM OF GENERALIZED FUNCTIONS OF SLOW GROWTH

In fact, for all $\varphi(\xi, \eta) \in \mathscr{S}(R^{n+m})$ we have

$$
\begin{aligned}
(F \mid f(x) \cdot g(y)], \varphi) &= (f(x) \cdot g(y), F[\varphi]) \\
&= (f(x), (g(y), F_\eta F_\xi[\varphi])) = (f(x), (F[g], F_\xi[\varphi])) \\
&= (f(x) \cdot F[g](\eta), F_\xi[\varphi]) = (F_x[f(x) \cdot F[g](\eta)], \varphi) \\
&= (F[g](\eta), (f(x), F_\xi[\varphi])) = (F[g](\eta), (F[f](\xi), \varphi)) \\
&= (F[f](\xi) \cdot F[g](\eta), \varphi)
\end{aligned}
$$

from which Eq. (15) follows.

(f) Analogous equations are valid for the Fourier transform F_x, for instance: If $f(x, y) \in \mathscr{S}'(R^{n+m})$, then

$$D_x^\alpha D_y^\beta F_x[f] = F_x[(ix)^\alpha D_y^\beta f], \qquad F_x[D_x^\alpha D_y^\beta f] = (-i\xi)^\alpha D_y^\beta F_x[f] \qquad (16)$$

4. Fourier Transform of Generalized Functions with Compact Support

THEOREM. *If f is a generalized function with compact support, then its Fourier transform belongs to the class θ_M and is represented by the equation*

$$F[f](\xi) = (f(x), \eta(x)e^{i(\xi, x)}) \qquad (17)$$

where η is any function belonging to $\mathscr{D}$ equal to 1 in the neighborhood of the support of f.

Proof. Taking into account Eq. (6) of Sec. 8.3 and (12) of Sec. 9.3, for all $\varphi \in \mathscr{S}$ we obtain

$$
\begin{aligned}
(D^\alpha F[f], \varphi) &= (-1)^{|\alpha|}(F[f], D^\alpha \varphi) = (-1)^{|\alpha|}(f, F[D^\alpha \varphi]) \\
&= (-1)^{|\alpha|}(f, \eta(x)(-ix)^\alpha F[\varphi]) \\
&= (f(x), \int \eta(x)(ix)^\alpha \varphi(\xi) e^{i(\xi, x)} \, d\xi)
\end{aligned}
$$

Noting now that

$$\eta(x)(ix)^\alpha \varphi(\xi) e^{i(\xi, x)} \in \mathscr{S}(R^{2n})$$

and using Eq. (15) of Sec. 8.5:

$$(f(x), \int \eta(x)(ix)^\alpha \varphi(\xi) e^{i(\xi, x)} \, d\xi) = \int (f, \eta(x)(ix)^\alpha e^{i(\xi, x)}) \varphi(\xi) \, d\xi$$

from the preceding equations we deduce the equation

$$(D^\alpha F[f], \varphi) = \int (f, \eta(x)(ix)^\alpha e^{i(\xi, x)}) \varphi(\xi) \, d\xi$$

from which it follows that

$$D^\alpha F[f](\xi) = (f, \eta(x)(ix)^\alpha e^{i(\xi,x)}) \tag{18}$$

As a result, when $\alpha = 0$, Eq. (17) follows.

From Eq. (18), as with the proof of the lemma of Sec. 7.1, we deduce that $D^\alpha F[f] \in C(R^n)$, so that $F[f] \in C^\infty(R^n)$. Further, according to L. Schwartz's theorem (cf. Sec. 8.2) there are numbers $C > 0$ and $p \geq 0$ (p being an integer) for which inequality (3) of Sec. 8.2 is valid. Applying this inequality to the right-hand side of Eq. (18), we obtain the estimate

$$\begin{aligned}
| D^\alpha F[f](\xi) | &= | (f, \eta(x)(ix)^\alpha e^{i(\xi,x)}) | \\
&\leq C \, \| \eta(x)(ix)^\alpha e^{i(\xi,x)} \|_p \\
&= C \sup_{|\beta| \leq p, x \in R^n} (1 + |x|)^p | D^\beta[\eta(x) x^\alpha e^{i(\xi,x)}] | \\
&\leq C_\alpha (1 + |\xi|)^p, \quad \xi \in R^n
\end{aligned}$$

from which it also follows that $F[f] \in \theta_M$ (cf. Sec. 8.1). The theorem is proved.

From Eq. (17), specifically, we obtain

$$\begin{aligned}
F[\delta(x - x_0)] &= (\delta(x - x_0), \eta(x) e^{i(\xi,x)}) \\
&= \eta(x_0) \exp[i(\xi, x_0)] = \exp[i(\xi, x_0)]
\end{aligned}$$

that is,

$$F[\delta(x - x_0)] = \exp[i(\xi, x_0)], \quad F[\delta] = 1 \tag{19}$$

From (19) we deduce

$$\delta = F^{-1}[1] = \frac{1}{(2\pi)^n} F[1]$$

so that

$$F[1] = (2\pi)^n \delta(\xi) \tag{20}$$

Finally, using Eqs. (11), (12), (19), and (20), we obtain

$$F[D^\alpha \delta] = (-i\xi)^\alpha F[\delta] = (-i\xi)^\alpha \tag{21}$$

$$F[x^\alpha] = (-i)^{|\alpha|} D^\alpha F[1] = (2\pi)^n (-i)^{|\alpha|} D^\alpha \delta(\xi) \tag{22}$$

§ 9. FOURIER TRANSFORM OF GENERALIZED FUNCTIONS OF SLOW GROWTH

5. Fourier Transform of a Convolution. Let $f \in \mathscr{S}'$ and g be a finite generalized function with compact support. Then

$$F[f * g] = F[g]F[f] \tag{23}$$

In fact, by virtue of Sec. 8.6, the convolution $f * g$ belongs to $\mathscr{S}'$ and appears in the form

$$(f * g, \varphi) = (f(x), (g(y), \eta(y)\varphi(x + y))), \qquad \varphi \in \mathscr{S}$$

where $\eta \in \mathscr{D}$ and $\eta = 1$, in the neighborhood of supp g. Taking this equation into account, for all $\varphi \in \mathscr{S}$ we obtain

$$(F[f * g], \varphi) = (f * g, F[\varphi])$$
$$= (f(x), (g(y), \eta(y) \int \varphi(\xi) \exp[i((x + y), \xi)] d\xi))$$

Since, according to the theorem of Sec. 9.4, $F[g] \in \mathscr{O}_M$, using Eqs. (15) of Sec. 8.5 and (17), we transform the equation obtained

$$(F[f * g], \varphi) = (f, \int (g, \eta(y)e^{i(\xi,y)})e^{i(\xi,x)}\varphi(\xi) d\xi$$
$$= (f, \int F[g](\xi)\varphi(\xi)e^{i(\xi,x)} d\xi) = (f, F[F[g]\varphi])$$
$$= (F[f], F[g]\varphi) = (F[g]F[f], \varphi)$$

from which Eq. (23) follows.

6. Examples, $n = 1.$

(a) $\quad F[\theta(R - |x|)] = \int_{-R}^{R} e^{i x \xi} dx = 2 \dfrac{\sin R\xi}{\xi} \tag{24}$

(b) $\quad F[e^{-\alpha^2 x^2}] = \dfrac{\sqrt{\pi}}{\alpha} \exp\left(-\dfrac{\xi^2}{4\alpha^2}\right) \tag{25}$

In fact,

$$F[e^{-\alpha^2 x^2}] = \int \exp(-\alpha^2 x^2 + i\xi x) dx = \dfrac{1}{\alpha} \int \exp\left(-\sigma^2 + i\dfrac{\xi}{\alpha}\sigma\right) d\sigma$$
$$= \dfrac{1}{\alpha} \exp\left(-\dfrac{\xi^2}{4\alpha^2}\right) \int \exp\left[-\left(\sigma + \dfrac{i\xi}{2\alpha}\right)^2\right] d\sigma$$
$$= \dfrac{1}{\alpha} \exp\left(-\dfrac{\xi^2}{4\alpha^2}\right) \int_{\operatorname{Im}\zeta = \xi/2\alpha} e^{-\zeta^2} d\zeta$$

It remains to prove that the line of integration $\operatorname{Im} \zeta = \xi/2\alpha$ in the integral obtained may be moved onto the real axis, that is, for all α

$$\int_{\operatorname{Im}\zeta=a} e^{-\zeta^2}\,d\zeta = \int_{-\infty}^{\infty} e^{-\sigma^2}\,d\sigma = \sqrt{\pi} \tag{26}$$

According to Cauchy's theorem for any $R > 0$ we have

$$\int_{c_R} e^{-\zeta^2}\,d\zeta = 0, \qquad \zeta = \sigma + i\tau \tag{27}$$

where the contour $c_R = c_R' \cup c_R'' \cup l_R^+ \cup l_R^-$ is shown in Fig. 21. But on

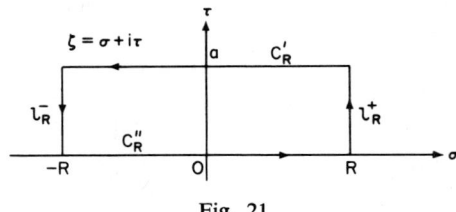

Fig. 21

the section $l_R^\pm = [0 \leq \tau \leq a, \sigma = \pm R]$

$$|e^{-\zeta^2}| = |\exp(-\sigma^2 + \tau^2 - 2i\sigma\tau)| = \exp(-R^2 + \tau^2) \xrightarrow{\tau \in [0,a]} 0, \qquad R \to \infty$$

and so

$$\lim_{R \to \infty} \left(\int_{l_R^+} + \int_{l_R^-}\right) e^{-\zeta^2}\,d\zeta = 0$$

from which, using Eq. (27), we obtain Eq. (26):

$$\lim_{R \to \infty} \int_{c_R} e^{-\zeta^2}\,d\zeta = \lim_{R \to \infty} \left(\int_{c_R'} + \int_{c_R''}\right) e^{-\zeta^2}\,d\zeta$$
$$= \int_{-\infty}^{\infty} e^{-\sigma^2}\,d\sigma - \int_{\tau=ia} e^{-\zeta^2}\,d\zeta = 0$$

(c) $$F[e^{ix^2}] = \sqrt{\pi}\,\exp\left[-\frac{i}{4}(\xi^2 - \pi)\right] \tag{28}$$

In fact, from the convergence of the improper integral (Fresnel's integral)

$$\int_{-\infty}^{\infty} e^{iy^2}\,dy = \sqrt{\pi}\,e^{i\pi/4}$$

§ 9. FOURIER TRANSFORM OF GENERALIZED FUNCTIONS OF SLOW GROWTH

follows the uniform convergence with respect to ξ over each finite interval of the improper integral

$$\int_{-\infty}^{\infty} \exp(ix^2 + ix\xi)\, dx = \lim_{\substack{N\to\infty \\ M\to\infty}} \int_{-M}^{N} \exp(ix^2 + ix\xi)\, dx$$

$$= \lim_{\substack{N\to\infty \\ M\to\infty}} \int_{-M}^{N} \exp\left[i\left(x + \frac{\xi}{2}\right)^2 - \frac{i}{4}\xi^2\right] dx$$

$$= \exp\left(-\frac{i}{4}\xi^2\right) \lim_{\substack{N\to\infty \\ M\to\infty}} \int_{-M+(\xi/2)}^{N+(\xi/2)} e^{iy^2}\, dy$$

$$= \exp\left(-\frac{i}{4}\xi^2\right) \int_{-\infty}^{\infty} e^{iy^2}\, dy$$

$$= \sqrt{\pi} \exp\left[-\frac{i}{4}(\xi^2 - \pi)\right]$$

Using this result, for all $\varphi \in \mathscr{D}$, $\operatorname{supp} \varphi \subset (-R, R)$ we obtain

$$(F[e^{ix^2}], \varphi) = (e^{ix^2}, F[\varphi]) = \int e^{ix^2} F[\varphi](x)\, dx$$

$$= \lim_{\substack{N\to\infty \\ M\to\infty}} \int_{-M}^{N} e^{ix^2} \int_{-R}^{R} \varphi(\xi) e^{ix\xi}\, d\xi\, dx$$

$$= \lim_{\substack{N\to\infty \\ M\to\infty}} \int_{-R}^{R} \varphi(\xi) \int_{-M}^{N} \exp(ix^2 + ix\xi)\, dx\, d\xi$$

$$= \int_{-R}^{R} \varphi(\xi) \lim_{\substack{N\to\infty \\ M\to\infty}} \int_{-M}^{N} \exp(ix^2 + ix\xi)\, dx\, d\xi$$

$$= \sqrt{\pi}\, e^{i\pi/4} \int \varphi(\xi) \exp\left(-\frac{i}{4}\xi^2\right) d\xi$$

from which, we conclude that Eq. (28) is valid over the test functions belonging to $\mathscr{D}$. But $\mathscr{D}$ is dense in $\mathscr{S}$ (cf. Sec. 8.1). Therefore, this equation is valid over the test functions belonging to $\mathscr{S}$.

(d)
$$F[\theta] = \pi\delta(\xi) + i\mathscr{P}\frac{1}{\xi} \qquad (29)$$

$$F[\theta(-x)] = \pi\delta(\xi) - i\mathscr{P}\frac{1}{\xi} \qquad (29')$$

In fact, for all $a > 0$ we have

$$F[\theta(x)e^{-ax}] = \int_0^\infty \exp(-ax + ix\xi)\, dx = \frac{i}{\xi + ia} \tag{30}$$

Since

$$\theta(x)e^{-ax} \to \theta(x), \quad a \to +0 \quad \text{in } \mathscr{S}'$$

then, proceeding to the limit as $a \to +0$ in Eq. (30), and using the continuity over $\mathscr{S}'$ of the Fourier transform (cf. Sec. 9.2), we deduce

$$F[\theta] = \frac{i}{\xi + i0} \tag{31}$$

Now applying Sokhotsky's formula (10) of Sec. 5.7, we obtain Eq. (29). Equation (29') is established analogously.

(e)
$$F\left[\mathscr{P}\frac{1}{|x|}\right] = -2C - 2\ln|\xi| \tag{32}$$

where C is Euler's constant

$$C = \int_0^1 \frac{1 - \cos u}{u}\, du - \int_1^\infty \frac{\cos u}{u}\, du$$

and the generalized function $\mathscr{P}(1/|x|)$ is as defined in Sec. 6.4, (b).

In fact, for all $\varphi \in \mathscr{S}$ we have

$$\left(F\left[\mathscr{P}\frac{1}{|x|}\right], \varphi\right) = \left(\mathscr{P}\frac{1}{|x|}, F[\varphi]\right)$$

$$= \int_{-1}^1 \frac{F[\varphi](x) - F[\varphi](0)}{|x|}\, dx + \int_{|x|>1} \frac{F[\varphi](x)}{|x|}\, dx$$

$$= \int_{-1}^1 \frac{1}{|x|} \int \varphi(\xi)(e^{ix\xi} - 1)\, d\xi\, dx + \int_{|x|>1} \frac{1}{|x|} \int \varphi(\xi) e^{ix\xi}\, d\xi\, dx$$

$$= 2\int_0^1 \int \varphi(\xi) \frac{\cos x\xi - 1}{x}\, d\xi\, dx + 2\int_1^\infty \int \varphi(\xi) \frac{\cos x\xi}{x}\, d\xi\, dx$$

$$= 2\int \varphi(\xi) \int_0^1 \frac{\cos x\xi - 1}{x}\, dx\, d\xi - 2\int_1^\infty \int \varphi'(\xi) \frac{\sin x\xi}{x^2}\, d\xi\, dx$$

$$= 2\int \varphi(\xi) \int_0^{|\xi|} \frac{\cos u - 1}{u}\, du\, d\xi - 2\int \varphi'(\xi) \int_1^\infty \frac{\sin x\xi}{x^2}\, dx\, d\xi$$

§ 9. FOURIER TRANSFORM OF GENERALIZED FUNCTIONS OF SLOW GROWTH

$$= 2\int \varphi(\xi)\left[\int_0^{|\xi|} \frac{\cos u - 1}{u}\,du + \frac{\partial}{\partial\xi}\int_1^\infty \frac{\sin x\xi}{x^2}\,dx\right]d\xi$$

$$= -2\int \varphi(\xi)(C + \ln|\xi|)\,d\xi$$

from which Eq. (32) follows.

(f) In Sec. 6.3, (c) the equation

$$\sum_{k=-\infty}^{\infty} \delta(x - 2\pi k) = \frac{1}{2\pi}\sum_{k=-\infty}^{\infty} e^{ikx} \qquad (33)$$

was established. It is not difficult to see that the series in Eq. (33) converges in $\mathscr{S}'$. Using Eq. (19), we shall rewrite Eq. (33) in the form

$$2\pi \sum_{k=-\infty}^{\infty} \delta(x - 2\pi k) = \sum_{k=-\infty}^{\infty} F[\delta(x - k)]$$

Applying this equation to $\varphi \in \mathscr{S}$, we obtain

$$2\pi\left(\sum_{k=-\infty}^{\infty} \delta(x - 2\pi k),\varphi\right) = 2\pi \sum_{k=-\infty}^{\infty} (\delta(x - 2\pi k),\varphi)$$

$$= 2\pi \sum_{k=-\infty}^{\infty} \varphi(2\pi k) = \left(\sum_{k=-\infty}^{\infty} F[\delta(x - k)],\varphi\right)$$

$$= \sum_{k=-\infty}^{\infty} (\delta(x - k), F[\varphi]) = \sum_{k=-\infty}^{\infty} F[\varphi](k)$$

that is,

$$2\pi \sum_{k=-\infty}^{\infty} \varphi(2\pi k) = \sum_{k=-\infty}^{\infty} F[\varphi](k) \qquad (34)$$

Equation (34) is known as *Poisson's summation formula*.
Supposing in Eq. (34) that

$$\varphi(x) = \exp\left(-\frac{tx^2}{4\pi^2}\right), \qquad F[\varphi](\xi) = \frac{2\pi\sqrt{\pi}}{\sqrt{t}}\exp\left(-\frac{\xi^2\pi^2}{t}\right), \qquad t > 0$$

we obtain

$$\sum_{k=-\infty}^{\infty} e^{-tk^2} = \sqrt{\frac{\pi}{t}}\sum_{k=-\infty}^{\infty}\exp\left(-\frac{k^2\pi^2}{t}\right) \qquad (35)$$

Formula (35) is used in the theory of elliptic functions.

7. Exercises. Use Eqs. (29) and (29') and the equation $\mathscr{P}(1/\xi^2) = -(\mathscr{P}(1/\xi))'$ to show that

(1) $F[\operatorname{sgn} x] = 2i\mathscr{P}\dfrac{1}{\xi}$, $F\left[\mathscr{P}\dfrac{1}{x}\right] = i\pi \operatorname{sgn} \xi$;

(2) $F\left[\mathscr{P}\dfrac{1}{x^2}\right] = -\pi\,|\,\xi\,|$, $F[|\,x\,|] = -2\mathscr{P}\dfrac{1}{\xi^2}$;

(3) $F[\theta(x)x] = -i\pi\delta'(\xi) - \mathscr{P}\dfrac{1}{\xi^2}$

8. Examples, $n \geq 2$. (a) Let the quadratic form

$$\sum_{i,j=1}^{n} a_{ij}x_i x_j = (Ax, x), \qquad A = (a_{ij})$$

be real and positive definite,

$$(Ax, x) \geq \sigma\,|\,x\,|^2, \qquad \sigma > 0$$

Then

$$F[e^{-(Ax,x)}] = \dfrac{\pi^{n/2}}{\sqrt{\det A}} \exp\left[-\dfrac{1}{4}(\xi, A^{-1}\xi)\right] \qquad (36)$$

To obtain formula (36) by means of a nonsingular real transformation $x = By$ we reduce the quadratic form (Ax, x) to a diagonal form

$$(Ax, x) = (ABy, By) = (B'ABy, y) = |\,y^2\,|$$

so that

$$A^{-1} = BB', \qquad \det A (\det B)^2 = 1$$

From this, using formula (25), we obtain

$$F[e^{-(Ax,x)}] = \int \exp[-(Ax, x) + i(\xi, x)]\,dx$$

$$= |\det B| \int \exp[-(ABy, By) + i(\xi, By)]\,dy$$

$$= \dfrac{1}{\sqrt{\det A}} \int \exp[-|\,y\,|^2 + i(B'\xi, y)]\,dy$$

$$= \dfrac{1}{\sqrt{\det A}} \prod_{j=1}^{n} \int \exp[-y_j^2 + i(B'\xi)_j y_j]\,dy_j$$

§9. FOURIER TRANSFORM OF GENERALIZED FUNCTIONS OF SLOW GROWTH

$$= \frac{\pi^{n/2}}{\sqrt{\det A}} \exp\left(-\frac{1}{4} |B'\xi|^2\right)$$

$$= \frac{\pi^{n/2}}{\sqrt{\det A}} \exp\left[-\frac{1}{4}(\xi, BB'\xi)\right] = \frac{\pi^{n/2}}{\sqrt{\det A}} \exp\left[-\frac{1}{4}(\xi, A^{-1}\xi)\right]$$

(b) Analogously, using formula (28), we obtain

$$F[\exp(i(Ax, x))] = \frac{\pi^{n/2}}{\sqrt{\det A}} \exp\left(i\frac{\pi n}{4}\right) \exp\left[-\frac{i}{4}(\xi, A^{-1}\xi)\right] \quad (37)$$

(c) Let $\delta_{S_R}(x)$ be a simple layer on the spherical surface S_R in R^3. Then

$$F[\delta_{S_R}] = 4\pi R \frac{\sin R|\xi|}{|\xi|} \quad (38)$$

In fact, since δ_{S_R} is a generalized function with compact support, then, applying formula (17), we obtain

$$F[\delta_{S_R}] = (\delta_{S_R}(x), \eta(x)e^{i(\xi,x)}) = \int_{S_R} \eta(x)e^{i(\xi,x)} \, dS_x$$

$$= R^2 \int_{S_1} e^{iR(\xi,s)} \, ds = R^2 \int_0^\pi \int_0^{2\pi} \exp(iR|\xi|\cos\theta) \sin\theta \, d\theta \, d\varphi$$

$$= 4\pi R \frac{\sin R|\xi|}{|\xi|}$$

(d) Let $n = 2$. We shall introduce the generalized function $\mathscr{P}(1/|x|^2)$,

$$\left(\mathscr{P}\frac{1}{|x|^2}, \varphi\right) = \int_{|x|<1} \frac{\varphi(x) - \varphi(0)}{|x|^2} \, dx + \int_{|x|>1} \frac{\varphi(x)}{|x|^2} \, dx \qquad \varphi \in \mathscr{S}$$

Then

$$F\left[\mathscr{P}\frac{1}{|x|^2}\right] = -2\pi \ln|\xi| - 2\pi C_0 \quad (39)$$

where

$$C_0 = \int_0^1 \frac{1 - J_0(u)}{u} \, du - \int_1^\infty \frac{J_0(u)}{u} \, du$$

and J_0 is a Bessel function.

In fact, for all $\varphi \in \mathscr{S}$ the sequence of equations

$$\left(F\left[\mathscr{P}\frac{1}{|x|^2}\right], \varphi\right) = \left(\mathscr{P}\frac{1}{|x|^2}, F[\varphi]\right)$$

$$= \int_{|x|<1} \frac{F[\varphi](x) - F[\varphi](0)}{|x|^2} dx + \int_{|x|>1} \frac{F[\varphi](x)}{|x|^2} dx$$

$$= \int_{|x|<1} \frac{1}{|x|^2} \int \varphi(\xi)(e^{i(x,\xi)} - 1) \, d\xi \, dx$$

$$+ \int_{|x|>1} \frac{1}{|x|^2} \int \varphi(\xi) e^{i(x,\xi)} \, d\xi \, dx$$

$$= \int_0^1 \frac{1}{r} \int \varphi(\xi) \int_0^{2\pi} (\exp(ir|\xi|\cos\theta) - 1) \, d\theta \, d\xi \, dr$$

$$+ \int_1^\infty \frac{1}{r} \int \varphi(\xi) \int_0^{2\pi} \exp(ir|\xi|\cos\theta) \, d\theta \, d\xi \, dr$$

$$= 2\pi \int_0^1 \frac{1}{r} \int \varphi(\xi) [J_0(r|\xi|) - 1] \, d\xi \, dr$$

$$+ 2\pi \int_1^\infty \frac{1}{r} \int \varphi(\xi) J_0(r|\xi|) \, d\xi \, dr$$

$$= 2\pi \int \varphi(\xi) \left[\int_0^1 \frac{J_0(r|\xi|) - 1}{r} \, dr + \int_1^\infty \frac{J_0(r|\xi|)}{r} \, dr \right] d\xi$$

$$= 2\pi \int \varphi(\xi) \left[\int_0^{|\xi|} \frac{J_0(u) - 1}{u} \, du + \int_{|\xi|}^\infty \frac{J_0(u)}{u} \, du \right] d\xi$$

$$= -2\pi \int \varphi(\xi) (C_0 + \ln|\xi|) \, d\xi$$

is true, and Eq. (39) follows from this.

(e) $$F\left[\frac{1}{z}\right] = \frac{2\pi i}{\zeta}, \qquad \zeta = \xi + i\eta \qquad (40)$$

Applying the Fourier transform to both sides of Eq. (37) of Sec. 6.5, we obtain

$$F\left[\frac{1}{2}\left(\frac{\partial}{\partial x} + i\frac{\partial}{\partial y}\right)\frac{1}{z}\right] = \frac{-i\zeta}{2} F\left[\frac{1}{z}\right] = \pi F[\delta] = \pi$$

Since $1/\zeta$ is a locally integrable function in R^2, then the last equation may

§9. FOURIER TRANSFORM OF GENERALIZED FUNCTIONS OF SLOW GROWTH

be divided by ζ in $\mathscr{S}'(R^2)$. As a result we shall obtain formula (40).

(f) $$F\left[\frac{\theta(R-|x|)}{\sqrt{R^2-|x|^2}}\right] = 2\pi \frac{\sin R|\xi|}{|\xi|}, \quad n=2 \qquad (41)$$

In fact,

$$F\left[\frac{\theta(R-|x|)}{\sqrt{R^2-|x|^2}}\right] = \int_{|x|<R} \frac{e^{i(\xi,x)}}{\sqrt{R^2-|x|^2}} dx$$

$$= \int_0^R \frac{r}{\sqrt{R^2-r^2}} \int_0^{2\pi} \exp(ir|\xi|\cos\varphi)\, d\varphi\, dr$$

$$= 2\pi \int_0^R \frac{rJ_0(r|\xi|)}{\sqrt{R^2-r^2}} dr$$

$$= 2\pi R \int_0^1 J_0(R|\xi|u)\frac{u\,du}{\sqrt{1-u^2}} = 2\pi \frac{\sin R|\xi|}{|\xi|}$$

Here we have used formula 6.554, (2) from the reference book of I. S. Gradshteyn and I. M. Ryzhik (*1*).

(g) $$F\left[\frac{1}{|x|^2}\right] = \frac{2\pi^2}{|\xi|}, \quad n=3 \qquad (42)$$

Establishing that the function $|x|^{-2}$ is locally integrable in R^3, for all $\varphi \in \mathscr{S}$ we obtain the following sequence of equations:

$$\left(F\left[\frac{1}{|x|^2}\right], \varphi\right) = \left(\frac{1}{|x|^2}, F[\varphi]\right) = \int \frac{1}{|x|^2} F[\varphi]\, dx$$

$$= \lim_{R\to\infty} \int_{|x|<R} \frac{1}{|x|^2} \int \varphi(\xi)e^{i(\xi,x)}\, d\xi\, dx$$

$$= \lim_{R\to\infty} \int \varphi(\xi) \int_{|x|<R} \frac{e^{i(\xi,x)}}{|x|^2} dx\, d\xi$$

$$= \lim_{R\to\infty} \int \varphi(\xi) \int_0^R \int_0^\pi \int_0^{2\pi} \frac{\exp(i|\xi|\varrho\cos\theta)}{\varrho^2} \varrho^2\, d\psi \sin\theta\, d\theta\, d\varrho\, d\xi$$

$$= 2\pi \lim_{R\to\infty} \int \varphi(\xi) \int_0^R \int_{-1}^1 \exp(i|\xi|\varrho\mu)\, d\mu\, d\varrho\, d\xi$$

$$= 4\pi \lim_{R\to\infty} \int \frac{\varphi(\xi)}{|\xi|} \int_0^R \frac{\sin|\xi|\varrho}{\varrho} d\varrho\, d\xi \qquad (43)$$

Since

$$\left| \xi \right| \left| \int_R^\infty \frac{\sin |\xi| \varrho}{\varrho} d\varrho \right|$$
$$= \left| \frac{\cos |\xi| R}{R} - \int_R^\infty \frac{\cos |\xi| \varrho}{\varrho^2} d\varrho \right| \leq \frac{1}{R} + \int_R^\infty \frac{d\varrho}{\varrho^2} = \frac{2}{R}$$

a transition to the limit is possible under the integral sign as $R \to \infty$ in the last term of Eqs. (43). As a result we obtain

$$\left(F\left[\frac{1}{|x|^2}\right], \varphi\right) = 4\pi \int \frac{\varphi(\xi)}{|\xi|^2} |\xi| \int_0^\infty \frac{\sin |\xi| \varrho}{\varrho} d\varrho \, d\xi = 2\pi^2 \int \frac{\varphi(\xi)}{|\xi|} d\xi$$

from which Eq. (42) follows.

CHAPTER

3

Fundamental Solutions and the Cauchy Problem

In this chapter the theory of generalized functions is applied to the solution of the Cauchy problem for the wave equation and for the equation for the conduction of heat. In this connection, the Cauchy problem is considered in a generalized context which allows one to include initial conditions involving momentarily acting sources (of the nature of simple and double layers over the surface $t = 0$). In this way, the Cauchy problem may be reduced to the problem of finding the (generalized) solution of the given equation (with a modified right-hand side) which becomes zero when $t < 0$. The problem is solved by a standard method—that of summing the perturbations produced by each point of the distributed source so that its solution appears in the form of a convolution of the fundamental solution together with the right-hand side.

§ 10. Fundamental Solutions of Linear Differential Operators

The Fourier transform is applied to construct the fundamental solutions of linear differential operators having constant coefficients. Naturally, only fundamental solutions of slow growth can be obtained by this method.

1. Generalized Solutions of Linear Differential Equations. Let

$$\sum_{|\alpha|=0}^{m} a_\alpha(x) D^\alpha u = f(x), \quad f \in \mathscr{D}' \tag{1}$$

be a linear differential equation of order m with coefficients $a_\alpha \in C^\infty(R^n)$. Introducing the differential operator

$$L(x, D) = \sum_{|\alpha|=0}^{m} a_\alpha(x) D^\alpha, \qquad D = \left(\frac{\partial}{\partial x_1}, \frac{\partial}{\partial x_2}, \ldots, \frac{\partial}{\partial x_n} \right)$$

we shall rewrite this equation in the form

$$L(x, D)u = f(x) \qquad (1')$$

Each generalized function $u \in \mathscr{D}'$ which satisfies this equation in the region G in a generalized sense, that is, for any $\varphi \in \mathscr{D}$, supp $\varphi \subset G$

$$(L(x, D)u, \varphi) = (f, \varphi) \qquad (2)$$

is known as the *generalized solution* of Eq. (1) in the region G. Equation (2) is equal in effect to the equation

$$(u, L^*(x, D)\varphi) = (f, \varphi), \qquad \varphi \in \mathscr{D}(G) \qquad (2')$$

where

$$L^*(x, D)\varphi = \sum_{|\alpha|=0}^{m} (-1)^{|\alpha|} D^\alpha(a_\alpha \varphi) \qquad (3)$$

In fact,

$$(L(x, D)u, \varphi) = \left(\sum_{|\alpha|=0}^{m} a_\alpha D^\alpha u, \varphi \right) = \sum_{|\alpha|=0}^{m} (a_\alpha D^\alpha u, \varphi)$$

$$= \sum_{|\alpha|=0}^{m} (D^\alpha u, a_\alpha \varphi) = \sum_{|\alpha|=0}^{m} (-1)^{|\alpha|} (u, D^\alpha(a_\alpha \varphi))$$

$$= \left(u, \sum_{|\alpha|=0}^{m} (-1)^{|\alpha|} D^\alpha(a_\alpha \varphi) \right) = (u, L^*(x, D)\varphi)$$

It is clear that every classical solution is also a generalized solution. We shall formulate the converse result as the following lemma.

LEMMA. *If the generalized solution $u(x)$ of Eq. (1) in the region G belongs to the class $C^m(G)$ and $f \in C(G)$, then it is also the classical solution of this equation in the region G.*

Proof. Since $u \in \mathscr{D}' \cap C^m(G)$, the classical and generalized derivatives of the function u up to and including the order m coincide in the region G (cf. Sec. 6.1). Since u is the generalized solution of Eq. (1) in the region G,

§ 10. FUNDAMENTAL SOLUTIONS OF LINEAR DIFFERENTIAL OPERATORS

then the function $L(x, D)u - f$ which is continuous in G vanishes in the region G in the sense of generalized functions. According to Du Bois Reymond's lemma (cf. Sec. 5.5), $L(x, D)u(x) - f(x) = 0$ at all points of the region G, so that u satisfies Eq. (1) in the region G in the classical sense. The lemma is proved.

2. Fundamental Solutions. Let L be an operator with constant coefficients $a_\alpha(x) = a_\alpha$:

$$L(D) = \sum_{|\alpha|=0}^{m} a_\alpha D^\alpha, \qquad L^*(D) = L(-D) \tag{4}$$

The generalized function $\mathscr{E} \in \mathscr{D}'$ which satisfies equation

$$L(D)\mathscr{E} = \delta(x) \tag{5}$$

in R^n is said to be the *fundamental solution* (*the function of influence*) of the differential operator $L(D)$.

The fundamental solution $\mathscr{E}(x)$ of the operator $L(D)$, generally speaking, is not unique; it is defined accurately as far as the term $\mathscr{E}_0(x)$, which is an arbitrary solution of the homogeneous equation $L(D)\mathscr{E}_0 = 0$.

In fact, the generalized function $\mathscr{E}(x) + \mathscr{E}_0(x)$ is also a fundamental solution of the operator $L(D)$,

$$L(D)(\mathscr{E} + \mathscr{E}_0) = L(D)\mathscr{E} + L(D)\mathscr{E}_0 = \delta(x)$$

LEMMA. *In order that the generalized function* $\mathscr{E} \in \mathscr{S}'$ *should be the fundamental solution of the operator $L(D)$, it is necessary and sufficient that its Fourier transform $F[\mathscr{E}]$ satisfy the equation*

$$L(-i\xi)F[\mathscr{E}] = 1 \tag{6}$$

where

$$L(\xi) = \sum_{|\alpha|=0}^{m} a_\alpha \xi^\alpha$$

Proof. Let $\mathscr{E} \in \mathscr{S}'$ be the fundamental solution of the operator $L(D)$. Applying the Fourier transform to both sides of Eq. (5), we obtain

$$F[L(D)\mathscr{E}] = F[\delta] = 1 \tag{7}$$

Taking Eq. (12) of Sec. 9.3 into account, we have

$$F[L(D)\mathscr{E}] = F\left[\sum_{|\alpha|=0}^{m} a_\alpha D^\alpha \mathscr{E}\right] = \sum_{|\alpha|=0}^{m} a_\alpha F[D^\alpha \mathscr{E}]$$

$$= \sum_{|\alpha|=0}^{m} a_\alpha (-i\xi)^\alpha F[\mathscr{E}] = L(-i\xi)F[\mathscr{E}] \qquad (8)$$

from which, and from (7), it follows that $F[\mathscr{E}]$ satisfies Eq. (6).

Conversely, if $\mathscr{E} \in \mathscr{S}'$ satisfies Eq. (6), then, by virtue of (8), $\mathscr{E}$ satisfies Eq. (7), from which it follows that $\mathscr{E}$ satisfies Eq. (5); that is, it is the fundamental solution of the operator $L(D)$. The lemma is proved.

The lemma which has just been proved reduces the problem of constructing fundamental solutions of slow growth for linear differential operators with constant coefficients to the solution in $\mathscr{S}'$ of algebraic equations of the form

$$P(\xi)X = 1 \qquad (9)$$

where P is an arbitrary polynomial.

As we see from Eq. (9), each of its solutions belonging to $\mathscr{D}'$ (if such solutions exist) must coincide with the function $1/P(\xi)$ outside the set N_P of the zeros of the polynomial $P(\xi)$

$$N_P = [\xi : P(\xi) = 0]$$

It follows from this that if $N_P \neq \emptyset$, the solution of Eq. (9) is not unique: the various solutions differ from each other by a generalized function with support in N_P. For instance, the generalized functions

$$\frac{1}{\xi + i0}, \quad \frac{1}{\xi - i0}, \quad \text{and} \quad \mathscr{P}\frac{1}{\xi}$$

which differ from each other by an expression of the form const $\delta(\xi)$ [cf. Sokhotsky's formulas (10) and (10') of Sec. 5.7], are different solutions of the equation $\xi X = 1$.

If the function $1/P(\xi)$ is locally integrable in R^n, then it (or more accurately, the regular functional defined by it) is the solution in $\mathscr{S}'$ of Eq. (9). However, if the function $1/P(\xi)$ is not locally integrable in R^n, there is the nontrivial problem of constructing a solution of Eq. (9) in $\mathscr{S}'$. It has been proved by L. Hörmander (2) that Eq. (9) is always soluble in $\mathscr{S}'$.

§ 10. FUNDAMENTAL SOLUTIONS OF LINEAR DIFFERENTIAL OPERATORS

We shall denote by $\text{reg}(1/P(\xi))$ any solution of Eq. (9) belonging to $\mathscr{S}'$. The construction of this solution depends essentially on the structure of the set N_P and can be carried out for each concrete polynomial P. In this way, Eq. (6) is always soluble in $\mathscr{S}'$ and

$$F[\mathscr{E}] = \text{reg}\,\frac{1}{L(-i\xi)}$$

is its solution.

Consequently, *each linear differential operator $L(D)$ with constant coefficients has a fundamental solution of slow growth, and this solution is given by the formula*

$$\mathscr{E} = F^{-1}\left[\text{reg}\,\frac{1}{L(-i\xi)}\right] = \frac{1}{(2\pi)^n}\,F\left[\text{reg}\,\frac{1}{L(i\xi)}\right] \qquad (10)$$

3. Equations with a Right-Hand Side. By means of the fundamental solution $\mathscr{E}(x)$ of the operator $L(D)$ it is possible to construct a solution of the equation

$$L(D)u = f(x) \qquad (11)$$

with an arbitrary right-hand side f. More accurately, the following theorem is valid.

THEOREM. *Let $f \in \mathscr{D}'$ be such that the convolution $\mathscr{E} * f$ exists in $\mathscr{D}'$. Then the solution of Eq. (11) exists in $\mathscr{D}'$ and is given by the formula*

$$u = \mathscr{E} * f \qquad (12)$$

This solution is unique in the class of generalized functions belonging to $\mathscr{D}'$ for which a convolution with $\mathscr{E}$ exists.

Proof. Using the formula for the differentiation of a convolution [cf. (22) of Sec. 7.6] and taking Eq. (5) into account, we obtain

$$L(D)(\mathscr{E} * f) = \sum_{|\alpha|=0}^{m} a_\alpha D^\alpha(\mathscr{E} * f)$$

$$= \left(\sum_{|\alpha|=0}^{m} a_\alpha D^\alpha \mathscr{E}\right) * f = L(D)\mathscr{E} * f = \delta * f = f$$

Therefore the formula $u = \mathscr{E} * f$ in fact gives the solution of Eq. (11). We shall prove the uniqueness of the solution of Eq. (11) in the class of the generalized functions belonging to $\mathscr{D}'$ for which a convolution with $\mathscr{E}$ exists in $\mathscr{D}'$. For this it is sufficient to establish that the corresponding homogeneous equation

$$L(D)u = 0$$

has only a zero solution in this class (cf. Sec. 1.9). But this is in fact so, by virtue of

$$u = u * \delta = u * L(D)\mathscr{E} = L(D)u * \mathscr{E} = 0$$

The theorem is proved.

COROLLARY. *If $u \in \mathscr{D}'$ and the convolution $u * \mathscr{E}$ exists in $\mathscr{D}'$, then the equation*

$$u = L(D)u * \mathscr{E} \tag{13}$$

is valid.

PHYSICAL SENSE OF THE SOLUTION $u = \mathscr{E} * f$. Let us represent the source $f(x)$ in the form of a "sum" of the point sources $f(\xi)\delta(x - \xi)$,

$$f(x) = \delta * f = \int f(\xi)\delta(x - \xi)\,d\xi$$

By virtue of (5), each point source $f(\xi)\delta(x - \xi)$ defines the influence $f(\xi)\mathscr{E}(x - \xi)$. Therefore the solution

$$u(x) = \mathscr{E} * f = \int f(\xi)\mathscr{E}(x - \xi)\,d\xi$$

is the *superposition of these influences.*

4. Method of Descent. Let us consider a linear differential equation with constant coefficients in the space R^{n+1} of the variables $(x, t) = (x_1, x_2, \ldots, x_n, t)$

$$L\left(D, \frac{\partial}{\partial t}\right)u = f(x) \cdot \delta(t), \qquad f \in \mathscr{D}'(R^n) \tag{14}$$

where

$$L\left(D, \frac{\partial}{\partial t}\right) = \sum_{q=1}^{p} \frac{\partial^q}{\partial t^q} L_q(D) + L_0(D)$$

and $L_q(D)$ are differential operators involving the variables x.

§ 10. FUNDAMENTAL SOLUTIONS OF LINEAR DIFFERENTIAL OPERATORS

Let the generalized function u belonging to $\mathscr{D}'(R^{n+1})$ allow continuation over functions of the form $\varphi(x)1(t)$, where $\varphi \in \mathscr{D}(R^n)$ in the following sense: Whatever may be the sequence of test functions $\eta_k(t)$, $k = 1, 2, \ldots$, belonging to $\mathscr{D}(R^1)$ and converging to 1 in R^1 (cf. Sec. 7.4), there is a limit

$$\lim_{k \to \infty} (u, \varphi(x)\eta_k(t)) = (u, \varphi(x)1(t)) \tag{15}$$

and this limit does not depend on the sequence $\{\eta_k\}$.

We shall denote the functional (15) by u_0,

$$(u_0, \varphi) = (u, \varphi(x)1(t)) = \lim_{k \to \infty} (u, \varphi(x)\eta_k(t)), \qquad \varphi \in \mathscr{D}(R^n) \tag{16}$$

Let us suppose, moreover, that the functional u_0 is continuous* over $\mathscr{D}(R^n)$. Then since the functional u_0 is linear, $u_0 \in \mathscr{D}'(R^n)$.

We shall give two examples of the construction of the continuation u_0.

(a) *Let the function $u(x, t)$ be such that the function $\int |u(x, t)| \, dt$ is locally integrable in R^n. Then $u_0(x)$ is a function locally integrable in R^n and is given by the integral*

$$u_0(x) = \int_{-\infty}^{\infty} u(x, t) \, dt \tag{17}$$

In fact, in this case the function $u(x, t)$ is locally integrable in R^{n+1} and, by virtue of Lebesgue's theorem (cf. Sec. 1.4), the limit (15) exists

$$\lim_{k \to \infty} (u, \varphi(x)\eta_k(t)) = \lim_{k \to \infty} \int u(x, t)\varphi(x)\eta_k(t) \, dx \, dt$$

$$= \int u(x, t)\varphi(x) \, dx \, dt = \int \varphi(x) \int_{-\infty}^{\infty} u(x, t) \, dt \, dx$$

for all $\varphi \in \mathscr{D}(R^n)$, does not depend on the sequence $\{\eta_k\}$, and defines a continuous functional over $\mathscr{D}(R^n)$. As a result, by virtue of (16), formula (17) follows.

(b) Let $u = f(x) \cdot \delta(t)$, where $f \in \mathscr{D}'(R^n)$. Then $u_0 = f$ by virtue of

$$(u_0, \varphi) = \lim_{k \to \infty} (u, \varphi(x)\eta_k(t)) = \lim_{k \to \infty} (f(x) \cdot \delta(t), \varphi(x)\eta_k(t))$$

$$= \lim_{k \to \infty} (f(x), \varphi(x)\eta_k(0)) = (f, \varphi), \qquad \varphi \in \mathscr{D}(R^n)$$

---

* The continuity of the functional u_0 follows from the completeness of the space $\mathscr{D}'$ (cf. Sec. 5.3).

THEOREM. *If the solution* $u \in \mathscr{D}'(R^{n+1})$ *of Eq.* (14) *allows the continuation* (16), *then the generalized function* u_0 *belonging to* $\mathscr{D}'(R^n)$ *satisfies the equation*

$$L_0(D)u_0 = f(x) \qquad (18)$$

Proof. Let $\eta_k(t)$, $k = 1, 2, \ldots$, be the sequence of functions belonging to $\mathscr{D}(R^1)$ which converges to 1 in R^1. Then for $q = 1, 2, \ldots$, the sequences of functions

$$\eta_k(t) + \eta_k^{(q)}(t), \qquad k = 1, 2, \ldots \qquad (19)$$

also converge to 1 in R^1 and therefore, for all φ belonging to $\mathscr{D}(R^n)$ (cf. Sec. 7.6)

$$\lim_{k \to \infty} (u, \varphi(x)\eta_k^{(q)}(t)) = \lim_{k \to \infty} (u, \varphi(x)[\eta_k(t) + \eta_k^{(q)}(t)])$$
$$- \lim_{k \to \infty} (u, \varphi(x)\eta_k(t)) = (u_0, \varphi) - (u_0, \varphi) = 0 \qquad (20)$$

Granting (20), we shall check that the generalized function u_0 satisfies Eq. (18):

$$(L_0(D)u_0, \varphi) = (u_0, L_0(-D)\varphi) = \lim_{k \to \infty} (u, L_0(-D)\varphi(x)\eta_k(t))$$
$$= \lim_{k \to \infty} \left(u, L_0(-D)\varphi(x)\eta_k(t) + \sum_{q=1}^{p} (-1)^q L_q(-D)\varphi(x)\eta_k^{(q)}(t)\right)$$
$$= \lim_{k \to \infty} \left(u, L\left(-D, -\frac{\partial}{\partial t}\right)\varphi(x)\eta_k(t)\right)$$
$$= \lim_{k \to \infty} \left(L\left(D, \frac{\partial}{\partial t}\right)u, \varphi(x)\eta_k(t)\right)$$
$$= \lim_{k \to \infty} (f(x) \cdot \delta(t), \varphi(x)\eta_k(t)) = \lim_{k \to \infty} (f(x), \varphi(x)\eta_k(0))$$
$$= (f, \varphi)$$

The theorem is proved.

The method which we have just set out for obtaining the solution $u_0(x)$ of Eq. (18) with n variables by means of the solution $u(x, t)$ of Eq. (14) with $n + 1$ variables is known as the *method of descent involving the variable t*.

The method of descent is especially useful in constructing fundamental solutions. In fact, applying the theorem which has just been proved to the

§ 10. FUNDAMENTAL SOLUTIONS OF LINEAR DIFFERENTIAL OPERATORS

case $f = \delta(x)$, we obtain: *If $\mathscr{E}(x, t)$, the fundamental solution of the operator $L(D, \partial/\partial t)$, allows the continuation* (16), *then the generalized function*

$$(\mathscr{E}_0, \varphi) = (\mathscr{E}, \varphi(x)1(t)), \qquad \varphi \in \mathscr{D}(R^n) \tag{21}$$

is a fundamental solution of the operator $L_0(D)$; specifically, if the function $\int |\mathscr{E}(x, t)|\, dt$ is locally integrable in R^n, then

$$\mathscr{E}_0(x) = \int_{-\infty}^{\infty} \mathscr{E}(x, t)\, dt \tag{22}$$

The fundamental solution satisfies the relation

$$\mathscr{E}_0(x) \cdot 1(t) = \mathscr{E} * \delta(x) \cdot 1(t)$$

The physical sense of this formula is that $\mathscr{E}_0(x)$ is a perturbation (not depending on t) from the source $\delta(x) \cdot 1(t)$ concentrated along the t axis (cf. Sec. 10.3).

5. Fundamental Solution of a Linear Differential Operator with Ordinary Derivatives

$$\frac{d^n \mathscr{E}}{dt^n} + a_1(t) \frac{d^{n-1}\mathscr{E}}{dt^{n-1}} + \cdots + a_n(t)\mathscr{E} = \delta(t)$$

In Sec. 6.3, (f) it has been shown that the fundamental solution of this operator is expressed by the formula

$$\mathscr{E}(t) = \theta(t) Z(t)$$

where $Z(t)$ satisfies the homogeneous equation $LZ = 0$ and the initial conditions

$$Z(0) = Z'(0) = \cdots = Z^{(n-2)}(0) = 0, \qquad Z^{(n-1)}(0) = 1$$

Specifically, the functions

$$\mathscr{E}_1(t) = \theta(t)e^{-at}, \qquad \mathscr{E}_2(t) = \theta(t)\frac{\sin at}{a} \tag{23}$$

are fundamental solutions, respectively, of the operators

$$\frac{d}{dt} + a, \qquad \frac{d^2}{dt^2} + a^2$$

6. Fundamental Solution of the Heat Conduction Operator

$$\frac{\partial \mathscr{E}}{\partial t} - a^2 \Delta \mathscr{E} = \delta(x, t) \tag{24}$$

In Sec. 6.5, (f) it has been shown that the solution of Eq. (24) is expressed by the formula

$$\mathscr{E}(x, t) = \frac{\theta(t)}{(2a\sqrt{\pi t})^n} \exp\left[-\frac{|x|^2}{4a^2 t}\right] \tag{25}$$

and consequently this function is the fundamental solution of the heat conduction operator.

We shall deduce formula (25) by means of the Fourier transform. Apply the Fourier transform F_x (cf. Sec. 9.2) to Eq. (24):

$$F_x\left[\frac{\partial \mathscr{E}}{\partial t}\right] - a^2 F_x[\Delta \mathscr{E}] = F_x[\delta(x, t)]$$

and use Eqs. (15) and (16) of Sec. 9.3:

$$F_x[\delta(x, t)] = F_x[\delta(x) \cdot \delta(t)] = F[\delta](\xi) \cdot \delta(t) = 1(\xi) \cdot \delta(t)$$

$$F_x\left[\frac{\partial \mathscr{E}}{\partial t}\right] = \frac{\partial}{\partial t} F_x[\mathscr{E}], \qquad F_x[\Delta \mathscr{E}] = -|\xi|^2 F_x[\mathscr{E}]$$

As a result, for the generalized function $\widetilde{\mathscr{E}}(\xi, t) = F_x[\mathscr{E}](\xi, t)$ we obtain the equation

$$\frac{\partial \widetilde{\mathscr{E}}(\xi, t)}{\partial t} + a^2 |\xi|^2 \widetilde{\mathscr{E}}(\xi, t) = 1(\xi) \cdot \delta(t) \tag{26}$$

Using Eq. (23) for $\mathscr{E}_1(t)$ and replacing a by $a^2 |\xi|^2$, we conclude that the solution in $\mathscr{S}'$ of Eq. (26) is the function

$$\widetilde{\mathscr{E}}(\xi, t) = \theta(t) \exp(-a^2 |\xi|^2 t)$$

From this, applying the inverse Fourier transform F_ξ^{-1} and using formula (36) of Sec. 9.8, we obtain Eq. (25):

$$\mathscr{E}(x, t) = F_\xi^{-1}[\widetilde{\mathscr{E}}(\xi, t)] = \frac{\theta(t)}{(2\pi)^n} \int \exp(-a^2 |\xi|^2 t - i(\xi, x))\, d\xi$$

$$= \frac{\theta(t)}{(2a\sqrt{\pi t})^n} \exp\left(-\frac{|x|^2}{4a^2 t}\right)$$

§ 10. FUNDAMENTAL SOLUTIONS OF LINEAR DIFFERENTIAL OPERATORS

7. Fundamental Solution of the Wave Operator

$$\Box_a \mathscr{E}_n = \delta(x, t) \tag{27}$$

Applying the Fourier transform F_x to Eq. (27) and acting as in the preceding section, instead of Eq. (26) for the generalized function $F_x[\mathscr{E}_n] = \tilde{\mathscr{E}}_n(\xi, t)$ we obtain the equation

$$\frac{\partial^2 \tilde{\mathscr{E}}_n(\xi, t)}{\partial t^2} + a^2 |\xi|^2 \tilde{\mathscr{E}}_n(\xi, t) = 1(\xi) \cdot \delta(t) \tag{28}$$

Using formula (23) for $\mathscr{E}_2(t)$ and replacing a by $a|\xi|$, we conclude that the solution in $\mathscr{S}'$ of Eq. (28) is the function

$$\tilde{\mathscr{E}}_n(\xi, t) = \theta(t) \frac{\sin a |\xi| t}{a |\xi|}$$

Consequently,

$$\mathscr{E}_n(x, t) = F_\xi^{-1}[\tilde{\mathscr{E}}_n(\xi, t)] = \theta(t) F_\xi^{-1}\left[\frac{\sin a |\xi| t}{a |\xi|}\right] \tag{29}$$

Let $n = 3$. Then from formula (38) of Sec. 9.8, we deduce that

$$F^{-1}\left[\frac{\sin a |\xi| t}{|\xi|}\right] = \frac{1}{4\pi a t} \delta_{S_{at}}(x)$$

from which, and from (29), we obtain

$$\mathscr{E}_3(x, t) = \frac{\theta(t)}{4\pi a^2 t} \delta_{S_{at}}(x) \equiv \frac{\theta(t)}{2\pi a} \delta(a^2 t^2 - |x|^2) \tag{30}$$

where the generalized function $\mathscr{E}_3$ acts according to the rule:

$$(\mathscr{E}_3, \varphi) = \frac{1}{4\pi a^2} \int_0^\infty (\delta_{S_{at}}, \varphi) \frac{dt}{t}$$

$$= \frac{1}{4\pi a^2} \int_0^\infty \frac{1}{t} \int_{S_{at}} \varphi(x, t) \, dS_x \, dt, \qquad \varphi \in \mathscr{S}(R^4) \tag{31}$$

Analogously, using formulas (24) of Sec. 9.6 and (41) of Sec. 9.8, we obtain [cf. Sec. 6.5, (g)]

$$\mathscr{E}_1(x, t) = \frac{1}{2a} \theta(at - |x|)$$

$$\mathscr{E}_2(x, t) = \frac{\theta(at - |x|)}{2\pi a \sqrt{a^2 t^2 - |x|^2}} \tag{32}$$

To obtain the fundamental solution $\mathscr{E}_2(x, t)$, $x = (x_1, x_2)$, we shall use the method of descent involving the variable x_3 (cf. Sec. 10.4). For this it

150   3. FUNDAMENTAL SOLUTIONS AND THE CAUCHY PROBLEM

is necessary to show that $\mathscr{E}_3(x, x_3, t)$ allows the continuation (16) over functions of the form $\varphi(x, t)1(x_3)$, where $\varphi \in \mathscr{D}(R^3)$, and that the functional

$$(\mathscr{E}_3(x, x_3, t), \varphi(x, t)1(x_3)), \quad \varphi \in \mathscr{D}(R^3)$$

is continuous over $\mathscr{D}(R^3)$.

Let $\eta_k \in \mathscr{D}(R^1)$ $k = 1, 2, \ldots$ and $\eta_k(x_3)$ tend to 1 in R^1. Then, using (31) for all $\varphi \in \mathscr{D}(R^3)$ we obtain

$$\lim_{k \to \infty} (\mathscr{E}_3, \varphi(x, t)\eta_k(x_3)) = \lim_{k \to \infty} \frac{1}{4\pi a^2} \int_0^\infty \frac{1}{t} \int_{S_{at}} \varphi(x, t)\eta_k(x_3) \, dS \, dt$$

$$= \frac{1}{4\pi a^2} \int_0^\infty \frac{1}{t} \int_{S_{at}} \varphi(x, t) \, dS \, dt = (\mathscr{E}_3, \varphi(x, t)1(x_3))$$

so that this limit exists which does not depend on the sequence $\{\eta_k\}$ and which defines a continuous functional over $\mathscr{D}(R^3)$. From this, and applying formula (21), we conclude that

$$(\mathscr{E}_2, \varphi) = (\mathscr{E}_3, \varphi(x, t)1(x_3)) = \frac{1}{4\pi a^2} \int_0^\infty \frac{1}{t} \int_{S_{at}} \varphi(x, t) \, dS \, dt \quad \varphi \in \mathscr{D}(R^3)$$

We shall transform the last integral. As φ does not depend on x_3, then, replacing the surface integral over the spherical surface $S_{at} = [|x|^2 + x_3^2 = a^2 t^2]$ by the doubled integral over the circle $|x| \leq at$ (Fig. 22), we obtain

$$(\mathscr{E}_2, \varphi) = \frac{1}{2\pi a} \int_0^\infty \int_{|x| \leq at} \frac{\varphi(x, t)}{(a^2 t^2 - |x|^2)^{1/2}} \, dx \, dt$$

$$= \frac{1}{2\pi a} \int \frac{\theta(at - |x|)}{(a^2 t^2 - |x|^2)^{1/2}} \varphi(x, t) \, dx \, dt$$

from which formula (32) follows for $\mathscr{E}_2$.

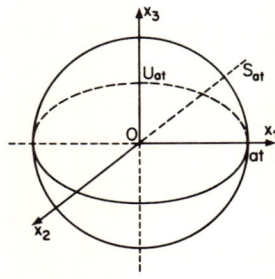

Fig. 22

§ 10. FUNDAMENTAL SOLUTIONS OF LINEAR DIFFERENTIAL OPERATORS

Analogously, using formula (22), by the method of descent involving x_2 we shall obtain formula (32) for the fundamental solution $\mathscr{E}_1(x, t)$:

$$\mathscr{E}_1(x, t) = \int_{-\infty}^{\infty} \mathscr{E}_2(x, x_2, t)\, dx_2$$

$$= \frac{1}{2\pi a} \int_{-\infty}^{\infty} \frac{\theta(at - \sqrt{x^2 + x_2^2})}{(a^2 t^2 - x^2 - x_2^2)^{1/2}}\, dx_2$$

$$= \frac{\theta(at - |x|)}{\pi a} \int_0^{(a^2 t^2 - x^2)^{1/2}} \frac{dx_2}{(a^2 t^2 - x^2 - x_2^2)^{1/2}}$$

$$= \frac{\theta(at - |x|)}{\pi a} \int_0^1 \frac{du}{\sqrt{1 - u^2}} = \frac{1}{2a}\theta(at - |x|)$$

8. Fundamental Solution of the Laplace Operator

$$\Delta \mathscr{E}_n = \delta(x) \tag{33}$$

In Sec. 6.5, (d) it has been shown that the functions

$$\mathscr{E}_2(x) = \frac{1}{2\pi} \ln |x|, \quad \mathscr{E}_n(x) = -\frac{1}{(n-2)\sigma_n}|x|^{-n+2}, \quad n \geq 3 \tag{34}$$

are fundamental solutions of the Laplace operator. We shall calculate these fundamental solutions by means of the Fourier transform. Applying the Fourier transform to Eq. (33) we obtain

$$-|\xi|^2 F[\mathscr{E}_n] = 1 \tag{35}$$

Let $n = 2$. Let us check that the generalized function $-\mathscr{P}(1/|\xi|^2)$ [cf. Sec. 9.8, (d)] satisfies Eq. (35). In fact,

$$\left(|\xi|^2 \mathscr{P}\frac{1}{|\xi|^2}, \varphi\right)$$

$$= \left(\mathscr{P}\frac{1}{|\xi|^2}, |\xi|^2 \varphi\right)$$

$$= \int_{|\xi|<1} \frac{|\xi|^2 \varphi(\xi) - |\xi|^2 \varphi(\xi)|_{\xi=0}}{|\xi|^2}\, d\xi + \int_{|\xi|>1} \frac{|\xi|^2 \varphi(\xi)}{|\xi|^2}\, d\xi$$

$$= \int \varphi(\xi)\, d\xi = (1, \varphi), \quad \varphi \in \mathscr{S}$$

Consequently, in agreement with the scheme of Sec. 10.2, we can suppose that

$$F[\mathscr{E}_2] = \operatorname{reg} \frac{1}{-|\xi|^2} = -\mathscr{P}\frac{1}{|\xi|^2}$$

From this, and using formula (39) of Sec. 9.8, we obtain

$$\mathscr{E}_2(x) = F^{-1}\left[-\mathscr{P}\frac{1}{|\xi|^2}\right] = -\frac{1}{4\pi^2}F\left[\frac{1}{|\xi|^2}\right]$$

$$= \frac{1}{2\pi}\ln|x| + \frac{C_0}{2\pi} \tag{36}$$

Since the constant satisfies the homogeneous Laplace equation, then, disregarding the term $C_0/2\pi$ in (36), we see that the fundamental solution $\mathscr{E}_2(x)$ may be chosen equal to $(1/2\pi)\ln|x|$.

Now let $n \geq 3$. In this case the function $-|\xi|^{-2}$ is locally integrable in R^n and so, in agreement with Sec. 10.2,

$$F[\mathscr{E}_n] = -\frac{1}{|\xi|^2}, \qquad \mathscr{E}_n = -F^{-1}\left[\frac{1}{|\xi|^2}\right]$$

From this, for $n = 3$, using formula (42) of Sec. 9.8, we obtain

$$\mathscr{E}_3(x) = -\frac{1}{4\pi|x|} \tag{37}$$

$\mathscr{E}_n(x)$ is calculated analogously for $n > 3$.

It is especially simple to construct $\mathscr{E}_n(x)$, $n \geq 3$, by the method of descent involving the variable t (cf. Sec. 10.4) from the fundamental solutions of the heat conduction operator or of the wave operator. For instance, using formula (22), from (25) when $a = 1$ we obtain formula (34):

$$\mathscr{E}_n(x) = -\int_{-\infty}^{\infty}\mathscr{E}(x, t)\,dt = -\int_0^{\infty}\frac{1}{(2\sqrt{\pi t})^n}\exp\left(-\frac{|x|^2}{4t}\right)dt$$

$$= -\frac{|x|^{-n+2}}{4\pi^{n/2}}\int_0^{\infty}e^{-u}u^{(n/2)-2}\,du = -\Gamma\left(\frac{n}{2}-1\right)\frac{|x|^{-n+2}}{4\pi^{n/2}}$$

$$= -\frac{1}{(n-2)\sigma_n}|x|^{-n+2}, \qquad n \geq 3$$

§ 10. FUNDAMENTAL SOLUTIONS OF LINEAR DIFFERENTIAL OPERATORS

9. Fundamental Solution of Helmholtz's Operator

$$(\Delta + k^2)\mathscr{E}_n = \delta(x) \tag{38}$$

It has been shown in Sec. 6.5, (e) that

$$\mathscr{E}_3(x) = -\frac{e^{ik|x|}}{4\pi|x|}, \qquad \bar{\mathscr{E}}_3(x) = -\frac{e^{-ik|x|}}{4\pi|x|} \tag{39}$$

are the fundamental solutions of Helmholtz's operator when $n = 3$. Formulas (39) are valid also when k is complex.

Let us calculate $\mathscr{E}_2(x)$ by means of the Fourier transform. From (38) we have

$$(-|\xi|^2 + k^2)F[\mathscr{E}_2] = 1 \tag{40}$$

We shall take the solution of Eq. (40) in the form

$$F[\mathscr{E}_2] = \lim_{\varepsilon \to +0} \frac{1}{k^2 + i\varepsilon - |\xi|^2} = \frac{1}{k^2 + i0 - |\xi|^2}$$

and so, since the Fourier transform is continuous,

$$\mathscr{E}_2(x) = F^{-1}\left[\frac{1}{k^2 + i0 - |\xi|^2}\right]$$

$$= \frac{1}{4\pi^2} \lim_{\varepsilon \to +0} \lim_{R \to \infty} \int_{|\xi|<R} \frac{e^{-i(\xi,x)}}{k^2 + i\varepsilon - |\xi|^2} d\xi$$

$$= \frac{1}{4\pi^2} \lim_{\varepsilon \to +0} \lim_{R \to \infty} \int_0^R \frac{\varrho}{k^2 + i\varepsilon - \varrho^2} \int_0^{2\pi} \exp(-i\varrho|x|\cos\varphi) \, d\varphi \, d\varrho$$

$$= \frac{1}{2\pi} \lim_{\varepsilon \to +0} \lim_{R \to \infty} \int_0^R \frac{\varrho J_0(\varrho|x|)}{k^2 + i\varepsilon - \varrho^2} d\varrho$$

$$= -\frac{1}{2\pi} \lim_{\varepsilon \to +0} K_0(-i\sqrt{k^2 + i\varepsilon}\,|x|)$$

$$= -\frac{1}{2\pi} K_0(-ik|x|) = -\frac{i}{4} H_0^{(1)}(k|x|)$$

Here we have used formula 6.532, (4) from the reference book of I. S.

Gradshteyn and I. M. Ryzhik (1); $H_0^{(j)}$, $j = 1, 2$, are Hankel functions. So the functions

$$\mathscr{E}_2(x) = -\frac{i}{4} H_0^{(1)}(k\,|\,x\,|), \qquad \bar{\mathscr{E}}_2(x) = \frac{i}{4} H_0^{(2)}(k\,|\,x\,|) \qquad (41)$$

are fundamental solutions of Helmholtz's operator when $n = 2$.

It is convenient to take fundamental solutions for $n = 1$ in the form (cf. Sec. 10.5)

$$\mathscr{E}_1(x) = \frac{\theta(x)}{k} \sin kx - \frac{\sin kx}{2k} + \frac{\cos kx}{2ik} = \frac{1}{2ik} e^{ik|x|} \qquad (42)$$

$$\bar{\mathscr{E}}_1(x) = -\frac{1}{2ik} e^{-ik|x|}$$

10. Fundamental Solution of the Cauchy–Riemann Operator

$$\frac{\partial}{\partial \bar{z}} \mathscr{E} = \delta(x, y) \qquad (43)$$

It has been shown in Sec. 6.5, (j) that

$$\mathscr{E}(x, y) = \frac{1}{\pi z} \qquad (44)$$

11. Fundamental Solution of the Transport Operator

$$\frac{1}{v} \frac{\partial \mathscr{E}_s}{\partial t} + (\mathbf{s}, \operatorname{grad} \mathscr{E}_s) + \alpha \mathscr{E}_s = \delta(x, t), \qquad |\mathbf{s}| = 1 \qquad (45)$$

Applying the Fourier transform F_x to Eq. (45) for the generalized function $F_x[\mathscr{E}_s] = \tilde{\mathscr{E}}_s(\xi, t)$ we obtain the equation

$$\frac{1}{v} \frac{\partial \tilde{\mathscr{E}}_s(\xi, t)}{\partial t} + [\alpha - i(\mathbf{s}, \xi)] \tilde{\mathscr{E}}_s(\xi, t) = 1(\xi) \cdot \delta(t) \qquad (46)$$

From this, and using the first of formulas (23), we conclude that the function

$$\tilde{\mathscr{E}}_s(\xi, t) = v\theta(t) \exp\{[i(\mathbf{s}, \xi) - \alpha]vt\}$$

is the solution belonging to $\mathscr{S}'$ of Eq. (46). Now applying the inverse Fourier transform F_ξ^{-1},

$$\mathscr{E}_s(x, t) = F_\xi^{-1}[\tilde{\mathscr{E}}_s(\xi, t)] = v\theta(t) e^{-\alpha vt} F^{-1}[\exp(i(\xi, \mathbf{s})vt)]$$

§ 10. FUNDAMENTAL SOLUTIONS OF LINEAR DIFFERENTIAL OPERATORS

and using formula (19) of Sec. 9.4 for $x_0 = v t \mathbf{s}$,

$$F^{-1}[\exp(i(\xi, \mathbf{s})vt)] = \delta(x - vt\mathbf{s})$$

we obtain the fundamental solution of the transport operator

$$\mathscr{E}_s(x, t) = v\theta(t)e^{-\alpha vt}\delta(x - vt\mathbf{s}) \tag{47}$$

To calculate the fundamental solution $\mathscr{E}_s^0(x)$ of the steady state transport operator

$$(\mathbf{s}, \operatorname{grad} \mathscr{E}_s^0) + \alpha \mathscr{E}_s^0 = \delta(x) \tag{48}$$

we shall use the method of descent involving the variable t (cf. Sec. 10.4). As a result, by virtue of (47), for all $\varphi \in \mathscr{D}(R^3)$ we obtain

$$(\mathscr{E}_s, \varphi(x)1(t)) = v \int_0^\infty e^{-\alpha vt}(\delta(x - vt\mathbf{s}), \varphi(x))\, dt$$

$$= v \int_0^\infty e^{-\alpha vt}\varphi(vt\mathbf{s})\, dt = \int_0^\infty e^{-\alpha u}\varphi(u\mathbf{s})\, du$$

$$= \left(\frac{e^{-\alpha|x|}}{|x|^2} \delta\!\left(\mathbf{s} - \frac{x}{|x|} \right), \varphi \right)$$

from which, by virtue of (21), it follows that

$$\mathscr{E}_s^0(x) = \frac{e^{-\alpha|x|}}{|x|^2} \delta\!\left(\mathbf{s} - \frac{x}{|x|} \right) \tag{49}$$

Specifically, from (49) we have

$$\frac{1}{4\pi} \int_{S_1} \mathscr{E}_s^0(x)\, ds = \frac{e^{-\alpha|x|}}{4\pi |x|^2} \tag{50}$$

12. Exercises. (a) Using formula (29), show that the generalized functions

$$\mathscr{E}_n(x, t) = \begin{cases} \dfrac{\theta(t)}{2\pi a}\left(\dfrac{d}{\pi a^2\, dt^2} \right)^{(n-3)/2} \delta(a^2 t^2 - |x|^2), & n \geq 3 \text{ odd} \\[2ex] \dfrac{(-1)^{(n/2)-1}}{2a} \pi^{-(n+1)/2} \Gamma\!\left(\dfrac{n-1}{2} \right) \dfrac{\theta(at - |x|)}{(a^2 t^2 - |x|^2)^{(n-1)/2}}, & n \geq 2 \text{ even} \end{cases} \tag{51}$$

are fundamental solutions of the wave operator $\square_a$ for $n \geq 2$.

(b) Prove that the fundamental solutions of the Klein–Gordon operator $\Box + m^2$ (cf. Sec. 2.8) are the generalized functions

$$D^r(x_0, x) = \frac{\theta(x_0)}{2\pi} \delta(x_0^2 - |x|^2)$$
$$- \frac{m}{4\pi} \theta(x_0 - |x|) \frac{J_1(m\sqrt{x_0^2 - |x|^2})}{\sqrt{x_0^2 - |x|^2}} \quad (52)$$

and $D^a(x_0, x) = D^r(-x_0, x)$; here J_1 is a Bessel function.

(c) Prove that the generalized functions

$$D^+(x_0, x) = \frac{1}{8\pi^3 i} F[\theta(\xi_0)\delta(\xi_0^2 - |\xi|^2 - m^2)]$$
$$D^-(x_0, x) = -\frac{1}{8\pi^3 i} F[\theta(-\xi_0)\delta(\xi_0^2 - |\xi|^2 - m^2)] \quad (53)$$

satisfy the Klein–Gordon equation and the result

$$D^+ + D^- = D^r - D^a$$

The generalized functions D^r, D^a, D^+, and D^- play a large part in quantum field theory [cf. N. N. Bogoliubov and D. V. Shirkov (*I*, Chap. II)].

(d) Using formula (40) of Sec. 2.8, show that the fourth-order matrix

$$\mathscr{E}(x_0, x) = -\left(i \sum_{k=0}^{3} \gamma^k \frac{\partial}{\partial x_k} + mI\right) D^r(x_0, x) \quad (54)$$

where D^r is defined in (52), is the fundamental solution of Dirac's operator (cf. Sec. 2.8),

$$\left(i \sum_{k=0}^{3} \gamma^k \frac{\partial}{\partial x_k} - mI\right) \mathscr{E} = \delta(x_0, x) I$$

(e) Using formula (37) of Sec. 9.8, show that the function

$$\mathscr{E}(x, t) = \frac{\theta(t)}{\hbar} \left(\frac{m}{2\pi \hbar t}\right)^{3/2} \exp\left[i\left(\frac{m}{2\hbar t} |x|^2 + \frac{3\pi}{4}\right)\right] \quad (55)$$

is the fundamental solution of Schrödinger's operator

$$i\hbar \frac{\partial}{\partial t} + \frac{\hbar^2}{2m} \Delta, \quad n = 3$$

§ 11. Retarded Potential

1. Properties of the Fundamental Solution of the Wave Operator. The fundamental solutions of the wave operator for $n = 1, 2,$ and 3 are the (generalized) functions [cf. formulas (32) and (30) of Sec. 10.7]:

$$\mathscr{E}_1(x, t) = \frac{1}{2a} \theta(at - |x|), \qquad \mathscr{E}_2(x, t) = \frac{\theta(at - |x|)}{2\pi a(a^2 t^2 - |x|^2)^{1/2}}$$

$$\mathscr{E}_3(x, t) = \frac{\theta(t)}{4\pi a^2 t} \delta_{S_{at}}(x) = \frac{\theta(t)}{2\pi a} \delta(a^2 t^2 - |x|^2)$$

The functions $\mathscr{E}_1$ and $\mathscr{E}_2$ are locally integrable, but the generalized function $\mathscr{E}_3$ acts on the test functions $\varphi \in \mathscr{D}(R^4)$ according to formula (31) of Sec. 10.7:

$$(\mathscr{E}_3, \varphi) = \frac{1}{4\pi a^2} \int_0^\infty \frac{1}{t} \int_{S_{at}} \varphi(x, t) \, dS \, dt$$

$$= \frac{1}{4\pi a^2} \int_{R^3} \frac{\varphi(x, |x|/a)}{|x|} \, dx \qquad (1)$$

The supports of the functions $\mathscr{E}_1$ and $\mathscr{E}_2$ coincide with the closure of the future light cone $\bar{\Gamma}^+$ (Fig. 17), but the support of the generalized function

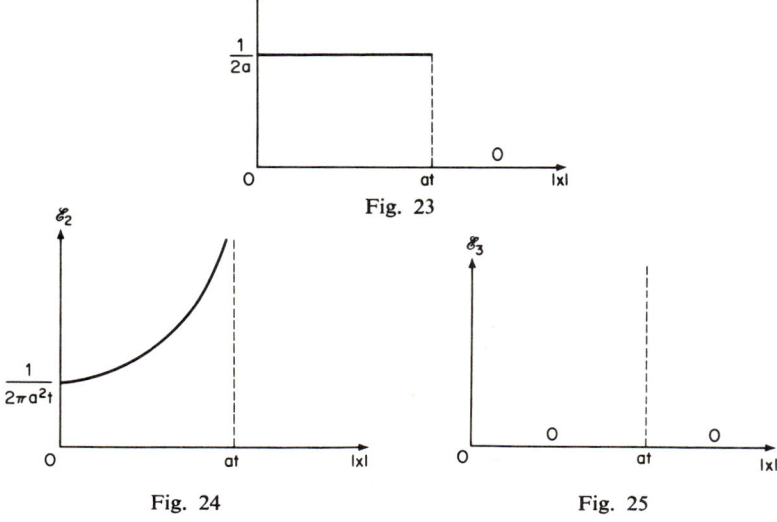

Fig. 23

Fig. 24

Fig. 25

$\mathscr{E}_3$ coincides with the boundary $[at = |x|]$ of this cone. Diagrams of the fundamental solutions $\mathscr{E}_1$, $\mathscr{E}_2$, and $\mathscr{E}_3$ at an instant of time t are shown schematically in Figs. 23–25.

Let $f(x, t) \in \mathscr{D}(R^{n+1})$ and $\varphi(x) \in \mathscr{D}(R^n)$.

Let us introduce the generalized function $(f(x, t), \varphi(x)) \in \mathscr{D}'(R^1)$, which obeys the formula

$$((f(x, t), \varphi(x)), \psi) = (f, \varphi\psi), \qquad \psi(t) \in \mathscr{D}(R^1) \qquad (2)$$

From this definition we obtain the following equation:

$$\left(\frac{\partial^k f(x, t)}{\partial t^k}, \varphi(x)\right) = \frac{d^k}{dt^k}(f(x, t), \varphi(x)), \qquad k = 1, 2, \ldots \qquad (3)$$

In fact, for all $\psi \in \mathscr{D}(R^1)$ we have

$$\left(\left(\frac{\partial^k f(x, t)}{\partial t^k}, \varphi(x)\right), \psi\right) = \left(\frac{\partial^k f}{\partial t^k}, \varphi\psi\right) = (-1)^k\left(f, \varphi\frac{d^k\psi}{dt^k}\right)$$

$$= (-1)^k\left((f(x, t), \varphi(x)), \frac{d^k\psi}{dt^k}\right)$$

$$= \left(\frac{d^k}{dt^k}(f(x, t), \varphi(x)), \psi\right)$$

from which Eqs. (3) follow.

We shall say that the generalized function $f(x, t)$ *belongs to the class* C^p, $0 \leq p \leq \infty$, *in the variable t in the interval* (a, b) *or* $[a, b]$ *if for any* $\varphi \in \mathscr{D}(R^n)$ *the generalized function* $(f(x, t), \varphi(x)) \in C^p(a, b)$ *or* $\in C^p([a, b])$ (cf. Sec. 5.4).

LEMMA. *The fundamental solutions* $\mathscr{E}_n(x, t)$, $n = 1, 2, 3$, *belong to the class* C^∞ *in the variable t in* $[0, \infty)$ *and satisfy the limiting results when* $t \to +0$

$$\mathscr{E}_n(x, t) \to 0, \qquad \frac{\partial \mathscr{E}_n(x, t)}{\partial t} \to \delta(x), \qquad \frac{\partial^2 \mathscr{E}_n(x, t)}{\partial t^2} \to 0 \quad \text{in} \quad \mathscr{D}'(R^n) \qquad (4)$$

Proof. Let $n = 3$ and $\varphi \in \mathscr{D}(R^3)$. It follows from (1) that

$$(\mathscr{E}_3(x, t), \varphi(x)) = \frac{\theta(t)}{4\pi a^2 t}\int_{S_{at}}\varphi(x)\,dS = \frac{\theta(t)t}{4\pi}\int_{S_1}\varphi(ats)\,ds \qquad (5)$$

Since the right-hand side of Eq. (5) is infinitely differentiable with respect

§ 11. RETARDED POTENTIAL

to t in $[0, \infty)$ (cf. Sec. 1.2), it follows that $\mathscr{E}_3$ belongs to the class C^∞ with respect to t in $[0, \infty)$. Moreover, it follows from (5) that

$$(\mathscr{E}_3(x, t), \varphi(x)) \to 0, \qquad t \to +0 \tag{6}$$

Further, using formula (3) for $f = \mathscr{E}_3$ and $k = 1, 2$, we shall obtain, when $t \to +0$,

$$\left(\frac{\partial \mathscr{E}_3(x, t)}{\partial t}, \varphi(x)\right) = \frac{d}{dt}\left[\frac{t}{4\pi}\int_{S_1}\varphi(ats)\,ds\right]$$

$$= \frac{1}{4\pi}\int_{S_1}\varphi(ats)\,ds + \frac{t}{4\pi}\frac{d}{dt}\int_{S_1}\varphi(ats)\,ds \to \varphi(0)$$

$$= (\delta, \varphi) \tag{7}$$

$$\left(\frac{\partial^2 \mathscr{E}_3(x, t)}{\partial t^2}, \varphi(x)\right) = \frac{d^2}{dt^2}\left[\frac{t}{4\pi}\int_{S_1}\varphi(ats)\,ds\right]$$

$$= \frac{1}{2\pi}\frac{d}{dt}\int_{S_1}\varphi(ats)\,ds + \frac{t}{4\pi}\frac{d^2}{dt^2}\int_{S_1}\varphi(ats)\,ds \to 0 \tag{8}$$

since the function

$$\int_{S_1}\varphi(ats)\,ds = \int_{S_1}\varphi(-ats)\,ds$$

is even and infinitely differentiable with respect to t, and so its first derivative when $t = 0$ is equal to zero. Since $\varphi \in \mathscr{D}(R^3)$ is arbitrary, the limiting results (6)–(8) are equivalent to results (4) when $n = 3$.

Now let $n = 2$ and 1, and $\varphi \in \mathscr{D}(R^n)$. Then when $t > 0$

$$(\mathscr{E}_2(x, t), \varphi(x)) = \frac{1}{2\pi a}\int_{U_{at}}\frac{\varphi(x)\,dx}{(a^2 t^2 - |x|^2)^{1/2}} = \frac{t}{2\pi}\int_{U_1}\frac{\varphi(at\eta)}{\sqrt{1 - |\eta|^2}}\,d\eta \tag{9}$$

$$(\mathscr{E}_1(x, t), \varphi(x)) = \frac{1}{2a}\int_{-at}^{at}\varphi(x)\,dx = \frac{t}{2}\int_{-1}^{1}\varphi(at\eta)\,d\eta \tag{10}$$

From this, as for the case $n = 3$, follow all the assertions of the lemma.

2. Additional Information about Convolutions. We shall give one more test when the convolution exists.

THEOREM. *Let the generalized functions f and g belonging to $\mathscr{D}'(R^{n+1})$ be such that $f(x, t) = 0$, $t < 0$, and supp $g \subset \bar{\Gamma}^+$. Then the convolution*

$f * g$ exists in $\mathscr{D}'(R^{n+1})$ and appears in the form

$$(f * g, \varphi) = (f(\xi, \tau') \cdot g(y, \tau), \eta(\tau')\eta(\tau)\eta(a^2\tau^2 - |y|^2)\varphi(\xi+y, \tau'+\tau))$$
$$\varphi \in \mathscr{D}(R^{n+1}) \tag{11}$$

where $\eta(\tau)$ is any function of the class $C^\infty(R^1)$ equal to 0 when $t < -\delta$ and to 1 when $t > -\varepsilon$ (δ and ε are any numbers, $\delta > \varepsilon > 0$). Then the convolution $f * g$ becomes zero when $t < 0$ and is continuous with respect to f and g separately: (1) if $f_k \to f$ as $k \to \infty$ in $\mathscr{D}'(R^{n+1})$, $f_k = 0$ for $t < 0$, then $f_k * g \to f * g$ as $k \to \infty$ in $\mathscr{D}'(R^{n+1})$; (2) if $g_k \to g$ as $k \to \infty$ in $\mathscr{D}'(R^{n+1})$, supp $g_k \subset \bar{\Gamma}^+$, then $f * g_k \to f * g$ as $k \to \infty$ in $\mathscr{D}'(R^{n+1})$.

Proof. Let $\varphi(x, t)$ be an arbitrary function belonging to $\mathscr{D}'(R^{n+1})$, let supp $\varphi \subset U_A$ and let $\eta_k(\xi, \tau'; y, \tau)$ for $k = 1, 2, \ldots$, be a sequence of functions belonging to $\mathscr{D}(R^{2n+2})$ and converging to 1 in R^{2n+2} (cf. Sec. 7.4). Then whenever k is sufficiently large

$$\psi_k \equiv \eta(\tau')\eta(\tau)\eta(a^2\tau^2 - |y|^2)\eta_k(\xi, \tau'; y, \tau)\varphi(\xi + y, \tau' + \tau)$$
$$= \eta(\tau')\eta(\tau)\eta(a^2\tau^2 - |y|^2)\varphi(\xi + y, \tau' + \tau) \equiv \psi \tag{12}$$

To prove Eq. (12) it is sufficient to establish that the function $\psi \in \mathscr{D}(R^{2n+2})$. But this follows from the fact that it is infinitely differentiable, while the set

$$[(\xi, \tau', y, \tau) : \tau' \geq -\delta, \tau \geq -\delta, a^2\tau^2 - |y|^2 \geq -\delta,$$
$$|y + \xi|^2 + (\tau' + \tau)^2 \leq A^2]$$

in which its support is contained, is bounded, since this set is contained

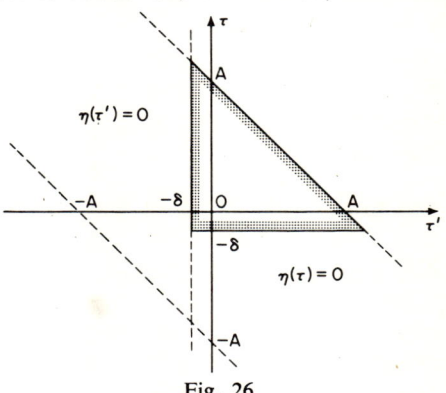

Fig. 26

§ 11. RETARDED POTENTIAL

in the bounded set

$$[-\delta \leq \tau' \leq A + \delta, -\delta \leq \tau \leq A + \delta, |y| \leq \sqrt{a^2(A+\delta)^2 + \delta},$$
$$|\xi| \leq \sqrt{a^2(A+\delta)^2 + \delta} + A] \quad \text{(Fig. 26)}$$

Moreover, by construction, $\eta(\tau') = 1$ in the neighborhood of the support of $f(\xi, \tau')$ and $\eta(\tau)\eta(a^2\tau^2 - |y|^2) = 1$ in the neighborhood of the support of $g(y, \tau)$. Consequently [cf. (13) of Sec. 5.9],

$$f(\xi, \tau') = \eta(\tau')f(\xi, \tau'), \qquad g(y, \tau) = \eta(\tau)\eta(a^2\tau^2 - |y|^2)g(y, \tau).$$

From these equations and Eq. (12), we see that formula (11) is valid:

$$(f * g, \varphi) = \lim_{k \to \infty} (f(\xi, \tau') \cdot g(y, \tau), \eta_k(\xi, \tau'; y, \tau)\varphi(\xi + y, \tau' + \tau))$$

$$= \lim_{k \to \infty} (f(\xi, \tau') \cdot g(y, \tau), \psi_k) = (f(\xi, \tau') \cdot g(y, \tau), \psi)$$

$$\varphi \in \mathscr{D}(R^{n+1})$$

We shall prove that the right-hand side of Eq. (11) defines a continuous functional over $\mathscr{D}(R^{n+1})$. In fact, let $\varphi_k \to \varphi$ as $k \to \infty$ in $\mathscr{D}(R^{n+1})$. Then

$$\psi_k \equiv \eta(\tau')\eta(\tau)\eta(a^2\tau^2 - |y|^2)\varphi_k(\xi + y, \tau' + \tau) \to \psi$$

as $k \to \infty$ in $\mathscr{D}(R^{2n+2})$, and so

$$(f * g, \varphi_k) = (f(\xi, \tau') \cdot g(y, \tau), \psi_k)$$
$$\to (f(\xi, \tau') \cdot g(y, \tau), \psi) = (f * g, \varphi), \quad k \to \infty$$

as was to be shown. In this way, $f * g \in \mathscr{D}'(R^{n+1})$.

We shall prove that $f * g = 0$ for $t < 0$. Let $\varphi(x, t) \in \mathscr{D}(R^{n+1})$ and $\operatorname{supp} \varphi \subset [t < 0]$. Since the support of φ is a compactum in R^{n+1}, then there will be a number $\delta_1 > 0$ such that $\operatorname{supp} \varphi \subset [t < -\delta_1]$. But then, choosing $\delta < \delta_1/2$, we shall obtain

$$\eta(\tau')\eta(\tau)\eta(a^2\tau^2 - |y|^2)\varphi(\xi + y, \tau' + \tau) = 0$$

from which, by virtue of (11), $(f * g, \varphi) = 0$, as was to be shown.

The continuity of the convolution $f * g$ with respect to f and g follows from Eq. (11) and from the continuity of the direct product $f(\xi, \tau') \cdot g(y, \tau)$ with respect to f and g [cf. Sec. 7.3, (a)]. For this an auxiliary function η may be chosen independent of k. The theorem is proved.

For $n = 0$ the theorem which has just been proved assumes the following form: *If the generalized functions $f(t)$ and $g(t)$ become zero when $t < 0$, then their convolution $f * g$ exists in $\mathscr{D}'(R^1)$, becomes zero when $t < 0$, and is expressed by the formula*

$$(f * g, \varphi) = (f(\tau') \cdot g(\tau), \quad \eta(\tau')\eta(\tau)\varphi(\tau' + \tau)), \qquad \varphi \in \mathscr{D}(R^1) \qquad (13)$$

We shall prove the following: *If $g(x, t) \in \mathscr{D}'(R^{n+1})$, supp $g \subset \bar{\Gamma}^+$, and $u(x) \in \mathscr{D}'(R^n)$, then*

$$g * u(x) \cdot \delta(t) = g(x, t) * u(x) \qquad (14)$$

and the generalized function $g(x, t) * u(x)$ acts according to the rule

$$(g(x, t) * u(x), \varphi)$$
$$= (g(y, t) \cdot u(\xi), \eta(a^2 t^2 - |y|^2)\varphi(y + \xi, t)), \varphi \in \mathscr{D}(R^{n+1}) \qquad (15)$$

In fact, supposing in formula (11) that $f = u(x) \cdot \delta(t)$, then for all $\varphi \in \mathscr{D}(R^{n+1})$ we shall obtain

$$(g * u(x) \cdot \delta(t), \varphi)$$
$$= (g(y, \tau) \cdot u(\xi) \cdot \delta(\tau'), \eta(\tau)\eta(\tau')\eta(a^2\tau^2 - |y|^2)\varphi(y+\xi, \tau+\tau'))$$
$$= (g(y, \tau) \cdot u(\xi), \eta(\tau)\eta(a^2\tau^2 - |y|^2)(\delta(\tau'), \eta(\tau')\varphi(y+\xi, \tau+\tau')))$$
$$= (g(y, \tau) \cdot u(\xi), \eta(\tau)\eta(a^2\tau^2 - |y|^2)\varphi(y + \xi, \tau)) \qquad (16)$$

Since the support of $g(y, \tau)$ is contained in the half-space $\tau \geq 0$, then, by virtue of Eq. (13) of Sec. 5.9, $g = \eta(\tau)g$. Moreover the function

$$\eta(a^2\tau^2 - |y|^2)\varphi(y + \xi, \tau) \in \mathscr{D}(R^{2n+1})$$

Therefore, continuing Eqs. (16) and taking (15) into account, we shall obtain Eq. (14):

$$(g * u(x) \cdot \delta(t), \varphi) = (\eta(\tau)g(y, \tau) \cdot u(\xi), \eta(a^2\tau^2 - |y|^2)\varphi(y + \xi, \tau))$$
$$= (g(y, \tau) \cdot u(\xi), \eta(a^2\tau^2 - |y|^2)\varphi(y + \xi, \tau))$$
$$= (g(x, t) * u(x), \varphi)$$

Now by means of formula (14), which we have just obtained, and the

We shall prove formula (20). Let $\varphi \in \mathscr{D}(R^4)$. Since f is a function locally integrable in R_x^4, we obtain from Eq. (18)

$$(V_3, \varphi) = (\mathscr{E}_3(y, \tau), \eta(\tau)\eta(a^2\tau^2 - |y|^2) \int f(\xi, \tau')\eta(\tau')\varphi(y+\xi, \tau+\tau')\,d\xi\,d\tau')$$
$$= (\mathscr{E}_3(y, \tau), \eta(\tau)\eta(a^2\tau^2 - |y|^2) \int f(x - y, t - \tau)\varphi(x, t)\,dx\,dt)$$

From this, using formula (1) and taking into account that $f = 0$ for $t < 0$, we deduce

$$(V_3, \varphi) = \frac{1}{4\pi a^2} \int_{R^3} \frac{\eta(|y|/a)\eta(0)}{|y|} \left[\int f\left(x - y, t - \frac{|y|}{a}\right)\varphi(x, t)\,dx\,dt \right] dy$$
$$= \frac{1}{4\pi a^2} \int \varphi(x, t) \int_{U_{at}} \frac{f(x - y, t - |y|/a)}{|y|}\,dy\,dx\,dt, \quad \varphi \in \mathscr{D}(R^4)$$

This means that the potential V_3 is a function locally integrable in R^4 and appears in the form

$$V_3(x, t) = \frac{1}{4\pi a^2} \int_{U_{at}} \frac{f(x - y, t - |y|/a)}{|y|}\,dy \qquad (21)$$

When we have completed the change of variables $x - y = \xi$ in this integral, we obtain Eq. (20).

Analogously, with corresponding simplifications, Eqs. (20') and (20'') are deduced for the potentials V_2 and V_1.

We shall associate with each point (x, t), $t > 0$, the open cone

$$\Gamma_0^-(x, t) = \Gamma^-(x, t) \cap [0 < \tau < t]$$

with vertex (x, t), base $U(x; at)$, and side surface $B(x, t)$ (Fig. 27); here $\Gamma^-(x, t)$ is a past light cone (cf. Sec. 3.3).

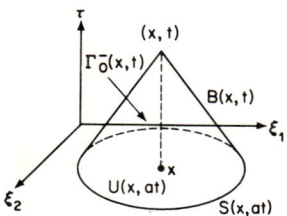

Fig. 27

§ 11. RETARDED POTENTIAL

THEOREM. *If $f \in C^2(t \geq 0)$ when $n = 3$ and 2, $f \in C^1(t \geq 0)$ for $n = 1$, then the potential $V_n \in C^2(t \geq 0)$ satisfies the estimate*

$$|V_3(x, t)| \leq \frac{t^2}{2} \max_{B(x,t)} |f(\xi, \tau)| \tag{22}$$

$$|V_n(x, t)| \leq \frac{t^2}{2} \max_{\bar{\Gamma}_0^-(x,t)} |f(\xi, \tau)|, \qquad n = 1, 2$$

and the initial conditions

$$V_n\bigg|_{t=0} = 0, \qquad \frac{\partial V_n}{\partial t}\bigg|_{t=0} = 0 \tag{23}$$

Proof. We shall prove the theorem for $n = 3$. The change of variables $y = at\eta$, $t > 0$, changes Eq. (21) to the form

$$V_3(x, t) = \frac{t^2}{4\pi} \int_{U_1} \frac{f(x + at\eta, t(1 - |\eta|))}{|\eta|} d\eta \tag{24}$$

Since $f \in C^2(t \geq 0)$ and the integrand in (24) has an integrable singularity, then $V_3 \in C^2(t \geq 0)$. From Eq. (24) there follows also the estimate (22) for the potential V_3:

$$|V_3(x, t)| \leq \frac{t^2}{4\pi} \max_{B(x,t)} |f(\xi, \tau)| \int_{U_1} \frac{d\eta}{|\eta|} = \frac{t^2}{2} \max_{B(x,t)} |f(\xi, \tau)|$$

Since $V_3 \in C^2(t \geq 0)$, then the initial conditions (23) follow from (22).

Now let $n = 2$. The change of variables $\xi = x + at\eta$, $\tau = t - \alpha t$ for $t > 0$, transforms Eq. (20') for the potential V_2 into the form

$$V_2(x, t) = \frac{t^2}{2\pi} \int_0^1 \int_{U_\alpha} \frac{f(x + at\eta; t - \alpha t)}{\sqrt{\alpha^2 - |\eta|^2}} d\eta \, d\alpha \tag{24'}$$

from which the required properties of this potential directly follow.

The properties of the potential V_1 follow from Eq. (20''). The theorem is proved.

Note. It is clear from formula (20) that the potential V_3 at the point x at an instant of time $t > 0$ is defined by the influence of the source $f(\xi, \tau)$ over the side surface $B(x, t) = [(\xi, \tau), |\xi - x| = a(t - \tau), 0 \leq \tau \leq t]$ of the cone $\Gamma_0^-(x, t)$. In other words, the retarded potential $V_3(x, t)$ is completely defined by the values of the source $f(\xi, \tau)$ in the sphere $\bar{u}(x; at)$, taken at early instants of time $\tau = t - |x - \xi|/a$; moreover, the time of retardation $(1/a)|x - \xi|$ is the time needed for a perturbation from the point ξ to reach the point x. On the other hand, it follows from Eqs. (20') and (20'') that the values of the potentials V_2 and V_1 at the point (x, t) for $t > 0$ are defined by the values of the source $f(\xi, \tau)$ in the closed cone $\bar{\Gamma}_0^-(x, t)$ itself. It is these differences of structure in retarded potentials which allow us to define the difference in the propagation of perturbations in a space, over a plane, and on a straight line (cf. Sec. 13).

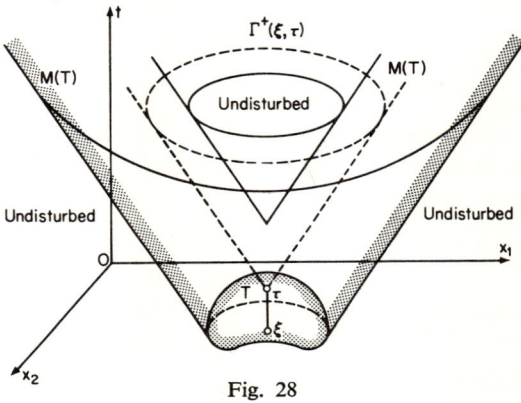

Fig. 28

Let the source f be concentrated over the closed set $T \subset R^{n+1}$. By virtue of what has just been said, when $n = 3$ the perturbation from T is spread over the union of the boundaries $a(t - \tau) = |x - \xi|$ of the future light cones $\Gamma^+(\xi, \tau)$ when their vertexes (ξ, τ) pass through the set T (Fig. 28);

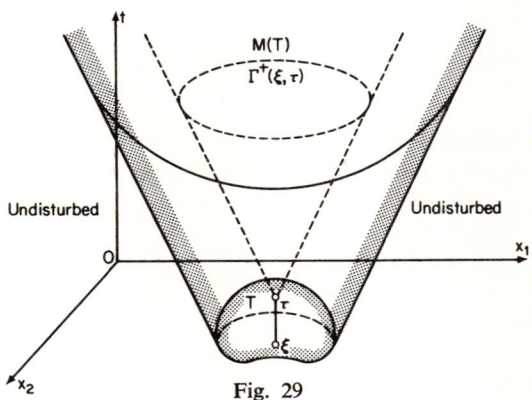

Fig. 29

when $n = 2$ and 1 this perturbation is spread over the union of the closed cones themselves $\bar{\Gamma}^+(\xi, \tau)$, $(\xi, \tau) \in T$ (Fig. 29). The set thus obtained, $M(T)$, is said to be the *region of influence* of the set T. It is clear that outside $M(T)$ there will be no disturbance.

4. Surface Retarded Potentials. If $f = u_1(x) \cdot \delta(t)$ or $f = u_0(x) \cdot \delta'(t)$, where u_0 and u_1 are arbitrary generalized functions belonging to $\mathscr{D}'(R^n)$, then their corresponding retarded potentials

$$V_n^{(0)} = \mathscr{E}_n * u_1(x) \cdot \delta(t), \qquad V_n^{(1)} = \mathscr{E}_n * u_0(x) \cdot \delta'(t), \qquad n = 1, 2, 3$$

are said to be *surface retarded potentials* (of simple and double layers with densities u_1 and u_0, respectively).

By virtue of Eqs. (14) and (17) of Sec. 11.2, the retarded potentials $V_n^{(0)}$ and $V_n^{(1)}$ appear in the form

$$V_n^{(0)} = \mathscr{E}_n(x, t) * u_1(x) \tag{25}$$

$$V_n^{(1)} = \frac{\partial \mathscr{E}_n(x, t)}{\partial t} * u_0(x) = \frac{\partial}{\partial t} [\mathscr{E}_n(x, t) * u_0(x)] \tag{26}$$

and the generalized function $\mathscr{E}_n(x, t) * u(x)$ acts according to Eq. (15).

LEMMA. *The surface retarded potentials $V_n^{(0)}$ and $V_n^{(1)}$ belong to the class C^∞ with respect to the variable t in $[0, \infty)$ and satisfy the initial conditions when $t \to +0$*

$$V_n^{(0)}(x, t) \to 0, \qquad \frac{\partial V_n^{(0)}(x, t)}{\partial t} \to u_1(x) \quad \text{in} \quad \mathscr{D}'(R^n) \tag{27}$$

$$V_n^{(1)}(x, t) \to u_0(x), \qquad \frac{\partial V_n^{(1)}(x, t)}{\partial t} \to 0 \quad \text{in} \quad \mathscr{D}'(R^n) \tag{28}$$

Proof. According to the lemma of Sec. 11.1, the generalized function $\mathscr{E}_n(x, t)$ belongs to the class C^∞ with respect to the variable t in $[0, \infty)$. Further, for each $t > 0$ the support of $\mathscr{E}_n(x, t)$ is contained in the sphere $\bar{U}_{at}$ and so is uniformly bounded in R^n for $t \to t_0 \geq 0$. Therefore, using the theorem of Sec. 7.5 concerning the continuity of a convolution in $\mathscr{D}'$, we conclude that for all $\varphi \in \mathscr{D}(R^n)$

$$\left(\frac{\partial^k \mathscr{E}_n(x, t)}{\partial t^k} * u_1(x), \varphi(x) \right) \in C[0, \infty), \qquad k = 0, 1, \ldots$$

From this, by virtue of Eqs. (3) and (17),

$$\frac{\partial^k}{\partial t^k} (\mathscr{E}_n(x, t) * u_1(x), \varphi) = \left(\frac{\partial^k}{\partial t^k} [\mathscr{E}_n(x, t) * u_1(x)], \varphi\right)$$

$$= \left(\frac{\partial^k \mathscr{E}_n(x, t)}{\partial t^k} * u_1(x), \varphi\right)$$

we deduce that $(\mathscr{E}_n(x, t) * u_1(x), \varphi) \in C^\infty[0, \infty)$. This means, by virtue of (25), that the potential $V_n^{(0)}(x, t)$ belongs to the class C^∞ with respect to t in $[0, \infty)$. Replacing u_1 by u_0, we deduce from (26) that the potential $V_n^{(1)}$ has the same property.

We shall prove the limiting results (27). Taking limiting results (4) into account and using the continuity of the convolution $\mathscr{E}_n(x, t) * u_1(x)$ in $\mathscr{D}'(R^n)$, we obtain when $t \to +0$

$$V_n^{(0)}(x, t) = \mathscr{E}_n(x, t) * u_1(x) \to 0 * u_1(x) = 0 \quad \text{in} \quad \mathscr{D}'(R^n)$$

$$\frac{\partial V_n^{(0)}(x, t)}{\partial t} = \frac{\partial}{\partial t} [\mathscr{E}_n(x, t) * u_1(x)] = \frac{\partial \mathscr{E}_n(x, t)}{\partial t} * u_1(x)$$

$$\to \delta * u_1 = u_1(x) \quad \text{in} \quad \mathscr{D}'(R^n)$$

Limiting results (28) are established analogously. The lemma is proved.

The other properties of the surface retarded potentials $V_n^{(0)}$ and $V_n^{(1)}$ essentially depend on the densities u_1 and u_0.

If u_1 is a function locally integrable in R^n, then the surface potential $V_n^{(0)}$ is a function locally integrable in R^{n+1} and is expressed by the formulas:

$$V_3^{(0)}(x, t) = \frac{\theta(t)}{4\pi a^2 t} \int_{S(x;at)} u_1(\xi)\, dS \tag{29}$$

$$V_2^{(0)}(x, t) = \frac{\theta(t)}{2\pi a} \int_{U(x;at)} \frac{u_1(\xi)\, d\xi}{(a^2 t^2 - |x - \xi|^2)^{1/2}} \tag{29'}$$

$$V_1^{(0)}(x, t) = \frac{\theta(t)}{2a} \int_{x-at}^{x+at} u_1(\xi)\, d\xi \tag{29''}$$

We shall establish formula (29). Since the function u_1 is locally integrable

§ 11. RETARDED POTENTIAL

in R^3, then, using formulas (25), (15), and (1), for all $\varphi \in \mathscr{D}(R^4)$, we obtain

$$(V_3^{(0)}, \varphi) = (\mathscr{E}_3(x, t) * u_1(x), \varphi)$$
$$= (\mathscr{E}_3(y, t), \eta(a^2 t^2 - |y|^2) \int u_1(\xi) \varphi(y + \xi, t) \, d\xi)$$
$$= \frac{1}{4\pi a^2} \int_0^\infty \frac{\eta(0)}{t} \int_{S_{at}} \int_{R^3} u_1(x - y) \varphi(x, t) \, dx \, dS_y \, dt$$
$$= \frac{1}{4\pi a^2} \int_0^\infty \int_{R^3} \frac{\varphi(x, t)}{t} \int_{S_{at}} u_1(x - y) \, dS_y \, dx \, dt$$

from which it follows that $V_3^{(0)}$ is locally integrable in R^4 and appears in the form [cf. formula (26) of Sec. 7.8]

$$V_3^{(0)}(x, t) = \frac{\theta(t)}{4\pi a^2 t} \int_{S_{at}} u_1(x - y) \, dS_y$$

Making the change of variables $x - y = \xi$ in this integral, we shall obtain formula (29). The Eqs. (29') and (29'') for the potentials $V_2^{(0)}$ and $V_1^{(0)}$ are deduced analogously, with corresponding simplifications.

THEOREM. *If $u_0 \in C^3(R^n)$, $u_1 \in C^2(R^n)$ for $n = 3$ and 2; $u_0 \in C^2(R^1)$, $u_1 \in C^1(R^1)$ for $n = 1$, then the potentials $V_n^{(0)}$ and $V_n^{(1)}$ belong to the class $C^2(t \geq 0)$ and satisfy the estimates:*

$$|V_3^{(0)}(x, t)| \leq t \max_{S(x; at)} |u_1(\xi)| \tag{30}$$

$$|V_n^{(0)}(x, t)| \leq t \max_{\bar{U}(x; at)} |u_1(\xi)|, \quad n = 1, 2 \tag{30'}$$

$$|V_3^{(1)}(x, t)| \leq \max_{S(x; at)} |u_0(\xi)| + at \max_{S(x; at)} |\operatorname{grad} u_0(\xi)| \tag{31}$$

$$|V_2^{(1)}(x, t)| \leq \max_{\bar{U}(x; at)} |u_0(\xi)| + at \max_{\bar{U}(x; at)} |\operatorname{grad} u_0(\xi)| \tag{31'}$$

$$|V_1^{(1)}(x, t)| \leq \max_{S(x; at)} |u_0(\xi)| \tag{31''}$$

and the initial conditions:

$$V_n^{(0)} \bigg|_{t=+0} = 0, \quad \frac{\partial V_n^{(0)}}{\partial t} \bigg|_{t=+0} = u_1(x) \tag{32}$$

$$V_n^{(1)} \bigg|_{t=+0} = u_0(x), \quad \frac{\partial V_n^{(1)}}{\partial t} \bigg|_{t=+0} = 0 \tag{33}$$

Proof. Let $n = 3$. When we have completed the change of variables $x - \xi = ats$ for $t > 0$ in formula (29), we obtain the equation

$$V_3^{(0)}(x, t) = \frac{\theta(t)t}{4\pi} \int_{S_1} u_1(x - ats)\, ds \tag{34}$$

from which it follows that $V_3^{(0)} \in C^2(t \geq 0)$, if $u_1 \in C^2(R^3)$, and satisfies (30). Differentiating formula (34) with respect to t and using (26), we obtain the representation for the potential $V_3^{(1)}$:

$$V_3^{(1)}(x, t) = \frac{\theta(t)}{4\pi} \int_{S_1} u_0(x - ats)\, ds - \frac{at\theta(t)}{4\pi} \int_{S_1} \frac{\partial u_0(x - ats)}{\partial s}\, ds$$

from which it follows that $V_3^{(1)} \in C^2(t \geq 0)$, if $u_0 \in C^3(R^3)$, and satisfies (31):

$$|V_3^{(1)}(x, t)| \leq \max_{|s|=1} |u_0(x - ats)| + at \max_{|s|=1} \left|\frac{\partial u_0(x - ats)}{\partial s}\right|$$

$$\leq \max_{S(x;at)} |u_0(\xi)| + at \max_{S(x;at)} |\operatorname{grad} u_0(x)|$$

When $n = 2$ the change of variables $\xi = x - at\eta$ for $t > 0$ transforms Eq. (29') for the potential $V_2^{(0)}$ into the form

$$V_2^{(0)}(x, t) = \frac{\theta(t)t}{2\pi} \int_{U_1} \frac{u_1(x - at\eta)}{\sqrt{1 - |\eta|^2}}\, d\eta$$

from which, and from (26), the required properties of smoothness and estimates (30') and (31') for the potentials $V_2^{(0)}$ and $V_2^{(1)}$ follow.

The corresponding properties of the potentials $V_1^{(0)}$ and $V_1^{(1)}$ follow from Eqs. (29'') and (26).

We shall now prove that the initial conditions (32) and (33) are valid. By virtue of (27) and (28), these conditions are satisfied in the sense of convergence in the space $\mathscr{D}'(R^n)$. But by what has just been proved, the functions $V_n^{(0)}(x, t)$ and $V_n^{(1)}(x, t)$ belong to the class $C^2(t \geq 0)$. Consequently, these functions satisfy conditions (32) and (33) in the usual sense. The theorem is proved.

Note. The equations (29), (29'), and (29'') follow formally from Eqs. (20), (20'), and (20''), if in them we set $f(\xi, \tau) = u_1(\xi) \cdot \delta(\tau)$ and "integrate" $\delta(\tau)$.

§ 12. The Cauchy Problem for the Wave Equation

1. The Cauchy Problem for the Ordinary Linear Differential Equation.
We shall first apply the theory of generalized functions to the solution of the Cauchy problem for the ordinary linear differential equation.

Since we propose only to outline this method, we shall consider only the simplest differential equations with constant coefficients.

Let us consider the Cauchy problem

$$u'' + a^2 u = f(t), \qquad u\big|_{t=+0} = u_0, \qquad u'\big|_{t=+0} = u_1 \tag{1}$$

where $f \in C(t \geq 0)$. We shall continue the solution $u(t)$ of this problem and the function $f(t)$ as zero when $t < 0$; we shall denote the continued functions by $\tilde{u}$ and $\tilde{f}$, respectively. Then [cf. Sec. 6.3, (a)] $\tilde{u}' = \{\tilde{u}'\} + u_0 \delta(t)$, $\tilde{u}'' = \{\tilde{u}''\} + u_0 \delta'(t) + u_1 \delta(t)$, and so the function $\tilde{u}$ satisfies in R^1 the equation

$$\tilde{u}'' + a^2 \tilde{u} = \tilde{f}(t) + u_0 \delta'(t) + u_1 \delta(t) \tag{2}$$

Let us construct a solution of Eq. (2). Since the fundamental solution $\mathscr{E}(t) = \theta(t)(\sin at)/a$ of the operator $u'' + a^2 u$ (cf. Sec. 10.5) becomes zero when $t < 0$, then its convolution with the right-hand side of Eq. (2) exists in $\mathscr{D}'(R^1)$ and becomes zero for $t < 0$ (cf. Sec. 11.2). So, by the theorem of Sec. 10.3, the solution of Eq. (2) exists and is unique in the class of generalized functions belonging to $\mathscr{D}'(R^1)$ which become zero when $t < 0$, and this solution is expressed by the convolution

$$\tilde{u} = \mathscr{E} * (\tilde{f} + u_0 \delta' + u_1 \delta) = \mathscr{E} * \tilde{f} + u_0 \mathscr{E}' + u_1 \mathscr{E} \tag{3}$$

Taking Eq. (15) of Sec. 7.4 into account, we write solution (3) in the form

$$\tilde{u}(t) = \frac{1}{a} \int_0^t f(\tau) \sin a(t-\tau) \, d\tau + u_0 \mathscr{E}'(t) + u_1 \mathscr{E}(t) \tag{4}$$

Since the solution of the Cauchy problem (1), continued as zero when $t < 0$, satisfies (2), and since such a solution for this equation is unique, then Eq. (4) when $t > 0$ in fact gives the solution of the Cauchy problem (1):

$$u(t) = \frac{1}{a} \int_0^t f(\tau) \sin a(t-\tau) \, d\tau + u_0 \cos at + u_1 \frac{\sin at}{a} \tag{5}$$

The Cauchy problem

$$u' + au = f(t), \quad u|_{t=+0} = u_0 \qquad (1')$$

is solved analogously. The corresponding equation (2) assumes the form

$$\tilde{u}' + a\tilde{u} = \tilde{f}(t) + u_0\delta(t) \qquad (2')$$

The solution of this equation is unique and is given by the formula

$$\tilde{u} = \mathscr{E} * (\tilde{f} + u_0\delta) = \mathscr{E} * \tilde{f} + u_0\mathscr{E} \qquad (3')$$

where $\mathscr{E}(t) = \theta(t)e^{-at}$ is the fundamental solution of the operator $u' + au$ (cf. Sec. 10.5). In this way, the solution of the Cauchy problem (1') is given by the formula

$$u(t) = \int_0^t f(\tau)e^{-a(t-\tau)}\,d\tau + u_0 e^{-at} \qquad (5')$$

analogous to formula (5).

2. Formulation of the Generalized Cauchy Problem for the Wave Equation. The scheme for solving the Cauchy problem which has been set out in the previous section for an ordinary linear differential equation of second order with constant coefficients is used to solve the Cauchy problem for the wave equation

$$\Box_a u = f(x, t) \qquad (6)$$

$$u|_{t=+0} = u_0(x), \quad \frac{\partial u}{\partial t}\bigg|_{t=+0} = u_1(x) \qquad (7)$$

We shall consider that $f \in C(t \geq 0)$, $u_0 \in C^1(R^n)$, and $u_1 \in C(R^n)$.

Let us suppose that there is a classical solution $u(x, t)$ of the Cauchy problem (6) and (7). This means that the function u of the class $C^2(t > 0) \cap C^1(t \geq 0)$ satisfies Eq. (6) for $t > 0$ and satisfies initial conditions (7) as $t \to +0$ (cf. Sec. 4.2).

We shall continue the functions u and f as zero for $t < 0$, supposing

$$\tilde{u} = \begin{cases} u, & t \geq 0 \\ 0, & t < 0, \end{cases} \quad \tilde{f} = \begin{cases} f, & t \geq 0 \\ 0, & t < 0 \end{cases}$$

We show that the function $\tilde{u}(x, t)$ satisfies in R^{n+1} the wave equation

$$\Box_a \tilde{u} = \tilde{f}(x, t) + u_0(x) \cdot \delta'(t) + u_1(x) \cdot \delta(t) \qquad (8)$$

§ 12. THE CAUCHY PROBLEM FOR THE WAVE EQUATION

In fact, for all $\varphi \in \mathscr{D}(R^{n+1})$ we have the sequence of equalities

$$(\Box_a \tilde{u}, \varphi) = (\tilde{u}, \Box_a \varphi) = \int_0^\infty \int_{R^n} u \Box_a \varphi \, dx \, dt$$

$$= \lim_{\varepsilon \to +0} \int_\varepsilon^\infty \int_{R^n} u \left(\frac{\partial^2 \varphi}{\partial t^2} - a^2 \Delta \varphi \right) dx \, dt$$

$$= \lim_{\varepsilon \to +0} \left[\int_\varepsilon^\infty \int_{R^n} \left(\frac{\partial^2 u}{\partial t^2} - a^2 \Delta u \right) \varphi \, dx \, dt \right.$$

$$\left. - \int_{R^n} \frac{\partial \varphi(x, \varepsilon)}{\partial t} u(x, \varepsilon) \, dx + \int_{R^n} \varphi(x, \varepsilon) \frac{\partial u(x, \varepsilon)}{\partial t} \, dx \right]$$

$$= \int_0^\infty \int_{R^n} f \varphi \, dx \, dt - \int_{R^n} \frac{\partial \varphi(x, 0)}{\partial t} u(x, 0) \, dx$$

$$+ \int_{R^n} \varphi(x, 0) \frac{\partial u(x, 0)}{\partial t} \, dx = \int_{R^{n+1}} \tilde{f} \, dx \, dt$$

$$- \int_{R^n} u_0(x) \frac{\partial \varphi(x, 0)}{\partial t} \, dx + \int_{R^n} u_1(x) \varphi(x, 0) \, dx$$

$$= (\tilde{f} + u_0(x) \cdot \delta'(t) + u_1(x) \cdot \delta(t), \varphi)$$

from which Eq. (8) follows.

As Eq. (8) shows, initial perturbations u_0 and u_1 for the function $\tilde{u}(x, t)$ play the role of the source $u_0(x) \cdot \delta'(t) + u_1(x) \cdot \delta(t)$ acting momentarily when $t = 0$. [In this case the double layer $u_0(x) \cdot \delta'(t)$ corresponds to the initial perturbation u_0, and the simple layer $u_1(x) \cdot \delta(t)$ corresponds to the initial perturbation u_1 over the plane $t = 0$.] Further, the classical solutions of the Cauchy problem (6) and (7) are contained among those solutions of Eq. (8) which become zero when $t < 0$. We may therefore say that the problem of finding (generalized) solutions of Eq. (8) which become zero when $t < 0$ is the generalized Cauchy problem for the wave equation. But in this case, in Eq. (8) $\tilde{f}$, u_0, and u_1 may be considered as generalized functions.

Let us therefore introduce the following definition. We shall say that the problem of finding the generalized function $u \in \mathscr{D}'(R^{n+1})$ which becomes zero when $t < 0$ and which satisfies the wave equation

$$\Box_a u = f(x, t) + u_0(x) \cdot \delta'(t) + u_1(x) \cdot \delta(t) \tag{9}$$

is the *generalized Cauchy problem* for the wave equation with the source $f \in \mathscr{D}'(R^{n+1})$ and the initial perturbations $u_0 \in \mathscr{D}'(R^n)$ and $u_1 \in \mathscr{D}'(R^n)$.

Equation (9) is equivalent to the following (cf. Sec. 10.1): For any $\varphi \in \mathscr{D}(R^{n+1})$ the equation

$$(u, \Box_a \varphi) = (f, \varphi) - \left(u_0, \frac{\partial \varphi(x, 0)}{\partial t}\right) + (u_1, \varphi(x, 0)) \tag{9'}$$

is valid.

It follows from Eq. (9) that f must become zero for $t < 0$ if the generalized Cauchy problem is to be solved. It will now be shown that this condition is sufficient as well as necessary.

3. Solution of the Generalized Cauchy Problem

THEOREM. *Let $f \in \mathscr{D}'(R^{n+1})$, $u_0 \in \mathscr{D}'(R^n)$, and $u_1 \in \mathscr{D}'(R^n)$, moreover $f = 0$ when $t < 0$. Then the solution of the corresponding generalized Cauchy problem exists, is unique, and appears in the form of a sum of three retarded potentials*

$$u = V_n + V_n^{(0)} + V_n^{(1)} \tag{10}$$

where

$$V_n = \mathscr{E}_n * f, \qquad V_n^{(0)} = \mathscr{E}_n(x, t) * u_1(x), \qquad V_n^{(1)} = \frac{\partial \mathscr{E}_n(x, t)}{\partial t} * u_0(x)$$

This solution depends continuously on f, u_0, and u_1 in $\mathscr{D}'$.

Proof. According to the conditions, the right-hand side of Eq. (9)

$$f(x, t) + u_0(x) \cdot \delta'(t) + u_1(x) \cdot \delta(t)$$

becomes zero when $t < 0$. Therefore, by the theorem of Sec. 11.2, its convolution with the fundamental solution $\mathscr{E}_n$ of the wave operator exists in $\mathscr{D}'(R^{n+1})$ and becomes zero when $t < 0$. By the theorem of Sec. 10.3, the solution of Eq. (9) exists and is unique in the class of generalized functions belonging to $\mathscr{D}'(R^{n+1})$ which become zero when $t < 0$ and this solution is expressed by the convolution

$$\begin{aligned} u &= \mathscr{E}_n * [f + u_0(x) \cdot \delta'(t) + u_1(x) \cdot \delta(t)] \\ &= \mathscr{E}_n * f + \mathscr{E}_n * u_0(x) \cdot \delta'(t) + \mathscr{E}_n * u_1(x) \cdot \delta(t) \end{aligned} \tag{11}$$

From this, using Eqs. (25) and (26) from Sec. 11.4, we obtain representation (10) for the required solution of the generalized Cauchy problem.

§ 12. THE CAUCHY PROBLEM FOR THE WAVE EQUATION

We shall prove the continuous dependence of this solution u on f, u_0 and u_1 in $\mathscr{D}'(R^{n+1})$. If $f_k = 0$, $t < 0$, and $f_k \to f$ as $k \to \infty$ in $\mathscr{D}'(R^{n+1})$, $u_{0k} \to u_0$, $u_{1k} \to u_1$ as $k \to \infty$ in $\mathscr{D}'(R^n)$, then by virtue of the continuity of the direct product [cf. Sec. 7.3, (a)], $f_k + u_{0k}(x) \cdot \delta'(t) + u_{1k}(x) \cdot \delta(t) \to f + u_0(x) \cdot \delta'(t) + u_1(x) \cdot \delta(t)$, $k \to \infty$ in $\mathscr{D}'(R^{n+1})$.

Therefore, using the continuity of the convolution (cf. the theorem of Sec. 11.2), from (11) we obtain

$$u_k = \mathscr{E}_n * [f_k + u_{0k}(x) \cdot \delta'(t) + u_{1k}(x) \cdot \delta(t)]$$
$$\to \mathscr{E}_n * [f + u_0(x) \cdot \delta'(t) + u_1(x) \cdot \delta(t)] = u$$

as

$$k \to \infty \text{ in } \mathscr{D}'(R^{n+1})$$

The theorem is proved.

COROLLARY 1. *If $u \in C^2(t \geq 0)$ and $u(x, t) = 0$ when $t < 0$, then the following equation is true*:

$$u(x, t) = V_n(x, t) + V_n^{(0)}(x, t) + V_n^{(1)}(x, t) \qquad (12)$$

where V_n is the retarded potential with density $\{\Box_a u\}$, and $V_n^{(0)}$ and $V_n^{(1)}$ are surface retarded potentials of simple and double layers with densities $[\partial u(x, +0)]/\partial t$ and $u(x, +0)$, respectively.

In fact, the function $u(x, t)$ is the solution of the following generalized Cauchy problem:

$$\Box_a u = \{\Box_a u\} + u(x, +0) \cdot \delta'(t) + \frac{\partial u(x, +0)}{\partial t} \cdot \delta(t)$$

(cf. Sec. 12.2) and therefore, according to the theorem of Sec. 12.3, appears in the form of a sum (12) of the three retarded potentials with densities as shown.

COROLLARY 2. *For $f = 0$ the solution $u(x, t)$ of the generalized Cauchy problem belongs to the class C^∞ with respect to the variable t in $[0, \infty)$ and satisfies the initial conditions (7) in the sense of a weak convergence*:

$$u(x, t) \to u_0(x), \quad \frac{\partial u(x, t)}{\partial t} \to u_1(x), \quad t \to +0 \text{ in } \mathscr{D}'(R^n) \qquad (13)$$

In fact, according to the lemma of Sec. 11.4, the potentials $V_n^{(0)}$ and $V_n^{(1)}$ belong to the class C^∞ with respect to t in $[0, \infty)$ and satisfy the initial conditions (27) and (28) from Sec. 11.4. Consequently, their sum $V_n^{(0)} + V_n^{(1)}$, which is, by virtue of (10), the solution of the generalized Cauchy problem for $f = 0$, belongs to the class C^∞ with respect to t in $[0, \infty)$ and satisfies the initial conditions (13).

4. The Solution of the Classical Cauchy Problem. From the theorems of Secs. 11.3, 11.4, and 12.3 we may make the following assertions concerning the solubility of the classical Cauchy problem for the wave equation.

Let $f \in C^2(t \geq 0)$, $u_0 \in C^3(R^n)$, and $u_1 \in C^2(R^n)$ when $n = 3, 2$; $f \in C^1(t \geq 0)$, $u_0 \in C^2(R^1)$, and $u_1 \in C^1(R^1)$ for $n = 1$. Then the classical solution of the Cauchy problem (6) and (7) exists, is unique, and is expressed by:

Kirchhoff's formula when $n = 3$

$$u(x, t) = \frac{1}{4\pi a^2} \int_{U(x;at)} \frac{f(\xi, t - |x - \xi|a)}{|x - \xi|} d\xi$$
$$+ \frac{1}{4\pi a^2 t} \int_{S(x;at)} u_1(\xi)\, dS + \frac{1}{4\pi a^2} \frac{\partial}{\partial t}\left[\frac{1}{t} \int_{S(x;at)} u_0(\xi)\, dS\right] \quad (14)$$

Poisson's formula when $n = 2$

$$u(x, t) = \frac{1}{2\pi a} \int_0^t \int_{U(x, a(t-\tau))} \frac{f(\xi, \tau)\, d\xi\, d\tau}{(a^2(t-\tau)^2 - |x - \xi|^2)^{1/2}}$$
$$+ \frac{1}{2\pi a} \int_{U(x;at)} \frac{u_1(\xi)\, d\xi}{(a^2 t^2 - |x - \xi|^2)^{1/2}}$$
$$+ \frac{1}{2\pi a} \frac{\partial}{\partial t} \int_{U(x;at)} \frac{u_0(\xi)\, d\xi}{(a^2 t^2 - |x - \xi|^2)^{1/2}} \quad (14')$$

d'Alembert's formula when $n = 1$

$$u(x, t) = \frac{1}{2a} \int_0^t \int_{x-a(t-\tau)}^{x+a(t-\tau)} f(\xi, \tau)\, d\xi\, d\tau + \frac{1}{2a} \int_{x-at}^{x+at} u_1(\xi)\, d\xi$$
$$+ \tfrac{1}{2}[u_0(x + at) + u_0(x - at)] \quad (14'')$$

This solution depends continuously on the data f, u_0, and u_1 of the Cauchy problem in the following sense: If these data are changed so that

$$|f - \tilde{f}| \leq \varepsilon, \quad |u_0 - \tilde{u}_0| \leq \varepsilon_0, \quad |u_1 - \tilde{u}_1| \leq \varepsilon_1, \quad |\operatorname{grad}(u_0 - \tilde{u}_0)| \leq \varepsilon_0'$$

§ 12. THE CAUCHY PROBLEM FOR THE WAVE EQUATION

(this latter inequality is necessary only for $n = 3$ and 2), then the corresponding solutions u and $\tilde{u}$ in any strip $0 \leq t \leq T$ satisfy the estimates:

$$|u(x,t) - \tilde{u}(x,t)| \leq \frac{T^2}{2}\varepsilon + T\varepsilon_1 + \varepsilon_0 + aT\varepsilon_0', \qquad n = 3, 2$$

$$|u(x,t) - \tilde{u}(x,t)| \leq \frac{T^2}{2}\varepsilon + T\varepsilon_1 + \varepsilon_0, \qquad n = 1$$

In summary, it may be said that the Cauchy problem for the wave equation is formulated correctly (cf. Sec. 4.6), and that $C^2(t \geq 0) \cap C^1(t \geq 0)$ is the correctness class for the classical Cauchy problem, and $\mathscr{D}'(R^{n+1})$ is the correctness class for the generalized Cauchy problem (cf. the theorem of Sec. 12.3).

Note. The method of solving the Cauchy problem for the wave equation which we have set out may be transferred without essential changes to the case involving any number of spatial variables n and also to problems where Cauchy data are given on an arbitrary spacelike surface* (cf. Sec. 4.3). Moreover, this method is applicable to the Cauchy problem for arbitrary equations of hyperbolic type with constant coefficients. This equation is characterized by the fact that it has a fundamental solution with a support included in a convex cone which does not contain a whole straight line [cf. L. Hörmander (*1*, Chap. V)]. This method was first used in a clear form by S. L. Sobolev (*2*) (in 1936) to solve the Cauchy problem for a hyperbolic equation of second order.

5. Exercises. (a) Using the fundamental solution of the Dirac operator [cf. Sec. 10.12, (d)], show that the solution of the Cauchy problem for Dirac's equation (cf. Sec. 2.8)

$$\left(i \sum_{k=0}^{3} \gamma^k \frac{\partial}{\partial x_k} - mI\right)\Psi = 0$$

$$\Psi|_{x_0=+0} = \Psi_0(x), \qquad \Psi_0 = (\psi_{01}, \psi_{02}, \psi_{03}, \psi_{04}), \qquad \psi_{0j} \in \mathscr{D}'(R^3)$$

is expressed by the formula

$$\Psi = \left(\sum_{k=0}^{3} \gamma^k \frac{\partial}{\partial x^k} - imI\right)D^r(x_0, x) * \gamma^0 \Psi_0(x)$$

---

* The smooth surface $\Sigma = [t = \sigma(x)]$ is said to be *spacelike* if at each of its points (x, t) the normal **n** lies in the cone $\Gamma^+(x, t)$ (cf. Secs. 3.3 and 4.2.).

The convolution of a matrix with a vector is defined by the usual rules, replacing the operation of multiplication by the operation of convolution $*$.

(b) Using the fundamental solution of the Klein–Gordon operator [cf. Sec. 10.12, (b)], show that the solution of the Cauchy problem for the Klein–Gordon equation (cf. Sec. 2.8) is expressed by the formula

$$u = D^r(x_0, x) * u_1(x) + \frac{\partial D^r(x_0, x)}{\partial x_0} * u_0(x)$$

(c) Show that the solution of the mixed problem

$$\Box_a u = 0, \quad u|_{t=+0} = \frac{\partial u}{\partial t}\bigg|_{t=+0} = 0, \quad u|_{x=+0} = \psi(t), \quad 0 < x, \quad t < \infty$$

is the function

$$u(x, t) = -2a^2 \frac{\partial \mathscr{E}_1(x, t)}{\partial x} * \psi(t) = \theta\left(t - \frac{x}{a}\right)\psi\left(t - \frac{x}{a}\right)$$

here $\psi \in C([0, \infty))$ and $\psi = 0$ for $t < 0$.

(d) Starting from Eqs. (20) and (20') of Sec. 11.3, show that the retarded potentials $V_n(x, t)$, $n = 3, 2$, belong to the class $C^2(t > 0)$ if the density $f \in C^1(t \geq 0)$.

§ 13. Wave Propagation

In this section we give a physical interpretation of the solution of the wave equation which we obtained in Sec. 12.

Let u_0 and u_1 be sufficiently smooth functions. Then the initial perturbation $u_0(x) \cdot \delta'(t) + u_1(x) \cdot \delta(t)$ produces, when $t > 0$, a perturbation $u(x, t)$ which is expressed by Eqs. (14), (14'), and (14'') of Sec. 12.4 when $f = 0$. We shall write these equations in the form

$$u(x, t) = (\mathscr{E}_n(x - \xi, t), u_1(\xi)) + \left(\frac{\partial \mathscr{E}_n(x - \xi, t)}{\partial t}, u_0(\xi)\right) \tag{1}$$

It is clear from formula (1) that the perturbation u at the point x at an instant of time $t > 0$ is a superposition of the elementary perturbations

$$u_1(\xi)\mathscr{E}_n(x - \xi, t) \quad \text{and} \quad u_0(\xi)\frac{\partial \mathscr{E}_n(x - \xi, t)}{\partial t}$$

produced, respectively, by the initial perturbations

$$u_1(\xi)\delta(x - \xi) \cdot \delta(t) \quad \text{and} \quad u_0(\xi)\delta(x - \xi) \cdot \delta'(t)$$

§ 13. WAVE PROPAGATION

concentrated at a separate point ξ (for this the point ξ naturally ranges over the set in which the initial perturbation is concentrated). This constitutes the principle of *superposition of waves* (cf. also Sec. 10.3).

Concrete realization of the principle of superposition depends essentially on the structure of the fundamental solution $\mathscr{E}_n(x, t)$ and, consequently, on the number n of space variables. This in its turn defines the differences in the character of wave propagation in space, in a plane, and on a straight line.

1. Wave Propagation in a Space. From the expression for the fundamental solution of the three-dimensional wave operator

$$\mathscr{E}_3(x, t) = \frac{\theta(t)}{4\pi a^2 t} \delta_{S_{at}}(x) \equiv \frac{\theta(t)}{2\pi a} \delta(a^2 t^2 - |x|^2) \qquad x = (x_1, x_2, x_3)$$

it follows that the perturbation $\mathscr{E}_3(x, t)$ from a momentarily acting point source $\delta(x) \cdot \delta(t)$ at an instant of time $t > 0$ will occupy a spherical surface of a radius at with a center at the point $x = 0$ (Figs. 30 and 25). This means that such a perturbation propagates in the form of a *spherical wave* $|x| = at$, moving with a speed a, and moreover, after the wave has passed, there will once more be no disturbance. In this case it is said that *Huygens' principle* takes place in a space.

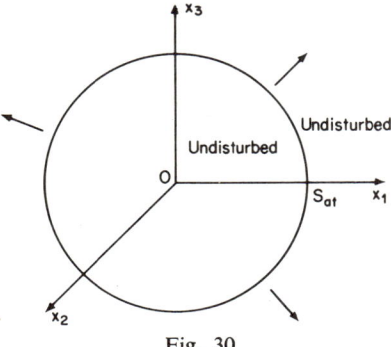

Fig. 30

Therefore, by virtue of the principle of superposition, it follows that the perturbation $u(x, t)$ for $t > 0$ from the arbitrary initial perturbation $u_0(x) \cdot \delta'(t) + u_1(x) \cdot \delta(t)$ is completely defined by the values $u_0(\xi)$ and $u_1(\xi)$ over the spherical surface $S(x; at)$, that is, over the points of the boundary of the base of the cone $\Gamma_0^-(x, t)$ (cf. Sec. 11.3 and Fig. 27).

Now let the initial perturbation be concentrated over the compactum K. By virtue of what has just been said, the point $x \notin K$, which the perturbation reaches at an instant of time $t_0 = d/a$, will be acted upon throughout the time $(D - d)/a$, where d and D are the minimum and maximum distances from the point x to the points of the set K (Fig. 31).

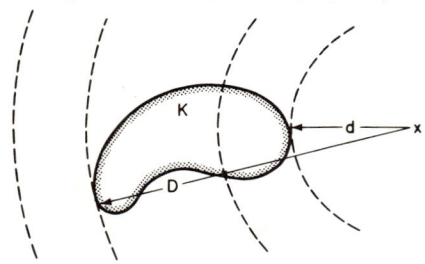

Fig. 31

For $t > D/a = t_1$ at the point x there will once more be no disturbance. In this way, at the instant of time t_0 the front edge of the wave passes through the point x; at the instant of time t_1 the rear edge of the wave passes through this point. At an instant of time t the front edge will be an external envelope of the spherical surfaces $S(\xi; at)$, when ξ passes through K, and the rear edge will be an internal envelope of these spheres (Fig. 32).

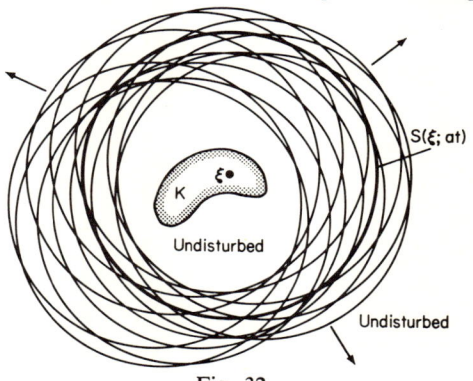

Fig. 32

In other words, at an instant of time t the perturbation propagates over the region enclosed between the front and the rear edges of the wave. When we examine this picture for all $t > 0$ we conclude that in the space of four variables (x, t) the perturbation from K will take place only on the union

§ 13. WAVE PROPAGATION

of the boundaries $|x - \xi| = at$ of the future light cones $\Gamma^+(\xi, 0)$, when their vertexes $(\xi, 0)$ pass through the compactum K in the plane $\tau = 0$. The set which is obtained in R^4 is the *region of influence* $M(K)$ of the compactum K (cf. Sec. 11.3 and Fig. 28).

2. Propagation of Waves over a Plane. From the equation for the fundamental solution of the two-dimensional wave operator

$$\mathscr{E}_2(x, t) = \frac{\theta(at - |x|)}{2\pi a(a^2 t^2 - |x|^2)^{1/2}}, \qquad x = (x_1, x_2)$$

it follows that the perturbation $\mathscr{E}_2(x, t)$ from the momentarily acting point source $\delta(x) \cdot \delta(t)$ at the instant of time $t > 0$ will occupy the circle $\bar{U}_{at}$ (Figs. 33 and 34). In this way the front edge of the wave S_{at} will be seen to move over the plane with speed a. However, in distinction from what happens in space, after the front edge the perturbation is observed to be present at all following instants of time, so that there is no rear edge. In this case, *wave diffusion* is said to take place. Huygens' principle is absent here, evidently.

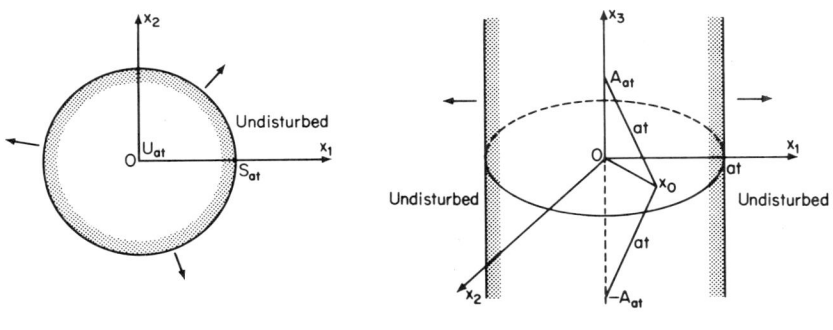

Fig. 33   Fig. 34

To understand why wave diffusion takes place over a plane, we note that the fundamental solution $\mathscr{E}_2(x, t)$ considered as a function of the variables (x, x_3, t) is a perturbation from the momentary source $\delta(x) \cdot 1(x_3) \cdot \delta(t)$ concentrated on the axis x_3 (cf. Secs. 10.4 and 10.7). From such a source in R^3 the perturbation propagates in the form of a *cylindrical wave* $x_1^2 + x_2^2 \leq a^2 t^2$, the front edge of which, $x_1^2 + x_2^2 = a^2 t^2$, moves with a speed a perpendicular to the axis x_3 (Fig. 34). After the front edge has passed, the perturbation is retained for an infinitely long time.

In fact, by virtue of Huygens' principle (cf. Sec. 13.1), a perturbation from the source $\delta(x) \cdot 1(x_3) \cdot \delta(t)$ will reach a given point $(x_0, 0) \in R^3$ at an instant of time $t > 0$ from those points of the sphere $|x - x_0|^2 + x_2^3 = a^2 t^2$ which lie along the axis x_3, that is, from the points (Fig. 34)

$$\pm A_{at} = \{0, \pm(a^2 t^2 - |x|^2)^{1/2}\}$$

It follows from this that when $t < |x_0|/a = t_0$ at the point $(x_0, 0)$ there will be no disturbance; at an instant of time t_0 the front edge of the wave will pass through this point (the perturbation will come from the point $\{0\}$); but at all succeeding instants of time $t > t_0$ similar perturbations from the points $\pm A_{at}$ will reach this point and so a composite perturbation distinct from zero will be observed at this point (the rear edge of the wave is absent).

From the presence of wave diffusion over a plane in the case of initial point perturbation $\delta(x) \cdot \delta(t)$, it follows that wave diffusion will also be observed for an arbitrary initial perturbation $u_0(x) \cdot \delta'(t) + u_1(x) \cdot \delta(t)$.

In fact, the corresponding perturbation $u(x, t)$ for $t > 0$, by virtue of the principle of superposition, is completely defined by the values $u_0(\xi)$ and $u_1(\xi)$ in the circle $\bar{U}(x; at)$, that is, on the base of the cone $\Gamma_0^-(x, t)$ (Fig. 27). Therefore, if the initial perturbation is concentrated, let us say, over the compactum K, then at an instant of time $t > 0$ the perturbation $u(x, t)$ propagates over a region which is a union of the circles $\bar{U}(\xi; at)$, when their centers ξ pass through K (Fig. 35). In this way, the front edge is observed, but the rear edge is absent.

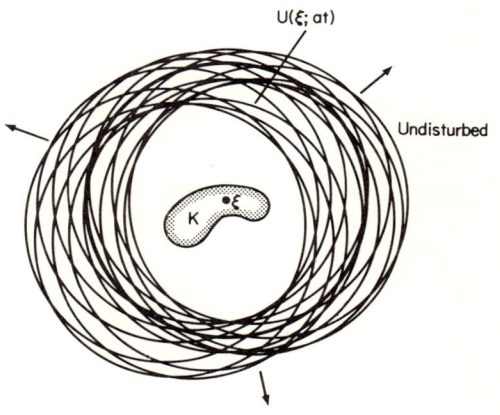

Fig. 35

§ 13. WAVE PROPAGATION

Therefore, the region of influence $M(K)$ of the compactum K is the union of the closed future light cones $\bar{\Gamma}^+(\xi, 0)$, when their vertexes $(\xi, 0)$ pass through K in the plane $\tau = 0$ (Fig. 29).

3. Wave Propagation on a Straight Line. From the form of the fundamental solution of the one-dimensional wave operator

$$\mathscr{E}_1(x, t) = \frac{1}{2a} \theta(at - |x|), \qquad x = x_1$$

it follows that the perturbation $\mathscr{E}_1(x, t)$ from a momentarily acting point source $\delta(x) \cdot \delta(t)$ at the instant of time $t > 0$ will occupy a segment $-at \leq x \leq at$ (Fig. 23). In this case two front edges $x = at$ and $x = -at$ will be observed, moving on a straight line with speed a to right and left, respectively. As in the case of a plane, a perturbation will be observed behind the front edge of the wave (in the case given it is constant and is equal to $1/2a$); that is, wave diffusion takes place.

To understand this occurrence, we shall interpret the fundamental solution $\mathscr{E}_1(x, t)$ three dimensionally. This solution is a perturbation from the momentary source $\delta(x_1) \cdot 1(x_2, x_3) \cdot \delta(t)$, concentrated over the plane $x_1 = 0$. From such a source in R^3 the perturbation propagates in the form of a *plane wave* $|x_1| \leq at$, the front edge of which, $|x_1| = at$, moves with a speed a, perpendicular to the plane $x_1 = 0$. We note that in this case the front edge consists of two planes, $x_1 = at$ and $x_1 = -at$, moving with a speed a to right and left, respectively, with respect to the plane $x_1 = 0$ (Fig. 36). After the front edge of the wave has passed, the perturbation is retained for an infinitely long time.

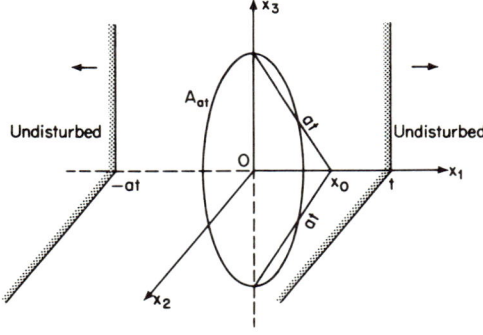

Fig. 36

In fact, by virtue of Huygens' principle, the perturbation from the source $\delta(x_1) \cdot 1(x_2, x_3) \cdot \delta(t)$ will reach a given point $(x_0, 0, 0) \in R^3$ at an instant of time $t > 0$ from those points of the sphere $|x_1 - x_0|^2 + x_2^2 + x_3^2 = a^2 t^2$ which lie on the plane $x_1 = 0$, that is, from the points of the circumference (Fig. 36),

$$A_{at} = [x_2^2 + x_3^2 = a^2 t^2 - x_0^2, \, x_1 = 0]$$

It follows from this that for $t < |x_0|/a = t_0$ at the point $(x_0, 0, 0)$ there will be no disturbance; at the instant of time t_0 the front edge of the wave will pass through this point (the perturbation will come from the point $\{0\}$); at all succeeding instants of time $t > t_0$ similar perturbations from the points of the circumference A_{at} will reach this point and therefore we shall observe at it a composite perturbation distinct from zero (the rear edge of the wave is absent).

Since there is wave diffusion on a straight line when there is an initial point perturbation $\delta(x) \cdot \delta(t)$, it follows that there is also wave diffusion for an arbitrary initial perturbation $u_1(x) \cdot \delta(t)$.

We shall now examine the momentary point source of the form $\delta(x) \cdot \delta'(t)$. According to the theorem of Sec. 12.3 this source produces the perturbation

$$\mathscr{E}_1(x, t) = \frac{\partial \mathscr{E}_1(x, t)}{\partial t} * \delta(x) = \frac{\partial \mathscr{E}_1(x, t)}{\partial t}$$

$$= \frac{1}{2a} \frac{\partial}{\partial t} \theta(at - |x|) = \tfrac{1}{2} \delta(at - |x|) \tag{2}$$

It is apparent from this that the perturbation $\mathscr{E}_1(x, t)$ at an instant of time $t > 0$ will be concentrated only at two points $x = \pm at$, so that after the front edge of the wave $|x| = at$ has passed there will once more be no disturbance. In this case Huygens' principle operates.

For an arbitrary initial perturbation of the form $u_0(x) \cdot \delta'(t)$ the perturbation $u(x, t)$ for $t > 0$ is completely defined by the values $u_0(\xi)$ at the points $x \pm at$, that is, at the points of the boundary of the base of the cone $\Gamma_0^-(x, t)$ (Fig. 27). This perturbation, by virtue of the theorem of Sec. 12.3, is given by the formula

$$u = \frac{\partial \mathscr{E}_1(x, t)}{\partial t} * u_0(x) = \mathscr{E}_1(x, t) * u_0(x)$$

§ 11. RETARDED POTENTIAL

rules of differentiation of a direct product [cf. Sec. 7.3, (c)] and of a convolution (cf. Sec. 7.6), for all $k = 1, 2, \ldots$, we obtain the equations

$$g * u(x) \cdot \delta^{(k)}(t) = \frac{\partial^k}{\partial t^k}[g(x, t) * u(x)] = \frac{\partial^k g(x, t)}{\partial t^k} * u(x) \quad (17)$$

3. Retarded Potential. Let the generalized function $f(x, t) \in \mathscr{D}'(R^{n+1})$ become zero in the half-space $t < 0$. The generalized function

$$V_n = \mathscr{E}_n * f$$

where $\mathscr{E}_n$ is the fundamental solution of the wave operator, is known as the *retarded potential with a density f*.

Since supp $\mathscr{E}_n \subset \bar{\Gamma}^+$, then, according to the theorem of Sec. 11.2, the retarded potential V_n exists in $\mathscr{D}'(R^{n+1})$ and appears in the form

$$(V_n, \varphi) = (\mathscr{E}_n(y, \tau) \cdot f(\xi, \tau'), \eta(\tau)\eta(\tau')$$
$$\times \eta(a^2\tau^2 - |y|^2)\varphi(y + \xi, \tau + \tau')), \quad \varphi \in \mathscr{D}(R^{n+1}), \quad (18)$$

where $\eta(\tau)$ is any function belonging to the class $C^\infty(R^1)$ which is equal to 0 for $\tau < -\delta$ and to 1 for $\tau > -\varepsilon$; δ and ε are any numbers, $\delta > \varepsilon > 0$. Moreover, according to the same theorem, the retarded potential $V_n(x, t)$ becomes zero when $t < 0$ and is continuously dependent on the density f in $\mathscr{D}'(R^{n+1})$. Finally, according to the theorem of Sec. 10.3, this potential satisfies the wave equation

$$\Box_a V_n = f \quad (19)$$

The other properties of the retarded potential V_n depend essentially on the properties of the density f.

If f is a function locally integrable in R^{n+1}, then V_n is a function locally integrable in R^{n+1} and is expressed by the formulas

$$V_3(x, t) = \frac{1}{4\pi a^2} \int_{U(x; at)} \frac{f(\xi, t - |x - \xi|/a)}{|x - \xi|} d\xi \quad (20)$$

$$V_2(x, t) = \frac{1}{2\pi a} \int_0^t \int_{S(x; a(t-\tau))} \frac{f(\xi, \tau) \, d\xi \, d\tau}{(a^2(t-\tau)^2 - |x - \xi|^2)^{1/2}} \quad (20')$$

$$V_1(x, t) = \frac{1}{2a} \int_0^t \int_{x-a(t-\tau)}^{x+a(t-\tau)} f(\xi, \tau) \, d\xi \, d\tau \quad (20'')$$

§ 13. WAVE PROPAGATION

From this, and taking Eqs. (2) into consideration, we obtain for $t > 0$

$$u(x, t) = \tfrac{1}{2}\delta(at - |x|) * u_0(x) = \tfrac{1}{2}u_0(x + at) + \tfrac{1}{2}u_0(x - at) \qquad (3)$$

The physical sense of Eq. (3) is that the initial perturbation $u_0(x) \cdot \delta'(t)$ for $t > 0$ disintegrates, as it were, into two similar perturbations $\tfrac{1}{2}u_0(x \pm at)$, each with half of the intensity (Fig. 37). For instance, the disintegration of

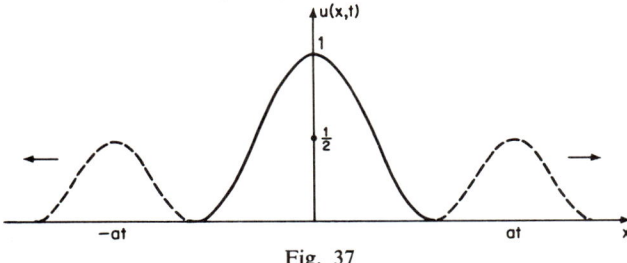

Fig. 37

the discontinuity of the step $u_0 = \theta(x)$, by virtue of Eq. (3), takes place in the manner shown in Fig. 38.

In correspondence with what has just been said, the regions of influence of the segment $K = [b, c]$ for the initial perturbations $u_1(x) \cdot \delta(t)$ and $u_0(x) \cdot \delta'(t)$ have the form which is shown in Figs. 39 and 40, respectively.

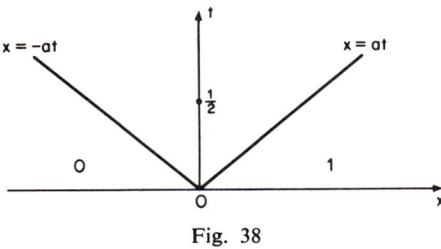

Fig. 38

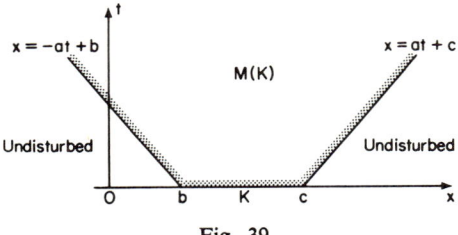

Fig. 39

In this way, for an initial perturbation $u_1(x) \cdot \delta(t)$ on a straight line wave diffusion takes place, and for an initial perturbation $u_0(x) \cdot \delta'(t)$ Huygens' principle applies.

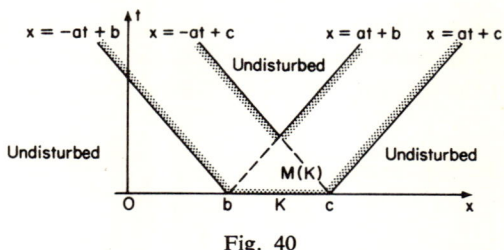

Fig. 40

4. Method of Propagating Waves. Taking the method of characteristics as our starting point, we shall set out another method of solving the classical Cauchy problem for a one-dimensional homogeneous wave equation, the method of propagating waves

$$\Box_a u = 0 \tag{4}$$

$$u\,|_{t=+0} = u_0(x), \qquad \frac{\partial u}{\partial t}\bigg|_{t=+0} = u_1(x) \tag{5}$$

First of all we shall prove the following lemma.

LEMMA. *In order that the function $u(x, t)$ of the class C^2 should be a solution of the wave equation* (4) *in a region, it is necessary and sufficient that in this region it appear in the form*

$$u(x, t) = f(x - at) + g(x + at) \tag{6}$$

where $f(\xi)$ and $g(\eta)$ are functions of the class C^2 in corresponding intervals of the transformed variables ξ and η.

Proof. Function (6) satisfies Eq. (4), since

$$\frac{\partial^2 u}{\partial t^2} = a^2 f''(x - at) + a^2 g''(x + at) = a^2 \frac{\partial^2 u}{\partial x^2}$$

Conversely, let the function $u(x, t)$ of the class C^2 satisfy Eq. (4) in some region. We shall present Eq. (4) in canonical form. In agreement with what has been said in Sec. 3.4, its differential equations describing

§ 13. WAVE PROPAGATION

the characteristics have the form

$$\frac{dx}{dt} = -a, \qquad \frac{dx}{dt} = a$$

and therefore the change of variables

$$\xi = x - at, \qquad \eta = x + at \tag{7}$$

reduces Eq. (4) to the canonical form

$$\frac{\partial^2 \tilde{u}}{\partial \xi \partial \eta} = 0$$

Integrating this equation with respect to ξ, we obtain

$$\frac{\partial \tilde{u}}{\partial \eta} = \chi(\eta)$$

where χ is an arbitrary function of the class C^1. Proceeding to integrate the equation we have just obtained with respect to η, we write the function $\tilde{u}$ in the form

$$\tilde{u}(\xi, \eta) = \int^{\eta} \chi(\eta') \, d\eta' + f(\xi) = f(\xi) + g(\eta) \tag{8}$$

where f and g are some functions of the class C^2. Changing to the old variables x and t by means of Eqs. (7), from (8) we deduce Eq. (6) for the solution $u(x, t)$. The lemma is proved.

PHYSICAL INTERPRETATION OF SOLUTION (6). The function $f(x - at)$ describes the perturbation which, from the point x_0 at an instant of time $t = 0$, reaches the point $x = x_0 + at$ at an instant of time t (Fig. 41). Therefore this function represents a wave moving to the right with the speed a. Analogously, the function $g(x + at)$ represents a wave moving to

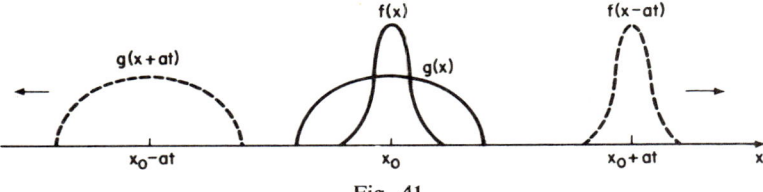

Fig. 41

the left with the speed a (Fig. 41). The general solution (6) of the wave equation (4) is a superposition of these two waves.

By means of representation (6) of the general solution of wave equation (4), the classical solution of the Cauchy problem (4) and (5) is constructed in the following way.

Let us suppose that the solution $u(x, t)$ of this problem exists. Then, according to the lemma of this section, this solution appears in the form (6), with the functions f and g belonging to the class $C^2(R^1)$. In order that the solution $u(x, t)$ should satisfy initial conditions (5), it is necessary that the functions f and g satisfy the equations

$$f(x) + g(x) = u_0(x), \qquad -af'(x) + ag'(x) = u_1(x)$$

that is,

$$f(\xi) + g(\xi) = u_0(\xi), \qquad g(\xi) - f(\xi) = \frac{1}{a}\int_0^\xi u_1(\xi')\,d\xi' + C \qquad (9)$$

where C is an arbitrary constant. Solving Eqs. (9) with respect to the unknown functions f and g,

$$f(\xi) = \tfrac{1}{2}u_0(\xi) - \frac{1}{2a}\int_0^\xi u_1(\xi')\,d\xi' - \frac{C}{2}$$

$$g(\eta) = \tfrac{1}{2}u_0(\eta) + \frac{1}{2a}\int_0^\eta u_1(\xi')\,d\xi' + \frac{C}{2}$$

and substituting the expressions we have obtained for f and g into formula (6), we obtain d'Alembert's formula (cf. Sec. 12.4)

$$u(x, t) = \tfrac{1}{2}[u_0(x + at) + u_0(x - at)] + \frac{1}{2a}\int_{x-at}^{x+at} u_1(\xi)\,d\xi \qquad (10)$$

We find by a direct check that d'Alembert's formula (10) does in fact give the classical solution of the Cauchy problem (4) and (5), if $u_0 \in C^2(R^1)$ and $u_1 \in C^1(R^1)$. This solution is unique (cf. Sec. 12.4).

5. Method of Reflections. A Semi-Infinite String. The solution of the Cauchy problem for Eq. (4) by the method of propagating waves, which has been set out in the previous section, enables us to solve several mixed problems for this equation. For clarity we shall consider the mixed problem

§ 13. WAVE PROPAGATION

(cf. Sec. 4.5) describing the vibration of a semi-infinite string $x \geq 0$ with a fixed left end

$$u\big|_{x=0} = 0 \tag{11}$$

We shall first prove that *each classical solution $u(x, t)$ of the wave equation (4) in the quadrant $x > 0$, $t > 0$ which satisfies condition (11) appears in the form*

$$u(x, t) = g(x + at) - g(-x + at), \qquad g \in C^2(R^1) \tag{12}$$

In fact, according to the lemma of Sec. 13.4, the solution $u(x, t)$ appears in the form (6), where $f(\xi) \in C^2(R^1)$ and $g(\eta) \in C^2(\eta > 0)$. From this, taking condition (11) into consideration, we obtain

$$0 = f(-at) + g(at)$$

from which Eq. (12) follows.

PHYSICAL INTERPRETATION OF THE SOLUTION (12). This solution is a superposition of two waves: the wave $g(x + at)$ moving with the speed a to the left, and the wave $-g(-x + at)$ moving with the same speed to the right. Let the wave $g(x + at)$ move along a semi-infinite string $x \geq 0$, fixed at the point $x = 0$. Then the wave $-g(-x + at)$ will move along the semiaxis $x \leq 0$ toward the wave $g(x + at)$ (Fig. 42). At a certain instant

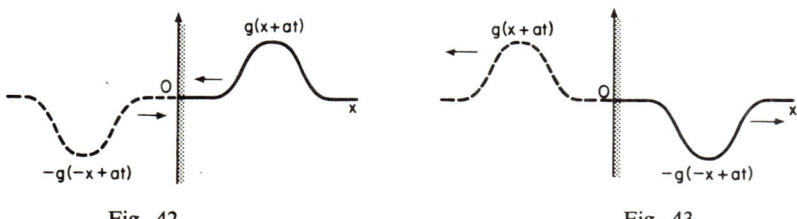

Fig. 42   Fig. 43

of time these waves will meet at the point $x = 0$ and, superposing one upon the other, will give a zero perturbation at this point. On further movement the wave $g(x + at)$ will be beyond the limits of the string, while the wave $-g(-x + at)$ will pass through onto the string itself. As a result, *the reflection of the wave $g(x + at)$ from the end of the string $x = 0$ with a change of sign of the function* (Fig. 43) *will be observed.*

We shall now construct a solution of the mixed problem (4)-(5)-(11). Each classical solution $u(x, t)$ of this problem, by virtue of (12), allows an odd continuation $\tilde{u}(x, t)$ along x of the class $C^2(R^2)$ and this continuation satisfies Eq. (4) in R^2. From this and from conditions (5) it follows that the solution $\tilde{u}(x, t)$ satisfies the initial conditions

$$\tilde{u}|_{t=+0} = \tilde{u}_0(x), \qquad \frac{\partial \tilde{u}}{\partial t}\bigg|_{t=+0} = \tilde{u}_1(x) \qquad (13)$$

where $\tilde{u}_0$ and $\tilde{u}_1$ are odd continuations of the functions u_0 and u_1, respectively. But the solution of such a Cauchy problem is unique and is presented by d'Alembert's formula (10) with a change of u_0 to $\tilde{u}_0$ and u_1 to $\tilde{u}_1$ if $\tilde{u} \in C^2(R^1)$ and $\tilde{u}_1 \in C^1(R^1)$. These latter conditions are satisfied if

$$u_0 \in C^2(x \geq 0), \qquad u_1 \in C^1(x \geq 0), \qquad u_0(0) = u_0''(0) = u_1(0) = 0 \qquad (14)$$

So if conditions (14) are satisfied, then the solution of the mixed problem (4)-(5)-(11) exists, is unique, and is given by the formula

$$u(x, t) = \tfrac{1}{2}[\tilde{u}_0(x + at) + \tilde{u}_0(x - at)]$$
$$+ \frac{1}{2a} \int_{x-at}^{x+at} \tilde{u}_1(\xi) \, d\xi, \qquad x \geq 0 \qquad (15)$$

Let $x - at \geq 0$. Then

$$\tilde{u}_0(x - at) = u_0(x - at), \qquad \tilde{u}_1(\xi) = u_1(\xi), \qquad \xi \geq x - at \geq 0$$

and formula (15) takes the form

$$u(x, t) = \tfrac{1}{2}[u_0(x + at) + u_0(x - at)]$$
$$+ \frac{1}{2a} \int_{x-at}^{x+at} u_1(\xi) \, d\xi, \qquad x \geq at \qquad (16)$$

Now let $x - at \leq 0$. In this case

$$\tilde{u}_0(x - at) = -u_0(-x + at), \qquad \tilde{u}_1(\xi) = -u_1(-\xi)$$
$$x - at \leq \xi \leq 0$$

and formula (15) takes the form

$$u(x, t) = \tfrac{1}{2}[u_0(x + at) - u_0(at - x)]$$
$$+ \frac{1}{2a} \int_{at-x}^{x+at} u_1(\xi) \, d\xi, \qquad 0 \leq x \leq at \qquad (17)$$

§ 13. WAVE PROPAGATION

As we see from formula (17), two waves reach the point (x, t), $0 \leq x \leq at$: a straight wave from the point $(x + at, 0)$ and a once-reflected wave from the point $(at - x, 0)$ [coinciding with the straight wave from the fictitious point $(x - at, 0)$; cf. Fig. 44].

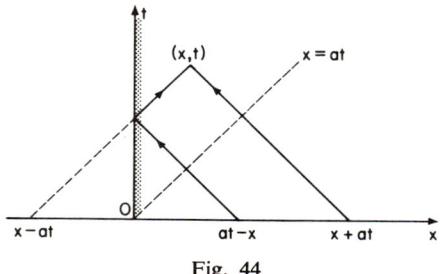

Fig. 44

The mixed problem for a semi-infinite string $x \geq 0$ with a free end is considered analogously:

$$\frac{\partial u}{\partial x}\bigg|_{x=0} = 0$$

Reflection of waves from the end of the string $x = 0$ takes place here too, but without change of sign.

6. Method of Reflections. A Finite String. We shall use the method of reflections which has been set out in the previous section to solve the mixed problem for a finite string $0 \leq x \leq l$ with fixed ends

$$u\big|_{x=0} = u\big|_{x=l} = 0 \tag{18}$$

We shall first prove that *each classical solution $u(x, t)$ of the wave equation (4) in the half-strip $0 < x < l$, $t > 0$, which satisfies conditions (18) appears in the form*

$$u(x, t) = g(x + at) - g(-x + at)$$
$$g(\xi + 2l) = g(\xi), \qquad g \in C^2(R^1) \tag{19}$$

In fact, according to the lemma of Sec. 13.4, the solution $u(x, t)$ appears in the form (6), where $f(\xi) \in C^2(\xi < l)$ and $g(\eta) \in C^2(\eta > 0)$. From this, and taking conditions (18) into account, we obtain

$$g(\xi) = -f(-\xi), \qquad f(l - \xi) = -g(l + \xi) \tag{20}$$

These results define the continuation of the functions f and g over the whole axis with conservation of the class C^2. In fact, the equation $g(\xi) = -f(-\xi)$ extends the function g onto the interval $(-l, \infty)$. But then the second of equations (20), written in the form $f(\eta) = -g(2l - \eta)$, extends the function f onto the interval $(-\infty, 3l)$; and so on. As a result of this continuation the functions f and g will belong to the class $C^2(R^1)$ and satisfy the relations (20). From this, Eq. (19) and the $2l$ periodicity of the function g follow:

$$-g(l + \xi) = f(l - \xi) = -g(-l + \xi)$$

Solution (19) shows that reflection of waves from both ends $x = 0$ and $x = l$ takes place with a change of sign. It follows that the movement of the string is periodic in time with a period $2l/a$ (Fig. 45).

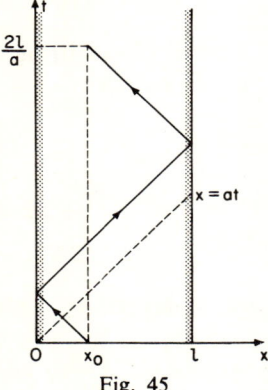

Fig. 45

We now construct a solution of the mixed problem (4)-(5)-(18). If the classical solution $u(x, t)$ of this problem exists, then, by virtue of (19), it allows an odd continuation $\tilde{u}(x, t)$ in x with respect to the points $x = 0$ and $x = l$, and this continuation belongs to the class $C^2(R^2)$ and satisfies Eq. (4) in R^2. From this and from conditions (5), it follows that the function $\tilde{u}(x, t)$ satisfies initial conditions (13) in which the functions $\tilde{u}_0$ and $\tilde{u}_1$ are the odd continuations of the functions u_0 and u_1, respectively, with regard to the points $x = 0$ and $x = l$.

Reasoning as in the previous section, we conclude that if the functions u_0 and u_1 satisfy the conditions

$$u_0 \in C^2([0, l]), \quad u_0(0) = u_0''(0) = u_0(l) = u_0''(l) = 0$$
$$u_1 \in C^1([0, l]), \quad u_1(0) = u_1(l) = 0 \tag{21}$$

§ 14. THE CAUCHY PROBLEM FOR THE EQUATION OF HEAT CONDUCTION

then the solution of the mixed problem (4)-(5)-(18) exists, is unique, and is given by the formula

$$u(x, t) = \tfrac{1}{2}[\tilde{u}_0(x+at)+\tilde{u}_0(x-at)] + \frac{1}{2a}\int_{x-at}^{x+at} \tilde{u}_1(\xi)\,d\xi, \qquad 0 \le x \le l \qquad (22)$$

Let the point (x, t) be placed as shown in Fig. 46. Then formula (22)

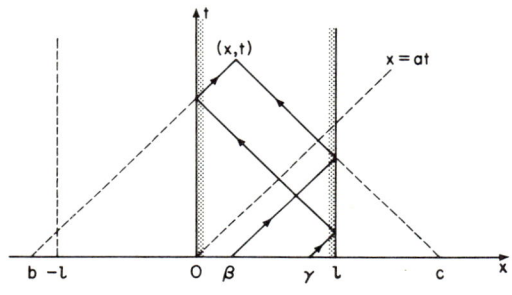

Fig. 46

at this point takes the form

$$u(x, t) = \tfrac{1}{2}[u_0(\gamma) - u_0(\beta)] - \frac{1}{2a}\int_{\beta}^{\gamma} u_1(\xi)\,d\xi \qquad (23)$$

In fact, using the rule of reflections, we have

$$\tilde{u}_0(b) = u_0(\gamma), \qquad \tilde{u}_0(c) = -u_0(\beta)$$

$$\int_b^c \tilde{u}_1(\xi)\,d\xi = \int_b^{-l} \tilde{u}_1(\xi)\,d\xi + \int_{l}^{c} \tilde{u}_1(\xi)\,d\xi$$

$$= \int_\gamma^l u_1(\xi)\,d\xi + \int_l^\beta u_1(\xi)\,d\xi = \int_\gamma^\beta u_1(\xi)\,d\xi$$

from which, and from (22), formula (23) follows. It shows that two waves reach the point (x, t): the once-reflected wave from the end $x = l$ is from the point β and the time-reflected wave from the ends $x = l$ and $x = 0$ is from the point γ (Fig. 46).

§ 14. The Cauchy Problem for the Equation of Heat Conduction

The solution of the Cauchy problem for the heat conduction equation is constructed in a way analogous to that shown in Sec. 12 for the solution of this problem for the wave equation.

1. Heat Potential. In Sec. 10.6 it was shown that the function

$$\mathscr{E}(x, t) = \frac{\theta(t)}{(2a\sqrt{\pi t})^n} \exp\left(-\frac{|x|^2}{4a^2 t}\right)$$

is the fundamental solution of the heat conduction operator. This function is nonnegative, becomes zero when $t < 0$, is infinitely differentiable for $(x, t) \neq (0, 0)$, and is locally integrable in R^{n+1}. Moreover [cf. Sec. 6.5, (f)],

$$\int \mathscr{E}(x, t)\, dx = 1, \qquad t > 0 \tag{1}$$

$$\mathscr{E}(x, t) \to \delta(x), \qquad t \to +0 \quad \text{in} \quad \mathscr{D}'(R^n) \tag{2}$$

The fundamental solution $\mathscr{E}(x, t)$ gives the temperature distribution from the momentary point source $\delta(x) \cdot \delta(t)$. Since $\mathscr{E}(x, t) > 0$ for all $t > 0$, and $x \in R^n$, it follows that heat is diffused with infinite velocity. But this is contrary to experience. Consequently, the heat conduction equation does not describe the processes of heat transfer with sufficient accuracy. A more exact description of the transfer processes (of heat and of particles) is given by the transport equation (cf. Sec. 2.4).

Let the generalized function $f \in \mathscr{D}'(R^{n+1})$ become zero for $t < 0$. The generalized function $V = \mathscr{E} * f$, where $\mathscr{E}$ is the fundamental solution of the heat conduction operator, is said to be the *heat potential with density f*.

If the heat potential V exists in $\mathscr{D}'(R^{n+1})$, then, by virtue of the theorem of Sec. 10.3, it satisfies the heat conduction equation

$$\frac{\partial V}{\partial t} = a^2\, \Delta V + f(x, t) \tag{3}$$

It follows from the theorem of Sec. 7.5 that if f is a generalized function with compact support and becomes zero when $t < 0$, then the heat potential is known to exist in $\mathscr{D}'(R^{n+1})$.

We shall distinguish another class of densities f for which the heat potential exists. Let $\mathscr{M}$ be a class of functions which become zero for $t < 0$ and which are bounded in each strip $0 \leq t \leq T$.

THEOREM. *If $f \in \mathscr{M}$, then the heat potential V with a density f exists in $\mathscr{M}$ and is expressed by the formula*

$$V(x, t) = \int_0^t \int_{R^n} \frac{f(\xi, \tau)}{[2a\sqrt{\pi(t-\tau)}]^n} \exp\left(-\frac{|x-\xi|^2}{4a^2(t-\tau)}\right) d\xi\, d\tau \tag{4}$$

§ 14. THE CAUCHY PROBLEM FOR THE EQUATION OF HEAT CONDUCTION

The potential V satisfies the estimate

$$|V(x, t)| \leq t \sup_{\substack{0 \leq \tau \leq t \\ \xi}} |f(\xi, \tau)|, \qquad t > 0 \tag{5}$$

and the initial condition: for any fixed $x \in R^n$

$$V(x, t) \to 0 \quad \text{as} \quad t \to +0 \tag{6}$$

Proof. Since the function $\mathscr{E}$ and f are locally integrable in R^{n+1}, then their convolution

$$\mathscr{E} * f = \int_0^t \int_{R^n} f(\xi, \tau) \mathscr{E}(x - \xi, t - \tau) \, d\xi \, d\tau$$

exists and is a locally integrable function in R^{n+1}, provided that the function

$$h(x, t) = \int_0^t \int_{R^n} |f(\xi, \tau)| \mathscr{E}(x - \xi, t - \tau) \, d\xi \, d\tau$$

is locally integrable in R^{n+1} (cf. Sec. 7.4). We shall check that this condition is satisfied. Since $h = 0$ when $t < 0$, then it is sufficient to establish that the function h satisfies Eq. (5) for $t > 0$. This follows from Eq. (1), by virtue of Fubini's theorem,

$$h(x, t) \leq \sup_{\substack{0 \leq \tau \leq t \\ \xi}} |f(\xi, \tau)| \int_0^t \int \mathscr{E}(x - \xi, t - \tau) \, d\xi \, d\tau$$

$$= t \sup_{\substack{0 \leq \tau \leq t \\ \xi}} |f(\xi, \tau)|, \qquad t > 0 \tag{7}$$

In this way, the heat potential $V = \mathscr{E} * f$ is represented by Eq. (4). Since $|V| \leq h$, then this potential becomes zero for $t < 0$ and, by virtue of (7), satisfies Eq. (5). This means that $V \in \mathscr{M}$. It follows from Eq. (5) that V satisfies the initial condition in the sense of (6). The theorem is proved.

2. Surface Heat Potential. The heat potential $V^{(0)}$ with a density $f = u_0(x) \cdot \delta(t)$ is known as the *surface heat potential* (of a simple layer with a density u_0),

$$V^{(0)} = \mathscr{E} * u_0(x) \cdot \delta(t) = \mathscr{E}(x, t) * u_0(x)$$

If u_0 is of compact support in R^n, then the surface heat potential $V^{(0)}$ is known to exist in $\mathscr{D}'(R^{n+1})$ (cf. Sec. 7.5).

The following theorem gives another test that the surface heat potential exists and also indicates some of its properties.

THEOREM. *If $u_0(x)$ is a function bounded in R^n, then the surface heat potential $V^{(0)}$ exists in $\mathcal{M}$, belongs to the class $C^\infty(t > 0)$, is represented by Poisson's integral*

$$V^{(0)}(x, t) = \frac{\theta(t)}{(2a\sqrt{\pi t})^n} \int_{R^n} u_0(\xi) \exp\left(-\frac{|x-\xi|^2}{4a^2 t}\right) d\xi \qquad (8)$$

and satisfies the inequality

$$|V^{(0)}(x, t)| \leq \sup_\xi |u_0(\xi)|, \qquad t > 0 \qquad (9)$$

If, moreover, the function $u_0(x)$ is continuous in R^n, then the potential $V^{(0)}$ satisfies the initial condition: for each $x \in R^n$

$$V^{(0)}(x, t) \to u_0(x), \qquad t \to +0 \qquad (10)$$

Proof. Since the function

$$h(x, t) = \int |u_0(\xi)| \, \mathscr{E}(x - \xi, t) \, d\xi$$

becomes zero for $t < 0$ and, for $t > 0$, by virtue of (1), satisfies Eq. (9):

$$h(x, t) \leq \sup_\xi |u_0(\xi)| \int \mathscr{E}(x - \xi, t) \, d\xi = \sup_\xi |u_0(\xi)|$$

then this function is locally integrable in R^{n+1}. Consequently, the surface heat potential $V^{(0)} = \mathscr{E}(x, t) * u_0(x)$ is represented by Eq. (8) (cf. Sec. 7.4):

$$V^{(0)}(x, t) = \int u_0(\xi) \mathscr{E}(x - \xi, t) \, d\xi \qquad (8')$$

becomes zero for $t < 0$, and, by virtue of the inequality $|V^{(0)}| \leq h$, satisfies Eq. (9). This means that $V^{(0)} \in \mathcal{M}$. Moreover, it follows from formula (8) that $V^{(0)} \in C^\infty(t > 0)$.

Now, let u_0 be a continuous function bounded in R^n. Taking limiting results (2)* into account, from formula (8') we deduce initial condition (10)

---

* Result (2), which has been proved for functions belonging to $\mathscr{D}(R^n)$ [cf. Sec. 6.5, (f)], is valid also for continuous functions bounded in R^n.

§ 14. THE CAUCHY PROBLEM FOR THE EQUATION OF HEAT CONDUCTION

for the potential $V^{(0)}$:

$$V^{(0)}(x, t) = (\mathscr{E}(x - \xi, t), u_0(\xi)) \to (\delta(x - \xi), u_0(\xi)) = u_0(x), \qquad t \to +0$$

The theorem is proved.

Note. Formula (8) follows formally from formula (4) if we set $f(\xi, \tau) = u_0(\xi) \cdot \delta(\tau)$ and $\delta(\tau)$ is "integrated."

3. Formulation of the Generalized Cauchy Problem for the Heat Conduction Equation. The scheme for solving the Cauchy problem which was set out in Sec. 12.1 for the ordinary linear first-order differential equation is also used in the solution of the Cauchy problem for the heat conduction equation

$$\frac{\partial u}{\partial t} = a^2 \, \Delta u + f(x, t) \tag{11}$$

$$u\big|_{t=+0} = u_0(x) \tag{12}$$

We shall consider that $f \in C(t \geq 0)$ and $u_0 \in C(R^n)$. Let us suppose that there is a classical solution $u(x, t)$ of this problem. This means that $u \in C^2(t > 0) \cap C(t \geq 0)$, and that Eq. (11) for $t > 0$ and initial condition (12) for $t \to +0$ are satisfied (cf. Sec. 4.2).

Continuing functions u and f as zero for $t < 0$, as in Sec. 12.2, we conclude that the continued functions $\tilde{u}$ and $\tilde{f}$ satisfy in R^{n+1} the heat conduction equation

$$\frac{\partial \tilde{u}}{\partial t} = a^2 \, \Delta \tilde{u} + \tilde{f}(x, t) + u_0(x) \cdot \delta(t) \tag{13}$$

As Eq. (13) shows, the initial distribution u_0 for the function $\tilde{u}(x, t)$ plays the role of a momentarily acting source $u_0(x) \cdot \delta(x)$ (of the type of a simple layer over the plane $t = 0$) and the classical solutions of the Cauchy problem (11)-(12) are contained among those solutions of Eq. (13) which become zero for $t < 0$. This enables us to introduce the following generalization of the Cauchy problem for the heat conduction equation.

We shall say that the problem of finding a generalized function $u \in \mathscr{D}'(R^{n+1})$ which becomes zero for $t < 0$ and which satisfies the heat conduction equation

$$\frac{\partial u}{\partial t} = a^2 \, \Delta u + f(x, t) + u_0(x) \cdot \delta(t) \tag{14}$$

is the *generalized Cauchy problem* for the equation of heat conduction with the source $f \in \mathscr{D}'(R^{n+1})$ and initial distribution $u_0 \in \mathscr{D}'(R^n)$. Equation (14) is equivalent to the following (cf. Sec. 10.1): For any $\varphi \in \mathscr{D}(R^{n+1})$ the equation

$$-\left(u, \frac{\partial \varphi}{\partial t}\right) = a^2(u, \Delta\varphi) + (f, \varphi) + (u_0, \varphi(x, 0)) \tag{14'}$$

is true. It follows from Eq. (14) that a necessary condition for the solvability of the generalized Cauchy problem is that f must become zero for $t < 0$.

4. Solution of the Cauchy Problem

THEOREM. *Let $f \in \mathscr{M}$ and u_0 be a function bounded in R^n. Then the solution of the corresponding generalized Cauchy problem exists and is unique in the class $\mathscr{M}$ and appears in the form of a sum of the two heat potentials*

$$u(x, t) = V(x, t) + V^{(0)}(x, t) \tag{15}$$

where the potentials V and $V^{(0)}$ are expressed by the equations (4) and (8). The solution depends continuously on f and u_0 in the following sense: If

$$|f - \tilde{f}| \leq \varepsilon, \qquad |u_0 - \tilde{u}_0| \leq \varepsilon_0$$

then the corresponding solutions u and $\tilde{u}$ in any strip $0 < t \leq T$ satisfy the inequality

$$|u(x, t) - \tilde{u}(x, t)| \leq T\varepsilon + \varepsilon_0 \tag{16}$$

If, moreover, $u \in C(R^n)$, then the solution $u(x, t)$ which has been constructed satisfies the initial condition: for each $x \in R^n$

$$u(x, t) \to u_0(x), \qquad t \to +0 \tag{17}$$

Proof. By virtue of the conditions of the theorem, the convolution of $\mathscr{E}$ with the right-hand side of Eq. (14) exists in $\mathscr{M}$ (cf. Secs. 14.1 and 14.2) and appears in the form of a sum (15) of the two heat potentials V and $V^{(0)}$ and these potentials are expressed by the formulas (4) and (8), respectively. In this way, according to the theorem of Sec. 10.3, formula (15) gives a solution of the generalized Cauchy problem for the heat conduction equation, and this solution is unique in the class $\mathscr{M}$. The continuous de-

§ 14. THE CAUCHY PROBLEM FOR THE EQUATION OF HEAT CONDUCTION

pendence of the solution on the data of the problem f and u_0 follows from estimates (5) and (9). Initial condition (17) follows from (6) and (10).

The theorem is proved.

Notes. (1) For $f = 0$ solution (15) is expressed by Poisson's integral

$$u(x, t) = \frac{1}{(2a\sqrt{\pi t})^n} \int_{R^n} u_0(\xi) \exp\left(-\frac{|x-\xi|^2}{4a^2 t}\right) d\xi \qquad (18)$$

This solution belongs to the class $C^\infty(t > 0)$ and therefore satisfies the homogeneous heat conduction equation for $t > 0$ in the classical sense (cf. Sec. 10.1).

If the function u_0 is continuous and bounded in R^n, then, using formula (18), it is easy to see that $u \in C(t \geq 0)$. Moreover, by the theorem just proved, this solution is unique, belongs to the class $\mathcal{M}$, satisfies initial condition (12), and depends continuously on u_0. So in this case Poisson's integral (18) gives the classical solution of the Cauchy problem for the heat conduction equation, and the classical formulation of this problem is correct, while the intersection $C^\infty(t > 0) \cap C(t \geq 0) \cap \mathcal{M}$ is the correctness class (cf. Sec. 4.6).

(2) The correctness of the solution of the Cauchy problem for the heat conduction equation may be established in a wider class, namely the class of functions which satisfy in any strip $0 \leq t \leq T$ the equation

$$|u(x, t)| \leq C_T \exp(a_T |x|^2)$$

This result belongs to A. N. Tikhonov and A. A. Samarsky (*1*).

5. Exercises. (a) Show that the solution of the mixed problem

$$\frac{\partial u}{\partial t} = a^2 \frac{\partial^2 u}{\partial x^2}, \qquad u\big|_{t=+0} = u_0(x), \qquad u\big|_{x=+0} = \psi(t)$$

is the function

$$u(x, t) = \mathscr{E}(x, t) * \tilde{u}_0(x) - 2a^2 \frac{\partial \mathscr{E}(x, t)}{\partial x} * \psi(t)$$

$$= \frac{1}{2a\sqrt{\pi t}} \int_0^\infty u_0(\xi) \left[\exp\left(-\frac{(x-\xi)^2}{4a^2 t}\right) - \exp\left(-\frac{(x+\xi)^2}{4a^2 t}\right)\right] d\xi$$

$$+ \frac{x}{2a\sqrt{\pi}} \int_0^t \frac{\psi(\tau)}{(t-\tau)^{3/2}} \exp\left(-\frac{x^2}{4a^2(t-\tau)}\right) d\tau$$

Here $u_0 \in C([0, \infty))$ is bounded and $\tilde{u}_0$ is its odd continuation

$$\psi \in C([0. \infty]), \qquad \psi = 0, \qquad t < 0$$

(b) Using the fundamental solution of Schrödinger's operator [cf. Sec. 10.12, (e)], show that the Cauchy problem for Schrödinger's equation (cf. Sec. 2.7) may be reduced to the integral equation

$$\psi(x, t) = \lambda \int_0^t \int \exp\left[i\frac{m|x - \xi|^2}{2\hbar(t - \tau)}\right] \frac{V(\xi, \tau)}{(t - \tau)^{3/2}} \psi(\xi, \tau) \, d\xi \, d\tau$$

$$+ \frac{\lambda \hbar i}{t^{3/2}} \int \exp\left(i\frac{m|x - \xi|^2}{2\hbar t}\right) \psi_0(\xi) \, d\xi,$$

$$\lambda = \frac{1}{\hbar}\left(\frac{m}{2\pi\hbar}\right)^{3/2} \exp\left(i\frac{3\pi}{4}\right)$$

(c) Using the fundamental solution of the transport operator (cf. Sec. 10.11), show that the Cauchy problem for the transport equation (cf. Sec. 2.4) may be reduced to the integral equation

$$\psi(x, \mathbf{s}, t) = \alpha h v \int_0^t \int_{S_1} \psi[x - v(t - \tau)\mathbf{s}, \mathbf{s}', \tau] \exp[-\alpha v(t - \tau)] \, d\mathbf{s}' \, d\tau$$

$$+ v \int_0^t F[x - v(t - \tau)\mathbf{s}, \mathbf{s}, \tau] \exp[-\alpha v(t - \tau)] \, d\tau$$

$$+ \psi_0(x - vt\mathbf{s}, \mathbf{s})e^{-\alpha vt}$$

CHAPTER

4

Integral Equations

Equations which contain an unknown function under the integral sign are known as *integral equations*.

Many problem in mathematical physics are reducible to linear integral equations of the form

$$\int_G \mathscr{K}(x, y)\varphi(y)\, dy = f(x) \tag{1}$$

$$\varphi(x) = \lambda \int_G \mathscr{K}(x, y)\varphi(y)\, dy + f(x) \tag{2}$$

with respect to the unknown function $\varphi(x)$ in the region $G \subset R^n$. Equations (1) and (2) are known as *Fredholm integral equations* of the first and second kind, respectively. The known functions $\mathscr{K}(x, y)$ and $f(x)$ are called the *kernel* and the *inhomogeneous term* (*free term*), respectively, of the integral equation; λ is a complex parameter.

We shall not consider Fredholm integral equations of the first kind here. The integral equation (2) for $f = 0$

$$\varphi(x) = \lambda \int_G \mathscr{K}(x, y)\varphi(y)\, dy \tag{3}$$

is known as a *homogeneous* Fredholm integral equation of the second kind corresponding to Eq. (2). Fredholm integral equations of the second kind

$$\psi(x) = \bar{\lambda} \int_G \mathcal{K}^*(x, y)\psi(y)\, dy + g(x) \tag{2*}$$

$$\psi(x) = \bar{\lambda} \int_G \mathcal{K}^*(x, y)\psi(y)\, dy \tag{3*}$$

where $\mathcal{K}^*(x, y) = \overline{\mathcal{K}(y, x)}$, are said to be *adjoint* to Eqs. (2) and (3), respectively. The kernel $\mathcal{K}^*(x, y)$ is known as the *Hermitian adjoint* kernel to the kernel $\mathcal{K}(x, y)$.

We shall write the integral equations (2), (3), (2*), and (3*) in abbreviated fashion by using the operator notation:

$$\varphi = \lambda K\varphi + f, \qquad \varphi = \lambda K\varphi$$
$$\psi = \bar{\lambda} K^*\psi + g, \qquad \psi = \bar{\lambda} K^*\psi$$

where the integral operators K and K^* are defined by the kernels $\mathcal{K}(x, y)$ and $\mathcal{K}^*(x, y)$, respectively (cf. Sec. 1.8):

$$(Kf)(x) = \int_G \mathcal{K}(x, y)f(y)\, dy$$

$$(K^*f)(x) = \int_G \mathcal{K}^*(x, y)f(y)\, dy$$

All the definitions and facts set out in Secs 1.8–1.10 are applicable to integral operators and equations. Moreover, the following definition is useful: The numbers λ (sometimes complex) for which the homogeneous integral equation (3) has nonzero solutions belonging to $\mathcal{L}_2(G)$ are known as the *eigenvalues* or *characteristic numbers* of the kernel $\mathcal{K}(x, y)$, and the solutions corresponding to these eigenvalues are the *eigenfunctions* of this kernel. In this way, the eigenvalues of the kernel $\mathcal{K}(x, y)$ and the eigenvalues of the operator K are mutually inverse and their eigenfunctions coincide.

§ 15. The Method of Successive Approximations

1. Integral Equations with a Continuous Kernel. Let us suppose that in integral equation (2) the region G is bounded in R^n, the function f is continuous over the closed region $\bar{G}$, and the kernel $\mathcal{K}(x, y)$ is continuous over $\bar{G} \times \bar{G}$ (we shall say that this type of kernel is *continuous*).

§ 15. THE METHOD OF SUCCESSIVE APPROXIMATIONS

We recall the definition of the norms in the spaces $\mathscr{L}_2(G)$ and $C(\bar{G})$ (cf. Secs. 1.3 and 1.5):

$$\|f\| = \left(\int_G |f(x)|^2 \, dx\right)^{1/2}, \quad f \in \mathscr{L}_2(G)$$

$$\|f\|_C = \max_{x \in \bar{G}} |f(x)|, \quad f \in C(\bar{G})$$

In order that the integral operator K with a continuous kernel $\mathscr{K}(x, y)$ should be zero in $\mathscr{L}_2(G)$, it is necessary and sufficient that $\mathscr{K}(x, y) = 0$ for $x \in G, y \in G$.

The sufficiency of the condition is evident, and its necessity follows from Du Bois Reymond's lemma (cf. Sec. 5.5): If for all $f \in \mathscr{L}_2(G)$

$$(Kf)(x) = \int_G \mathscr{K}(x, y) f(y) \, dy = 0, \quad x \in G$$

then $\mathscr{K}(x, y) = 0$ for $x \in G, y \in G$.

LEMMA. *The integral operator K with a continuous kernel $\mathscr{K}(x, y)$ maps $\mathscr{L}_2(G)$ into $C(\bar{G})$ [and, therefore, $C(\bar{G})$ into $C(\bar{G})$ and $\mathscr{L}_2(G)$ into $\mathscr{L}_2(G)$] and is bounded; moreover*

$$\|Kf\|_C \le M\sqrt{V} \, \|f\|, \quad f \in \mathscr{L}_2(G) \tag{4}$$

$$\|Kf\|_C \le MV \, \|f\|_C, \quad f \in C(\bar{G}) \tag{5}$$

$$\|Kf\| \le MV \, \|f\|, \quad f \in \mathscr{L}_2(G) \tag{6}$$

where

$$M = \max_{x \in \bar{G}, y \in \bar{G}} |\mathscr{K}(x, y)|, \quad V = \int_G dy$$

Proof. Let $f \in \mathscr{L}_2(G)$. Then f is an absolutely integrable function over G (cf. Sec. 1.5) and, since the kernel $\mathscr{K}(x, y)$ is continuous over $\bar{G} \times \bar{G}$, the function $(Kf)(x)$ is continuous over $\bar{G}$. Therefore the operator K maps $\mathscr{L}_2(G)$ into $C(\bar{G})$ and, by virtue of the Cauchy–Buniakowski inequality, is bounded:

$$\|Kf\|_C = \max_{x \in \bar{G}} |(Kf)(x)| = \max_{x \in \bar{G}} \left|\int_G \mathscr{K}(x, y) f(y) \, dy\right|$$

$$\le \max_{x \in \bar{G}} \left(\int |\mathscr{K}(x, y)|^2 \, dy\right)^{1/2} \left(\int_G |f(y)|^2 \, dy\right)^{1/2} \le M\sqrt{V} \, \|f\|$$

Inequalities (5) and (6) are proved analogously and more simply. The lemma is proved.

We shall seek the solution of Eq. (2) by the method of successive approximations, supposing that $\varphi^{(0)}(x) = f(x)$,

$$\varphi^{(p)}(x) = \lambda \int_G \mathcal{K}(x, y)\varphi^{(p-1)}(y)\, dy + f(x) \equiv \lambda K\varphi^{(p-1)} + f, \qquad p = 1, 2, \ldots \tag{7}$$

We shall prove that

$$\varphi^{(p)} = \sum_{k=0}^{p} \lambda^k K^k f, \qquad p = 0, 1, \ldots \tag{8}$$

where K^k denotes a power of the operator K (cf. Sec. 1.8).

In fact, for $p = 0$ formula (8) is true: $\varphi^{(0)} = f$. Supposing that this formula is true for p, and replacing p by $p + 1$ in the recurrence relation (7), we obtain formula (8) for $p + 1$:

$$\varphi^{(p+1)} = \lambda K\varphi^{(p)} + f = \lambda K \sum_{k=0}^{p} \lambda^k K^k f + f$$

$$= f + \sum_{k=0}^{p} \lambda^{k+1} K^{k+1} f = \sum_{k=0}^{p+1} \lambda^k K^k f$$

In this way, formula (8) is seen to be true for all p.

The functions $(K^p f)(x)$, $p = 0, 1, \ldots$, are said to be *iterates* of the function f.

According to the lemma of Sec. 15.1, the iterates of f are continuous over $\bar{G}$ and, by virtue of (5), satisfy the inequality

$$\| K^p f \|_C = \| K(K^{p-1} f) \|_C \leq MV \| K^{p-1} f \|_C$$
$$\leq (MV)^2 \| K^{p-2} f \|_C \leq \cdots \leq (MV)^p \| f \|_C$$

that is,

$$\| K^p f \|_C \leq (MV)^p \| f \|_C, \qquad p = 0, 1, \ldots \tag{9}$$

It follows from this result that the series

$$\sum_{k=0}^{\infty} \lambda^k (K^k f)(x), \qquad x \in \bar{G} \tag{10}$$

which is known as *Neumann's series*, is majorized by the numerical series

$$\| f \|_C \sum_{k=0}^{\infty} |\lambda|^k (MV)^k = \frac{\| f \|_C}{1 - |\lambda| MV} \tag{11}$$

§ 15. THE METHOD OF SUCCESSIVE APPROXIMATIONS

converging in the circle $|\lambda| < 1/MV$. Therefore for these λ the series (10) converges regularly on $x \in \bar{G}$, and on account of this it defines a function $\varphi(x)$ which is continuous over $\bar{G}$. This means by virtue of (8), that the successive approximations $\varphi^{(p)}(x)$ as $p \to \infty$ converge uniformly to the function $\varphi(x)$:

$$\varphi^{(p)}(x) \xrightarrow{x \in \bar{G}} \varphi(x) = \sum_{k=0}^{\infty} \lambda^k (K^k f)(x), \qquad p \to \infty \tag{12}$$

moreover, by virtue of (11), the estimate

$$\|\varphi\|_C \leq \frac{\|f\|_C}{1 - |\lambda| MV} \tag{13}$$

is valid.

We shall prove that the function $\varphi(x)$ satisfies integral equation (2). In fact, passing to the limit as $p \to \infty$ in the recurrence relation (7), and using the uniform convergence of the sequence $\varphi^{(p)}(x)$ to $\varphi(x)$ over $\bar{G}$, we obtain

$$\varphi(x) = \lim_{p \to \infty} \varphi^{(p)}(x) = \lambda \int_G \mathcal{K}(x, y) \lim_{p \to \infty} \varphi^{(p-1)}(y) \, dy + f(x)$$
$$= \lambda \int_G \mathcal{K}(x, y) \varphi(y) \, dy + f(x)$$

We shall prove the uniqueness of the solution of Eq. (2) in the class $\mathscr{L}_2(G)$ if $|\lambda| < 1/MV$. For this it is sufficient to show that the homogeneous equation (3) has only a zero solution in this class (cf. Sec. 1.9). In fact, if $\varphi_0 \in \mathscr{L}_2(G)$ is the solution of Eq. (3), $\varphi_0 = \lambda K \varphi_0$, then, according to the lemma of Sec. 15.1,

$$\|\varphi_0\| \leq |\lambda| MV \|\varphi_0\|$$

from which, due to the inequality $|\lambda| MV < 1$, it follows that $\|\varphi_0\|_C = 0$; that is, $\varphi_0 = 0$, as was to be shown.

We now summarize these results in the following theorem.

THEOREM. *Each Fredholm integral equation* (2) *with a continuous kernel* $\mathcal{K}(x, y)$ *for which* $|\lambda| < 1/MV$ *has a unique solution* φ *in the class* $C(\bar{G})$ *for any inhomogeneous term* $f \in C(\bar{G})$. *This solution appears in the form of the Neumann series* (12), *which is regularly convergent in* $\bar{G}$ *and satisfies Eq.* (13). *In other words, the inverse operator* $(I - \lambda K)^{-1}$ *exists and is bounded in the circle* $|\lambda| < 1/MV$.

Note. The method of successive approximations may be used for an approximate solution of integral equation (2) for sufficiently small $|\lambda|$.

2. Iterated Kernels. Resolvents.

First we shall test the validity of the equation

$$(Kf, g) = (f, K^*g), \qquad f \text{ and } g \in \mathscr{L}_2(G) \tag{14}$$

In fact, if f and $g \in \mathscr{L}_2(G)$, then by the lemma of Sec. 15.1, Kf and $K^*g \in \mathscr{L}_2(G)$, and therefore

$$(Kf, g) = \int_G (Kf)\bar{g}\, dx = \int_G \left[\int_G \mathscr{K}(x,y) f(y)\, dy \right] \bar{g}(x)\, dx$$

$$= \int_G f(y) \left[\int_G \mathscr{K}(x,y) \bar{g}(x)\, dx \right] dy = \int_G f \overline{K^*g}\, dy = (f, K^*g)$$

LEMMA. *If K_i, $i = 1, 2$, are integral operators with continuous kernels $K_i(x, y)$, respectively then the operator $K_3 = K_2 K_1$ is an integral operator with the continuous kernel*

$$\mathscr{K}_3(x, y) = \int_G \mathscr{K}_2(x, y') \mathscr{K}_1(y', y)\, dy' \tag{15}$$

The following result is true:

$$K_3^* = (K_2 K_1)^* = K_1^* K_2^* \tag{16}$$

Proof. For all $f \in \mathscr{L}_2(G)$ we have

$$(K_3 f)(x) = (K_2 K_1 f)(x) = \int_G \mathscr{K}_2(x, y') \int_G \mathscr{K}_1(y', y) f(y)\, dy\, dy'$$

$$= \int_G \left[\int_G \mathscr{K}_2(x, y') \mathscr{K}_1(y', y)\, dy' \right] f(y)\, dy$$

from which formula (14) follows. It is clear that the kernel $\mathscr{K}_3(x, y)$ is continuous for $x \in \bar{G}$, $y \in \bar{G}$.

Taking Eq. (14) into account, for all f and $g \in \mathscr{L}_2(G)$ we obtain

$$(f, K_3^* g) = (K_3 f, g) = (K_2 K_1 f, g) = (K_1 f, K^* g) = (f, K_1^* K_2^* g)$$

that is,

$$(f, K_3^* g - K_1^* K_2^* g) = 0$$

and so $K_3^* g = K_1^* K_2^* g$, which is equivalent to Eq. (16). The lemma is proved.

§ 15. THE METHOD OF SUCCESSIVE APPROXIMATIONS

It follows from the lemma which has just been proved that the operators $K^p = K(K^{p-1}) = (K^{p-1})K$, $p = 2, 3, \ldots$, are integral operators and that their kernels $\mathscr{K}_p(x, y)$ are continuous and satisfy the recurrence relations: $\mathscr{K}_1(x, y) = \mathscr{K}(x, y)$,

$$\mathscr{K}_p(x, y) = \int_G \mathscr{K}(x, y')\mathscr{K}_{p-1}(y', y)\, dy'$$
$$= \int_G \mathscr{K}_{p-1}(x, y')\mathscr{K}(y', y)\, dy' \qquad (17)$$

The kernels $\mathscr{K}_p(x, y)$ are known as *iterated kernels* of the kernel $\mathscr{K}(x, y)$.

It follows from the recurrence relations (17) that the iterated kernels satisfy the inequality

$$|\mathscr{K}_p(x, y)| \leq M^p V^{p-1}, \qquad p = 1, 2, \ldots \qquad (18)$$

It follows from (18) that the series

$$\sum_{k=0}^{\infty} \lambda^k \mathscr{K}_{k+1}(x, y), \qquad x \in \bar{G}, \qquad y \in \bar{G} \qquad (19)$$

is majorized by the numerical series

$$\sum_{k=0}^{\infty} |\lambda|^k M^{k+1} V^k$$

converging in the circle $|\lambda| < 1/MV$. Therefore the series (19) converges regularly for $x \in \bar{G}$, $y \in \bar{G}$, and $|\lambda| \leq 1/MV - \varepsilon$ for any $\varepsilon > 0$. Therefore its sum is continuous in $\bar{G} \times \bar{G} \times U_{1/MV}$ and analytic in λ in the circle $|\lambda| < 1/MV$. We shall denote the sum of the series (19) by $\mathscr{R}(x, y; \lambda)$:

$$\mathscr{R}(x, y; \lambda) = \sum_{k=0}^{\infty} \lambda^k \mathscr{K}_{k+1}(x, y)$$

The function $\mathscr{R}(x, y; \lambda)$ is known as the *resolvent* of the kernel $\mathscr{K}(x, y)$.

THEOREM. *The solution φ of the integral equation (2) with the continuous kernel $\mathscr{K}(x, y)$ is unique in the class $C(\bar{G})$ for $|\lambda| < 1/MV$, and for any $f \in C(\bar{G})$ is represented in terms of the resolvent $\mathscr{R}(x, y; \lambda)$ of the kernel $\mathscr{K}(x, y)$ by the equation*

$$\varphi(x) = f(x) + \lambda \int_G \mathscr{R}(x, y; \lambda) f(y)\, dy \qquad (20)$$

In other words, the following operator equation is valid:

$$(I - \lambda K)^{-1} = I + \lambda R, \qquad |\lambda| < \frac{1}{MV} \qquad (21)$$

where R is the integral operator with the kernel $\mathscr{R}(x, y; \lambda)$.

Proof. By the theorem of Sec. 15.1, the solution φ of Eq. (2) is unique in the class $C(\bar{G})$ for $|\lambda| < 1/MV$, and for any $f \in C(\bar{G})$ is represented in the form of the uniformly converging Neumann series (12). If we substitute in this series the expressions for the iterates $K^k f$ in terms of the iterated kernels $\mathscr{K}_k(x, y)$ and use the uniform convergence of the series (19) for the resolvent $\mathscr{R}(x, y; \lambda)$, we obtain formula (20):

$$\varphi(x) = \int_G \left[\lambda \sum_{k=0}^{\infty} \lambda^k \mathscr{K}_{k+1}(x, y) \right] f(y) \, dy + f(x)$$
$$= \lambda \int_G \mathscr{R}(x, y; \lambda) f(y) \, dy + f(x)$$

The theorem is proved.

We shall prove that the iterated kernels $(\mathscr{K}^*)_p(x, y)$ and the resolvent $\mathscr{R}_*(x, y; \lambda)$ of the Hermitian conjugate kernel $\mathscr{K}^*(x, y)$ may be expressed in terms of the iterated kernels $\mathscr{K}_p(x, y)$ and the resolvent of the original kernel $\mathscr{K}(x, y)$ by the equations

$$(\mathscr{K}^*)_p(x, y) = \mathscr{K}_p^*(x, y), \qquad p = 1, 2, \ldots \qquad (22)$$

$$\mathscr{R}_*(x, y; \lambda) = \overline{\mathscr{R}}(y, x; \bar{\lambda}), \qquad |\lambda| < \frac{1}{MV} \qquad (23)$$

Equation (22) follows from Eq. (16), in accordance with which

$$(K^*)^p = (K^p)^*, \qquad p = 1, 2, \ldots$$

Since $|\mathscr{K}^*(x, y)| = |\mathscr{K}(y, x)| \leq M$, then, according to what has just been proved, series (19) for the resolvent $\mathscr{R}_*(x, y; \lambda)$ of the kernel $\mathscr{K}^*(x, y)$ converges for $x \in \bar{G}$, $y \in \bar{G}$, $|\lambda| < 1/MV$. From this, using Eq. (22), we obtain formula (23):

$$\mathscr{R}_*(x, y; \lambda) = \sum_{k=0}^{\infty} \lambda^k (\mathscr{K}^*)_{k+1}(x, y)$$
$$= \sum_{k=0}^{\infty} \lambda^k \mathscr{K}_{k+1}^*(x, y) = \sum_{k=0}^{\infty} \lambda^k \overline{\mathscr{K}}_{k+1}(y, x)$$
$$= \overline{\sum_{k=0}^{\infty} \bar{\lambda}^k \mathscr{K}_{k+1}(y, x)} = \overline{\mathscr{R}}(y, x; \bar{\lambda})$$

§ 15. THE METHOD OF SUCCESSIVE APPROXIMATIONS

From (23) we obtain

$$\mathscr{R}_*(x, y; \bar{\lambda}) = \overline{\mathscr{R}}(y, x; \lambda) = \mathscr{R}^*(x, y; \lambda), \qquad |\lambda| < \frac{1}{MV}$$

and so, according to Eq. (21), formula

$$(I - \bar{\lambda}K^*)^{-1} = I + \bar{\lambda}R^*, \qquad |\lambda| < \frac{1}{MV} \qquad (21^*)$$

is true.

Note. It can be proved that the resolvent $\mathscr{R}(x, y; \lambda)$ of the continuous kernel $\mathscr{K}(x, y)$ allows a meromorphic continuation over the whole plane of the complex variable λ and that the eigenvalues of the kernel $\mathscr{K}(x, y)$ are the poles of the resolvent. This proposition will be proved later for degenerate and for Hermitian kernels.

3. Volterra Integral Equations. Let $n = 1$, the region G be the interval $(0, a)$, and the kernel $\mathscr{K}(x, y)$ become zero in the triangle $0 < x < y < a$. This kernel is known as the *Volterra kernel*. The integral equations (1) and (2) with Volterra kernels take the form

$$\int_0^x \mathscr{K}(x, y)\varphi(y)\,dy = f(x) \qquad (24)$$
$$\varphi(x) = \lambda \int_0^x \mathscr{K}(x, y)\varphi(y)\,dy + f(x)$$

and are known as *Volterra integral equations* of first and second kind, respectively.

Volterra integral equations of the first kind are reducible by differentiation to equations of the second kind

$$\mathscr{K}(x, x)\varphi(x) + \int_0^x \frac{\partial \mathscr{K}(x, y)}{\partial x}\varphi(y)\,dy = f'(x)$$

if $\mathscr{K}(x, y)$ and $\partial \mathscr{K}(x, y)/\partial x$ are continuous for $0 \leq y \leq x \leq a$, $\mathscr{K}(x, x) \neq 0$ for $x \in [0, a]$, $f \in C^1([0, a])$, and $f(0) = 0$. We shall not consider Volterra integral equations of the first kind here.

Let us suppose that in the integral equation (24), $f \in C([0, a])$ and the kernel $\mathscr{K}(x, y)$ is continuous in the closed triangle $0 \leq y \leq x \leq a$. In this case $|\mathscr{K}(x, y)| \leq M$ and the integral operator

$$(Kf)(x) = \int_0^x \mathscr{K}(x, y)f(y)\,dy$$

maps $C([0, a])$ into $C([0, a])$.

As with Fredholm equations (cf. Sec. 15.1), we shall define successive approximations $\varphi^{(p)}$ according to the formula $\varphi^{(0)} = f$,

$$\varphi^{(p)} = \sum_{k=0}^{p} \lambda^k K^k f = \lambda K \varphi^{(p-1)} + f, \qquad p = 1, 2, \ldots \qquad (25)$$

The iterates $K^p f \in C[0, a]$ also satisfy the result

$$|(K^p f)(x)| \leq \|f\|_C \frac{(Mx)^p}{p!}, \qquad x \in [0, a], \qquad p = 0, 1, \ldots \qquad (26)$$

We shall prove result (26) by induction with respect to p. Equation (26) is true for $p = 0$. Supposing that it is true for $p - 1$, we shall prove it for p:

$$|(K^p f)(x)| = |K(K^{p-1} f)| = \left| \int_0^x \mathcal{K}(x, y)(K^{p-1} f)(y) \, dy \right|$$

$$\leq M \|f\|_C M^{p-1} \int_0^x \frac{y^{p-1}}{(p-1)!} \, dy = \|f\|_C \frac{(Mx)^p}{p!}$$

It follows from result (26) that the Neumann series (10) is majorized over $[0, a]$ by the converging numerical series

$$\|f\|_C \sum_{k=0}^{\infty} |\lambda|^k \frac{(Ma)^k}{k!} = \|f\|_C e^{|\lambda| Ma} \qquad (27)$$

and so converges regularly with respect to x in $[0, a]$ for any λ and defines a continuous function $\varphi(x)$. In this way, by virtue of (25), the successive approximations $\varphi^{(p)}$ as $p \to \infty$ uniformly tend to the function φ:

$$\varphi^{(p)}(x) \xrightarrow{x \in [0, a]} \varphi(x) = \sum_{k=0}^{\infty} \lambda^k (K^k f)(x), \qquad p \to \infty \qquad (28)$$

Because of this, by virtue of (27), the inequality

$$\|\varphi\|_C \leq \|f\|_C e^{|\lambda| Ma} \qquad (29)$$

is true.

Proceeding to the limit as $p \to \infty$ in the recurrence relation (25) and using the uniform convergence of the sequence $\varphi^{(p)}$ to φ over $[0, a]$, we find that the function $\varphi(x)$ which has been constructed satisfies the integral equation (24).

§ 15. THE METHOD OF SUCCESSIVE APPROXIMATIONS

We shall prove that the solution of Eq. (24) is unique in the class $C([0, a])$ for any λ. It is sufficient to show that the corresponding homogeneous equation has only a zero solution in this class. In fact, if φ_0 is the solution of the homogeneous equation (24), $\varphi_0 = \lambda K \varphi_0$, then,

$$\varphi_0 = \lambda K(\lambda K \varphi_0) = \lambda^2 K^2 \varphi_0 = \cdots = \lambda^p K^p \varphi_0, \qquad p = 1, 2, \ldots$$

Applying result (26) to these equations gives,

$$|\varphi_0(x)| = |\lambda^p K^p \varphi_0| \le |\lambda|^p \|\varphi_0\|_C \frac{(Mx)^p}{p!}, \qquad p = 1, 2, \ldots$$

and allowing p to tend to infinity, we obtain $\varphi_0(x) = 0$ for $x \in [0, a]$, as was to be shown.

We now formulate the results obtained as the following theorem.

THEOREM. *Each Volterra integral equation* (24) *with a continuous kernel $\mathcal{K}(x, y)$ has a unique solution φ in the class $C([0, a])$ for any λ and for any inhomogeneous term $f \in C([0, a])$. This solution is the regularly convergent Neumann series* (28) *and satisfies inequality* (29).

COROLLARY. *Volterra's continuous kernel does not have eigenvalues.*

4. Integral Equations with a Polar Kernel.

The kernel

$$\mathcal{K}(x, y) = \frac{\mathcal{H}(x, y)}{|x - y|^\alpha}, \qquad \alpha < n$$

where $\mathcal{H}(x, y) \in C(\bar{G} \times \bar{G})$, is known as a *polar kernel*; if $\alpha < n/2$, then $\mathcal{K}(x, y)$ is known as a *weakly polar kernel*.

In order that the kernel $\mathcal{K}(x, y)$ should be polar, it is necessary and sufficient that it be continuous for $x \ne y$, where $x \in \bar{G}$, $y \in \bar{G}$, and that it satisfy the estimate

$$|\mathcal{K}(x, y)| \le \frac{A}{|x - y|^\alpha}, \qquad \alpha < n, \qquad x \in G, \qquad y \in G$$

In fact, the necessity of the condition is evident, and its sufficiency follows from the equation

$$\mathcal{K}(x, y) = \frac{\mathcal{K}(x, y) |x - y|^{\alpha + \varepsilon}}{|x - y|^{\alpha + \varepsilon}}, \qquad 0 < \varepsilon < n - \alpha$$

where the function

$$\mathcal{H}(x, y) = \mathcal{K}(x, y) | x - y |^{\alpha+\varepsilon}$$

is continuous over $\bar{G} \times \bar{G}$.

LEMMA 1. *The integral operator K with the polar kernel $\mathcal{K}(x, y)$ maps $C(\bar{G})$ into $C(\bar{G})$, $\mathcal{L}_2(G)$ into $\mathcal{L}_2(G)$, and is bounded*:

$$\| Kf \|_C \leq N \| f \|_C, \qquad f \in C(\bar{G}) \tag{30}$$

$$\| Kf \| \leq \sqrt{NN^*} \| f \|, \qquad f \in \mathcal{L}_2(G) \tag{31}$$

where

$$N = \max_{x \in \bar{G}} \int_G | \mathcal{H}(x, y) | \, dy, \qquad N^* = \max_{x \in \bar{G}} \int_G | \mathcal{H}^*(x, y) | \, dy$$

Proof. Let $f \in C(\bar{G})$. By virtue of the uniform convergence of the integral over $x \in \bar{G}$

$$\int_G \frac{\mathcal{H}(x, y)}{| x - y |^\alpha} f(y) \, dy = \int_G \mathcal{K}(x, y) f(y) \, dy = (Kf)(x)$$

the operator K maps $C(\bar{G})$ into $C(\bar{G})$ and inequality (30) is valid:

$$\| Kf \|_C = \max_{x \in G} \left| \int_G \mathcal{K}(x, y) f(y) \, dy \right|$$

$$\leq \| f \|_C \max_{x \in G} \int | \mathcal{K}(x, y) | \, dy = N \| f \|_C$$

Let $f \in \mathcal{L}_2(G)$. Using the Cauchy–Buniakowski inequality, we obtain

$$\| Kf \|^2 = \int_G | Kf |^2 \, dx = \int_G \left| \int_G \mathcal{K}(x, y) f(y) \, dy \right|^2 dx$$

$$\leq \int_G \left[\int_G \sqrt{| \mathcal{K}(x, y) |} \sqrt{| \mathcal{K}(x, y) |} \, | f(y) | \, dy \right]^2 dx$$

$$\leq \int_G \int_G | \mathcal{K}(x, y') | \, dy' \int_G | \mathcal{K}(x, y) | \, | f(y) |^2 \, dy \, dx$$

$$\leq NN^* \int_G | f(y) |^2 \, dy = NN^* \| f \|^2$$

from which it follows that the operator K maps $\mathcal{L}_2(G)$ into $\mathcal{L}_2(G)$, and inequality (31) is true. The lemma is proved.

Using Lemma 1, which has just been proved, and repeating the arguments

§ 15. THE METHOD OF SUCCESSIVE APPROXIMATIONS

of Sec. 15.1, we conclude that the theorem of Sec. 15.1 remains valid for the integral equation (2) with the polar kernel $\mathcal{K}(x, y)$ if MV is replaced by N: *If* $|\lambda| < 1/N$, *then in the class* $C(\bar{G})$ *there is a unique solution for any* $f \in C(\bar{G})$, *and this solution is the Neumann series which is regularly convergent over* $\bar{G}$.

LEMMA 2. *If* $\mathcal{K}_i(x\ y)$ *are the polar kernels*

$$|\mathcal{K}_i(x, y)| \leq \frac{A_i}{|x - y|^{\alpha_i}}, \qquad \alpha_i < n, \qquad i = 1, 2$$

then the kernel

$$\mathcal{K}_3(x, y) = \int_G \mathcal{K}_2(x, y') \mathcal{K}_1(y', y)\, dy'$$

is also polar, moreover

$$|\mathcal{K}_3(x, y)| \leq \begin{cases} \dfrac{A_3}{|x - y|^{\alpha_1 + \alpha_2 - n}}, & \text{if } \alpha_1 + \alpha_2 > n \\ A_4 |\ln|x - y|| + A_5, & \text{if } \alpha_1 + \alpha_2 = n \end{cases} \tag{32}$$

$\mathcal{K}_3(x, y)$ *is continuous over* $\bar{G} \times \bar{G}$ *if* $\alpha_1 + \alpha_2 < n$.

Proof. Representing the polar kernels $\mathcal{K}_i(x, y)$ in the form

$$\mathcal{K}_i(x, y) = \frac{\mathcal{H}_i(x, y)}{|x - y|^{\alpha_i + \varepsilon}}, \qquad 0 < \varepsilon < n - \alpha_i$$

where $\mathcal{H}_i(x, y)$ are continuous functions over $\bar{G} \times \bar{G}$, we express the kernel $\mathcal{K}_3(x, y)$ in the form

$$\mathcal{K}_3(x, y) = \int_G \frac{\mathcal{H}_2(x, y')\mathcal{H}_1(y', y)}{|x - y'|^{\alpha_2 + \varepsilon} |y' - y|^{\alpha_1 + \varepsilon}}\, dy'$$

Since this integral converges uniformly for $x \in \bar{G}$, $y \in \bar{G}$ and $|x - y| \geq \delta$ for any $\delta > 0$, we conclude that the kernel $\mathcal{K}_3(x, y)$ is continuous for $x \neq y$, $x \in \bar{G}$, $y \in \bar{G}$. If then $\alpha_1 + \alpha_2 < n$, for $\varepsilon < \frac{1}{2}(n - \alpha_1 - \alpha_2)$ and $\varepsilon > 0$, this integral converges uniformly for $x \in \bar{G}$, $y \in \bar{G}$, and therefore the kernel $\mathcal{K}_3(x, y)$ is continuous over $\bar{G} \times \bar{G}$.

So, to prove the lemma, we must establish results (32) for $\alpha_1 + \alpha_2 \geq n$. Taking into account the results for the kernels $\mathcal{K}_i(x, y)$, we have

$$|\mathcal{K}_3(x, y)| \leq A_1 A_2 \int_G \frac{dy'}{|x - y'|^{\alpha_2} |y' - y|^{\alpha_1}}, \qquad x \in G, \qquad y \in G$$

Changing the variable of integration to $\eta = x - y'$, and enlarging the region of integration to the sphere U_D, where D is the diameter of the region G (Fig. 47), we deduce the equation

$$|\mathcal{K}_3(x,y)| \leq A_1 A_2 \int_{U_D} \frac{d\eta}{|\eta|^{\alpha_2} |x-y-\eta|^{\alpha_1}}$$

Writing

$$|x-y| = r, \quad \frac{x-y}{r} = s, \quad |s| = 1$$

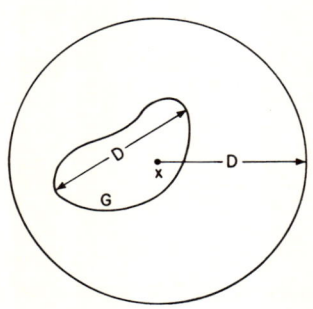

Fig. 47

and in the last integral making the change of variable $\eta = r\xi$, $d\eta = r^n d\xi$, we obtain

$$|\mathcal{K}_3(x,y)| \leq A_1 A_2 r^{n-\alpha_1-\alpha_2} \int_{U_{D/r}} \frac{d\xi}{|\xi|^{\alpha_2} |s-\xi|^{\alpha_1}}$$

$$= A_1 A_2 r^{n-\alpha_1-\alpha_2} \bigg(\int_{|\xi| \leq 2} \frac{d\xi}{|\xi|^{\alpha_2} |s-\xi|^{\alpha_1}}$$

$$+ \int_{2 < |\xi| < D/r} \frac{d\xi}{|\xi|^{\alpha_2} |s-\xi|^{\alpha_1}} \bigg) \qquad (33)$$

Since $|s| = 1$, the integral in the first term is of bounded magnitude

$$\int_{|\xi| \leq 2} \frac{d\xi}{|\xi|^{\alpha_2} |s-\xi|^{\alpha_1}} \leq A_6 \qquad (34)$$

Taking the inequality into account

$$|\xi - s| \geq |\xi| - |s| = |\xi| - 1 \geq \tfrac{1}{2} |\xi|$$

§ 15. THE METHOD OF SUCCESSIVE APPROXIMATIONS

we shall evaluate the second integral when $|\xi| > 2$,

$$\int_{2<|\xi|<D/r} \frac{d\xi}{|\xi|^{\alpha_2}|s-\xi|^{\alpha_1}} \leq 2^{\alpha_1} \int_{1<|\xi|<D/r} \frac{d\xi}{|\xi|^{\alpha_1+\alpha_2}} = 2^{\alpha_1}\sigma_n \int_1^{D/r} \varrho^{n-1-\alpha_1-\alpha_2}\,d\varrho$$

$$\leq 2^{\alpha_1}\sigma_n \begin{cases} \dfrac{1}{\alpha_1+\alpha_2-n}, & \text{if } \alpha_1+\alpha_2 > n \\[1em] \ln \dfrac{D}{r}, & \text{if } \alpha_1+\alpha_2 = n \end{cases} \tag{35}$$

Equations (32) follow from Eq. (33)–(35). The lemma is proved.

It follows from Lemma 2, which has just been proved, that all the iterated kernels $\mathcal{H}_p(x,y)$ of the polar kernel $\mathcal{H}(x,y)$ are polar and satisfy the estimates

$$|\mathcal{H}_p(x,y)| \leq \begin{cases} A_p|x-y|^{-p\alpha+(p-1)n}, & \text{if } p\alpha-(p-1)n > 0 \\ A_p|\ln|x-y|| + B_p, & \text{if } p\alpha-(p-1)n = 0 \end{cases} \tag{36}$$

Beginning with the number $p_0 = [n/(n-\alpha)] + 1$, the iterated kernels $\mathcal{H}_p(x,y)$ are continuous. (Here $[t]$ denotes the integral part of the number $t \geq 0$.)

From this, using Lemma 1 of this section and reasoning as in Sec. 15.2, we deduce that the resolvent of the polar kernel $\mathcal{H}(x,y)$

$$\mathcal{R}(x,y;\lambda) = \sum_{k=0}^{\infty} \lambda^k \mathcal{H}_{k+1}(x,y)$$
$$= \mathcal{R}_1(x,y;\lambda) + \mathcal{R}_2(x,y;\lambda) \tag{37}$$

is the sum of two terms: of the polar term

$$\mathcal{R}_1(x,y;\lambda) = \sum_{k=0}^{p_0-2} \lambda^k \mathcal{H}_{k+1}(x,y)$$

and the continuous term

$$\mathcal{R}_2(x,y;\lambda) = \sum_{k=p_0-1}^{\infty} \lambda^k \mathcal{H}_{k+1}(x,y) \tag{38}$$

For this series (38) converges uniformly for $x \in \bar{G}$, $y \in \bar{G}$, and $|\lambda| \leq 1/N - \varepsilon$ for any $\varepsilon > 0$, defining a continuous function $\mathcal{R}_2(x,y;\lambda)$ for $x \in \bar{G}$, $y \in \bar{G}$, $|\lambda| < 1/N$, and an analytic function with respect to λ in the circle $|\lambda| < 1/N$.

It follows from what has been said that the *theorem of Sec. 15.2 remains valid for the integral equation* (2) *with a polar kernel* $\mathcal{K}(x, y)$ *on the condition that* $|\lambda| < 1/N$. Further, the formulas (22), (23), and (21*) for $(\mathcal{K}_p^*)(x, y)$, $\mathcal{R}_*(x, y; \lambda)$, and $(I - \bar{\lambda}K^*)^{-1}$ may also be retained, provided $|\lambda| < 1/N$ and $|\lambda| < 1/N^*$.

5. Exercises. (a) Prove that if $|\lambda| MV < 1$, the resolvent $\mathcal{R}(x, y; \lambda)$ of the continuous kernel $\mathcal{K}(x, y)$ satisfies the Fredholm integral equation

$$\mathcal{R}(x, y; \lambda) = \lambda \int_G \mathcal{K}(x, y') \mathcal{R}(y', y; \lambda) \, dy' + \mathcal{K}(x, y)$$

(b) Let the kernel $\mathcal{K}(x, y)$ of the Fredholm integral equation (2) belong to $\mathcal{L}_2(G \times G)$. Using Eq. (28) of Sec. 1.8, establish the convergence in $\mathcal{L}_2(G)$ of the method of successive approximations for any $f \in \mathcal{L}_2(G)$, provided $|\lambda| C < 1$.

(c) Prove that the resolvent of the Volterra kernel is analytic in the whole plane of the complex variable λ (an entire function).

(d) Let $\mathcal{K} \in C(x \geq 0)$, $\mathcal{K}(x) = 0$ for $x < 0$. Prove that the generalized function

$$\mathcal{E}(x) = \delta(x) + \mathcal{R}(x), \qquad \mathcal{R} = \sum_{k=1}^{\infty} \underbrace{\mathcal{K} * \mathcal{K} * \cdots * \mathcal{K}}_{k \text{ times}}$$

is a fundamental solution of the Volterra operator of the second kind with the kernel $\mathcal{K}(x - y)$:

$$\mathcal{E} - \mathcal{K} * \mathcal{E} = \delta$$

For this the series for $\mathcal{R}(x)$ converges uniformly in each finite interval and satisfies the Volterra integral equation

$$\mathcal{R}(x) = \int_0^x \mathcal{K}(x - y) \mathcal{R}(y) \, dy + \mathcal{K}(x), \qquad x \geq 0$$

The function $\mathcal{R}(x - y)$ is the resolvent of the kernel $\mathcal{K}(x - y)$ for $\lambda = 1$.

(e) Using Exercise (c) of Sec. 7.9 for $\alpha = 0$ and $\beta = 1 - \alpha$, check that the function

$$f(x) = \frac{\sin \pi \alpha}{\pi} \int_0^x \frac{g'(y) \, dy}{(x - y)^{1-\alpha}}$$

satisfies Abel's integral equation

$$\int_0^x \frac{f(y)}{(x - y)^\alpha} \, dy = g(x), \qquad g(0) = 0, \qquad g \in C^1(x \geq 0), \qquad 0 < \alpha < 1$$

(f) Using fundamental solutions (23) of Sec. 10.5, establish that the Cauchy problems

$$u' + au = \varrho u, \qquad u|_{t=+0} = u_0, \qquad \varrho \in C(t \geq 0)$$

$$u'' + a^2 u = \varrho u, \qquad u|_{t=+0} = u_0, \qquad u'|_{t=+0} = u_1$$

are equivalent, respectively, to the following Volterra integral equations of the second kind

$$u(t) = \int_0^t e^{-a(t-\tau)} \varrho(\tau) u(\tau) \, d\tau + u_0 e^{-at}, \qquad t \geq 0$$

$$u(t) = \frac{1}{a} \int_0^t \sin a(t-\tau) \varrho(\tau) u(\tau) \, d\tau + u_0 \cos at + \frac{u_1}{a} \sin at, \qquad t \geq 0$$

(g) Prove that for $|\lambda| < 1$, Milne's integral equation

$$\varphi(x) = \lambda \int_0^\infty \mathscr{H}(x-y) \varphi(y) \, dy, \qquad \mathscr{H}(\xi) = \frac{1}{2} \int_{|\xi|}^\infty \frac{e^{-t}}{t} \, dt$$

has a unique solution $\varphi = 0$ in the class of bounded functions on $[0, \infty)$.

§ 16. Fredholm's Theorems

In this section Fredholm's theorems concerning solubility will be proved for the Fredholm integral equation

$$\varphi = \lambda K \varphi + f \tag{1}$$

with a continuous kernel $\mathscr{H}(x, y)$ and for the equation adjoint to it

$$\psi = \bar{\lambda} K^* \psi + g \tag{1*}$$

1. Integral Equations with a Degenerate Kernel. The kernel

$$\mathscr{H}(x, y) = \sum_{i=1}^N f_i(x) g_i(y) \tag{2}$$

where f and $g \in C(\bar{G})$, is known as a *degenerate kernel*.

Without loss of generality, it may be assumed that the systems of functions $\{f_i, 1 \leq i \leq N\}$ and $\{g_i, 1 \leq i \leq N\}$ are linearly independent. In fact, if this is not so, then for instance

$$f_N(x) = c_1 f_1(x) + \cdots + c_{N-1} f_{N-1}(x)$$

and the kernel $\mathscr{H}(x, y)$, by virtue of (2), assumes the form

$$\mathscr{H}(x, y) = \sum_{i=1}^{N-1} f_i(x) g_i(y) + \sum_{i=1}^{N-1} c_i f_i(x) g_N(y) = \sum_{i=1}^{N-1} f_i(x) g_i^*(y)$$

Arguing in a similar fashion, by means of a finite number of steps, we reach the situation in which there are linearly independent systems of functions $\{f_i\}$ and $\{g_i\}$ in Eq. (2).

Let us consider the Fredholm integral equation with a degenerate kernel (2)

$$\varphi(x) = \lambda \sum_{i=1}^{N} f_i(x) \int_G g_i(y)\varphi(y)\,dy + f(x) \qquad (3)$$

and the equation adjoint to it

$$\psi(x) = \bar{\lambda} \sum_{i=1}^{N} \bar{g}_i(x) \int_G \bar{f}_i(y)\psi(y)\,dy + g(x) \qquad (3^*)$$

Solutions φ and ψ of the integral equations (3) and (3*) will be sought in the class $C(\bar{G})$.

We shall show that these equations are reducible to systems of linear algebraic equations, and so may be investigated and solved by known methods of linear algebra.

We shall write Eq. (3) in the form

$$\varphi(x) = \lambda \sum_{i=1}^{N} c_i f_i(x) + f(x) \qquad (4)$$

where

$$c_i = \int_G \varphi(y) g_i(y)\,dy = (\varphi, \bar{g}_i) \qquad (5)$$

are unknown numbers. Multiplying Eq. (4) by $g_k(x)$, integrating over the region G, and using (5), we obtain the following system of linear algebraic equations defining the numbers c_k:

$$c_k = \lambda \sum_{i=1}^{N} c_i \int_G g_k(x) f_i(x)\,dx + \int_G g_k(x) f(x)\,dx \qquad (6)$$

Setting

$$\alpha_{ki} = \int_G g_k(x) f_i(x)\,dx, \qquad a_k = \int_G f(x) g_k(x)\,dx = (f, \bar{g}_k) \qquad (7)$$

we write system (6) in the form

$$c_k = \lambda \sum_{i=1}^{N} \alpha_{ki} c_i + a_k, \qquad k = 1, 2, \ldots, N \qquad (8)$$

§ 16. FREDHOLM'S THEOREMS

Introducing the matrix A and the vectors $\mathbf{c}$ and $\mathbf{a}$,

$$A = (\alpha_{ki}), \quad \mathbf{c} = (c_1, c_2, \ldots, c_N), \quad \mathbf{a} = (a_1, a_2, \ldots, a_N)$$

we may present system (8) in the matrix form

$$\mathbf{c} = \lambda A \mathbf{c} + \mathbf{a} \tag{9}$$

We shall prove that the integral equation (3) and the algebraic equation (9) are equivalent. In fact, if $\varphi \in C(\bar{G})$ is the solution of Eq. (3), then, as we have just shown, the numbers $c_i = (\varphi, \bar{g}_i)$, $i = 1, 2, \ldots, N$, satisfy system (8). Conversely, if the numbers c_i, $i = 1, 2, \ldots, N$, satisfy system (8), then the function $\varphi(x)$, constructed according to formula (4), is continuous over $\bar{G}$ and, by virtue of (7), satisfies Eq. (3):

$$\varphi(x) - \lambda \sum_{i=1}^{N} f_i(x) \int_G g_i(y) \varphi(y) \, dy - f(x)$$

$$= \lambda \sum_{i=1}^{N} c_i f_i(x) + f(x) - \lambda \sum_{i=1}^{N} f_i(x) \int_G g_i(y)$$

$$\times \left[\lambda \sum_{k=1}^{N} c_k f_k(y) + f(y) \right] dy - f(x)$$

$$= \lambda \sum_{i=1}^{N} f_i(x) \left(c_i - \lambda \sum_{k=1}^{N} c_k \alpha_{ik} - a_i \right) = 0$$

We shall denote by $D(\lambda)$ the determinant of system (9),

$$D(\lambda) = \det(I - \lambda A) \tag{10}$$

and by $M_{ki}(\lambda)$ the cofactors of the matrix $I - \lambda A$. It is clear that $D(\lambda)$ and $M_{ki}(\lambda)$ are polynomials in λ, and that $D(\lambda) \not\equiv 0$, since $D(0) = \det I = 1$.

Let the (complex) number λ be such that $D(\lambda) \neq 0$. According to Cramer's theorem the solution of the algebraic system (9) is unique and can be expressed by the equation

$$c_k = \frac{1}{D(\lambda)} \sum_{i=1}^{N} M_{ki}(\lambda) a_i, \quad k = 1, 2, \ldots, N \tag{11}$$

Substituting this solution (11) into formula (4) and recalling the definition of the numbers a_k, we obtain the solution of the integral equation (3) for

$D(\lambda) \neq 0$ in the form

$$\varphi(x) = \frac{\lambda}{D(\lambda)} \sum_{i,k=1}^{N} M_{ik}(\lambda) f_i(x) \int_G g_k(y) f(y)\, dy + f(x) \tag{12}$$

On the other hand, according to the theorem of Sec. 15.2, for sufficiently small λ [and then $D(\lambda) \neq 0$], this solution is expressed in terms of the resolvent $\mathscr{R}(x, y; \lambda)$ by Eq. (20) of Sec. 15.2. Therefore,

$$\mathscr{R}(x, y; \lambda) = \frac{1}{D(\lambda)} \sum_{i,k=1}^{N} M_{ik}(\lambda) f_i(x) g_k(y) \tag{13}$$

In this way, the resolvent $\mathscr{R}(x, y; \lambda)$ of the degenerate kernel is seen to be a rational function and so allows a meromorphic continuation over the whole plane of the complex variable λ (cf. note of Sec. 15.2).

2. Fredholm's Theorems for Integral Equations with a Degenerate Kernel. In the previous section we have constructed in an explicit form the solution of an integral equation with a degenerate kernel. Here we shall continue our investigation of these equations and shall formulate the conditions for their solubility.

As with Eq. (3), we change Eq. (3*), which is adjoint to it, to an equivalent system of linear algebraic equations. We have

$$\psi(x) = \bar{\lambda} \sum_{i=1}^{N} d_i \bar{g}_i(x) + g(x) \tag{4*}$$

where $d_i = (\psi, f_i)$ are unknown numbers. The corresponding system of linear algebraic equations, equivalent to Eq. (3*), has the form

$$d_k = \bar{\lambda} \sum_{i=1}^{N} \beta_{ki} d_i + b_k, \qquad k = 1, 2, \ldots, N \tag{8*}$$

where

$$\beta_{ki} = \int_G \bar{f}_k(x) \bar{g}_i(x)\, dx = \bar{\alpha}_{ik}, \qquad b_k = (g, f_k) \tag{7*}$$

In this way, the system (8*) is adjoint to system (8):

$$\mathbf{d} = \bar{\lambda} A^* \mathbf{d} + \mathbf{b} \tag{9*}$$

where

$$A^* = (\beta_{ki}) = (\bar{\alpha}_{ik}) = \bar{A}', \quad \mathbf{d} = (d_1, d_2, \ldots, d_N), \quad \mathbf{b} = (b_1, b_2, \ldots, b_N)$$

§ 16. FREDHOLM'S THEOREMS

From courses on linear algebra it is known that the determinant and rank of a matrix and its transpose coincide. Therefore, by virtue of (10),

$$\det(I - \bar{\lambda}A^*) = \det(I - \bar{\lambda}\bar{A}') = \overline{\det(I - \lambda A')} = \overline{D(\lambda)}$$
$$\operatorname{rang}(I - \bar{\lambda}A^*) = \operatorname{rang}(\overline{I - \lambda A'}) = \operatorname{rang}(I - \lambda A) = q \tag{14}$$

There are the following alternatives.

I. $D(\lambda) \neq 0$. Then $q = N$ and the systems (9) and (9*) are uniquely soluble for any **a** and **b**. Consequently, Eqs. (3) and (3*) are also uniquely soluble for any f and g and these solutions are given by expressions (4) and (4*), respectively.

II. $D(\lambda) = 0$. Then $q < N$ and, by virtue of (14), the homogeneous systems (9) and (9*) each have $N - q$ linearly independent solutions:

$$\mathbf{c}^{(s)} = (c_1^{(s)}, c_2^{(s)}, \ldots, c_N^{(s)}), \quad \mathbf{d}^{(s)} = (d_1^{(s)}, d_2^{(s)}, \ldots, d_N^{(s)}), \quad s = 1, 2, \ldots, N-q$$

Homogeneous integral equations (3) and (3*) will also each have $N - q$ linearly independent solutions, defined by expressions (4) and (4*), respectively:

$$\varphi_s(x) = \lambda \sum_{i=1}^{N} c_i^{(s)} f_i(x), \quad \psi_s(x) = \bar{\lambda} \sum_{i=1}^{N} d_i^{(s)} \bar{g}_i(x) \tag{15}$$
$$s = 1, 2, \ldots, N - q$$

We shall prove the linear independence of these systems of solutions $\{\varphi_s, 1 \leq s \leq N - q\}$ and $\{\psi_s, 1 \leq s \leq N - q\}$. Let there be numbers p_s, $s = 1, 2, \ldots, N - q$, such that

$$\sum_{s=1}^{N-q} p_s \varphi_s(x) = 0, \quad x \in G$$

that is, by virtue of (15)

$$\sum_{i=1}^{N} f_i(x) \sum_{s=1}^{N-q} c_i^{(s)} p_s = 0, \quad x \in G$$

From this, by virtue of the linear independence of the system of functions $\{f_i, 1 \leq i \leq N\}$, follow the equations

$$\sum_{s=1}^{N-q} c_i^{(s)} p_s = 0, \quad i = 1, 2, \ldots, N$$

Since the system of vectors $\{\mathbf{c}^{(s)}, 1 \leq s \leq N - q\}$ is linearly independent in R^N, it follows from the last equations that $p_s = 0$, $s = 1, 2, \ldots, N - q$, which also proves the linear independence of the system of solutions $\{\varphi_s\}$. The linear independence of the system of solutions $\{\psi_s\}$ is established analogously.

Further, to solve the system (9) for $D(\lambda) = 0$ it is necessary and sufficient to satisfy the following orthogonality conditions:

$$(\mathbf{a}, \mathbf{d}^{(s)}) = \sum_{i=1}^{N} a_i \bar{d}_i^{(s)} = 0, \qquad s = 1, 2, \ldots, N - q \tag{16}$$

Conditions (16) are equivalent to the conditions

$$(f, \psi_s) = \int_G f(x) \bar{\psi}_s(x)\, dx = 0, \qquad s = 1, 2, \ldots, N - q$$

since, by virtue of (15) and (7),

$$\int_G f(x) \bar{\psi}_s(x)\, dx = \lambda \sum_{i=1}^{N} \int_G f(x) g_i(x)\, dx\, \bar{d}_i^{(s)} = \lambda \sum_{i=1}^{N} a_i \bar{d}_i^{(s)} = \lambda(\mathbf{a}, \mathbf{d}^{(s)})$$

So the following theorems, which are known as *Fredholm's theorems*, have been proved.

THEOREM 1. *If $D(\lambda) \neq 0$, then Eq. (3) and its adjoint equation (3*) are uniquely soluble for any inhomogeneous terms f and g.*

THEOREM 2. *If $D(\lambda) = 0$, then the homogeneous equations (3) and (3*) have a like number of linearly independent solutions, equal to $N - q$ in number, where q is the rank of the matrix $I - \lambda A$.*

THEOREM 3. *If $D(\lambda) = 0$, then for Eq. (3) to have a solution it is necessary and sufficient that the inhomogeneous term f be orthogonal to all the solutions ψ_s, $s = 1, 2, \ldots, N - q$, of the adjoint homogeneous equation (3*).*

It follows from Theorems 1 and 2 that the eigenvalues of the degenerate kernel coincide with the roots of the polynomial $D(\lambda)$, and therefore are finite in number. Moreover from formula (13) for the resolvent it follows that the eigenvalues of the degenerate kernel coincide with the poles of its resolvent (cf. Sec. 15.2, note).

Note. It may happen that the functions f_i and g_i in Eq. (2) of the degenerate kernel depend on the complex parameter λ n the following way:

§ 16. FREDHOLM'S THEOREMS

Let $f_i(x; \lambda)$ and $g_i(x; \lambda)$ be continuous with respect to (x, λ) in $\bar{G} \times U_\omega$ and analytic in λ in the circle U_ω. *In this case Fredholm's theorems 1–3 remain valid on condition that $|\lambda| < \omega$.*

We shall prove that the *determinant* $D(\lambda)$ *is an analytic function in the circle* $|\lambda| < \omega$. In fact, the elements of the matrix A, calculated according to formula (7),

$$\alpha_{ki}(\lambda) = \int_G g_k(x; \lambda) f_i(x; \lambda) \, dx$$

are analytic functions in the circle $|\lambda| < \omega$. Therefore, by virtue of (10), $D(\lambda)$ is an analytic function in this circle, and $D(\lambda) \neq 0$.

3. Fredholm's Theorems for Integral Equations with a Continuous Kernel. The Fredholm theorems for integral equations with a degenerate kernel that have been proved in the previous section may be extended to integral equations with an arbitrary continuous kernel. The essence of the proof lies in the fact that a continuous kernel appears in the form of a sum of a degenerate kernel and of a sufficiently small continuous kernel. This makes it possible, using the results of Sec. 15 concerning the solubility of integral equations with a small kernel, to reduce the corresponding integral equation to an integral equation with a degenerate kernel, for which Fredholm's theorems have already been established. We may then deduce that Fredholm's theorems are also valid for integral equations with a continuous kernel.

So let the kernel $\mathscr{H}(x, y)$ be continuous over $\bar{G} \times \bar{G}$. According to Weierstrass's theorem (cf. Sec. 1.3), it may be approximated with arbitrary accuracy by polynomials; that is, for any $\varepsilon > 0$ there is a polynomial

$$\mathscr{P}(x, y) = \sum_{|\alpha+\beta| \leq N} a_{\alpha\beta} x^\alpha y^\beta \tag{17}$$

such that

$$|\mathscr{H}(x, y) - \mathscr{P}(x, y)| < \varepsilon, \qquad x \in \bar{G}, \quad y \in \bar{G}$$

In this way, the kernel $\mathscr{H}(x, y)$ appears in the form

$$\mathscr{H}(x, y) = \mathscr{P}(x, y) + \eta(x, y) \tag{18}$$

where $\mathscr{P}(x, y)$ is a degenerate kernel (polynomial) and $\eta(x, y)$ is a small continuous kernel, $|\eta(x, y)| < \varepsilon$ for $x \in \bar{G}$, $y \in \bar{G}$.

By virtue of (18), Fredholm's integral equation takes the form

$$\varphi = \lambda P \varphi + \lambda \eta \varphi + f \tag{19}$$

where P and η are integral operators with the kernels $\mathscr{P}(x, y)$ and $\eta(x, y)$, respectively, and $P + \eta = K$.

We shall show that for $|\lambda| < 1/\varepsilon V$ in the class $C(\bar{G})$, the integral equation (19) is equivalent to an integral equation with a degenerate kernel. For this we shall introduce a new unknown function $\Phi(x)$ according to the formula

$$\Phi = \varphi - \lambda \eta \varphi \tag{20}$$

According to the theorem of Sec. 15.2, the function φ is *expressible uniquely* in terms of Φ by means of the expression

$$\varphi = (I - \lambda \eta)^{-1} \Phi = (I + \lambda R) \Phi \tag{21}$$

where R is an integral operator with a kernel $\mathscr{R}(x, y; \lambda)$, which is the resolvent of the kernel $\eta(x, y)$. By virtue of (20) and (21), Eq. (19) takes the following equivalent form:

$$\Phi = \lambda P(I + \lambda R)\Phi + f = \lambda T\Phi + f \tag{22}$$

where

$$T = P + \lambda PR \tag{23}$$

We recall that the resolvent $\mathscr{R}(x, y; \lambda)$ is continuous with respect to $(x, y; \lambda)$ in $\bar{G} \times \bar{G} \times U_{1/\varepsilon V}$, and analytic in λ in the circle $|\lambda| < 1/\varepsilon V$ (cf. Sec. 15.2). From this, taking the lemma of Sec. 15.2 into account, we conclude that the operator T is an integral with a continuous kernel

$$\mathscr{E}(x, y; \lambda) = \mathscr{P}(x, y) + \lambda \int_G \mathscr{P}(x, y') \mathscr{R}(y', y; \lambda) \, dy'$$

Moreover, it follows from (17) that the kernel $\mathscr{E}(x, y; \lambda)$ is degenerate and analytic in λ in the circle $|\lambda| < 1/\varepsilon V$.

We shall now transform the adjoint integral equation (1*). By virtue of (18), $K^* = P^* + \eta^*$, and therefore Eq. (1*) takes the form

$$(I - \bar{\lambda} \eta^*) \psi = \bar{\lambda} P^* \psi + g \tag{19*}$$

Applying the operator $(I - \bar{\lambda} \eta^*)^{-1}$ to Eq. (19*) and using Eq. (21*) of Sec. 15.2, we have

$$(I - \bar{\lambda} \eta^*)^{-1} = I + \bar{\lambda} R^*, \qquad |\lambda| < \frac{1}{\varepsilon V}$$

§ 16. FREDHOLM'S THEOREMS

and we reduce this to the equivalent equation

$$\psi = (I - \bar{\lambda}\eta^*)^{-1}(\bar{\lambda}P^*\psi + g) = (I + \bar{\lambda}R^*)(\bar{\lambda}P^*\psi + g)$$
$$= \bar{\lambda}(P^* + \bar{\lambda}R^*P^*)\psi + (I + \bar{\lambda}R^*)g \tag{24}$$

Setting

$$Q = (I + \bar{\lambda}R^*)g, \qquad g = (I - \bar{\lambda}\eta^*)Q \tag{25}$$

and considering that, in accordance with Eqs. (16) of Sec. 15.2 and (23) above,

$$P^* + \bar{\lambda}R^*P^* = (P + \lambda PR)^* = T^*$$

we shall write Eq. (24) in the form

$$\psi = \bar{\lambda}T^*\psi + Q \tag{22*}$$

In this way, for $|\lambda| < 1/\varepsilon V$ in the class $C(\bar{G})$ the integral equation (1) is equivalent to integral equation (22) with the degenerate kernel $\mathscr{E}(x, y; \lambda)$, which is analytic in the circle $|\lambda| < 1/\varepsilon V$, while Eq. (1*) adjoint to Eq. (1) is equivalent to Eq. (22*), which is adjoint to Eq. (22). But for Eqs. (22) and (22*) Fredholm's theorems 1–3 are valid and the determinant $D(\lambda)$ is an analytic function in the circle $|\lambda| < 1/\varepsilon V$ (cf. note of Sec. 16.2). Hence, as these equations are equivalent to the initial equations (1) and (1*), we obtain the following Fredholm theorems for integral equations with a continuous kernel The set of these theorems is known as *Fredholm's alternative theorems*.

FREDHOLM'S ALTERNATIVE THEOREMS. *If the integral equation* (1) *with a continuous kernel has a solution in* $C(\bar{G})$ *for any inhomogeneous term* $f \in C(\bar{G})$ *then its adjoint equation* (1*) *has a solution in* $C(\bar{G})$ *for any inhomogeneous term* $g \in C(\bar{G})$, *and these solutions are unique* (Fredholm's first theorem).

If the integral equation (1) *has no solution in* $C(\bar{G})$ *for any inhomogeneous term* f, *then*

(1) *the homogeneous equations* (1) *and* (1*) *have an equal (finite) number of linearly independent solutions* (Fredholm's second theorem);

(2) *for Eq.* (1) *to have a solution it is necessary and sufficient that the inhomogeneous term* f *be orthogonal to all the solutions of the adjoint homogeneous equation* (2*) (Fredholm's third theorem).

Proof. For $\lambda = 0$ Fredholm's alternative theorems are obviously valid. Therefore we shall consider that $\lambda \neq 0$, and as in previous constructions we shall choose $\varepsilon < 1/|\lambda|V$.

Let Eq. (1) have a solution in $C(\bar{G})$ for any $f \in C(\bar{G})$. Then Eq. (22) with a degenerate kernel which is equivalent to (1) will also have a solution in $C(\bar{G})$ for any f. From this, using Theorem 3 of Sec. 16.2, we conclude that $D(\lambda) \neq 0$. But then, according to Theorem 1 of Sec. 16.2, Eq. (22) and the adjoint equation (22*) have unique solutions for any f and Q belonging to $C(\bar{G})$. But the functions Q and g are expressible in a unique manner by the formulas (25). Therefore, the equivalent equations (1) and (1*) have unique solutions in $C(\bar{G})$ for any f and g. Fredholm's first theorem is proved.

If Eq. (1) has no solution in $C(\bar{G})$ for any f, then the equivalent equation (22) with a degenerate kernel also has no solution in $C(\bar{G})$ for any f. From this, by Theorem 1 of Sec. 16.2, we conclude that $D(\lambda) = 0$. But then, by Theorem 2 of Sec. 16.2, the homogeneous equations (22) and (22*) have an equal (finite) number of linearly independent solutions in $C(\bar{G})$. Consequently the homogeneous equations (1) and (1*) (which are equivalent to them) also have an equal (finite) number of linearly independent solutions in $C(\bar{G})$. Fredholm's second theorem is proved.

Moreover, by Theorem 3 of Sec. 16.2, for the solubility of Eq. (22) for $D(\lambda) = 0$ it is necessary and sufficient that the inhomogeneous term f should be orthogonal to all solutions of the adjoint homogeneous equation (22*). But the solutions ψ of the equivalent homogeneous equations (1*) and (22*), like the right-hand sides f of the equivalent equations (1) and (22), are the same. Therefore, for the solubility of Eq. (1) in the case we are considering, it is necessary and sufficient that the inhomogeneous term f be orthogonal to all solutions of the adjoint homogeneous equation (1*). Fredholm's third theorem is proved.

FREDHOLM'S FOURTH THEOREM. *In each circle $|\lambda| \leq R$ there is only a finite number of eigenvalues of the kernel $\mathcal{K}(x, y)$.*

Proof. We shall choose $\varepsilon = 1/(R+1)V$. Then for $|\lambda| < R+1$ we have $|\lambda| < 1/\varepsilon V$. Therefore for $|\lambda| < R+1$, the homogeneous equations (1) and (22) are equivalent. Consequently, in the circle $|\lambda| < R+1$ the eigenvalues of the kernel $\mathcal{K}(x, y)$ coincide with the roots of the equation $D(\lambda) = 0$ (cf. Sec. 16.2). Since the kernel $\mathcal{E}(x, y; \lambda)$ is analytic in λ in the

§ 15. FREDHOLM'S THEOREMS

circle $|\lambda| < R + 1$, then $D(\lambda)$ is an analytic function in this circle (cf. note of Sec. 16.2). From this, since analytic functions are unique, we conclude that in the circle $|\lambda| \leq R$ there is only a finite number of roots of the equation $D(\lambda) = 0$, and this means that the kernel $\mathscr{K}(x, y)$ can have only a finite number of eigenvalues. The theorem is proved.

4. Consequences of Fredholm's Theorems. It follows from Fredholm's fourth theorem that *the set of eigenvalues of a continuous kernel is no more than countable and does not have finite limit points.* (This set may also be empty as, for instance, for Volterra's kernel; cf. Sec. 15.3.)

Moreover, it follows from Fredholm's second theorem that *the multiplicity of each eigenvalue is finite.*

Therefore, all the eigenvalues of the kernel $\mathscr{K}(x, y)$ may be enumerated in ascending order of their modulus,

$$|\lambda_1| \leq |\lambda_2| \leq \cdots \qquad (26)$$

where λ_k is repeated in this series according to its multiplicity. We denote the corresponding eigenfunctions by $\varphi_1, \varphi_2, \ldots$, so that for each eigenvalue λ_k belonging to (26) there is one and only one, corresponding eigenfunction φ_k,

$$\varphi_k = \lambda_k K \varphi_k, \qquad k = 1, 2, \ldots \qquad (27)$$

According to Fredholm's second theorem, $\bar{\lambda}_1, \bar{\lambda}_2, \ldots$, are all eigenvalues of the kernel $\mathscr{K}^*(x, y)$; moreover the multiplicities of λ_k and $\bar{\lambda}_k$ are equal. We shall denote the corresponding eigenfunctions by ψ_k,

$$\psi_k = \bar{\lambda}_k K^* \psi_k, \qquad k = 1, 2, \ldots \qquad (27^*)$$

The eigenfunctions φ_k and ψ_k are continuous over $\bar{G}$.

We shall prove that if $\lambda_k \neq \lambda_i$, then

$$(\varphi_k, \psi_i) = 0 \qquad (28)$$

Taking Eq. (14) of Sec. 15.2 into account, from (27) and (27*) we obtain

$$(\varphi_k, \psi_i) = (\varphi_k, \bar{\lambda}_i K^* \psi_i) = \lambda_i (K \varphi_k, \psi_i) = \frac{\lambda_i}{\lambda_k} (\varphi_k, \psi_i)$$

from which, since $\lambda_k \neq \lambda_i$, Eqs. (28) follow.

We note that λ_k^p and φ_k, $k = 1, 2, \ldots$, are *eigenvalues and corresponding eigenfunctions of the iterated kernel* $\mathcal{K}_p(x, y)$.

This assertion follows from Eqs. (27), by which

$$\varphi_k = \lambda_k^p K^p \varphi_k, \qquad k = 1, 2, \ldots \tag{29}$$

Conversely, *if μ and φ are the eigenvalue and the corresponding eigenfunction of the iterated kernel* $\mathcal{K}_p(x, y)$, *then at least one of the roots λ_j, $j = 1, 2, \ldots, p$, of the equation $\lambda^p = \mu$ is an eigenvalue of the initial kernel* $\mathcal{K}(x, y)$.

This assertion follows from the equation

$$(\mu K^p - I)\varphi = (-1)^{p-1}(\lambda_1 K - I)(\lambda_2 K - I) \cdots (\lambda_p K - I)\varphi = 0 \tag{30}$$

In fact, if

$$\psi = (\lambda_2 K - I) \cdots (\lambda_p K - I)\varphi \neq 0 \tag{31}$$

then, by virtue of (30), $(\lambda_1 K - I)\psi = 0$, and therefore λ_1 is an eigenvalue of the kernel $\mathcal{K}(x, y)$. If then $\psi = 0$, that is, by virtue of (31),

$$(\lambda_2 K - I) \cdots (\lambda_p K - I)\varphi = 0$$

then, repeating the previous argument, we shall obtain: either λ_2 is an eigenvalue of the kernel $\mathcal{K}(x, y)$ or

$$(\lambda_3 K - I) \cdots (\lambda_p K - I)\varphi = 0 \qquad \text{etc.}$$

We shall now restate Fredholm's alternative theorems in terms of eigenvalues and eigenfunctions.

If $\lambda \neq \lambda_k$, $k = 1, 2, \ldots$, then the integral equations (1) and (1*) are uniquely soluble for any inhomogeneous terms.

If $\lambda = \lambda_k$, then for the solubility of Eq. (1), it is necessary and sufficient that

$$(f, \psi_{k+i}) = 0, \qquad i = 0, 1, \ldots, r_k - 1 \tag{32}$$

where $\psi_k, \psi_{k+1}, \ldots, \psi_{k+r_k-1}$ are eigenfunctions, corresponding to the eigenvalue $\bar{\lambda}_k$ of the kernel $\mathcal{K}^*(x, y)$, and r_k is the multiplicity of λ_k and $\bar{\lambda}_k$.

Note. This process of reducing Eq. (1) to the integral equation (22) with a degenerate kernel indicates that Eq. (1) may be approximately solved for any λ as follows:

(1) The kernel $\mathcal{K}(x, y)$ is approximated by the polynomial $\mathcal{P}(x, y)$ (or by any other degenerate kernel); (2) for a small kernel $\eta(x, y) = \mathcal{K}(x, y) - \mathcal{P}(x, y)$, by the method of Sec. 15.2, the resolvent $\mathcal{R}(x, y; \lambda)$ is constructed approximately; (3) the integral equation (22) is constructed with a degenerate kernel $\mathcal{C}(x, y; \lambda)$; (4) the solution Φ of Eq. (22) is constructed by the method of Sec. 16.1; and, finally, (5) the solution φ of Eq. (1) is found using formula (21).

5. Fredholm's Theorems for Integral Equations with a Polar Kernel.
We shall extend Fredholm's theorems to include integral equations with a polar kernel (cf. Sec. 15.4)

$$\mathcal{K}(x, y) = \frac{\mathcal{H}(x, y)}{|x - y|^\alpha}, \qquad \alpha < n$$

where $\mathcal{H}(x, y)$ is a continuous kernel.

We shall prove that for any $\varepsilon > 0$ there is a degenerate kernel $\mathcal{P}(x, y)$ such that

$$\max_{x \in \bar{G}} \int_G |\mathcal{K}(x, y) - \mathcal{P}(x, y)|\, dy < \varepsilon \tag{33}$$

$$\max_{x \in \bar{G}} \int_G |\mathcal{K}^*(x, y) - \mathcal{P}^*(x, y)|\, dy < \varepsilon \tag{33*}$$

In fact, the kernel

$$\mathcal{L}(x, y) = \begin{cases} \mathcal{K}(x, y), & |x - y| \geq \dfrac{1}{N} \\ \mathcal{H}(x, y) N^\alpha, & |x - y| < \dfrac{1}{N} \end{cases}$$

is continuous and for a sufficiently large N

$$\int_G |\mathcal{K}(x, y) - \mathcal{L}(x, y)|\, dy$$

$$= \int_{U(x;1/N)} |\mathcal{H}(x, y)| \left(\frac{1}{|x - y|^\alpha} - N^\alpha \right) dy$$

$$\leq \int_{U(x;1/N)} \frac{|\mathcal{H}(x, y)|}{|x - y|^\alpha}\, dy \leq \max_{\bar{G} \times \bar{G}} |\mathcal{H}(x, y)| \int_{U(x;1/N)} \frac{dy}{|x - y|^\alpha}$$

$$= c \int_{U_{1/N}} \frac{d\xi}{|\xi|^\alpha} = c\sigma_n \int_0^{1/N} \varrho^{n-1-\alpha}\, d\varrho = \frac{c\sigma_n}{n - \alpha} \frac{1}{N^{n-\alpha}} < \frac{\varepsilon}{2}, \qquad x \in \bar{G}$$

and analogously

$$\int_G |\mathcal{K}^*(x,y) - \mathcal{L}^*(x,y)|\,dy = \int_G |\mathcal{K}(y,x) - \mathcal{L}(y,x)|\,dy < \varepsilon/2, \quad x \in \bar{G}$$

Moreover, we shall approximate the continuous kernel $\mathcal{L}(x,y)$ by the degenerate kernel $\mathcal{P}(x,y)$ (cf. Sec. 16.3)

$$|\mathcal{L}(x,y) - \mathcal{P}(x,y)| < \frac{\varepsilon}{2V}, \quad x \in \bar{G}, \quad y \in \bar{G}$$

It follows that the polar kernel $\mathcal{K}(x,y)$ may be approximated by a degenerate kernel in the sense of (33) and (33*):

$$\max_{x \in \bar{G}} \int_G |\mathcal{K}(x,y) - \mathcal{P}(x,y)|\,dy \leq \max_{x \in \bar{G}} \int_G |\mathcal{K}(x,y) - \mathcal{L}(x,y)|\,dy$$
$$+ \max_{x \in \bar{G}} \int_G |\mathcal{L}(x,y) - \mathcal{P}(x,y)|\,dy < \frac{\varepsilon}{2} + \frac{\varepsilon}{2V}\int_G dy = \varepsilon$$

Equation (33*) is established analogously.

So for any $\varepsilon > 0$ the polar kernel $\mathcal{K}(x,y)$ may be represented in the form $\mathcal{K}(x,y) = \mathcal{P}(x,y) + \eta(x,y)$, where $\mathcal{P}(x,y)$ is a degenerate kernel and $\eta(x,y)$ is a small polar kernel which, by virtue of (33) and (33*), satisfies the estimates

$$\max_{x \in \bar{G}} \int_G |\eta(x,y)|\,dy < \varepsilon, \quad \max_{x \in \bar{G}} \int_G |\eta^*(x,y)|\,dy < \varepsilon$$

Repeating the arguments of Secs. 16.3 and 16.4 and using the results of Sec. 15.4 concerning the solubility of integral equations with a small polar kernel, we conclude that *all Fredholm's theorems and their consequences are carried over to integral equations with a polar kernel.*

We note that *all eigenfunctions of the polar kernel $\mathcal{K}(x,y)$ which belong to $\mathcal{L}_2(G)$ belong to $C(\bar{G})$.*

In fact, if $\varphi_0 = \lambda_0 K \varphi_0$, $\varphi_0 \in \mathcal{L}_2(G)$, then $\varphi_0 = \lambda_0^p K^p \varphi_0$. But for a sufficiently large p the kernel $\mathcal{K}_p(x,y)$ of the integral operator K^p is continuous (cf. Sec. 15.4). Then, by the lemma of Sec. 15.1, $\varphi_0 = \lambda_0^p K^p \varphi_0 \in C(\bar{G})$, as was to be shown.

Note. Fredholm's theorems remain valid for integral equations with a polar kernel over a bounded piecewise smooth surface S:

$$\varphi(x) = \lambda \int_S \frac{\mathcal{H}(x,y)}{|x-y|^\alpha} \varphi(y)\,dS_y + f(x)$$

§ 17. INTEGRAL EQUATIONS WITH AN HERMITIAN KERNEL

where the kernel $\mathscr{H}(x, y)$ is uniformly continuous over $S \times S$ and the exponent α is smaller than the dimension of the surface S [cf. I. G. Petrovsky (2, Sec. 8)].

6. Exercises. (a) Prove that if $\mathscr{H}(t)$ is a continuous 2π-periodic function and
$$\int_{-\pi}^{\pi} \mathscr{H}(t) e^{ikt} \, dt \neq 0, \quad \text{with} \quad k \quad \text{an integer}$$
then
$$\lambda_k = \frac{1}{\int_{-\pi}^{\pi} \mathscr{H}(t) e^{ikt} \, dt} \quad \text{and} \quad \varphi_k(x) = e^{-ikx}$$
are the eigenvalue and the corresponding eigenfunction of the kernel $\mathscr{H}(x - y)$, $-\pi < x, y < \pi$.

(b) Prove that if $\mathscr{H}(t)$ is an absolutely integrable function over R^1 and $F[\mathscr{H}](\mu) \neq 0$, then
$$\lambda = \frac{1}{F[\mathscr{H}](\mu)} \quad \text{and} \quad \varphi(x) = e^{-i\mu x}$$
are the eigenvalue and the corresponding eigenfunction of the kernel $\mathscr{H}(x - y)$, $-\infty < x, y < \infty$.

(c) Prove that $\lambda = \sqrt{2/\pi}$ is an eigenvalue of the kernel $\cos xy$, $0 < x, y < \infty$, and that the corresponding eigenfunction is
$$\varphi(x) = f(x) + \sqrt{\frac{2}{\pi}} \int_0^\infty \cos xy f(y) \, dy$$
where $f(x)$ is any function belonging to $\mathscr{L}_2(0, \infty)$.

We note that for integral equations with kernels, as in Exercises (b) and (c) above, Fredholm's theorems are not valid (the regions of integration in them are not bounded!).

§ 17. Integral Equations with an Hermitian Kernel

The kernel $\mathscr{H}(x, y)$ is said to be *Hermitian* if it coincides with its Hermitian adjoint kernel, $\mathscr{H}(x, y) = \mathscr{H}^*(x, y)$.

The corresponding integral equation
$$\varphi(x) = \lambda \int_G \mathscr{H}(x, y) \varphi(y) \, dy + f(x) \tag{1}$$
for real λ coincides with its adjoint equation, since $K^* = K$. It is convenient to consider this equation in the space $\mathscr{L}_2(G)$.

1. Integral Operators with an Hermitian Continuous Kernel. Let K be an integral operator with an Hermitian continuous kernel $\mathcal{H}(x, y)$. This operator maps $\mathscr{L}_2(G)$ into $\mathscr{L}_2(G)$ (c.f. Sec. 15.1) and is Hermitian (cf. Secs. 15.2 and 1.10),

$$(Kf, g) = (f, Kg), \qquad f, g \in \mathscr{L}_2(G) \tag{2}$$

Conversely, if the integral operator K with a continuous kernel $\mathcal{H}(x, y)$ is Hermitian, then this kernel is also Hermitian.

In fact, it follows from Eq. (2) that the kernel $\mathcal{H}(x, y) = \mathcal{H}^*(x, y)$ is Hermitian (cf. Sec. 15.2).

From Eq. (22) of Sec. 15.2 it follows that *all iterated kernels $\mathcal{H}_p(x, y)$ of the Hermitian continuous kernel $\mathcal{H}(x, y)$ are Hermitian,*

$$\mathcal{H}_p^*(x, y) = (\mathcal{H}^*)_p(x, y) = \mathcal{H}_p(x, y)$$

LEMMA. *The integral operator K with a continuous kernel $\mathcal{H}(x, y)$ maps each bounded set in $\mathscr{L}_2(G)$ into a set bounded in $C(\bar{G})$ consisting of equicontinuous* functions over $\bar{G}$.*

Proof. Let B be a bounded set in $\mathscr{L}_2(G)$: $\|f\| \leq A$, $f \in B$. By the lemma of Sec. 15.1 the operator K maps the set B into a set bounded in $C(\bar{G})$: $\|Kf\|_C \leq M\sqrt{V}A$, $f \in B$. Moreover, since the kernel $\mathcal{H}(x, y)$ is uniformly continuous over $\bar{G} \times \bar{G}$, then for any $\varepsilon > 0$ there is a number $\delta > 0$ such that

$$|\mathcal{H}(x', y) - \mathcal{H}(x'', y)| < \frac{\varepsilon}{\sqrt{V}A}$$

if $|x' - x''| < \delta$, x', x'', and $y \in \bar{G}$. From this, using inequality (4) of Sec. 15.1 and changing $\mathcal{H}(x, y)$ to $\mathcal{H}(x', y) - \mathcal{H}(x'', y)$, for all $f \in B$ we obtain

$$|(Kf)(x') - (Kf)(x'')| = \left| \int_G [\mathcal{H}(x', y) - \mathcal{H}(x'', y)] f(y) \, dy \right|$$

$$\leq \frac{\varepsilon}{\sqrt{V}A} \sqrt{V} \|f\| \leq \varepsilon$$

---

* A definition of a set of equicontinuous functions is given in Sec. 1.3.

§ 17. INTEGRAL EQUATIONS WITH AN HERMITIAN KERNEL

if $|x' - x''| < \delta$, $x', x'' \in \bar{G}$. This means that the set $\{(Kf)(x), f \in B\}$ consists of equicontinuous functions over $\bar{G}$. The lemma is proved.

2. Arzelà's Lemma. *If the infinite set B is bounded in $C(K)$, where K is a compactum, and if it consists of equicontinuous functions over K, then a sequence may be chosen from it which converges in $C(K)$.*

Proof. As we know, a set of points with rational coordinates is countable. Therefore all such points of the set K may be enumerated: $x_1, x_2, \ldots$. Since, by assumption, the set of numbers $\{f(x_1), f \in B\}$ is bounded, then using the Bolzano–Weierstrass theorem (cf. Sec. 1.1), we may choose from it a convergent sequence $f_k^{(1)}(x_1)$, $k = 1, 2, \ldots$. Moreover, since the set of numbers $\{f_k^{(1)}(x_2)\}$ is bounded, we may choose from it a convergent subsequence $f_k^{(2)}(x_2)$, $k = 1, 2, \ldots$, and so on.

We now consider the diagonal sequence $f_k(x) = f_k^{(k)}(x)$, $k = 1, 2, \ldots$, of functions from the set B. For any point x_i the numerical sequence $f_k(x_i)$, $k = 1, 2, \ldots$, converges, since by construction for $k \geq i$ this sequence is contained in the convergent sequence $f_k^{(i)}(x_i)$, $k = 1, 2, \ldots$.

We shall now prove that the sequence f_k, $k = 1, 2, \ldots$, converges uniformly over K. Let $\varepsilon > 0$. Since this sequence consists of equicontinuous functions over K, then there will be a number δ such that for $k = 1, 2, \ldots$

$$|f_k(x) - f_k(x')| < \varepsilon/3 \qquad (3)$$

if $|x - x'| < \delta$, x and $x' \in K$. Since K is a bounded set, then from the set of points $x_1, x_2, \ldots$, it is possible to choose a finite number of them, $x_1, x_2, \ldots, x_l$, $l = l(\varepsilon)$, so that for any point $x \in K$ there would be a point x_i, $1 \leq i \leq l$, such that $|x - x_i| < \delta$. If we recall that the sequence $f_k(x)$, $k = 1, 2, \ldots$, converges over the points $x_1, x_2, \ldots, x_l$, we may conclude that there will be a number $N = N(\varepsilon)$ such that

$$|f_k(x_i) - f_p(x_i)| < \varepsilon/3, \qquad k, p \geq N, \qquad i = 1, 2, \ldots, l \qquad (4)$$

Let x be an arbitrary point of the set K. Choosing a point x_i, $1 \leq i \leq l$, such that $|x - x_i| < \delta$, by virtue of (3) and (4), we obtain

$$|f_k(x) - f_p(x)| \leq |f_k(x) - f_k(x_i)| + |f_k(x_i) - f_p(x_i)|$$

$$+ |f_p(x_i) - f_p(x)| < \frac{\varepsilon}{3} + \frac{\varepsilon}{3} + \frac{\varepsilon}{3} = \varepsilon, \qquad k, p \geq N$$

where N does not depend on x. This means that the sequence f_k, $k = 1$, $2, \ldots$, converges in itself in $C(K)$. According to the Cauchy theorem (cf. Sec. 1.3) this sequence converges in $C(K)$ to a certain function belonging to $C(K)$. The lemma is proved.

Note. Arzelà's lemma expresses the property of compactness of any set bounded in $C(K)$ and consisting of functions equicontinuous over K. The lemma of Sec. 17.1 asserts that an integral operator with a continuous kernel maps each bounded set from $\mathscr{L}_2(G)$ into a compact set in $C(\bar{G})$. Each operator which has this property is said to be *completely continuous from* $\mathscr{L}_2(G)$ *into* $C(\bar{G})$.

3. Integral Equations with an Hermitian Continuous Kernel. Not every kernel which is not identically zero has eigenvalues; for instance, as was shown in Sec. 15.3, Volterra continuous kernels do not have such numbers. Nevertheless, the following theorem is valid.

THEOREM. *Each Hermitian continuous kernel $\mathscr{K}(x, y) \not\equiv 0$ has at least one eigenvalue, and the eigenvalue λ_1 with the smallest modulus satisfies the variational principle*

$$\frac{1}{|\lambda_1|} = \sup_{f \in \mathscr{L}_2(G)} \frac{\|Kf\|}{\|f\|} \tag{5}$$

Proof. We shall denote by ν the supremum of the functional $\|Kf\|$ over the set of functions f belonging to $\mathscr{L}_2(G)$ with a unit norm,

$$\nu = \sup_{\|f\|=1} \|Kf\| \tag{6}$$

It follows from Eq. (6) of Sec. 15.1 that over the functions of this set $\|Kf\| \leq MV$, and so $\nu \leq MV$. Moreover, it is obvious that $\nu \geq 0$. We shall prove that $\nu > 0$. In fact, for $\nu = 0$, by virtue of (6), we should have $\|Kf\| = 0$, that is, $Kf = 0$ for all $f \in \mathscr{L}_2(G)$ and therefore $\mathscr{K}(x, y) \equiv 0$, $x \in G$, $y \in G$, (cf. Sec. 15.1), despite what has been supposed.

From the definition of the supremun ν it follows that there is a sequence f_k, $k = 1, 2, \ldots$, $\|f_k\| = 1$ such that

$$\|Kf_k\| \to \nu \quad \text{as} \quad k \to \infty \tag{7}$$

§ 17. INTEGRAL EQUATIONS WITH AN HERMITIAN KERNEL

moreover the inequality

$$\| K^2 f \| = \left\| K\left(\frac{Kf}{\| Kf \|}\right) \right\| \| Kf \| \leq \nu \| Kf \|, \qquad f \in \mathscr{L}_2(G) \tag{8}$$

is valid.

We shall now prove that

$$K^2 f_k - \nu^2 f_k \to 0 \quad \text{as} \quad k \to \infty \quad \text{in} \quad \mathscr{L}_2(G) \tag{9}$$

In fact, using (2), (8), and (7), we obtain

$$\| K^2 f_k - \nu^2 f_k \|^2 = (K^2 f_k - \nu^2 f_k, K^2 f_k - \nu^2 f_k)$$
$$= (K^2 f_k, K^2 f_k) + \nu^4 (f_k, f_k) - \nu^2 (f_k, K^2 f_k) - \nu^2 (K^2 f_k, f_k)$$
$$= \| K^2 f_k \|^2 + \nu^4 - 2\nu^2 (Kf_k, Kf_k)$$
$$\leq \nu^2 \| Kf_k \|^2 + \nu^4 - 2\nu^2 \| Kf_k \|^2 = \nu^4 - \nu^2 \| Kf_k \|^2 \to 0$$
$$\text{as} \quad k \to \infty$$

which is equivalent to limiting result (9).

According to the lemma of Sec. 17.1, the sequence of functions Kf_k, $k = 1, 2, \ldots$, is bounded in $C(\bar{G})$ and consists of equicontinuous functions over $\bar{G}$. But then, according to Arzelà's lemma (cf. Sec. 17.2), there is a subsequence $\psi_i = Kf_{k_i}$, $i = 1, 2, \ldots$, convergent in $C(\bar{G})$ to the function $\psi \in C(\bar{G})$, $\| \psi - \psi_i \|_C \to 0$ as $i \to \infty$. From this, using Eqs. (4) and (5) of Sec. 15.1 and result (9), we obtain

$$\| K^2 \psi - \nu^2 \psi \|_C \leq \| K^2(\psi - \psi_i) \|_C + \nu^2 \| \psi - \psi_i \|_C$$
$$+ \| K^2 \psi_i - \nu^2 \psi_i \|_C \leq MV \| K(\psi - \psi_i) \|_C + \nu^2 \| \psi - \psi_i \|_C$$
$$+ \| K(K^2 f_{k_i} - \nu^2 f_{k_i}) \|_C \leq (M^2 V^2 + \nu^2) \| \psi - \psi_i \|_C$$
$$+ M\sqrt{V} \| K^2 f_{k_i} - \nu^2 f_{k_i} \| \to 0, \qquad i \to \infty$$

and therefore

$$K^2 \psi = \nu^2 \psi$$

It is clear that $\psi \equiv 0$.

So the function ψ is the eigenfunction of the kernel $\mathscr{K}_2(x, y)$ that corresponds to the eigenvalue $1/\nu^2$. But then at least one of the numbers $\pm(1/\nu)$ is the eigenvalue of the kernel $\mathscr{K}(x, y)$ (cf. Sec. 16.4). In this way, the eigenvalue λ_1 has absolute value equal to $1/\nu$ and therefore, by virtue of (6), satisfies the variational principle (5).

We now need only to prove the fact that λ_1 is the eigenvalue of the kernel $\mathcal{K}(x, y)$ with the smallest modulus. In fact, if λ_0 and φ_0 are the eigenvalue and the corresponding eigenfunction, $\lambda_0 K\varphi_0 = \varphi_0$, then, by virtue of (5)

$$\frac{1}{|\lambda_1|} = \sup_{f \in \mathscr{L}_2(G)} \frac{\|Kf\|}{\|f\|} \geq \frac{\|K\varphi_0\|}{\|\varphi_0\|} = \frac{1}{|\lambda_0|}$$

and so $|\lambda_1| \leq |\lambda_0|$. The theorem is proved.

As established in Sec. 17.1, the integral operator K with an Hermitian continuous kernel $\mathcal{K}(x, y)$ is Hermitian. According to the theorem of Sec. 1.10, the eigenvalues of the kernel $\mathcal{K}(x, y)$ are real, while the eigenfunctions corresponding to the various eigenvalues are orthogonal. Moreover, according to Fredholm's fourth theorem, the set of eigenvalues is no more than countable, while according to Fredholm's second theorem, the multiplicity of each eigenvalue is finite. Therefore the system of eigenfunctions of the operator K is no more than countable and this system may be chosen to be orthonormal (cf. Sec. 1.10).

Taking this theorem and Fredholm's theorems (cf. Sec. 16.3) into account, for integral equations with an Hermitian continuous kernel $\mathcal{K}(x, y) \neq 0$ we obtain the following statements:

The set of eigenvalues $\{\lambda_k\}$ is not empty; it is located on the real axis; it is no more than countable and does not have finite limit points; each eigenvalue has a finite multiplicity; the system of eigenfunctions $\{\varphi_k\}$ may be chosen to be orthonormal,

$$(\varphi_k, \varphi_i) = \delta_{ki} \tag{10}$$

If $\lambda \neq \lambda_k$, $k = 1, 2, \ldots$, then Eq. (1) is uniquely soluble for any inhomogeneous term $f \in C(\bar{G})$. If $\lambda = \lambda_k$, then for the solubility of Eq. (1) it is necessary and sufficient that

$$(f, \varphi_{k+i}) = 0, \quad i = 0, 1, \ldots, r_k - 1 \tag{11}$$

where $\varphi_k, \varphi_{k+1}, \ldots, \varphi_{k+r_k-1}$ are the eigenfunctions corresponding to the eigenvalues λ_k and r_k is the multiplicity of λ_k.

4. Integral Equations with an Hermitian Polar Kernel. *All the results established in Sec. 17.3 for integral equations with an Hermitian continuous kernel remain valid for integral equations with an Hermitian polar kernel.*

§ 18. THE HILBERT-SCHMIDT THEOREM AND ITS COROLLARIES

In fact, for these integral equations Fredholm's theorems, and their corollaries, are valid (cf. Sec. 16.5).

Moreover, for the Hermitian polar kernel $\mathcal{H}(x, y)$ all the iterated kernels $\mathcal{H}_p(x, y)$ are Hermitian and polar, and for $p \geq p_0 = [n/(n - \alpha)] + 1$ these kernels are continuous (cf. Sec. 15.4).

It now remains to extend the theorem of Sec. 17.3 to include Hermitian polar kernels $\mathcal{H}(x, y) \neq 0$. We shall write

$$v = \sup_{\|f\|=1} \| Kf \| \tag{12}$$

Then, by virtue of $\mathcal{H}(x, y) \neq 0$ and inequality (31) of Sec. 15.4, $0 < v \leq N = N^*$. As in the proof of the theorem of Sec. 17.3, it follows from (12) that there is a sequence f_k, $k = 1, 2, \ldots$, $\|f_k\| = 1$ such that

$$K^2 f_k - v^2 f_k \to 0 \quad \text{as} \quad k \to \infty \quad \text{in} \quad \mathcal{L}_2(G)$$

From this, applying (31) of Sec. 15.4, for $p = 1, 2, \ldots$, we obtain

$$\| K^{2p} f_k - v^{2p} f_k \| = \| (K^{2p-2} + v^2 K^{2p-4} + \cdots + v^{2p-2} I)(K^2 f_k - v^2 f_k) \|$$
$$\leq (N^{2p-2} + v^2 N^{2p-4} + \cdots + v^{2p-2}) \| K^2 f_k - v^2 f_k \| \to 0$$
$$\text{as} \quad k \to \infty$$

that is,

$$K^{2p} f_k - v^{2p} f_k \to 0 \quad \text{as} \quad k \to \infty \quad \text{in} \quad \mathcal{L}_2(G) \tag{13}$$

But for $2p \geq p_0$ the kernel $\mathcal{H}_{2p}(x, y)$ is continuous. Therefore, as in the proof of the theorem of Sec. 17.3, it follows from the limiting result (13) that $1/v^{2p}$ is the eigenvalue of the kernel $\mathcal{H}_{2p}(x, y)$. But then since the kernel $\mathcal{H}(x, y)$ in Hermitian, and therefore all its eigenvalues are real, at least one of the numbers $\pm(1/v)$ is an eigenvalue λ_1 of this kernel (cf. Sec. 16.4). From this and from (12) it follows that variational principle (5) is valid for the eigenvalue λ_1. Clearly λ_1 is the eigenvalue of the kernel $\mathcal{H}(x, y)$ having the smallest modulus. This completes our extension of the theorem of Sec. 17.3 to Hermitian polar kernels.

§ 18. The Hilbert-Schmidt Theorem and its Corollaries

1. The Hilbert-Schmidt Theorem for an Hermitian Continuous Kernel. Let $\lambda_1, \lambda_2, \ldots$, be the eigenvalues of the Hermitian continuous kernel $\mathcal{H}(x, y) \neq 0$, arranged in order of size of their modulus, $|\lambda_1| \leq |\lambda_2| \leq \ldots$, and let $\varphi_1, \varphi_2, \ldots,$ be the corresponding orthonormal eigenfunctions, $(\varphi_k, \varphi_i) = \delta_{ki}$.

As we know, the eigenvalues λ_k are real, and the eigenfunctions $\varphi_k(x)$ are continuous over $\bar{G}$; the set $\{\lambda_k\}$ is either finite or countable; in the latter case $|\lambda_k| \to \infty$ as $k \to \infty$. Further, by virtue of the theorem of Sec. 17.3, the inequality

$$\|Kf\| \leq \frac{1}{|\lambda_1|} \|f\|, \quad f \in \mathscr{L}_2(G) \tag{1}$$

is valid.

We shall note another inequality*

$$\sum_{k=1}^{\infty} \frac{|\varphi_k(x)|^2}{\lambda_k^2} \leq \int_G |\mathscr{K}(x, y)|^2 \, dy, \quad x \in \bar{G} \tag{2}$$

[Below, in Sec. 18.2, it will be shown that in inequality (2) the equality sign in fact occurs.]

Inequality (2) for a fixed $x \in \bar{G}$ is Bessel's inequality (cf. Sec. 1.6) for the function $\mathscr{K}(x, y)$, whose Fourier coefficients, according to the orthonormal system $\{\varphi_k(y)\}$, are equal to

$$(\mathscr{K}, \varphi_k) = \int_G \mathscr{K}(x, y) \bar{\varphi}_k(y) \, dy = \overline{K \varphi_k} = \frac{1}{\lambda_k} \bar{\varphi}_k(x)$$

We shall introduce the sequence of Hermitian continuous kernels

$$\mathscr{K}^{(p)}(x, y) = \mathscr{K}(x, y) - \sum_{i=1}^{p} \frac{\varphi_i(x) \bar{\varphi}_i(y)}{\lambda_i}, \quad p = 1, 2, \ldots \tag{3}$$

The corresponding integral Hermitian operators $K^{(p)}$ obey the formula

$$K^{(p)}f = Kf - \sum_{i=1}^{p} \frac{(f, \varphi_i)}{\lambda_i} \varphi_i, \quad f \in \mathscr{L}_2(G) \tag{4}$$

We shall prove that $\lambda_{p+1}, \lambda_{p+2}, \ldots,$ and $\varphi_{p+1}, \varphi_{p+2}, \ldots,$ comprise all the eigenvalues and eigenfunctions of the kernel $\mathscr{K}^{(p)}(x, y)$.

In fact, by virtue of (4), we have

$$K^{(p)} \varphi_k = K \varphi_k - \sum_{i=1}^{p} \frac{(\varphi_k, \varphi_i)}{\lambda_i} \varphi_i = K \varphi_k = \frac{1}{\lambda_k} \varphi_k, \quad k \geq p + 1$$

* If the kernel $\mathscr{K}(x, y)$ has a finite number of eigenvalues $\lambda_1, \lambda_2, \ldots, \lambda_N$, then we shall assume that $\lambda_k = \infty$ for $k > N$.

§ 18. THE HILBERT-SCHMIDT THEOREM AND ITS COROLLARIES

so that λ_k and φ_k, $k \geq p+1$, are in fact the eigenvalues and the eigenfunctions of the kernel $\mathcal{K}^{(p)}(x, y)$. Conversely, let λ_0 and φ_0 be the eigenvalue and the corresponding eigenfunction of the kernel $\mathcal{K}^{(p)}(x, y)$; that is, by virtue of (4),

$$\varphi_0 = \lambda_0 K^{(p)}\varphi_0 = \lambda_0 K\varphi_0 - \lambda_0 \sum_{i=1}^{p} \frac{(\varphi_0, \varphi_i)}{\lambda_i} \varphi_i \qquad (5)$$

From this, for $k = 1, 2, \ldots, p$ we obtain

$$(\varphi_0, \varphi_k) = \lambda_0(K\varphi_0, \varphi_k) - \lambda_0 \sum_{i=1}^{p} \frac{(\varphi_0, \varphi_i)(\varphi_i, \varphi_k)}{\lambda_i}$$

$$= \lambda_0(\varphi_0, K\varphi_k) - \lambda_0 \sum_{i=1}^{p} \frac{(\varphi_0, \varphi_i)}{\lambda_i} \delta_{ik}$$

$$= \frac{\lambda_0}{\lambda_k}(\varphi_0, \varphi_k) - \frac{\lambda_0}{\lambda_k}(\varphi_0, \varphi_k) = 0$$

and so, by virtue of (5), $\varphi_0 = \lambda_0 K\varphi_0$. In this way, λ_0 and φ_0 are the eigenvalue and the corresponding eigenfunction of the kernel $\mathcal{K}(x, y)$. Since φ_0 is orthogonal to all the eigenfunctions $\varphi_1, \varphi_2, \ldots, \varphi_p$, then it follows that λ_0 coincides with one of the eigenvalues $\lambda_{p+1}, \lambda_{p+2}, \ldots,$ and so φ_0 may be considered equal to φ_k for some $k \geq p+1$.

In this way, λ_{p+1} is the eigenvalue of the kernel $\mathcal{K}^{(p)}(x, y)$ having the smallest modulus. Applying inequality (1) to this kernel and taking (4) into consideration, we obtain the inequality

$$\| K^{(p)}f \| = \left\| Kf - \sum_{i=1}^{p} \frac{(f, \varphi_i)}{\lambda_i} \varphi_i \right\| \leq \frac{\| f \|}{| \lambda_{p+1} |}, \quad f \in \mathcal{L}_2(G) \qquad (6)$$

$$p = 1, 2, \ldots$$

Let the Hermitian kernel $\mathcal{K}(x, y)$ have a finite number of eigenvalues: $\lambda_1, \lambda_2, \ldots, \lambda_N$. By what has been proved, the Hermitian kernel $\mathcal{K}^{(N)}(x, y)$ does not have eigenvalues, and so, by the theorem of Sec. 17.3, $\mathcal{K}^{(N)}(x, y) = 0$, so that, by virtue of (3)

$$\mathcal{K}(x, y) = \sum_{i=1}^{N} \frac{\varphi_i(x)\bar{\varphi}_i(y)}{\lambda_i} \qquad (7)$$

that is, the kernel $\mathcal{K}(x, y)$ is degenerate.

From this, recalling that a degenerate kernel always has a finite number of eigenvalues (cf. Sec. 16.2), we deduce the following result: *In order that an Hermitian continuous kernel should be degenerate, it is necessary and sufficient that it have a finite number of eigenvalues.*

We shall say that a function $f(x)$ is *sourcewise* represented by the kernel $\mathcal{H}(x, y)$ if there is a function $h \in \mathscr{L}_2(G)$ such that

$$f(x) = \int_G \mathcal{H}(x, y)h(y)\,dy, \qquad x \in G \tag{8}$$

HILBERT–SCHMIDT THEOREM. *If the function $f(x)$ is sourcewise represented by the Hermitian continuous kernel $\mathcal{H}(x, y)$, $f = Kh$, then its Fourier series involving the eigenfunctions of the kernel $\mathcal{H}(x, y)$ converges regularly (and therefore uniformly) over $\bar{G}$ to this function,*

$$f(x) = \sum_{k=1}^{\infty} (f, \varphi_k)\varphi_k(x) = \sum_{k=1}^{\infty} \frac{(h, \varphi_k)}{\lambda_k} \varphi_k(x) \tag{9}$$

Proof. Since $f = Kh$, $h \in \mathscr{L}_2(G)$, then, according to the lemma of Sec. 15.1, $f \in C(\bar{G})$ and the Fourier coefficients of the functions f and h involving the eigenfunctions $\{\varphi_k\}$ of the kernel $\mathcal{H}(x, y)$ are linked by the relation

$$(f, \varphi_k) = (Kh, \varphi_k) = (h, K\varphi_k) = \frac{(h, \varphi_k)}{\lambda_k} \tag{10}$$

If the kernel $\mathcal{H}(x, y)$ has a finite number of eigenvalues, then, by virtue of (7),

$$f(x) = Kh = \sum_{k=1}^{N} \frac{(h, \varphi_k)}{\lambda_k} \varphi_k(x)$$

and the Hilbert–Schmidt theorem is proved.

Now let the kernel $\mathcal{H}(x, y)$ have an infinite number of eigenvalues. In this case $|\lambda_k| \to \infty$ as $k \to \infty$. So, by virtue of (6) and (10), series (9) converges to f in $\mathscr{L}_2(G)$:

$$\left\| f - \sum_{k=1}^{p} (f, \varphi_k)\varphi_k \right\| = \left\| Kh - \sum_{k=1}^{p} \frac{(h, \varphi_k)}{\lambda_k} \varphi_k \right\| \leq \frac{\|h\|}{|\lambda_{p+1}|} \to 0, \qquad p \to \infty$$

It remains for us to prove that the series (9) converges regularly over $\bar{G}$. Using the Cauchy–Buniakowski inequality and inequality (2), for all p

§ 18: THE HILBERT-SCHMIDT THEOREM AND ITS COROLLARIES

and q we obtain

$$\sum_{k=p}^{q} |(h, \varphi_k)| \left|\frac{\varphi_k(x)}{\lambda_k}\right| \leq \left[\sum_{k=p}^{q} |(h, \varphi_k)|^2\right]^{1/2} \left[\sum_{k=p}^{q} \frac{|\varphi_k(x)|^2}{\lambda_k^2}\right]^{1/2}$$

$$\leq \left[\sum_{k=p}^{q} |(h, \varphi_k)|^2\right]^{1/2} \left[\int_G |\mathscr{K}(x, y)|^2 \, dy\right]^{1/2}$$

$$\leq M\sqrt{V} \left[\sum_{k=p}^{q} |(h, \varphi_k)|^2\right]^{1/2}, \qquad x \in \bar{G} \quad (11)$$

By virtue of Bessel's inequality

$$\sum_{k=1}^{\infty} |(h, \varphi_k)|^2 \leq \|h\|^2$$

the right-hand side of inequality (11) tends to zero as $p, q \to \infty$. This also means that the series (9) converges regularly over $\bar{G}$. The theorem is proved.

We add some corollaries of the Hilbert–Schmidt theorem.

2. Bilinear Expansion of Iterated Kernels. We shall prove that *the iterated kernel $\mathscr{K}_p(x, y)$ of the Hermitian continuous kernel $\mathscr{K}(x, y)$ can be expanded into a bilinear series involving the eigenfunctions of this kernel*

$$\mathscr{K}_p(x, y) = \sum_{k=1}^{\infty} \frac{\varphi_k(x)\bar{\varphi}_k(y)}{\lambda_k^p}, \qquad p = 2, 3, \ldots \quad (12)$$

regularly converging over $\bar{G} \times \bar{G}$.

By virtue of formula (17) of Sec. 15.2, for each $y \in \bar{G}$ the kernel $\mathscr{K}_p(x, y)$ is sourcewise represented by the kernel $\mathscr{K}(x, y)$ and so, by the Hilbert–Schmidt theorem, it can be expanded into a regularly converging Fourier series involving the eigenfunctions of this kernel

$$\mathscr{K}_p(x, y) = \sum_{k=1}^{\infty} (\mathscr{K}_p(x, y), \varphi_k)\varphi_k(x)$$

Since the kernel $\mathscr{K}_p(x, y)$ is Hermitian then

$$(\mathscr{K}_p(x, y), \varphi_k) = \int_G \mathscr{K}_p(x, y)\bar{\varphi}_k(x) \, dx$$

$$= \int_G \bar{\mathscr{K}}_p(y, x)\bar{\varphi}_k(x) \, dx = \overline{(K^p\varphi_k)(y)} = \frac{\bar{\varphi}_k(y)}{\lambda_k^p}, \qquad p \geq 1 \quad (13)$$

In this way, Eq. (12) is proved and the series in (12) is seen to converge regularly in $x \in \bar{G}$ for each $y \in \bar{G}$.

Specifically, if in formula (12) we set $p = 2$, $x = y$, and take into account that, by virtue of (17) of Sec. 15.2,

$$\mathcal{K}_2(x, x) = \int_G \mathcal{K}(x, y') \mathcal{K}(y', x) \, dy'$$
$$= \int_G \mathcal{K}(x, y') \bar{\mathcal{K}}(x, y') \, dy' = \int_G |\mathcal{K}(x, y)|^2 \, dy$$

we obtain the equality

$$\sum_{k=1}^{\infty} \frac{|\varphi_k(x)|^2}{\lambda_k^2} = \int_G |\mathcal{K}(x, y)|^2 \, dy \qquad (14)$$

It follows from Dini's lemma (cf. Sec. 1.3) that series (14) converges uniformly over $\bar{G}$. From this, using the Cauchy–Buniakowski inequality, we have

$$\sum_{k=1}^{\infty} \frac{|\varphi_k(x) \bar{\varphi}_k(y)|}{\lambda_k^p} \leq \frac{1}{|\lambda_1|^{p-2}} \left[\sum_{k=1}^{\infty} \frac{|\varphi_k(x)|^2}{\lambda_k^2} \sum_{k=1}^{\infty} \frac{|\varphi_k(y)|^2}{\lambda_k^2} \right]^{1/2}$$

and so we conclude that the series (12) converges regularly over $\bar{G} \times \bar{G}$.

Integrating the uniformly convergent series (14) termwise, and taking the normalization of the eigenfunctions into account, we obtain the formula

$$\sum_{k=1}^{\infty} \frac{1}{\lambda_k^2} = \int_G \int_G |\mathcal{K}(x, y)|^2 \, dx \, dy \qquad (15)$$

3. Bilinear Expansion of an Hermitian Continuous Kernel.

We shall investigate the convergence of series (12) for $p = 1$, and, in particular, we shall prove that the Hermitian continuous kernel $\mathcal{K}(x, y)$ may be expanded into a bilinear series in terms of its own eigenfunctions

$$\mathcal{K}(x, y) = \sum_{k=1}^{\infty} \frac{\varphi_k(x) \bar{\varphi}_k(y)}{\lambda_k} \qquad (16)$$

which converges in $\mathcal{L}_2(G)$ uniformly with respect to $y \in \bar{G}$, that is,

$$\left\| \mathcal{K}(x, y) - \sum_{k=1}^{p} \frac{\varphi_k(x) \bar{\varphi}_k(y)}{\lambda_k} \right\| \xrightarrow{y \in \bar{G}} 0, \quad p \to \infty \qquad (17)$$

§ 18. THE HILBERT-SCHMIDT THEOREM AND ITS COROLLARIES

Equation (13) for $p = 1$ shows that for each $y \in \bar{G}$, the Fourier coefficients of the kernel $\mathcal{K}(x, y)$ in terms of the orthonormal system $\{\varphi_k(x)\}$ are equal to $\bar{\varphi}_k(y)/\lambda_k$. Therefore, using Eq. (16) of Sec. 1.6, we obtain the equation

$$\left\| \mathcal{K}(x, y) - \sum_{k=1}^{p} \frac{\varphi_k(x)\bar{\varphi}_k(y)}{\lambda_k} \right\|^2 = \int_G |\mathcal{K}(x, y)|^2 \, dx - \sum_{k=1}^{p} \frac{|\varphi_k(y)|^2}{\lambda_k^2}, \quad y \in \bar{G}$$

from which, by virtue of the uniform convergence of series (14), we conclude that the bilinear series (16) converges to the kernel $\mathcal{K}(x, y)$ in the sense of (17).

It follows from (17), specifically, that series (16) converges to the kernel $\mathcal{K}(x, y)$ in $\mathcal{L}_2(G \times G)$, that is,

$$\int_G \int_G \left| \mathcal{K}(x, y) - \sum_{k=1}^{p} \frac{\varphi_k(x)\bar{\varphi}_k(y)}{\lambda_k} \right|^2 dx \, dy \to 0, \quad p \to \infty \tag{18}$$

For the bilinear form (Kf, g) we shall prove the formula

$$(Kf, g) = \sum_{k=1}^{\infty} \frac{(f, \varphi_k)\overline{(g, \varphi_k)}}{\lambda_k}, \quad f, g \in \mathcal{L}_2(G) \tag{19}$$

In fact, since $f \in \mathcal{L}_2(G)$, then, by the Hilbert–Schmidt theorem,

$$(Kf)(x) = \sum_{k=1}^{\infty} \frac{(f, \varphi_k)}{\lambda_k} \varphi_k(x)$$

and this series converges uniformly over $\bar{G}$. Multiplying this series by a function $\bar{g}$ belonging to $\mathcal{L}_2(G)$ (and, consequently, absolutely integrable over $\bar{G}$; cf. Sec. 1.5) and integrating it termwise over the region G, we obtain Eq. (19):

$$(Kf, g) = \int_G Kf\bar{g} \, dx = \sum_{k=1}^{\infty} \frac{(f, \varphi_k)}{\lambda_k} \int_G \varphi_k(x)\bar{g}(x) \, dx$$

$$= \sum_{k=1}^{\infty} \frac{(f, \varphi_k)\overline{(g, \varphi_k)}}{\lambda_k}$$

Setting $f = g$ in Eq. (19) we obtain a representation of the quadratic form (Kf, f) in the form

$$(Kf, f) = \sum_{k=1}^{\infty} \frac{|(f, \varphi_k)|^2}{\lambda_k}, \quad f \in \mathcal{L}(G) \tag{20}$$

Formula (20) is a generalization of the formula for the reduction to principal axes of a quadratic form involving a finite number of variables.

4. Solution of a Nonhomogeneous Integral Equation with an Hermitian Continuous Kernel.
We construct a formula for the solution of a nonhomogeneous integral equation

$$\varphi = \lambda K\varphi + f \tag{21}$$

involving an Hermitian continuous kernel $\mathcal{K}(x, y)$.

If $\lambda \neq \lambda_k$, $k = 1, 2, \ldots$, and $f \in C(\bar{G})$, then the (unique) solution φ of the integral equation (21) appears in the form of a series uniformly convergent over $\bar{G}$ (by Schmidt's formula)

$$\varphi(x) = \lambda \sum_{k=1}^{\infty} \frac{(f, \varphi_k)}{\lambda_k - \lambda} \varphi_k(x) + f(x) \tag{22}$$

In fact, for $\lambda \neq \lambda_k$, $k = 1, 2, \ldots$, the solution of integral equation (21) exists and is unique in $C(\bar{G})$ for any inhomogeneous term $f \in C(\bar{G})$ (cf. Sec. 16.3). According to the Hilbert–Schmidt theorem, the function $K\varphi$ can be expanded into a uniformly convergent Fourier series involving the eigenfunctions of the kernel $\mathcal{K}(x, y)$. Therefore

$$\varphi = \lambda K\varphi + f = \lambda \sum_{k=1}^{\infty} \frac{(\varphi, \varphi_k)}{\lambda_k} \varphi_k + f \tag{23}$$

We shall calculate the Fourier coefficients (φ, φ_k). From Eq. (21) we have

$$(\varphi, \varphi_k) = \lambda(K\varphi, \varphi_k) + (f, \varphi_k) = \lambda(\varphi, K\varphi_k) + (f, \varphi_k) = \frac{\lambda}{\lambda_k} (\varphi, \varphi_k) + (f, \varphi_k)$$

and, therefore,

$$(\varphi, \varphi_k) = \frac{\lambda_k}{\lambda_k - \lambda} (f, \varphi_k), \qquad k = 1, 2, \ldots$$

from which, by virtue of (23), Schmidt's formula (22) follows.

According to the Hilbert–Schmidt theorem

$$(Kf)(x) = \sum_{k=1}^{\infty} \frac{(f, \varphi_k)}{\lambda_k} \varphi_k(x)$$

and the series converges uniformly over $\bar{G}$. Therefore Schmidt's formula (22)

§ 18. THE HILBERT-SCHMIDT THEOREM AND ITS COROLLARIES

assumes the form

$$\varphi(x) = \lambda \sum_{k=1}^{\infty} \frac{(f, \varphi_k)}{\lambda_k} \varphi_k(x) + \lambda^2 \sum_{k=1}^{\infty} \frac{(f, \varphi_k)}{\lambda_k(\lambda_k - \lambda)} \varphi_k(x) + f(x)$$

$$= \lambda \int_G \mathcal{K}(x, y) f(y) \, dy + \lambda^2 \sum_{k=1}^{\infty} \frac{(f, \varphi_k)}{\lambda_k(\lambda_k - \lambda)} \varphi_k(x) + f(x) \quad (24)$$

Moreover, since the bilinear series (12) converges regularly for $p = 2$, it follows that the bilinear series

$$\sum_{k=1}^{\infty} \frac{\varphi_k(x) \bar{\varphi}_k(y)}{\lambda_k(\lambda_k - \lambda)}$$

converges uniformly and its sum is a continuous function for $x \in \bar{G}$, $y \in \bar{G}$, $\lambda \neq \lambda_k$, $k = 1, 2, \ldots$, and is meromorphic in λ with simple poles λ_k. Therefore, for $\lambda \neq \lambda_k$, $k = 1, 2, \ldots$, in formula (24) we may change the order of summation and integration, as a result of which we obtain

$$\varphi(x) = \lambda \int_G \left[\mathcal{K}(x, y) + \lambda \sum_{k=1}^{\infty} \frac{\varphi_k(x) \bar{\varphi}_k(y)}{\lambda_k(\lambda_k - \lambda)} \right] f(y) \, dy + f(x) \quad (25)$$

On the other hand, by the theorem of Sec. 15.2, for small λ the solution of Eq. (21) is expressed by the resolvent $\mathcal{R}(x, y; \lambda)$ of the kernel $\mathcal{K}(x, y)$ in terms of formula (20) of Sec. 15.2. Therefore

$$\mathcal{R}(x, y; \lambda) = \mathcal{K}(x, y) + \lambda \sum_{k=1}^{\infty} \frac{\varphi_k(x) \bar{\varphi}_k(y)}{\lambda_k(\lambda_k - \lambda)} \quad (26)$$

In this way, the resolvent $\mathcal{R}(x, y; \lambda)$ of the Hermitian continuous kernel $\mathcal{K}(x, y)$ allows a meromorphic continuation over the whole plane of the complex variable λ with simple poles λ_k and with the residues

$$- \sum_{i=0}^{r_k - 1} \varphi_{k+i}(x) \bar{\varphi}_{k+i}(y) \quad (27)$$

where $\varphi_k, \varphi_{k+1}, \ldots, \varphi_{k+r_k-1}$ are eigenfunctions of the kernel $\mathcal{K}(x, y)$, corresponding to λ_k, and r_k is the multiplicity of λ_k (cf. note of Sec. 15.2).

Using Eq. (16), we write formula (26) in the form

$$\mathcal{R}(x, y; \lambda) = \sum_{k=1}^{\infty} \frac{\varphi_k(x) \bar{\varphi}_k(y)}{\lambda_k - \lambda} \quad (28)$$

and this bilinear series converges in $\mathcal{L}_2(G \times G)$ (cf. Sec. 18.3).

Note. Formula (22) remains valid for $\lambda = \lambda_j$ if, in agreement with Fredholm's third theory,

$$(f, \varphi_{j+i}) = 0, \qquad i = 0, 1, \ldots, r_j - 1$$

In this case the solution of Eq. (21) is not unique and its general solution, expressed by Eq. (38) of Sec. 1.9, is given by the formula

$$\varphi(x) = \lambda_j \sum_{\substack{k=1 \\ \lambda_k \neq \lambda_j}}^{\infty} \frac{(f, \varphi_k)}{\lambda_k - \lambda_j} \varphi_k(x) + f(x) + \sum_{i=0}^{r_j-1} c_i \varphi_{j+i}(x) \qquad (29)$$

where c_i are arbitrary constants.

5. Positive Kernels. The kernel $\mathscr{K}(x, y)$ is said to be positive if the corresponding operator K is positive (cf. Sec. 1.10), that is,

$$(Kf, f) \geq 0, \qquad f \in \mathscr{L}_2(G)$$

Each positive kernel $\mathscr{K}(x, y)$ is Hermitian.

In fact, since the operator K is Hermitian (cf. Sec. 1.10), then its kernel $\mathscr{K}(x, y)$ is also Hermitian (cf. Sec. 17.1).

In order that the Hermitian continuous kernel $\mathscr{K}(x, y)$ should be positive, it is necessary and sufficient that all its eigenvalues λ_k be positive.

In fact, if $\lambda_k > 0$, then, by virtue of (20), $(Kf, f) \geq 0$ for $f \in \mathscr{L}_2(G)$, so that the kernel $\mathscr{K}(x, y)$ is positive. Conversely, if the kernel $\mathscr{K}(x, y)$ is positive, then

$$\frac{1}{\lambda_k} = (K\varphi_k, \varphi_k) \geq 0, \qquad \text{that is} \quad \lambda_k > 0$$

If $\mathscr{K}(x, y)$ is a positive continuous kernel, then the following variational principle is valid:

$$\frac{1}{\lambda_k} = \sup_{\substack{f \in \mathscr{L}_2(G) \\ (f, \varphi_i) = 0, i = 1, 2, \ldots, k-1}} \frac{(Kf, f)}{\|f\|^2}, \qquad k = 1, 2, \ldots \qquad (30)$$

and also, the supremum in (30) is attained over any eigenfunction corresponding to the eigenvalue λ_k.

In fact, using formula (20) and taking into account the inequality $\lambda_i \geq \lambda_k > 0$, $i \geq k$, for all $f \in \mathscr{L}_2(G)$ such that $(f, \varphi_i) = 0$, $i = 1, 2, \ldots$,

§ 18. THE HILBERT–SCHMIDT THEOREM AND ITS COROLLARIES

$k - 1$, we obtain

$$\frac{(Kf,f)}{\|f\|^2} = \frac{1}{\|f\|^2} \sum_{i=1}^{\infty} \frac{|(f,\varphi_i)|^2}{\lambda_i} \leq \frac{1}{\lambda_k \|f\|^2} \sum_{i=k}^{\infty} |(f,\varphi_i)|^2$$

and consequently, by virtue of Bessel's inequality, the inequality

$$\frac{(Kf,f)}{\|f\|^2} \leq \frac{1}{\lambda_k} \tag{31}$$

is valid. On the other hand, for $f = \varphi_k$ we have

$$\frac{(K\varphi_k, \varphi_k)}{\|\varphi_k\|^2} = \frac{1}{\lambda_k} \tag{32}$$

Inequality (31) and Eq. (32) establish the validity of variational principle (30).

Setting $k = 1$ in (30), we obtain

$$\frac{1}{\lambda_1} = \sup_{f \in \mathscr{L}_2(G)} \frac{(Kf,f)}{\|f\|^2} \tag{33}$$

6. Extension of the Hilbert–Schmidt Theory to Integral Equations with an Hermitian Polar Kernel. *The Hilbert–Schmidt theorem and its corollaries, which have been established in this section for integral equations with an Hermitian continuous kernel, are carried over to integral equations with an Hermitian weakly polar kernel* (cf. Sec. 15.4)

$$\mathscr{K}(x,y) = \frac{\mathscr{H}(x,y)}{|x-y|^\alpha}, \quad \alpha < \frac{n}{2}, \quad \mathscr{H}^*(x,y) = \mathscr{H}(x,y)$$

In fact, for these kernels the results of Sec. 17 are valid. Therefore, as analysis of the proof of the Hilbert–Schmidt theorem shows, to extend this theorem to weakly polar kernels it is sufficient to establish the following lemma.

LEMMA. *The integral operator K with a weakly polar kernel $\mathscr{K}(x,y)$ maps $\mathscr{L}_2(G)$ into $C(\bar{G})$ and is bounded*

$$\|Kf\|_C \leq L \|f\| \tag{34}$$

where

$$L^2 = \max_{x \in \bar{G}} \int_G |\mathscr{K}(x, y)|^2 \, dy$$

Proof. Let $f \in \mathscr{L}_2(G)$. Using the Cauchy–Buniakowski inequality, for all $x \in \bar{G}$ we have

$$|(Kf)(x)| = \left|\int_G \mathscr{K}(x, y) f(y) \, dy\right|$$
$$\leq \left[\int_G |\mathscr{K}(x, y)|^2 \, dy\right]^{1/2} \|f\| \leq L \|f\| \qquad (35)$$

We shall prove that the function $(Kf)(x)$ is continuous over $\bar{G}$. Take an $\varepsilon > 0$. Then there is a function $f_\varepsilon \in C(\bar{G})$ such that

$$\|f - f_\varepsilon\| \leq \varepsilon/L$$

(cf. Sec. 1.5); according to Lemma 1 of Sec. 15.4, $Kf_\varepsilon \in C(\bar{G})$ and, by virtue of inequality (35),

$$|(Kf)(x) - (Kf_\varepsilon)(x)| = |K(f - f_\varepsilon)| \leq L \|f - f_\varepsilon\| \leq L\varepsilon/L = \varepsilon, \qquad x \in \bar{G}$$

In this way, the function Kf with any arbitrary degree of accuracy may be uniformly approximated by continuous functions over $\bar{G}$. Therefore $Kf \in C(\bar{G})$. Consequently the operator K maps $\mathscr{L}_2(G)$ into $C(\bar{G})$. In this case inequality (34) follows from inequality (35). The lemma is proved.

Now let the Hermitian kernel $\mathscr{K}(x, y)$ be polar with $\alpha < n$. For such kernels the results of Sec. 17 are valid. Therefore, as follows from the proof of the Hilbert–Schmidt theorem, *series (9) converges in $\mathscr{L}_2(G)$*. Considering now that for $p \geq p_1 = [n/2(n - \alpha)] + 1$, the iterated kernels $\mathscr{K}_p(x, y)$ are Hermitian and weakly polar (cf. Sec. 15.4 and 17.4), we conclude that *the bilinear series (12) converges regularly for $p \geq 2p_1$. Further, Eqs. (19) and (20), and consequently all the results of Sec. 18.5, are retained. Schmidt's formula (22) remains valid with a change from uniform convergence to convergence in $\mathscr{L}_2(G)$.*

Note. We shall consider the integral equation

$$\varphi(x) = \lambda \int_G \varrho(y) \mathscr{K}(x, y) \varphi(y) \, dy + f(x) \qquad (36)$$

where the kernel $\mathscr{K}(x, y)$ is Hermitian and the weight function $\varrho(y)$ is a

§ 18. THE HILBERT-SCHMIDT THEOREM AND ITS COROLLARIES

positive and continuous function over $\bar{G}$. The change of the unknown function to $\psi = \sqrt{\varrho}\,\varphi$ transforms Eq. (36) to the equivalent integral equation

$$\psi(x) = \lambda \int_G \sqrt{\varrho(x)\varrho(y)}\,\mathscr{H}(x, y)\psi(y)\,dy + \sqrt{\varrho(x)}f(x)$$

with an Hermitian kernel $\sqrt{\varrho(x)\varrho(y)}\,\mathscr{H}(x, y)$. Changing to the original equation (36), we see that the Hilbert–Schmidt theory may be carried over without change to the integral equation (36) with a non-Hermitian kernel $\varrho(y)\mathscr{H}(x, y)$, provided it is understood to be in the space $\mathscr{L}_2(G; \varrho)$ with the scalar product $(f, g)_\varrho$ (cf. note of Sec. 1.7).

7. Jentsch's Theorem. Many problems of mathematical physics are reduced to integral equations with a real Hermitian kernel. Such kernels are called *symmetrical*; they satisfy the equation $\mathscr{H}(x, y) = \mathscr{H}(y, x)$.

The eigenfunctions of the symmetrical kernel $\mathscr{H}(x, y)$ may be chosen to be real.

In fact, if $\varphi_0 = \varphi_1 + i\varphi_2$ is the eigenfunction of the kernel $\mathscr{H}(x, y)$ corresponding to the eigenvalue λ_0,

$$\varphi_0 = \varphi_1 + i\varphi_2 = \lambda_0 K\varphi_0 = \lambda_0 K\varphi_1 + i\lambda_0 K\varphi_2$$

then, since $\mathscr{H}(x, y)$ and λ_0 are real, we conclude that the nonzero real and imaginary parts φ_1 and φ_2 of the function φ_0 are also eigenfunctions corresponding to λ_0,

$$\varphi_1 = \lambda_0 K\varphi_1, \qquad \varphi_2 = \lambda_0 K\varphi_2$$

The kernel $\mathscr{H}(x, y)$ is known as a kernel of *positive type** if $\mathscr{H}(x, y) > 0$, $x \in G$, $y \in G$.

Evidently if the kernel $\mathscr{H}(x, y)$ is of positive type, then all its iterated kernels $\mathscr{H}_p(x, y)$ are of positive type.

JENTSCH'S THEOREM. *If the symmetrical polar kernel $\mathscr{H}(x, y)$ is of positive type, then its eigenvalue λ_1 with the smallest modulus is positive and simple; the corresponding eigenfunction $\varphi(x)$ is positive in G.*

Proof. Let λ_1 be the (real) eigenvalue with the smallest modulus of the symmetrical (polar) kernel $\mathscr{H}(x, y)$ of positive type and let φ_1 be an arbitrary

---

* As distinct from a positive kernel (cf. Sec. 18.5).

(real) eigenfunction corresponding to λ_1; $\varphi_1 = \lambda_1 K\varphi_1$. Then λ_1^2 is the smallest eigenvalue of the positive polar kernel $\mathscr{K}_2(x, y)$ and φ_1 is the eigenfunction corresponding to λ_1^2; $\varphi_1 = \lambda_1^2 K^2 \varphi_1$.

We shall prove that $\varphi_1(x)$ *cannot change sign in the region G*, that is,

$$|\varphi_1(x)\varphi_1(y)| = \varphi_1(x)\varphi_1(y), \qquad x \in G, \qquad y \in G$$

In fact, in the opposite case, by virtue of the continuity of the function $\varphi_1(x)$ (cf. Sec. 16.5), there would be neighborhoods $U(x'; r) \subset G$ and $U(y'; \varrho) \subset G$ such that

$$|\varphi_1(x)||\varphi_1(y)| > \varphi_1(x)\varphi_1(y), \qquad x \in U(x'; r), \qquad y \in U(y'; \varrho)$$

and so, by virtue of the condition $\mathscr{K}_2(x, y) > 0$,

$$\frac{(K^2|\varphi_1|, |\varphi_1|)}{\||\varphi_1|\|^2} = \frac{1}{\|\varphi_1\|^2} \int_G \int_G \mathscr{K}_2(x, y) |\varphi_1(x)| |\varphi_1(y)| \, dx \, dy$$

$$> \frac{1}{\|\varphi_1\|^2} \int_G \int_G \mathscr{K}_2(x, y) \varphi_1(x) \varphi_1(y) \, dx \, dy$$

$$= \frac{(K^2\varphi_1, \varphi_1)}{\|\varphi_1\|^2} = \frac{1}{\lambda_1^2}$$

which contradicts the variational principle (33).

We shall prove that the function $\varphi_1(x)$ *cannot become zero in the region G and therefore can be chosen positive in G*.

In fact, in the opposite case there will be a point $x' \in G$ such that

$$\varphi_1(x') = \lambda_1^2 \int_G \mathscr{K}_2(x', y) \varphi_1(y) \, dy = 0$$

from which, by virtue of the condition $\mathscr{K}_2(x, y) > 0$, the contradiction follows: $\varphi_1(y) = 0$, $y \in G$.

Since $\varphi_1(x)$ is positive, it follows that λ_1 is positive, as $\mathscr{K}(x, y) > 0$ and $\lambda_1 = K\varphi_1/\varphi_1 > 0$.

We shall prove that λ_1 is a *simple eigenvalue*.

In fact, if there were a real eigenfunction φ_2 linearly independent of φ_1 and corresponding to λ_1, then for all real c the linear combination $\varphi_1 + c\varphi_2$ also would be a real eigenfunction corresponding to λ_1 and therefore, by what has been proved, it could not become zero in the region G. As c is arbitrary, this is impossible. The theorem is proved.

§ 18. THE HILBERT-SCHMIDT THEOREM AND ITS COROLLARIES   251

Note. Jentsch's theorem is valid for any polar kernel of positive type (without supposing it symmetrical). The corresponding theorem is valid also for matrices with positive elements; in this case it is known as Perron's theorem.

8. Kellogg's Method. To find approximately the eigenvalue λ_1 having the smallest modulus and the corresponding eigenfunctions of the Hermitian polar kernel $\mathcal{K}(x, y)$, Kellogg's method of successive approximations is used. Suppose that the real function $\varphi^{(0)}$ belonging to $\mathscr{L}_2(G)$ is not orthogonal to all the eigenfunctions corresponding to λ_1. We shall construct the sequences

$$\varphi_{(p)}(x) = \frac{\varphi^{(p)}(x)}{\|\varphi^{(p)}\|}, \qquad \lambda_{(p)} = \frac{\|\varphi^{(p-1)}\|}{\|\varphi^{(p)}\|}, \qquad p = 1, 2, \ldots \qquad (37)$$

where $\varphi^{(p)} = K^p \varphi^{(0)}$ are the iterates of the function $\varphi^{(0)}$ (cf. Sec. 15.1). The terms $\lambda_{(p)}$ and $\varphi_{(p)}(x)$ of these sequences are taken as approximations to $|\lambda_1|$ and to the corresponding eigenfunction $\varphi_1(x)$.

We shall give the basis of Kellogg's method for integral equations with a symmetrical weakly polar kernel of positive type. According to Jentsch's theorem for such kernels, λ_1 is positive and simple, so that $0 < \lambda_1 < |\lambda_2| \leq \cdots$; the corresponding eigenfunction $\varphi_1(x)$ is positive for $x \in G$; the eigenfunctions $\varphi_k(x)$ are real (cf. Sec. 18.7).

THEOREM. *Let $\mathcal{K}(x, y)$ be a symmetrical weakly polar kernel of positive type. Then for any function $\varphi^{(0)}(x) \geq 0$ with $\|\varphi^{(0)}\| = 1$, the sequence $\{\lambda_{(p)}\}$ converges in a monotonically decreasing fashion toward λ_1, the sequence $\{\varphi_{(p)}\}$ converges toward φ_1 in $\mathscr{L}_2(G)$ and in $C(\bar{G})$, and the estimates*

$$0 \leq \lambda_{(p)} - \lambda_1 \leq \frac{\lambda_1}{2}\left(\frac{\lambda_1}{\lambda_2}\right)^{2p-2} \frac{1 - c_1^2}{c_1^2}, \qquad p = 2, 3, \ldots \qquad (38)$$

$$\|\varphi_{(p)} - \varphi_1\| \leq \left(\frac{\lambda_1}{\lambda_2}\right)^p \frac{\sqrt{1 - c_1^2}}{c_1}, \qquad p = 1, 2, \ldots \qquad (39)$$

$$\|\varphi_{(p)} - \varphi_1\|_C \leq L\lambda_2\left(\frac{\lambda_1}{\lambda_2}\right)^p \frac{\sqrt{1 - c_1^2}}{c_1}, \qquad p = 2, 3, \ldots \qquad (40)$$

are valid, where

$$c_1 = (\varphi^{(0)}, \varphi_1), \qquad L^2 = \max_{x \in \bar{G}} \int_G |\mathcal{K}(x, y)|^2 \, dy$$

Proof. According to the Hilbert–Schmidt theorem (cf. Sec. 18.1 and 18.6) we have

$$\varphi^{(p)}(x) = K^p \varphi^{(0)} = \sum_{k=1}^{\infty} \frac{c_k}{\lambda_k^p} \varphi_k(x), \qquad p = 1, 2, \ldots \qquad (41)$$

where, by virtue of Bessel's inequality,

$$\sum_{k=1}^{\infty} c_k^2 \leq \| \varphi^{(0)} \|^2 = 1, \qquad c_k = (\varphi^{(0)}, \varphi_k), \qquad c_1 > 0 \qquad (42)$$

and the series (41) converges uniformly for $x \in \bar{G}$.

From Eqs. (41), by virtue of $(\varphi_i, \varphi_k) = \delta_{ik}$, there follow the equations

$$\| \varphi^{(p)} \|^2 = (\varphi^{(p)}, \varphi^{(p)}) = \sum_{k=1}^{\infty} \frac{c_k^2}{\lambda_k^{2p}}, \qquad p = 1, 2, \ldots \qquad (43)$$

We shall prove that the sequence $\lambda_{(p)}$, $p = 1, 2, \ldots$, is monotonic decreasing, and that $\lambda_{(p)} \geq \lambda_1$.

In fact, using the Cauchy–Buniakowski inequality, we obtain

$$\| \varphi^{(p)} \|^2 = (\varphi^{(p)}, \varphi^{(p)}) = (K\varphi^{(p-1)}, \varphi^{(p)}) = (\varphi^{(p-1)}, K\varphi^{(p)})$$
$$= (\varphi^{(p-1)}, \varphi^{(p+1)}) \leq \| \varphi^{(p-1)} \| \, \| \varphi^{(p+1)} \|$$

from which, together with (37), follow the inequalities

$$\lambda_{(p+1)} = \frac{\| \varphi^{(p)} \|}{\| \varphi^{(p+1)} \|} \leq \frac{\| \varphi^{(p-1)} \|}{\| \varphi^{(p)} \|} = \lambda_{(p)}, \qquad p = 1, 2, \ldots$$

Further, from the variational principle (5) of Sec. 17.3 (cf. also Sec. 17.4) we deduce

$$\lambda_{(p)} = \frac{\| \varphi^{(p-1)} \|}{\| \varphi^{(p)} \|} = \frac{\| \varphi^{(p-1)} \|}{\| K\varphi^{(p-1)} \|} \geq \inf_{f \in \mathscr{L}_2(G)} \frac{\| f \|}{\| Kf \|} = \lambda_1, \qquad p = 1, 2, \ldots$$

as was to be shown.

Taking Eqs. (43) into account, we obtain

$$\lambda_{(p)} - \lambda_1 = \frac{\| \varphi^{(p-1)} \|}{\| \varphi^{(p)} \|} - \lambda_1$$
$$= \lambda_1 \left[\frac{1 + \sum_{k=2}^{\infty} (c_k/c_1)^2 (\lambda_1/\lambda_k)^{2p-2}}{1 + \sum_{k=2}^{\infty} (c_k/c_1)^2 (\lambda_1/\lambda_k)^{2p}} \right]^{1/2} - \lambda_1, \qquad p = 2, 3, \ldots \qquad (44)$$

§ 18. THE HILBERT-SCHMIDT THEOREM AND ITS COROLLARIES

We note the inequalities valid for $x \geq y \geq 0$:

$$\sqrt{\frac{1+x}{1+y}} - 1 \leq \tfrac{1}{2}(x-y), \qquad 1 + \frac{1+x}{1+y} - \frac{2}{\sqrt{1+y}} \leq x \qquad (45)$$

Applying the first of inequalities (45) with

$$x = \sum_{k=2}^{\infty} \left(\frac{c_k}{c_1}\right)^2 \left(\frac{\lambda_1}{\lambda_k}\right)^{2p-2} \quad \text{and} \quad y = \sum_{k=2}^{\infty} \left(\frac{c_k}{c_1}\right)^2 \left(\frac{\lambda_1}{\lambda_k}\right)^{2p}$$

to the right-hand sides of Eqs. (44), and using (42), we obtain inequalities (38)

$$0 \leq \lambda_{(p)} - \lambda_1 \leq \frac{\lambda_1}{2} \sum_{k=2}^{\infty} \left(\frac{c_k}{c_1}\right)^2 \left(\frac{\lambda_1}{\lambda_k}\right)^{2p-2} \left(1 - \frac{\lambda_1^2}{\lambda_k^2}\right)$$

$$\leq \frac{\lambda_1}{2} \left(\frac{\lambda_1}{\lambda_2}\right)^{2p-2} \sum_{k=2}^{\infty} \left(\frac{c_k}{c_1}\right)^2 \leq \frac{\lambda_1}{2} \left(\frac{\lambda_1}{\lambda_2}\right)^{2p-2} \frac{1-c_1^2}{c_1^2}$$

We shall prove the equations

$$\left\| \frac{\varphi^{(p)}}{\lambda_1^q \, \|\varphi^{(p+q)}\|} - \varphi_1 \right\| \leq \left(\frac{\lambda_1}{\lambda_2}\right)^p \frac{\sqrt{1-c_1^2}}{c_1} \qquad (46)$$

$$p = 1, 2, \ldots, \qquad q = 0, 1, \ldots$$

Using Eqs. (41) and (43) we obtain

$$\left\| \frac{\varphi^{(p)}}{\lambda_1^q \, \|\varphi^{(p+q)}\|} - \varphi_1 \right\|^2$$

$$= \left\| \left(\frac{c_1}{\lambda_1^{q+p} \, \|\varphi^{(p+q)}\|} - 1 \right) \varphi_1 + \frac{1}{\lambda_1^q \, \|\varphi^{(p+q)}\|} \sum_{k=2}^{\infty} \frac{c_k}{\lambda_k^p} \varphi_k \right\|^2$$

$$= \left(\frac{c_1}{\lambda_1^{p+q} \, \|\varphi^{(p+q)}\|} - 1 \right)^2 + \frac{1}{\lambda_1^{2q} \, \|\varphi^{(p+q)}\|^2} \sum_{k=2}^{\infty} \frac{c_k^2}{\lambda_k^{2p}}$$

$$= 1 + \frac{1 + \sum_{k=2}^{\infty} (c_k/c_1)^2 (\lambda_1/\lambda_k)^{2p}}{1 + \sum_{k=2}^{\infty} (c_k/c_1)^2 (\lambda_1/\lambda_k)^{2p+2q}}$$

$$- \frac{2}{[1 + \sum_{k=2}^{\infty} (c_k/c_1)^2 (\lambda_1/\lambda_k)^{2p+2q}]^{1/2}} \qquad (47)$$

Applying the second of inequalities (45) with

$$x = \sum_{k=2}^{\infty} \left(\frac{c_k}{c_1}\right)^2 \left(\frac{\lambda_1}{\lambda_k}\right)^{2p} \quad \text{and} \quad y = \sum_{k=2}^{\infty} \left(\frac{c_k}{c_1}\right)^2 \left(\frac{\lambda_1}{\lambda_k}\right)^{2p+2q}$$

to the right-hand side of Eqs. (47), and taking (42) into consideration, we obtain inequalities (46):

$$\left\| \frac{\varphi^{(p)}}{\lambda_1^q \, \|\varphi^{(p+q)}\|} - \varphi_1 \right\|^2 \le \sum_{k=2}^{\infty} \left(\frac{c_k}{c_1}\right)^2 \left(\frac{\lambda_1}{\lambda_k}\right)^{2p}$$

$$\le \left(\frac{\lambda_1}{\lambda_2}\right)^{2p} \sum_{k=1}^{\infty} \left(\frac{c_k}{c_1}\right)^2 \le \left(\frac{\lambda_1}{\lambda_2}\right)^{2p} \frac{1 - c_1^2}{c_1^2}$$

Inequalities (39) follow from inequalities (46) by setting $q = 0$, by virtue of (37). We shall prove inequalities (40). Applying inequality (34) of Sec. 18.6, we obtain

$$\|\varphi_{(p)} - \varphi_1\|_C = \left\| \frac{\varphi^{(p)}}{\|\varphi^{(p)}\|} - \varphi_1 \right\|_C = \left\| K\left(\frac{\varphi^{(p-1)}}{\|\varphi^{(p)}\|} - \lambda_1 \varphi_1 \right) \right\|_C$$

$$\le L\lambda_1 \left\| \frac{\varphi^{(p-1)}}{\lambda_1 \, \|\varphi^{(p)}\|} - \varphi_1 \right\|$$

Applying inequality (46), for $q = 1$ with p replaced by $p - 1$, to the right-hand side of the inequality we have just obtained, we obtain estimates (40). The theorem is proved.

Note. The theorem which has just been proved concerning the convergence of Kellogg's method is also valid for symmetrical polar kernels of positive type. In this case estimates (40) occur for $p \ge 2p_1$ (cf. Sec. 18.6).

9. Exercises. (a) Prove: If the kernel $\mathscr{K}(x, y)$ is polar and Hermitian and $f \in C(\bar{G})$, then the method of successive approximations of Sec. 15.1 converges in $C(\bar{G})$ for $|\lambda| < |\lambda_1|$ with an error $O(|\lambda/\lambda_1|^p)$.

(b) Prove Mercer's Theorem. *If the continuous kernel $\mathscr{K}(x, y)$ is positive, then its bilinear series converges regularly over $\bar{G} \times \bar{G}$.*

(c) For Peierl's integral equation

$$\varphi(x) = \lambda \int_G \mathscr{K}(|x - y|)\varphi(y) \, dy, \qquad \mathscr{K}(\xi) = \frac{e^{-\xi}}{4\pi \xi^2}$$

prove the result $\lambda_1(1 - e^{-D}) \ge 1$, where D is the diameter of the region $G \subset R^3$.

CHAPTER

5

Boundary Value Problems for Elliptic Equations

In this chapter we shall study boundary value problems for equations of elliptic type. The region G is considered to be bounded, if this point is not mentioned specifically, and its boundary S is a piecewise smooth surface.

§ 19. The Eigenvalue Problem

1. Formulation of the Eigenvalue Problem. We shall consider the following linear homogeneous boundary value problem for an equation of elliptic type (cf. Sec. 4.4):

$$-\operatorname{div}(p \operatorname{grad} u) + qu = \lambda u, \quad x \in G \quad (1)$$

$$\left. \alpha u + \beta \frac{\partial u}{\partial n} \right|_S = 0 \quad (2)$$

We shall suppose (cf. Secs. 4.1 and 4.4)

$$\left. \begin{array}{l} p \in C^1(\bar{G}), \quad q \in C(\bar{G}), \quad p(x) > 0, \quad q(x) \geq 0, \quad x \in \bar{G} \\ \alpha \in C(S), \quad \beta \in C(S), \quad \alpha(x) \geq 0, \quad \beta(x) \geq 0 \\ \alpha(x) + \beta(x) > 0, \quad x \in S \end{array} \right\} \quad (3)$$

Let S_0 be that part of S on which $\alpha(x) > 0$ and $\beta(x) > 0$ are simultaneously true.

The problem (1)–(2) is to find a function $u(x)$ of the class $C^2(G) \cap C^1(\bar{G})$ which satisfies Eq. (1) in the region G and the boundary conditions (2) on the boundary S. Obviously problem (1)–(2) always has a zero solution, and this solution is of no interest. We must therefore consider problem (1)–(2) as an eigenvalue problem (cf. Sec. 1.9) for the operator

$$L = -\operatorname{div}(p \operatorname{grad}) + q$$

All the functions $f(x)$ of the class $C^2(G) \cap C^1(\bar{G})$ satisfying the boundary condition (2) and the condition that $Lf \in \mathscr{L}_2(G)$ will be related to the domain of definition $\mathscr{M}_L$ of the operator L (cf. Sec. 1.8). p.20

So the problem (1)–(2) is to find those values λ (the eigenvalues of the operator L) for which the equation

$$Lu = \lambda u \qquad (4)$$

has a nonzero solution $u(x)$ belonging to the domain of definition $\mathscr{M}_L$ (the eigenfunctions which correspond to this eigenvalue).

2. Green's Formulas. If $u \in C^2(G) \cap C^1(\bar{G})$ and $v \in C^1(\bar{G})$, then *Green's first formula* holds:

$$\int_G vLu\,dx = \int_G p \sum_{i=1}^n \frac{\partial v}{\partial x_i}\frac{\partial u}{\partial x_i}\,dx - \int_S pv\frac{\partial u}{\partial \mathbf{n}}\,dS + \int_G quv\,dx \qquad (5)$$

To prove Eq. (5) we shall take an arbitrary region G' with a piecewise smooth boundary S', strictly lying in the region G (Fig. 48). Since $u \in C^2(G)$, then $u \in C^2(\bar{G}')$ and so

$$\int_{G'} vLu\,dx = \int_{G'} v[-\operatorname{div}(p \operatorname{grad} u) + qu]\,dx$$

$$= -\int_{G'} \operatorname{div}(pv \operatorname{grad} u)\,dx + \int_{G'} p\sum_{i=1}^n \frac{\partial v}{\partial x_i}\frac{\partial u}{\partial x_i}\,dx + \int_{G'} quv\,dx$$

Now using the Gauss–Ostrogradski formula, we obtain

$$\int_{G'} vLu\,dx = \int_{G'} p \sum_{i=1}^n \frac{\partial v}{\partial x_i}\frac{\partial u}{\partial x_i}\,dx - \int_{S'} pv\frac{\partial u}{\partial \mathbf{n}}\,dS' + \int_{G'} quv\,dx$$

Allowing G' to tend to G in this equation, and using the fact that u and $v \in C^1(\bar{G})$, we conclude that the limit of the right-hand side exists, so that

§ 19. THE EIGENVALUE PROBLEM

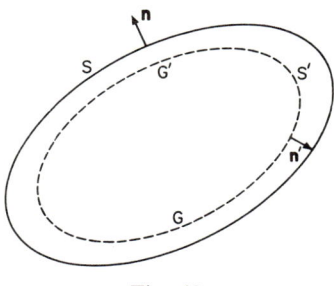

Fig. 48

there is a limit of the left-hand side and Eq. (5) is true. For this the integral on the left in (5) must be understood to be nonsingular.

If u and $v \in C^2(G) \cap C^1(\bar{G})$, then *Green's second formula* holds:

$$\int_G (vLu - uLv)\, dx = \int_S p\left(u \frac{\partial v}{\partial n} - v \frac{\partial u}{\partial n}\right) dS \tag{6}$$

To prove formula (6) we interchange u and v in Green's first formula (5)

$$\int_G uLv\, dx = \int_G p \sum_{i=1}^n \frac{\partial u}{\partial x_i} \frac{\partial v}{\partial x_i}\, dx - \int_S pu \frac{\partial v}{\partial n}\, dS + \int_G quv\, dx$$

and subtract the equation obtained from Eq. (5). As a result we obtain Green's second formula (6).

Specifically, for $p = 1$, $q = 0$ Green's formulas (5) and (6) are transformed into the following [cf. formula (22) of Sec. 6.5]:

$$\int_G v\, \Delta u\, dx = -\int_G \sum_{i=1}^n \frac{\partial v}{\partial x_i} \frac{\partial u}{\partial x_i}\, dx + \int_S v \frac{\partial u}{\partial n}\, dS \tag{7}$$

$$\int_G (v\, \Delta u - u\, \Delta v)\, dx = \int_S \left(v \frac{\partial u}{\partial n} - u \frac{\partial v}{\partial n}\right) dS \tag{8}$$

3. Properties of the Operator L. *The operator L is Hermitian (self adjoint in the sense of Lagrange),*

$$(Lf, g) = (f, Lg), \qquad f, g \in \mathscr{M}_L \tag{9}$$

In fact, since the functions f and $\bar{g}$ belong to the region $\mathscr{M}_L$, then $Lf \in \mathscr{L}_2(G)$ and $L\bar{g} = \overline{Lg} \in \mathscr{L}_2(G)$ and Green's second formula (6) with

5. BOUNDARY VALUE PROBLEMS FOR ELLIPTIC EQUATIONS

$u = f$ and $v = \bar{g}$ takes the form

$$\int_G (\bar{g}Lf - f\overline{Lg})\, dx = (Lf, g) - (f, Lg)$$
$$= \int_S p\left(f\frac{\partial \bar{g}}{\partial \mathbf{n}} - \bar{g}\frac{\partial f}{\partial \mathbf{n}}\right) dS \quad (10)$$

Moreover the functions f and $\bar{g}$ satisfy boundary condition (2):

$$\alpha f + \beta \frac{\partial f}{\partial \mathbf{n}}\bigg|_S = 0, \quad \alpha \bar{g} + \beta \frac{\partial \bar{g}}{\partial \mathbf{n}}\bigg|_S = 0 \quad (11)$$

By supposition (3), $\alpha + \beta > 0$ over S. Therefore the homogeneous system of linear algebraic equations (11) has a nonzero solution (α, β) and so its determinant is equal to zero, that is,

$$\begin{vmatrix} f & \dfrac{\partial f}{\partial \mathbf{n}} \\ \bar{g} & \dfrac{\partial \bar{g}}{\partial \mathbf{n}} \end{vmatrix}_S = f\frac{\partial \bar{g}}{\partial \mathbf{n}} - \bar{g}\frac{\partial f}{\partial \mathbf{n}}\bigg|_S = 0$$

Taking this equation into consideration, we obtain Eq. (9) from Eq. (10), and this means that the operator L is Hermitian (cf. Sec. 1.10).

Let $f \in \mathcal{M}_L$. Setting $u = f$ and $v = \bar{f}$ in Green's first formula (5) and taking into consideration that $Lf \in \mathscr{L}_2(G)$, we obtain

$$(Lf, f) = \int_G p\,|\operatorname{grad} f|^2\, dx - \int_S pf\frac{\partial \bar{f}}{\partial \mathbf{n}}\, dS + \int_G q\,|f|^2\, dx \quad (12)$$

It follows from boundary condition (2) that

$$\frac{\partial f}{\partial \mathbf{n}} = -\frac{\alpha}{\beta} f, \quad \text{if} \quad \beta(x) > 0, \quad x \in S$$
$$f = 0, \quad \text{if} \quad \beta(x) = 0, \quad x \in S$$

Substituting these results into Eq. (12), we obtain an expression for quadratic form

$$(Lf, f) = \int_G (p\,|\operatorname{grad} f|^2 + q\,|f|^2)\, dx$$
$$+ \int_{S_0} p\frac{\alpha}{\beta}\,|f|^2\, dS, \quad f \in \mathcal{M}_L \quad (13)$$

where S_0 is that part of S on which $\alpha(x) > 0$ and $\beta(x) > 0$.

§ 19. THE EIGENVALUE PROBLEM

The quadratic form $(Lf, f), f \in \mathcal{M}_L$, is known as the *energy integral*.

By virtue of suppositions (3), all three terms in the right-hand side of (13) are nonnegative. Therefore, disregarding the second and third terms and underestimating the first term, we obtain the inequality

$$(Lf, f) \geq \int_G p \,|\,\mathrm{grad}\, f\,|^2 \, dx \geq \min_{x \in \bar{G}} p(x) \int_G |\,\mathrm{grad}\, f\,|^2 \, dx$$

that is,

$$(Lf, f) \geq p_0 \,\|\,|\,\mathrm{grad}\, f\,|\,\|^2, \qquad f \in \mathcal{M}_L \tag{14}$$

where $p_0 = \min p(x)$; since the function p is continuous and *positive* over $\bar{G}$, $p_0 > 0$.

It follows from inequality (14) that the operator L is positive (cf. Sec. 1.10), that is,

$$(Lf, f) \geq 0, \qquad f \in \mathcal{M}_L \tag{15}$$

From this, specifically, it follows once more that the operator L is Hermitian (cf. Sec. 1.10).

4. Properties of Eigenvalues and Eigenfunctions of the Operator L

All eigenvalues of the operator L are nonnegative. This statement is true because the operator is positive (cf. Sec. 1.10).

The eigenfunctions of the operator L corresponding to the different eigenvalues are orthogonal. This statement is true because the operator is Hermitian (cf. Sec. 1.10).

The eigenfunctions of the operator L may be chosen to be real. This statement is true because the operator L is real (cf. Sec. 18.7). In fact, let λ_0 be an eigenvalue and u_0 the corresponding eigenfunction of the operator L,

$$Lu_0 = \lambda_0 u_0, \qquad u_0 \in \mathcal{M}_L \tag{16}$$

Then, separating the real and imaginary parts in Eq. (16), we find that the real and imaginary parts of the eigenfunction $u_0 = u_1 + iu_2$ which are distinct from zero are also eigenfunctions corresponding to the eigenvalue λ_0, $Lu_j = \lambda_0 u_j$, $j = 1, 2$.

LEMMA. *In order that $\lambda = 0$ be an eigenvalue of the operator L, it is necessary and sufficient that $q = 0$ and $\alpha = 0$. For this $\lambda = 0$ is a simple eigenvalue and $u_0 = $ const is the corresponding eigenfunction.*

Proof. Necessity. Let $\lambda = 0$ be the eigenvalue of the operator L and let u_0 be the corresponding eigenfunction, so that $Lu_0 = 0$, $u_0 \in \mathcal{M}_L$. Applying Eq. (13) to the function u_0, we obtain

$$0 = (Lu_0, u_0) = \int_G (p \,|\, \mathrm{grad}\, u_0\,|^2 + q\,|\, u_0\,|^2)\, dx + \int_{S_0} p\, \frac{\alpha}{\beta}\, |\, u_0\,|^2\, dS$$

from which, taking supposition (3) into account, we deduce

$$p\, \mathrm{grad}\, u_0 = 0, \qquad qu_0 = 0, \qquad x \in G$$

that is, $u_0 = \mathrm{const}$ and $q = 0$. It follows from boundary condition (2) for the eigenfunction $u_0 = \mathrm{const}$ that $\alpha = 0$. The necessity of the conditions is proved. Moreover it is established that $u_0 = \mathrm{const}$ is a unique eigenfunction corresponding to the eigenvalue $\lambda = 0$; that is, this eigenvalue is simple.

Sufficiency. If $q = 0$ and $\alpha = 0$, then, by virtue of (3), $\beta > 0$ and the problem (1)-(2) becomes the following:

$$-\mathrm{div}(p\, \mathrm{grad}\, u) = \lambda u, \qquad \left.\frac{\partial u}{\partial \mathbf{n}}\right|_S = 0$$

for which $u_0 = \mathrm{const}$ is the eigenfunction corresponding to the eigenvalue $\lambda = 0$. The lemma is proved.

We shall suppose that in boundary condition (2) either $\beta = 0$ or 1; that is, that this condition has the form

$$\text{either } \left. u \right|_S = 0, \quad \text{or} \quad \left.\frac{\partial u}{\partial \mathbf{n}} + \alpha u\right|_S = 0, \qquad \alpha \geq 0 \qquad (17)$$

Then, if the boundary S of the region G is a sufficiently smooth surface and the coefficients $p > 0$, $q \geq 0$, and $\alpha \geq 0$ are sufficiently smooth functions, the following theorem holds:

THEOREM 1. *The set of eigenvalues of the operator L is countable and does not have finite limit points; each eigenvalue has a finite multiplicity. Each function belonging to $\mathcal{M}_L$ may be expanded into a regularly converging Fourier series involving the eigenfunctions of the operator L.*

This theorem will be proved for two specific cases: (1) for the Sturm–Liouville problem (cf. Sec. 20); and (2) for the Dirichlet problem (cf. Sec. 24). Proof of this theorem in its general form is contained in Miranda's book (*1*).

§ 19. THE EIGENVALUE PROBLEM

On the basis of this theorem and of previous statements, all the eigenvalues of the operator L may be enumerated in the order of their magnitude

$$0 \le \lambda_1 \le \lambda_2 \le \ldots, \qquad \lambda_k \to \infty, \qquad k \to \infty \qquad (18)$$

repeating λ_k in this series as many times as its multiplicity. We shall denote the corresponding eigenfunctions by $u_1, u_2, \ldots$, so that in series (18) one and only one eigenfunction u_k corresponds to each eigenvalue λ_k,

$$Lu_k = \lambda_k u_k, \qquad k = 1, 2, \ldots, \qquad u_k \in \mathcal{M}_L$$

Here the eigenfunctions $\{u_k\}$ may be chosen to be real and orthonormal (cf. Sec. 1.10) so that

$$(Lu_k, u_i) = \lambda_k(u_k, u_i) = \lambda_k \delta_{ki} \qquad (19)$$

Moreover, each function f belonging to $\mathcal{M}_L$ may be expanded into Fourier series in terms of the orthonormal system $\{u_k\}$,

$$f(x) = \sum_{k=1}^{\infty} (f, u_k) u_k(x) \qquad (20)$$

and this series converges regularly over $\bar{G}$. But by the lemma of Sec. 1.5, the set $\mathcal{M}_L$ is dense in $\mathscr{L}_2(G)$. From this and from the theorem of Sec. 1.7 we obtain the following theorem.

THEOREM 2. *The system of eigenfunctions of the operator L is complete in $\mathscr{L}_2(G)$.*

Multiplying the series (20) scalarly on the left by Lf, for all f belonging to $\mathcal{M}_L$, we obtain the formula for the energy integral

$$(Lf, f) = \sum_{k=1}^{\infty} (f, \lambda_k u_k)\overline{(f, u_k)} = \sum_{k=1}^{\infty} \overline{(f, u_k)}(Lf, u_k)$$

$$= \sum_{k=1}^{\infty} (f, \lambda_k u_k)\overline{(f, u_k)} = \sum_{k=1}^{\infty} \lambda_k |(f, u_k)|^2 \qquad (21)$$

Now we shall establish the following variational principle (cf. Sec. 18.5):

$$\lambda_k = \inf_{\substack{f \in \mathcal{M}_L \\ (f, u_i) = 0, \, i = 1, 2, \ldots k-1}} \frac{(Lf, f)}{\|f\|^2}, \qquad k = 1, 2, \ldots \qquad (22)$$

and the infimum in (22) is attained for any eigenfunction corresponding to the eigenvalue λ_k.

In fact, using Eq. (21) for the quadratic form (Lf,f), and taking inequalities (18) into account: $\lambda_i \geq \lambda_k \geq 0$, $i \geq k$, for all $f \in \mathcal{M}_L$ such that $(f, u_i) = 0$, $i = 1, 2, \ldots, k-1$, we obtain

$$(Lf,f) = \sum_{i=k}^{\infty} \lambda_i |(f, u_i)|^2 \geq \lambda_k \sum_{i=k}^{\infty} |(f, u_i)|^2$$

But, by virtue of Theorem 2 of this subsection, Parseval's equation (cf. Sec. 1.6) is true:

$$\sum_{i=1}^{\infty} |(f, u_i)|^2 = \sum_{i=k}^{\infty} |(f, u_i)|^2 = \|f\|^2$$

and so

$$\lambda_k \leq \frac{(Lf,f)}{\|f\|^2}$$

On the other hand, for $f = u_k$, by virtue of (19) we have

$$\frac{(Lu_k, u_k)}{\|u_k\|^2} = \lambda_k, \qquad (u_k, u_i) = 0, \qquad i = 1, 2, \ldots, k-1$$

This establishes the validity of the variational principle (22).

Setting $k = 1$ in (22) we obtain, specifically,

$$\lambda_1 = \inf_{f \in \mathcal{M}_L} \frac{(Lf,f)}{\|f\|^2}$$

Applying formula (21) to the functions

$$\eta_p = f - \sum_{i=1}^{p} (f, u_i) u_i, \qquad p = 1, 2, \ldots$$

belonging to $\mathcal{M}_L$ and taking into consideration that

$$(\eta_p, u_k) = \left(f - \sum_{i=1}^{p} (f, u_i) u_i, u_k \right) = \begin{cases} 0, & k = 1, 2, \ldots, p \\ (f, u_k), & k = p+1, \ldots \end{cases}$$

we obtain

$$(L\eta_p, \eta_p) = \sum_{k=p+1}^{\infty} \lambda_k |(f, u_k)|^2$$

§ 19. THE EIGENVALUE PROBLEM

From this and from the convergence of the series (21) it follows that

$$(L\eta_p, \eta_p) \to 0, \qquad p \to \infty \qquad (23)$$

Applying inequality (14) to the functions η_p and taking (23) into account, we obtain, as $p \to \infty$,

$$\| |\operatorname{grad} \eta_p| \|^2 = \left\| \left| \operatorname{grad} f - \sum_{i=1}^{p} (f, u_i) \operatorname{grad} u_i \right| \right\|^2 \leq \frac{1}{p_0} (L\eta_p, \eta_p) \to 0$$

The result obtained shows that

$$\operatorname{grad} f(x) = \sum_{k=1}^{\infty} (f, u_k) \operatorname{grad} u_k(x) \qquad (24)$$

and the series (24) converges to $\operatorname{grad} f$ in $\mathscr{L}_2(G)$.

So we obtain the following theorem.

THEOREM 3. *If $f \in \mathscr{M}_L$, then the series (20) may be differentiated term by term once with respect to x_i, $i = 1, 2, \ldots, n$, and the series (24) obtained will converge to $\partial f/\partial x_i$ in $\mathscr{L}_2(G)$.*

Note. The results obtained may be extended to the boundary value problem involving the eigenvalues

$$Lu = \lambda \varrho u, \qquad \alpha u + \beta \left. \frac{\partial u}{\partial \mathbf{n}} \right|_S = 0$$

where the weight $\varrho(x) > 0$ is a continuous function over $\bar{G}$ if this problem is considered in the space $\mathscr{L}_2(G; \varrho)$ (cf. note of Sec. 1.7).

5. The Fourier Method (Separation of Variables). The Fourier method may be used to define eigenvalues and eigenfunctions of a many-dimensional elliptic operator which permits separation of its variables. The essence of this method is as follows. We divide the independent variables into two groups, $x = (x_1, x_2, \ldots, x_n)$ and $y = (y_1, y_2, \ldots, y_m)$, and let $G \subset R^n$ be the region of variation of x and $D \subset R^m$ be the region of variation of y. We shall use S and Γ to denote the boundaries of the regions G and D, respectively. Then $(S \times \bar{D}) \cup (\Gamma \times \bar{G})$ is the boundary of the region $G \times D \subset R^{n+m}$.

In the region $G \times D$ we shall examine the following boundary value problem involving the eigenvalues of an equation of elliptic type

$$Lu + Mu = \lambda u \qquad (25)$$

$$\alpha u + \beta \left.\frac{\partial u}{\partial \mathbf{n}}\right|_{S\times \bar{D}} = 0, \qquad \gamma u + \delta \left.\frac{\partial u}{\partial \mathbf{n}}\right|_{\Gamma \times \bar{G}} = 0 \qquad (26)$$

where L and M are elliptic operators not depending on y and x, respectively; the functions α, β do not depend on y and the functions γ, δ do not depend on x.

We shall seek the eigenfunctions of problem (25)-(26) in the form of the product $X(x)Y(y)$,

$$u(x, y) = X(x)Y(y) \qquad (27)$$

Substituting this expression into Eq. (25), we obtain

$$Y(y)LX(x) + X(x)MY(y) = \lambda X(x)Y(y)$$

from which

$$\frac{LX(x)}{X(x)} = \lambda - \frac{MY(y)}{Y(y)} \qquad (28)$$

The left-hand side of Eq. (28) does not depend on y, nor the right-hand side on x. Therefore these expressions do not depend either on x or on y; that is, they are equal to a constant. Denoting this constant by μ and writing $\nu = \lambda - \mu$, from (28) we obtain two equations:

$$LX = \mu X \qquad (29)$$

$$MY = \nu Y \qquad (30)$$

In this way, Eq. (25) has split into the two equations (29) and (30), or as it is said, *the variables have been separated*; in addition, an unknown parameter μ has appeared.

To deduce the boundary conditions for the functions $X(x)$ and $Y(y)$ we shall substitute the product $X(x)Y(y)$ in the boundary conditions (26). As a result, after abbreviation, we obtain

$$\alpha X + \beta \left.\frac{\partial X}{\partial \mathbf{n}}\right|_{S} = 0 \qquad (31)$$

$$\gamma Y + \delta \left.\frac{\partial Y}{\partial \mathbf{n}}\right|_{\Gamma} = 0 \qquad (32)$$

§ 19. THE EIGENVALUE PROBLEM

So the boundary value problem involving the eigenvalues (25)-(26) becomes two boundary value problems involving the eigenvalues (29)-(31) and (30)-(32) with a smaller number of independent variables. We shall denote by μ_k, $X_k(x)$, $k = 1, 2, \ldots$, and ν_j, $Y_j(y)$, $j = 1, 2, \ldots$, all the eigenvalues and the eigenfunctions of the operators L and M, respectively. By virtue of (27),

$$\lambda_{kj} = \mu_k + \nu_j, \qquad u_{kj}(x, y) = X_k(x) Y_j(y), \qquad k, j = 1, 2, \ldots \qquad (33)$$

are the eigenvalues and the eigenfunctions of the initial boundary value problem (25)-(26).

Note. Let the orthonormal systems of eigenfunctions $\{X_k\}$ and $\{Y_j\}$ be complete in $\mathscr{L}_2(G)$ and $\mathscr{L}_2(D)$, respectively (cf. Sec. 19.4). Then by the lemma of Sec. 1.7 the system of eigenfunctions $\{X_k Y_j\}$ is orthonormal and complete in $\mathscr{L}_2(G \times D)$. In this case Eqs. (33) give all the eigenvalues and eigenfunctions of the boundary value problem (25)-(26).

6. Examples. (a) Let us consider a boundary value problem involving the eigenvalues for a rectangle $\varPi = (0, l) \times (0, m)$ with the boundary L (Fig. 49)

$$-\frac{\partial^2 u}{\partial x^2} - \frac{\partial^2 u}{\partial y^2} = \lambda u, \qquad u|_L = 0 \qquad (34)$$

In accordance with the scheme set out in Sec. 19.5, this problem may be divided into two one-dimensional boundary value problems:

$$-X'' = \mu X, \qquad X(0) = X(l) = 0 \qquad (35)$$

$$-Y'' = \nu Y, \qquad Y(0) = Y(m) = 0 \qquad (36)$$

The eigenvalues and eigenfunctions of these boundary value problems are easily calculated.

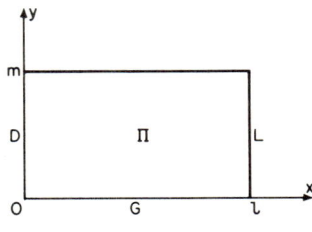

Fig. 49

We shall do this for problem (35). We write out the general solution of the differential equation (35)

$$X(x) = c_1 \sin \sqrt{\mu} x + c_2 \cos \sqrt{\mu} x$$

and select the arbitrary constants c_1 and c_2 and the parameter μ so as to satisfy the boundary conditions (35) and the normalization condition $\|X\| = 1$. For this it is necessary to put $c_2 = 0$ and $\sqrt{\mu} l = k\pi$, $k = \pm 1, \pm 2, \ldots$, so that

$$X(x) = c_1 \sin \frac{k\pi x}{l}$$

The normalization condition

$$1 = \int_0^l X^2(x)\, dx = c_1^2 \int_0^l \sin^2 \frac{k\pi x}{l}\, dx = \frac{l}{2} c_1^2$$

gives $c_1 = \sqrt{2/l}$ and, therefore,

$$\mu_k = \left(\frac{k\pi}{l}\right)^2, \qquad X_k(x) = \sqrt{\frac{2}{l}} \sin \frac{k\pi x}{l}, \qquad k = 1, 2, \ldots \qquad (37)$$

It follows from the construction just made that problem (35) has no other eigenfunctions. The system of eigenfunctions (37) is complete in $\mathscr{L}_2(0, l)$ (cf. Sec. 20.3).

Analogously for problem (36) we have

$$\nu_j = \left(\frac{j\pi}{m}\right)^2, \qquad Y_j(y) = \sqrt{\frac{2}{m}} \sin \frac{j\pi y}{m}, \qquad j = 1, 2, \ldots \qquad (38)$$

From (37) and (38), in accordance with Eqs. (33), we obtain the following eigenvalues and eigenfunctions of the boundary value problem (34):

$$\lambda_{kj} = \pi^2 \left(\frac{k^2}{l^2} + \frac{j^2}{m^2}\right), \qquad u_{kj}(x, y) = \frac{2}{\sqrt{lm}} \sin \frac{k\pi x}{l} \sin \frac{j\pi y}{m} \qquad (39)$$

$$k, j = 1, 2, \ldots$$

By virtue of the note of Sec. 19.5, problem (34) has no other eigenvalues and eigenfunctions. We remark that the eigenvalues λ_{kj} may be repeated; that is, $\lambda_{kj} = \lambda_{k_0 j_0}$ for a certain set of numbers (k, j). The number of such

§ 19. THE EIGENVALUE PROBLEM

repetitions, which is equal to the integral number of solutions of the equation,

$$\frac{k^2}{l^2} + \frac{j^2}{m^2} = \frac{k_0^2}{l^2} + \frac{j_0^2}{m^2}$$

gives the multiplicity of the eigenvalue $\lambda_{k_0 j_0}$.

(b) Let us consider the boundary value problem involving the eigenvalues for the circle U_R (Fig. 50):

$$-\Delta u = \lambda u, \qquad u|_{S_R} = 0 \qquad (40)$$

It is convenient to solve this problem in polar coordinates $x = r \cos \varphi$,

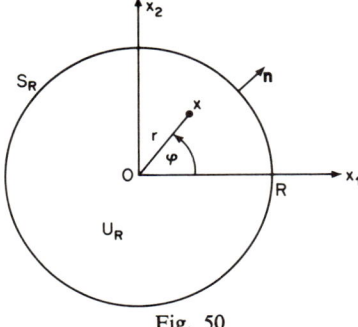

Fig. 50

$y = r \sin \varphi$, $0 \leq r < R$, $0 \leq \varphi \leq 2\pi$. In these coordinates problem (40) for the function $\tilde{u}(r, \varphi) = u(r \cos \varphi, r \sin \varphi)$ takes the form (cf. Sec. 3.2)

$$-\frac{1}{r}\frac{\partial}{\partial r}\left(r \frac{\partial \tilde{u}}{\partial r}\right) - \frac{1}{r^2}\frac{\partial^2 \tilde{u}}{\partial \varphi^2} = \lambda \tilde{u}, \qquad \tilde{u}(R, \varphi) = 0 \qquad (41)$$

The boundary condition for $r = 0$ must be added to the boundary condition for $r = R$. The boundary condition for $r = 0$ is that the function $\tilde{u}$ must be bounded in the neighborhood of the point $r = 0$. Moreover the function $\tilde{u}$ must obviously be 2π-periodic with respect to φ.

Applying the Fourier method to problem (41) (cf. Sec. 19.5), for the function $\tilde{u}(r, \varphi) = \mathscr{R}(r)\Phi(\varphi)$ we obtain two one-dimensional boundary value problems:

$$-\Phi'' = \mu \Phi, \qquad \Phi(\varphi) = \Phi(\varphi + 2\pi) \qquad (42)$$

$$r(r\mathscr{R}')' + (\lambda r^2 - \mu)\mathscr{R} = 0, \qquad |\mathscr{R}(0)| \neq \infty, \qquad \mathscr{R}(R) = 0 \qquad (43)$$

5. BOUNDARY VALUE PROBLEMS FOR ELLIPTIC EQUATIONS

The eigenvalues and eigenfunctions of problem (42) are easily calculated,

$$\mu_k = k^2, \qquad \Phi_k(\varphi) = \frac{1}{\sqrt{2\pi}} e^{ik\varphi}, \qquad k = 0, 1, \ldots \tag{44}$$

Moreover, Eq. (43) is Bessel's equation. A bounded solution $\mathscr{R}(r)$ of this equation, for $\mu = k^2$, is expressed by the Bessel function* $J_k(\sqrt{\lambda} r)$. To obtain the eigenvalues λ, it is necessary to use the second boundary condition (43), namely, $J_k(\sqrt{\lambda} R) = 0$; that is, $\sqrt{\lambda} R = \mu_{kj}$, where μ_{kj}, $j = 1, 2, \ldots$, are the positive roots of the Bessel function $J_k(\mu)$. From this it follows that

$$\lambda_{kj} = \frac{\mu_{kj}^2}{R^2}, \qquad \mathscr{R}_{kj}(r) = c_{kj} J_k\left(\mu_{kj} \frac{r}{R}\right), \qquad j = 1, 2, \ldots \tag{45}$$

where

$$c_{kj} = \frac{1}{\left[\int_0^R r J_k^2(\mu_{kj}(r/R)) \, dr\right]^{1/2}} = \frac{\sqrt{2}}{R \, |J_k'(\mu_{kj})|}$$

are the eigenvalues and eigenfunctions of the boundary value problem (43) for $\mu = k^2$.

From (44) and (45), according to Eqs. (33) we obtain the following eigenvalues and eigenfunctions of the boundary value problem (41), and hence of (40),[†]

$$\lambda_{kj} = \frac{\mu_{kj}^2}{R^2}, \qquad u_{kj}(x) = \frac{J_k(\mu_{kj}(r/R)) e^{ik\varphi}}{\sqrt{\pi} R \, |J_k'(\mu_{kj})|}$$
$$k = 0, 1, \ldots, \qquad j = 1, 2, \ldots \tag{46}$$

---

* The elements of the theory of Bessel functions may be found in the books of V. I. Smirnov (3, Chap. 6), A. N. Tikhonov and A. A. Samarsky (1, supplement), and V. Ya. Arsenin (1, Chap. XI).

[†] Strictly speaking, it has only been established so far that the functions u_{kj} satisfy Eq. (40) for $\lambda = \lambda_{kj}$ in the circle $|x| < R$ with the point $\{0\}$ being excluded. But from the equation

$$u_{kj}(x) = J_k\left(\mu_{kj} \frac{r}{R}\right) e^{ik\varphi} = \left(\frac{x_1 + ix_2}{2R}\right)^k \sum_{p=0}^{\infty} \frac{(-1)^p (|x|/2R)^{2p} \mu_{kj}^{2p+k}}{\Gamma(p+k+1)\Gamma(p+1)}$$

it follows that $u_{kj} \in C^\infty(U_R)$, and therefore they satisfy Eq. (40) at the point $x = 0$ also.

§ 19. THE EIGENVALUE PROBLEM

For each fixed $k = 0, 1, \ldots$, the eigenvalues λ_{kj}, $j = 1, 2, \ldots$, are different. Therefore the eigenfunctions u_{kj}, $j = 1, 2, \ldots$, are orthogonal in $\mathscr{L}_2(U_R)$ (cf. Sec. 19.4). It follows from this that the system of Bessel functions (45) is orthogonal in the space $\mathscr{L}_2((0, R); r)$.

Moreover, for each $k = 0, 1, \ldots$, the system of functions (45) is complete in $\mathscr{L}_2((0, R); r)$.* From this, by the lemma of Sec. 1.7, it follows that the system of eigenfunctions defined by Eq. (46) is orthonormal and complete in $\mathscr{L}_2(U_R)$ and therefore problem (40) has no other eigenvalues or eigenfunctions.

Analogous results are valid for the boundary value problem

$$-\Delta u = \lambda u, \qquad \left.\frac{\partial u}{\partial n} + \alpha u\right|_{S_R} = 0, \qquad \alpha \geq 0$$

7. Physical Sense of Eigenvalues and Eigenfunctions. For $p = 1$ and $\beta = 0$ the problem (1)-(2) involving the eigenvalues takes the form

$$-\Delta u + q(x)u = \lambda u, \qquad u|_S = 0 \qquad (47)$$

As is known,† the eigenvalues of problem (47) define the energy levels of a quantum particle moving in a external force field with a potential

$$V(x) = \begin{cases} q(x), & x \in \bar{G} \\ +\infty, & x \notin \bar{G} \end{cases}$$

The corresponding eigenfunctions are wave functions of the steady state Schrödinger operator (cf. Sec. 2.7),

$$-\Delta u + V(x)u = \lambda u \qquad (48)$$

As we show in Sec. 28, the eigenvalues of the operator L define the fundamental frequencies of the vibrations of bounded regions (of volumes, membranes, strings, rods, etc.) and the corresponding eigenfunctions define the modes of harmonic vibration.

The smallest eigenvalue of the steady state transport operator (cf. Sec. 2.4) also determines the criticality of a nuclear reactor, and the corresponding eigenfunction determines the density of the neutrons in a reactor in a critical condition.

---

* See A. N. Tikhonov and A. A. Samarsky (*I*, supplement).
† Compare, for instance, L. D. Landau and E. M. Lifschitz (*I*, Chap. III).

8. Uniqueness of the Solution of an Inhomogeneous Boundary Value Problem.

We shall consider the inhomogeneous boundary value problem

$$Lu = f, \qquad \alpha u + \beta \frac{\partial u}{\partial \mathbf{n}}\bigg|_S = \nu \qquad (49)$$

belonging to the class $C^2(G) \cap C^1(\bar{G})$.

If $q \not\equiv 0$ or $\alpha \not\equiv 0$, then the solution of the boundary value problem (49) is unique in the class $C^2(G) \cap C^1(\bar{G})$.

In fact, if $\tilde{u} \in C^2(G) \cap C^1(\bar{G})$ is another solution of problem (49), then the difference $\eta = u - \tilde{u}$ belongs to $\mathcal{M}_L$ and satisfies the homogeneous equation $L\eta = 0$; that is, it is an eigenfunction of the operator L corresponding to the eigenvalue $\lambda = 0$. But then, by the lemma of Sec. 19.4, $q \equiv 0$ and $\alpha \equiv 0$, despite the proposition. Therefore $\eta = u - \tilde{u} = 0$, as was to be shown.

9. Exercises.

(a) Prove the following maximum principle: If the function $u(x)$ of the class $C^2(G) \cap C(\bar{G})$ satisfies in the region G the differential inequality $-\Delta u + q(x)u \leq 0$, $q \geq 0$, then either $u \leq 0$ on $\bar{G}$ or $u(x)$ takes its (positive) maximum on $\bar{G}$ on the boundary S.

(b) Using (a), prove: If the function $u \in C^2(G) \cap C(\bar{G})$ is the solution of the boundary value problem

$$-\Delta u + q(x)u = F(x), \qquad u|_S = \nu(x) \qquad (50)$$

then the inequality

$$\|u\|_{C(\bar{G})} \leq \frac{\|F\|_{C(\bar{G})}}{q_0} + \|\nu\|_{C(S)}, \qquad q_0 = \min_{x \in \bar{G}} q(x)$$

is valid.

(c) Using (b), prove that the solution of problem (50) is unique in the class $C^2(G) \cap C(\bar{G})$ and depends continuously on F and ν in terms of the norm C (by the condition that $q_0 > 0$).

§ 20. The Sturm–Liouville Problem

For $n = 1$ the problem involving the eigenvalues (1)-(2) of Sec. 19.1 is known as the *Sturm–Liouville problem*,

$$Lu \equiv -(pu')' + qu = \lambda u, \qquad 0 < x < l \qquad (1)$$

$$h_1 u(0) - h_2 u'(0) = 0, \qquad H_1 u(l) + H_2 u'(l) = 0 \qquad (2)$$

§ 20. THE STURM-LIOUVILLE PROBLEM

According to the conditions (3) of Sec. 19.1, we consider

$$p \in C^1([0, l]), \quad q \in C([0, l]), \quad p(x) > 0, \quad q(x) \geq 0$$
$$h_1 \geq 0, \quad h_2 \geq 0, \quad H_1 \geq 0, \quad H_2 \geq 0, \quad h_1 + h_2 > 0, \quad H_1 + H_2 > 0$$

We recall that the domain of definition $\mathscr{M}_L$ of the operator L consists of the functions $u(x)$ of the class $C^2(0, l) \cap C^1([0, l])$, $u'' \in \mathscr{L}_2(0, l)$, satisfying the boundary conditions (2).

Expression (13) of Sec. 19.3 for the quadratic form (Lf, f), $f \in \mathscr{M}_L$, takes the following form:

$$(Lf, f) = \int_0^l (p|f'|^2 + q|f|^2)\, dx + \frac{h_1}{h_2} p(0)\, |f(0)|^2 + \frac{H_1}{H_2} p(l)\, |f(l)|^2$$

(the last terms to be excluded if $h_2 = 0$ or $H_2 = 0$, respectively).

1. Green's Function. Let us suppose that $\lambda = 0$ is not an eigenvalue of the operator L; this means, by virtue of the lemma of Sec. 19.4, that either $q \neq 0$, or $h_1 \neq 0$, or $H_1 \neq 0$.

We shall consider the boundary value problem

$$Lu \equiv -(pu')' + qu = f(x), \quad u \in \mathscr{M}_L \tag{3}$$

where $f \in C(0, l) \cap \mathscr{L}_2(0, l)$. Since $\lambda = 0$ is not an eigenvalue of the operator L, the solution of the boundary value problem (3) in the class $\mathscr{M}_L$ is unique. We shall construct the solution of this problem.

Let v_1 and v_2 be nonzero (real) solutions of the homogeneous equation $Lv = 0$, satisfying the conditions

$$h_1 v_1(0) - h_2 v_1'(0) = 0, \quad H_1 v_2(l) + H_2 v_2'(l) = 0 \tag{4}$$

It follows from the theory of ordinary differential equations that such solutions always exist and belong to the class $C^2([0, l])$. The solutions v_1 and v_2 are linearly independent. In fact, in the opposite case $v_1(x) = cv_2(x)$ and therefore, by virtue of (4), the solution v_1 also satisfies the second boundary condition (2). This means that v_1 is an eigenfunction of the operator L corresponding to the eigenvalue $\lambda = 0$, despite the supposition. Therefore the Wronskian determinant

$$w(x) = \begin{vmatrix} v_1(x) & v_2(x) \\ v_1'(x) & v_2'(x) \end{vmatrix} \neq 0, \quad x \in [0, l]$$

Moreover, the Ostrogradski–Liouville identity will hold:

$$p(x)w(x) = p(0)w(0), \qquad x \in [0, l] \tag{5}$$

[cf., for instance, L. S. Pontryagin (*1*, Chap. 3)].

We shall seek the solution of problem (3) by the method of variation of parameters,

$$u(x) = C_1(x)v_1(x) + C_2(x)v_2(x) \tag{6}$$

In accordance with this method, the functions C_1 and C_2 must satisfy the system of linear differential equations

$$C_1'v_1 + C_2'v_2 = 0, \qquad C_1'v_1' + C_2'v_2' = -\frac{f}{p} \tag{7}$$

with the determinant $w(x) \neq 0$. When we have solved this system and used identity (5), we shall obtain

$$
\begin{aligned}
C_1' &= \frac{1}{w} \begin{vmatrix} 0 & v_2 \\ -\dfrac{f}{p} & v_2' \end{vmatrix} = \frac{f(x)v_2(x)}{p(0)w(0)} \\
C_2' &= \frac{1}{w} \begin{vmatrix} v_1 & 0 \\ v_1' & -\dfrac{f}{p} \end{vmatrix} = -\frac{f(x)v_1(x)}{p(0)w(0)}
\end{aligned}
\tag{8}
$$

To satisfy boundary conditions (2), we put $C_2(0) = 0$, $C_1(l) = 0$, since, by virtue of (4) and (7),

$$h_1[C_1(0)v_1(0) + C_2(0)v_2(0)] - h_2[C_1(0)v_1'(0) + C_2(0)v_2'(0)]$$
$$= C_1(0)[h_1v_1(0) - h_2v_1'(0)] + C_2(0)[h_1v_2(0) - h_2v_2'(0)] = 0$$

and, analogously, for the end $x = l$. Integrating (8) using the conditions $C_1(l) = 0$, $C_2(0) = 0$, we have

$$C_1(x) = -\frac{1}{p(0)w(0)} \int_x^l f(y)v_2(y)\, dy$$

$$C_2(x) = -\frac{1}{p(0)w(0)} \int_0^x f(y)v_1(y)\, dy$$

If we substitute these expressions into (6), we find the required solution of

§ 20. THE STURM-LIOUVILLE PROBLEM

problem (3) in the form

$$u(x) = -\frac{1}{p(0)w(0)} \left[v_2(x) \int_0^x f(y)v_1(y)\, dy + v_1(x) \int_x^l f(y)v_2(y)\, dy \right]$$

or

$$u(x) = \int_0^l \mathscr{G}(x, y) f(y)\, dy \qquad (9)$$

where

$$\mathscr{G}(x, y) = -\frac{1}{p(0)w(0)} \begin{cases} v_1(x)v_2(y), & 0 \le x \le y \\ v_2(x)v_1(y), & y \le x \le l \end{cases} \qquad (10)$$

The function $\mathscr{G}(x, y)$ is known as the *Green's function* of the boundary value problem (3), or of the operator L.

Therefore the following result has been proved.

LEMMA. *If $\lambda = 0$ is not an eigenvalue of the operator L, then the solution of the boundary value problem (3) exists, is unique, and is expressed by Eq. (9).*

Equation (10) gives us the following properties of the Green's function $\mathscr{G}(x, y)$.

(1) It is real and continuous in the closed square $\bar{\Pi} = [0, l] \times [0, l]$ and belongs to the class C^2 in the closed triangles $[0 \le x \le y \le l]$ and $[0 \le y \le x \le l]$ (Fig. 51).

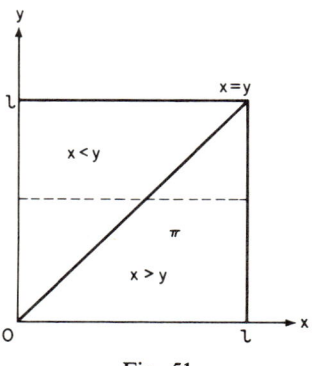

Fig. 51

(2) It is symmetrical

$$\mathscr{G}(x, y) = \mathscr{G}(y, x), \qquad (x, y) \in \bar{\Pi}$$

(3) On the diagonal $x = y$ the jump in the derivative $\partial \mathscr{G}/\partial x$ is equal to

$-[1/p(y)]$, that is,

$$\frac{\partial \mathscr{G}(y+0, y)}{\partial x} - \frac{\partial \mathscr{G}(y-0, y)}{\partial x} = -\frac{1}{p(y)}, \quad y \in (0, l)$$

(4) Away from the diagonal $x = y$ it satisfies the homogeneous equation

$$L_x \mathscr{G}(x, y) = 0, \quad x \neq y, \quad (x, y) \in \bar{\Pi}$$

(5) On the sides of the square Π it satisfies the boundary conditions (2):

$$h_1 \mathscr{G}(0, y) - h_2 \frac{\partial \mathscr{G}(0, y)}{\partial x} = H_1 \mathscr{G}(l, y) + H_2 \frac{\partial \mathscr{G}(l, y)}{\partial x} = 0, \quad y \in [0, l]$$

Note. It follows from properties (1), (3), and (4) that for each $y \in (0, l)$ the Green's function $\mathscr{G}(x, y)$ satisfies, in a generalized sense (cf. Sec. 10.1), the equation

$$L_x \mathscr{G}(x, y) = \delta(x - y), \quad x \in (0, l)$$

Therefore $\mathscr{G}(x, y)$ is a perturbation generated by a point source of intensity 1, situated at the point y. So the Green's function $\mathscr{G}(x, y)$ is a natural generalization of the fundamental solution (cf. Sec. 10.2) of equations with variable coefficients which are subject to boundary conditions.

Example. The Green's function of the boundary value problem

$$-u'' = f(x), \quad u(0) = u(1) = 0$$

has the form

$$\mathscr{G}(x, y) = \begin{cases} x(1-y), & 0 \leq x \leq y \\ (1-x)y, & y \leq x \leq 1 \end{cases}$$

2. Reduction of the Sturm–Liouville Problem to an Integral Equation.

We shall show that the Sturm–Liouville problem may be reduced to a Fredholm integral equation with a real, symmetrical, and continuous kernel $\mathscr{G}(x, y)$.

THEOREM. *The boundary value problem*

$$Lu = \lambda u + f, \quad u \in \mathscr{M}_L, \quad f \in C(0, l) \cap \mathscr{L}_2(0, l) \tag{11}$$

for which $\lambda = 0$ is not an eigenvalue of the operator L is equivalent to the

§ 20. THE STURM-LIOUVILLE PROBLEM

integral equation

$$u(x) = \lambda \int_0^l \mathscr{G}(x, y)u(y)\, dy + \int_0^l \mathscr{G}(x, y)f(y)\, dy, \qquad u \in C([0, l]) \quad (12)$$

where $\mathscr{G}(x, y)$ is the Green's function of the operator L.

Proof. If $u(x)$ is the solution of the boundary value problem (11), then applying the lemma of Sec. 20.1 with a change of f to $\lambda u + f$, we obtain

$$u(x) = \int_0^l \mathscr{G}(x, y)[\lambda u(y) + f(y)]\, dy$$

that is, the function $u(x)$ satisfies integral equation (12).

Conversely, let the function $u_0 \in C([0, l])$ satisfy the integral equation (12). We shall consider the boundary value problem

$$Lu = \lambda u_0 + f, \qquad u \in \mathscr{M}_L$$

By the lemma of Sec. 20.1 the unique solution of this problem is given by the equation

$$u(x) = \int_0^l \mathscr{G}(x, y)[\lambda u_0(y) + f(y)]\, dy = u_0(x)$$

from which it follows that $u_0 \in \mathscr{M}_L$ and satisfies the equation

$$Lu_0 = \lambda u_0 + f$$

that is, u_0 is the solution of the boundary value problem (11). The theorem is proved.

When $f = 0$ the boundary value problem (11) is the Sturm–Liouville problem, and therefore the Sturm–Liouville problem (1)–(2) is equivalent to a problem involving the eigenvalues of the homogeneous integral equation

$$u(x) = \lambda \int_0^l \mathscr{G}(x, y)u(y)\, dy \qquad (13)$$

provided that $\lambda = 0$ is not an eigenvalue of the operator L.

We shall now eliminate the assumption that $\lambda = 0$ is not an eigenvalue of the operator L. For this we note that, by virtue of the lemma of Sec. 19.4,

$\mu = 0$ is not an eigenvalue of the Sturm–Liouville problem

$$L_1 u \equiv -(pu')' + (q+1)u = \mu u \tag{14}$$

$$h_1 u(0) - h_2 u'(0) = H_1 u(l) + H_2 u'(l) = 0 \tag{15}$$

But $\mathscr{M}_L = \mathscr{M}_{L_1}$ and therefore problem (14)-(15) is equivalent to problem (1)-(2) for $\mu = \lambda + 1$.

Therefore the Sturm–Liouville problem (1)-(2) is equivalent to the integral equation

$$u(x) = (\lambda + 1) \int_0^l \mathscr{G}_1(x, y) u(y)\, dy \tag{16}$$

where $\mathscr{G}_1(x, y)$ is the Green's function of the operator L_1.

3. Properties of Eigenvalues and Eigenfunctions. We have now established that the Sturm–Liouville problem (1)-(2) is equivalent to the problem involving the eigenvalues of the homogeneous integral equation (16) with a symmetrical (and therefore Hermitian) continuous kernel $\mathscr{G}_1(x, y)$. For this the eigenvalues λ of the problem (1)-(2) are linked with the characteristic numbers μ of the kernel $\mathscr{G}_1(x, y)$ by the equation $\mu = \lambda + 1$, and the eigenfunctions corresponding to them coincide. Therefore all the statements of the theory of integral equations with a symmetrical continuous kernel which were developed in Sec. 17 and 18 are also valid for the Sturm–Liouville problem. Specifically, *the set of eigenvalues $\{\lambda_k\}$ of this problem is not empty; it is no more than countable and does not have finite limit points; the eigenvalues are real and of finite multiplicity; the eigenfunctions may be chosen real and orthonormal.*

But the Sturm–Liouville problem has a number of specific properties. We shall note some of them.

Eigenvalues are nonnegative. This statement is proved in Sec. 19.4.

The set of eigenvalues is countable. In fact, if this set were finite $\{\lambda_1, \lambda_2, \ldots, \lambda_N\}$, then the kernel $\mathscr{G}_1(x, y)$ would have to be represented in the form (cf. Sec. 18.1)

$$\mathscr{G}_1(x, y) = \sum_{k=1}^N \frac{u_k(x) u_k(y)}{\lambda_k + 1} \tag{17}$$

But $u_k \in C^2([0, l])$ and therefore Eq. (17) contradicts property (2) of the Green's function $\mathscr{G}_1(x, y)$. This contradiction proves our assertion.

§ 20. THE STURM–LIOUVILLE PROBLEM

Each eigenvalue is simple. In fact, let u_1 and u_2 be eigenfunctions corresponding to the eigenvalue λ_0. This means that these functions satisfy Eq. (1) for $\lambda = \lambda_0$ and satisfy boundary conditions (2). From the first boundary condition (2)

$$h_1 u_1(0) - h_2 u_1'(0) = 0, \qquad h_1 u_2(0) - h_2 u_2'(0) = 0$$

it follows, by virtue of the supposition that $h_1 + h_2 > 0$, that

$$\begin{vmatrix} u_1(0) & -u'(0) \\ u_2(0) & -u_2'(0) \end{vmatrix} = - \begin{vmatrix} u_1(0) & u_2(0) \\ u_1'(0) & u_2'(0) \end{vmatrix} = 0$$

that is, the Wronskian determinant of the solutions $u_1(x)$ and $u_2(x)$ of Eq. (1) for $\lambda = \lambda_0$ becomes zero at the point $x = 0$. So these solutions are linearly dependent. This also means that λ_0 is a simple eigenvalue of the Sturm–Liouville problem (1)-(2).

THEOREM (V. A. Steklov). *Each function f belonging to $\mathcal{M}_L$ may be expanded into a regularly convergent Fourier series involving the eigenfunctions $\{u_k\}$ of the Sturm–Liouville problem,*

$$f(x) = \sum_{k=1}^{\infty} (f, u_k) u_k(x) \tag{18}$$

Proof. Since $f \in \mathcal{M}_L$, then

$$L_1 f = Lf + f = h \in C(0, l) \cap \mathscr{L}_2(0, l)$$

Evidently $\mathcal{M}_{L_1} = \mathcal{M}_L$ and so $f \in \mathcal{M}_{L_1}$. In this way we see that the function f is a solution of the boundary value problem

$$L_1 f = h, \qquad f \in \mathcal{M}_{L_1}$$

and, by the lemma of Sec. 19.4, $\lambda = 0$ is not an eigenvalue of the operator L_1. We shall denote the Green's function of the operator L_1 by $\mathscr{G}_1(x, y)$. By the lemma of Sec. 20.1, the function f can be expressed by the integral

$$f(x) = \int_0^l \mathscr{G}_1(x, y) h(y) \, dy$$

that is, it can be represented sourcewise by the Hermitian continuous kernel $\mathscr{G}_1(x, y)$. By the Hilbert–Schmidt theorem (cf. Sec. 18.1), the function f

can be expanded into a regularly convergent Fourier series involving the eigenfunctions of the kernel $\mathscr{G}_1(x, y)$. But the eigenfunctions of the kernel $\mathscr{G}_1(x, y)$ coincide with the eigenfunctions of the operator L_1, which in their turn coincide with the eigenfunctions $\{u_k\}$ of the operator L. The theorem is proved.

So Theorem 1 of Sec. 19.4 and its corollaries are true for the Sturm–Liouville problem. Specifically, the system of eigenfunctions of the Sturm–Liouville problem is complete in $\mathscr{L}_2(0, l)$.

4. Finding Eigenvalues and Eigenfunctions. We shall set out the process of calculating eigenvalues and eigenfunctions of the Sturm–Liouville problem (1)-(2). Let $u_1(x; \lambda)$ and $u_2(x; \lambda)$ be the solutions of Eq. (1), satisfying the initial conditions:

$$u_1(0; \lambda) = 1, \quad u_1'(0; \lambda) = 0; \quad u_2(0; \lambda) = 0, \quad u_2'(0; \lambda) = 1$$

Then the function

$$u(x; \lambda) = h_2 u_1(x; \lambda) + h_1 u_2(x; \lambda) \tag{19}$$

satisfies Eq. (1) and the first of the boundary conditions (2). To satisfy the second of the boundary conditions (2), it is necessary to require

$$H_1 h_2 u_1(l; \lambda) + H_1 h_1 u_2(l; \lambda) + H_2 h_2 u_1'(l; \lambda) + H_2 h_1 u_2'(l; \lambda) = 0$$

The roots $\lambda_1, \lambda_2, \ldots$, of the transcendental equation obtained give all the eigenvalues of the Sturm–Liouville problem (1)-(2). The corresponding eigenfunctions u_k are defined according to Eq. (19) for $\lambda = \lambda_k$,

$$u_k(x) = u(x; \lambda_k) = h_2 u_1(x; \lambda_k) + h_1 u_2(x; \lambda_k), \quad k = 1, 2, \ldots$$

§ 21. Harmonic Functions

In this section we shall study the main properties of harmonic functions.

The real-valued function $u(x)$ of the class $C^2(G)$ is said to be *harmonic in the region G* if it satisfies Laplace's equation $\Delta u = 0$ in this region.

For $n = 1$ harmonic functions are reduced to linear functions and so their theory is of no interest; we shall therefore consider in future that $n \geq 2$.

§ 21. HARMONIC FUNCTIONS

The fundamental solution of the Laplace operator (cf. Sec. 10.8)

$$\mathscr{E}_2(x) = \frac{1}{2\pi} \ln |x|, \quad n = 2$$

$$\mathscr{E}_n(x) = -\frac{1}{(n-2)\sigma_n} |x|^{-n+2}, \quad n \geq 3$$

is a nontrivial example of a harmonic function for $x \neq 0$.

1. Green's Formula. If $u \in C^2(\bar{G})$ and $u(x) = 0$, $x \notin \bar{G}$, then for $x \notin S$ the following Green's formula is true:

$$u(x) = -\frac{1}{(n-2)\sigma_n} \int_G \frac{\Delta u(y)}{|x-y|^{n-2}} dy$$

$$+ \frac{1}{(n-2)\sigma_n} \int_S \left[\frac{1}{|x-y|^{n-2}} \frac{\partial u(y)}{\partial n} - u(y) \frac{\partial}{\partial n_y} \frac{1}{|x-y|^{n-2}} \right] dS_y \quad (1)$$

$$u(x) = -\frac{1}{2\pi} \int_G \Delta u(y) \ln \frac{1}{|x-y|} dy$$

$$+ \frac{1}{2\pi} \int_S \left[\ln \frac{1}{|x-y|} \frac{\partial u(y)}{\partial n} - u(y) \frac{\partial}{\partial n_y} \ln \frac{1}{|x-y|} \right] dS_y, \quad n = 2$$

In other words, the function u appears in the form of a sum of three Newtonian (logarithmic) potentials

$$u(x) = V_n(x) + V_n^{(0)}(x) + V_n^{(1)}(x) \quad (2)$$

where (we shall consider for the sake of definition that $n \geq 3$)

$$V_n(x) = \mathscr{E}_n * \{\Delta u\} = -\frac{1}{(n-2)\sigma_n} \int_G \frac{\Delta u(y)}{|x-y|^{n-2}} dy$$

is the volume potential with a density $-[1/(n-2)\sigma_n]\{\Delta u\}$;

$$V_n^{(0)}(x) = -\mathscr{E}_n * \left(\frac{\partial u}{\partial n} \delta_S \right) = \frac{1}{(n-2)\sigma_n} \int_S \frac{1}{|x-y|^{n-2}} \frac{\partial u(y)}{\partial n} dS_y$$

is the simple layer potential over S with a surface density $[1/(n-2)\sigma_n] \times (\partial u/\partial \mathbf{n})$;

$$V_n^{(1)}(x) = -\mathscr{E}_n * \frac{\partial}{\partial \mathbf{n}} (u\delta_S) = \frac{-1}{(n-2)\sigma_n} \int_S u(y) \frac{\partial}{\partial n_y} \frac{1}{|x-y|^{n-2}} dS_y$$

is the double layer potential over S with a surface density $-[1/(n-2)\sigma_n]u$.

We shall prove Green's formula (1) for $n \geq 3$. In Sec. 6.5, (c), formula (21) was deduced:

$$\Delta u = \{\Delta u\} - \frac{\partial u}{\partial \mathbf{n}} \delta_S - \frac{\partial}{\partial \mathbf{n}} (u\delta_S) \qquad (3)$$

Since the function u has compact support, its convolution with the fundamental solution $\mathscr{E}_n$ of the Laplace operator exists (cf. Sec. 7.5). Therefore, applying formula (13) of Sec. 10.3 and using Eq. (3), for the function u we obtain the representation

$$\begin{aligned} u &= \mathscr{E}_n * \Delta u = \mathscr{E}_n * \{\Delta u\} - \mathscr{E}_n * \left(\frac{\partial u}{\partial \mathbf{n}} \delta_S\right) - \mathscr{E}_n * \frac{\partial}{\partial \mathbf{n}} (u\delta_S) \\ &= \frac{1}{(n-2)\sigma_n} \left[-\frac{1}{|x|^{n-2}} * \{\Delta u\} \right. \\ &\quad + \frac{1}{|x|^{n-2}} * \left(\frac{\partial u}{\partial \mathbf{n}} \delta_S\right) + \frac{1}{|x|^{n-2}} * \frac{\partial}{\partial \mathbf{n}} (u\delta_S) \right] \end{aligned} \qquad (4)$$

From this, using the definition of Newtonian potentials and Eqs. (29), (32), and (33) of Sec. 7.8, we obtain Green's formula (1) for $n \geq 3$. The case $n = 2$ is considered analogously.

Green's formula (1) is also valid for functions u of the class $C^2(G)$ $\cap \, C^1(\bar{G})$ if the integral in it involving the region G is understood to be improper (cf. Sec. 19.2). (This integral may not converge absolutely and so may not be the volume potential V_n.)

For proof we shall apply Green's formula (1) to each subregion $G' \Subset G$ with a piecewise smooth boundary and we shall proceed to the limit as $G' \to G$. Using the supposed smoothness of the function u, we see that Green's formula (1) is valid in this case too (cf. Sec. 19.2).

For the function u of the class $C^1(\bar{G})$, harmonic in the region G, Green's formula (1) takes the following form:

$$\begin{aligned} u(x) = \frac{1}{(n-2)\sigma_n} \int_S \left[\frac{1}{|x-y|^{n-2}} \frac{\partial u(y)}{\partial \mathbf{n}} \right. \\ \left. - u(y) \frac{\partial}{\partial \mathbf{n}_y} \frac{1}{|x-y|^{n-2}} \right] dS_y, \qquad n \geq 3 \end{aligned} \qquad (5)$$

$$\begin{aligned} u(x) = \frac{1}{2\pi} \int_S \left[\ln \frac{1}{|x-y|} \frac{\partial u(y)}{\partial \mathbf{n}} \right. \\ \left. - u(y) \frac{\partial}{\partial \mathbf{n}_y} \ln \frac{1}{|x-y|} \right] dS_y, \qquad n = 2 \end{aligned}$$

§ 21. HARMONIC FUNCTIONS

The surface potentials $V_n^{(0)}(x)$ and $V_n^{(1)}(x)$ may be repeatedly differentiated under the integral sign *an infinite number of times outside S* and these potentials are *harmonic functions outside S*. It follows from this and from formula (5) that *each harmonic function is infinitely differentiable.**

Note. Green's formula (5) expresses the values of a harmonic function in a region in terms of its values and the values of its normal derivative over the boundary of this region. This formula is analogous to Cauchy's formula for analytic functions. It is easy to see a further analogy between Green's formula in form (2) and the similar formula (12) of Sec. 12.3 for the wave equation.

2. Extension of Green's Formulas. Let the boundary S of the region G be a surface of the class C^1 (cf. Sec. 1.1) and let the function $u \in C^1(G)$. We shall say that the function u has a *correct normal derivative* $\partial u/\partial \mathbf{n}$ over S if uniformly for all $x \in S$ there is a limit of the normal derivative $\partial u(x')/\partial \mathbf{n}_x$ as $x' \to x$ with $x' \in G$, $x' \in -\mathbf{n}_x$. This limit will be denoted by $\partial u/\partial \mathbf{n} \equiv \partial u(x)/\partial \mathbf{n}_x$.

It follows from this definition that the correct normal derivative is continuous over S, if it exists.

Let S be a surface of the class C^2 (cf. Sec. 1.1). At each point $x \in S$ we shall measure along the interior normal $-\mathbf{n}_x$ a segment of constant length δ. The set of the ends x' of these segments is described by the equation

$$x' = x - \delta \mathbf{n}_x \qquad (6)$$

For a sufficiently small δ this set forms a closed surface of the class C^1, which we shall denote by S_δ and shall call a surface *parallel to the surface S* (Fig. 52).

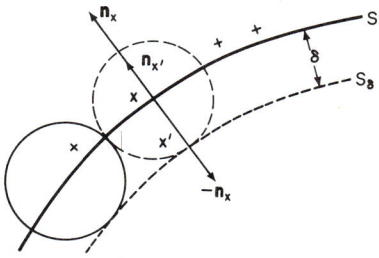

Fig. 52

---

* And each one is even analytic.

The normal $\mathbf{n}_{x'}$ at the point $x' = x - \delta \mathbf{n}_x \in S_\delta$ is directed along the normal $\mathbf{n}_x$ where $x \in S$.

This statement follows from the fact that the surface S_δ is an envelope of a family of spherical surfaces

$$(x_1 - x_1')^2 + (x_2 - x_2')^2 + \cdots + (x_n - x_n')^2 = \delta^2 \tag{7}$$

having their centers x located on the surface S. We shall prove this fact. Let a segment U of the surface S be given by the equation $x_n = z(x_1, \ldots, x_{n-1})$. If we differentiate the equation of the family (7) involving the parameters $x_1, \ldots, x_{n-1}$, in order to define the envelope of this family covering the segment U, we obtain the conditions:

$$x_k - x_k' + (x_n - x_n') \frac{\partial z}{\partial x_k} = 0, \quad k = 1, 2, \ldots, n-1 \tag{8}$$

If we now note that over U the vector $\mathbf{n}_x$ is proportional to the vector $(-(\partial z/\partial x_1), \ldots, -(\partial z/\partial x_{n-1}), 1)$, we can deduce from Eqs. (8) and (7) the equation of the surface S_δ in the form (6).

LEMMA. *Let the boundary S of the region G be a surface of the class C^2 and let the function u belonging to $C^1(G)$ have a correct normal derivative $\partial u/\partial \mathbf{n}$ over S. Then for any $f \in C(\bar{G})$ the following equation is true*:

$$\lim_{\delta \to 0} \int_{S_\delta} f(x') \frac{\partial u(x')}{\partial \mathbf{n}_{x'}} dS_{x'} = \int_S f(x) \frac{\partial u(x)}{\partial \mathbf{n}_x} dS_x \tag{9}$$

where S_δ are surfaces parallel to S.

Proof. Since the normals $\mathbf{n}_x$ and $\mathbf{n}_{x'}$ at the points $x \in S$ and $x' = x - \delta \mathbf{n}_x \in S_\delta$ are similarly directed, then

$$f(x') \frac{\partial u(x')}{\partial \mathbf{n}_{x'}} = f(x') \frac{\partial u(x')}{\partial \mathbf{n}_x} \xrightarrow{x \in S} f(x) \frac{\partial u(x)}{\partial \mathbf{n}_x} \tag{10}$$

$$x' \to x, \quad x' \in -\mathbf{n}_x$$

by virtue of the definition of the correct normal derivative and the continuity of the function f over $\bar{G}$. Equation (9) follows from the limiting result (10). The lemma is proved.

It follows from the lemma that *Green's formulas (7) and (8) of Sec. 19.2 and (1) of Sec. 21.1 remain valid if S is a surface of the class C^2, and the*

§ 21. HARMONIC FUNCTIONS

functions $u, v \in C^2(G) \cap C(\bar{G})$ have correct normal derivatives over S and $\Delta u, \Delta v \in \mathscr{L}_2(G)$.

In fact, we shall apply Green's formulas to any subregion bounded by the surface S_δ parallel to S. Proceeding to the limit as $\delta \to 0$ in these formulas and using the limiting result (9), we see that Green's formulas are valid under the conditions we have formulated.

3. Theorem of the Mean Value. We shall first prove the following statement: *If the function $u \in C^1(\bar{G})$ is harmonic in the region G, then*

$$\int_S \frac{\partial u}{\partial n} \, dS = 0 \tag{11}$$

Equation (11) follows from Green's first formula (7) of Sec. 19.2 when $v = 1$.

THEOREM OF THE MEAN VALUE. *If the function $u(x)$ is harmonic in the sphere U_R and continuous on $\bar{U}_R$, then its value in the center of this sphere is equal to the mean value over the spherical surface S_R,*

$$u(0) = \frac{1}{\sigma_n R^{n-1}} \int_{S_R} u(x) \, dS = \frac{1}{\sigma_n} \int_{S_1} u(Rs) \, ds \tag{12}$$

Proof. Applying Green's formula (5) for the point $x = 0$ to any sphere $|x| < \varrho$, $\varrho < R$, and using Eq. (11), for $n \geq 3$ we obtain Eq. (12):

$$u(0) = \frac{1}{(n-2)\sigma_n} \left[\frac{1}{\varrho^{n-2}} \int_{S_\varrho} \frac{\partial u(y)}{\partial n} \, dS - \int_{S_\varrho} u(y) \frac{\partial}{\partial n} \frac{1}{|y|^{n-2}} \, dS \right]$$

$$= \frac{1}{\sigma_n \varrho^{n-1}} \int_{S_\varrho} u(y) \, dS$$

Since the function $u(x)$ is continuous on the closed sphere $\bar{U}_R$, Eq. (12) is also retained when $\varrho \to R$. The case $n = 2$ is considered analogously. The theorem is proved.

4. The Maximum Principle. Using the theorem of the mean value, we shall establish the following maximum principle for harmonic functions.

THEOREM. *If the function $u(x) \neq$ const is harmonic in the region G and continuous on $\bar{G}$, then it cannot assume its minimum and maximum values in*

the region G, that is,

$$\min_{x \in S} u(x) < u(x) < \max_{x \in S} u(x), \quad x \in G \tag{13}$$

Proof. Assume the contrary to be true and let the function $u(x)$ assume its maximum value M at a point $x_0 \in G$,

$$M = u(x_0) = \max_{x \in \bar{G}} u(x) \tag{14}$$

Since x_0 is an interior point of the region G, then there is a sphere $U(x_0; r_0)$ of largest radius r_0, contained in G (Fig. 53).

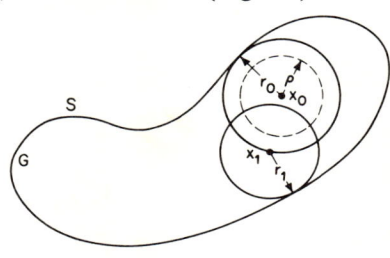

Fig. 53

We shall prove that

$$u(x) \equiv M, \quad x \in \bar{U}(x_0; r_0) \tag{15}$$

It follows from (14) that

$$u(x) \leq M = u(x_0), \quad x \in U(x_0; r_0) \tag{16}$$

If at a point $x' \in \bar{U}(x_0; r_0)$ there were a $u(x') < M$, then, according to continuity, the inequality $u(x) < M$ would also hold in a neighborhood of this point. But then, applying the formula of the mean value (12) to the spherical surface $S(x_0; \varrho)$, where $\varrho = |x' - x_0|$, and using inequality (16), we obtain

$$u(x_0) = \frac{1}{\sigma_n \varrho^{n-1}} \int_{S(x_0,\varrho)} u(x) \, dS < \frac{M}{\sigma_n \varrho^{n-1}} \int_{S(x_0,\varrho)} dS = M$$

which contradicts (14). So identity (15) is established.

We shall now take an arbitrary point $x_1 \in G$ lying on the boundary of the sphere $\bar{U}(x_0; r_0)$ (Fig. 53). By what has been proved $u(x_1) = M$. Applying our previous arguments to the point x_1, we conclude that $u(x) \equiv M$ in the largest sphere $U(x_1; r_1) \subset \bar{G}$, and so on. In this way the whole region G becomes exhausted and so $u(x) \equiv M$ for $x \in G$, despite what has been supposed.

This contradiction shows that our original supposition is untrue; so the function $u(x)$ cannot assume its maximum value in the region G. From this, replacing u by $-u$, we conclude that the function $u(x)$ cannot assume its minimum value in the region G either. The theorem is proved.

5. Corollaries of the Maximum Principle

(a) *If the function $u \in C(\bar{G})$ is harmonic in G, then*

$$|u(x)| \leq \max_{x \in S} |u(x)|, \quad x \in \bar{G} \tag{17}$$

Specifically, if $u|_S = 0$, then $u(x) \equiv 0$, $x \in G$.

This assertion follows from inequality (13),

$$-\max_{x \in S} |u(x)| \leq \min_{x \in S} u(x) \leq u(x) \leq \max_{x \in S} u(x) \leq \max_{x \in S} |u(x)|, \quad x \in \bar{G}$$

(b) *If the function $u \in C(\bar{G}_1)$ is harmonic in the region $G_1 = R^n \setminus \bar{G}$ and $u(\infty) = 0$, then*

$$|u(x)| \leq \max_{x \in S} |u(x)|, \quad x \in \bar{G}_1 \tag{18}$$

Specifically, if $u|_S = 0$ and $u(\infty) = 0$, then $u(x) \equiv 0$ for $x \in G_1$.

In fact, let the sphere U_R contain the region $\bar{G}$. Then $S \cup S_R$ is the boundary of the region $Q_R = G_1 \cap U_R$ (Fig. 54). Applying inequality (17) to this region, we obtain

$$|u(x)| \leq \max_{x \in S \cup S_R} |u(x)|, \quad x \in \bar{Q}_R \tag{19}$$

But by assumption $u(x) \to 0$ as $|x| \to \infty$, that is

$$\max_{x \in S_R} |u(x)| \to 0, \quad R \to \infty$$

Therefore, proceeding to the limit as $R \to \infty$ in inequality (19), we obtain inequality (18).

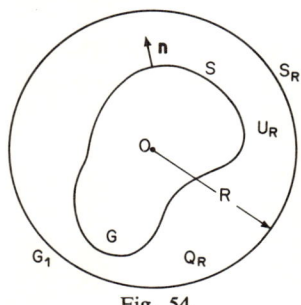

Fig. 54

(c) *If the sequence of functions $u_1, u_2, \ldots$, which are harmonic in the region G and continuous on $\bar{G}$, converges uniformly over the boundary S, then it also converges uniformly over $\bar{G}$.*

This statement follows from inequality (17):

$$|u_p(x) - u_q(x)| \le \max_{x \in S} |u_p(x) - u_q(x)| \to 0, \qquad p, q \to \infty, \qquad x \in \bar{G}$$

An analogous result is valid for the region $G_1 = R^n \setminus \bar{G}$ on condition that $u_k(\infty) = 0$.

6. Removal of the Singularities of a Harmonic Function.

The following theorem concerning the removal of singularities is valid for harmonic functions. It is analogous to a corresponding theorem for analytic functions.

THEOREM. *If the function $u(x)$ is harmonic in the region $G \setminus \{0\}$ and satisfies the condition*

$$u(x) = o(|\mathscr{E}_n(x)|), \qquad x \to 0 \tag{20}$$

then it can be continued harmonically to the point $\{0\}$.

Proof. Let $U_R \Subset G$. We introduce the function $\tilde{u}(x)$ equal to $u(x)$ in $\bar{U}_R$ and to zero outside $\bar{U}_R$. This function is locally integrable and, by virtue of (3) of Sec. 21.1, the functional

$$\Delta \tilde{u} + \frac{\partial u}{\partial \mathbf{n}} \delta_{S_R} + \frac{\partial}{\partial \mathbf{n}} (u \delta_{S_R}) \tag{21}$$

becomes zero over all test functions equal to zero in the neighborhood of the point $\{0\}$. This means that the generalized function (21) is either equal to zero or its support is the point $\{0\}$. Then, by the theorem of Sec. 8.4, this generalized function can be represented in the form of a finite combination of derivatives of $\delta(x)$, that is,

$$\Delta \tilde{u} = -\frac{\partial u}{\partial n}\delta_{S_R} - \frac{\partial}{\partial n}(u\delta_{S_R}) + \sum_{|\alpha|=0}^{m} c_\alpha D^\alpha \delta \qquad (22)$$

Since the function $\tilde{u}$ has compact support, its convolution with the fundamental solution $\mathscr{E}_n$ exists (cf. Sec. 7.5). So, applying formula (13) of Sec. 10.3, from (22) we obtain the equation

$$\tilde{u} = \mathscr{E}_n * \Delta \tilde{u} = -\mathscr{E}_n * \left(\frac{\partial u}{\partial n}\delta_{S_R}\right) - \mathscr{E}_n * \frac{\partial}{\partial n}(u\delta_{S_R})$$
$$+ \sum_{|\alpha|=0}^{m} c_\alpha(\mathscr{E}_n * D^\alpha \delta) = V_n^{(0)}(x) + V_n^{(1)}(x) + \sum_{|\alpha|=0}^{m} c_\alpha D^\alpha \mathscr{E}_n(x) \qquad (23)$$

Since the surface potentials $V_n^{(0)}$ and $V_n^{(1)}$ are harmonic functions in the sphere U_R (cf. Sec. 21.1), it follows from (23) and from condition (20) that all $c_\alpha = 0$, so that the function

$$u(x) = V_n^{(0)}(x) + V_n^{(1)}(x)$$

is harmonic in the sphere U_R. The theorem is proved.

7. Generalized Harmonic Functions. The real-valued function $u \in C(G)$ is said to be a *generalized harmonic function* in the region G if it satisfies Laplace's equation in this region, that is,

$$(\Delta u, \varphi) = \int u(x)\, \Delta\varphi(x)\, dx = 0, \qquad \varphi \in \mathscr{D}(G) \qquad (24)$$

Obviously each harmonic function is a generalized harmonic function, and the following converse theorem is also true:

THEOREM. *Each generalized harmonic function $u(x)$ in the region G is infinitely differentiable and therefore is harmonic in that region.*

Proof. In view of the local character of the theorem, we may consider that $u \in C(\bar{G})$. We continue the function u as zero outside $\bar{G}$ and let $\tilde{u}$ be a continued function. Applying Eq. (13) of Sec. 10.3, we obtain the equation

$$\tilde{u} = \Delta \tilde{u} * \mathscr{E}_n \qquad (25)$$

where $\mathscr{E}_n$ is a fundamental solution of the Laplace operator. Since $\Delta\tilde{u} = \Delta u = 0$ for $x \in G$ and $\Delta\tilde{u} = \Delta 0 = 0$ for $x \in G_1$, then supp $\Delta\tilde{u} \subset S$. So, by the theorem of Sec. 7.5, for the convolution $\Delta\tilde{u} * \mathscr{E}_n$ we have the representation

$$(\Delta\tilde{u} * \mathscr{E}_n, \varphi) = (\Delta\tilde{u}(y) \cdot \mathscr{E}_n(\xi), \eta(y)\varphi(y + \xi))$$
$$= (\Delta\tilde{u}(y), \eta(y) \int \mathscr{E}_n(\xi)\varphi(y + \xi)\, d\xi)$$
$$= (\Delta\tilde{u}(y), \eta(y) \int \mathscr{E}_n(x - y)\varphi(x)\, dx), \qquad \varphi \in \mathscr{D} \qquad (26)$$

where η is an arbitrary function belonging to $\mathscr{D}$ with a support in a neighborhood of S.

Let $G' \Subset G$. We choose in (26) an auxiliary function η such that supp $\eta \cap \bar{G}' = \varnothing$ (Fig. 55). Since the fundamental solution $\mathscr{E}_n(x - y)$ is an

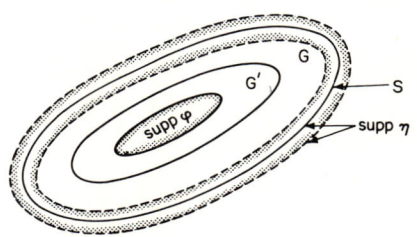

Fig. 55

infinitely differentiable function for $x \neq y$, then for the chosen η and all $\varphi \in \mathscr{D}(G')$

$$\eta(y)\mathscr{E}_n(x - y)\varphi(x) \in \mathscr{D}(R^{2n})$$

Now applying Eq. (14) of Sec. 7.3, (f) to the right-hand side of Eq. (26), we obtain

$$(\Delta\tilde{u} * \mathscr{E}_n, \varphi) = \int \varphi(x)(\Delta\tilde{u}(y), \eta(y)\mathscr{E}_n(x - y))\, dx, \qquad \varphi \in \mathscr{D}(G')$$

from which, by virtue of (25), there follows the equation:

$$u(x) = (\Delta\tilde{u}(y), \eta(y)\mathscr{E}_n(x - y)), \qquad x \in G'$$

From this, as in the proof of the lemma of Sec. 7.1, we deduce that $u \in C^\infty(G')$. Since the region $G' \Subset G$ is arbitrary, it follows from this that $u \in C^\infty(G)$. Therefore the function $u(x)$ satisfies Laplace's equation in the region G in the classical sense (cf. Sec. 10.1); that is, it is harmonic in G. The theorem is proved.

§ 21. HARMONIC FUNCTIONS

8. Further Properties of Harmonic Functions. It has been established in Sec. 21.7 that the generalized harmonic property and the harmonic property are equivalent, and we shall now consider two corollaries of this.

(a) *If the sequence $u_1, u_2, \ldots,$ of functions which are harmonic in the region G converges weakly (specifically, uniformly over each compactum $K \subset G$, or monotonically) to the function $u \in C(G)$, that is,*

$$\int u_k(x)\varphi(x)\,dx \to \int u(x)\varphi(x)\,dx, \qquad k \to \infty, \qquad \varphi \in \mathscr{D}(G) \tag{27}$$

then u is a harmonic function in G.

In fact, each function of the sequence $\{u_k\}$ satisfies the integral condition (24). But then, by virtue of (27), the limit function $u(x)$ belonging to $C(G)$ will also satisfy Eq. (24); that is, it is a generalized harmonic function and therefore a harmonic function in the region G.

(b) *If the function $u \in C(G)$ is such that for each point $x \in G$ there is a number $r_0 = r_0(x) > 0$ such that for all $r < r_0$ the mean value condition is satisfied:*

$$u(x) = \frac{1}{\sigma_n r^{n-1}} \int_{S_r} u(x - y)\, dS_y \tag{28}$$

then $u(x)$ is a harmonic function in the region G.

In proving this we may consider that $u \in C(\bar{G})$; let $\tilde{u}$ be the function u continued as zero outside $\bar{G}$. We take $G' \Subset G$. By the Heine–Borel lemma (cf. Sec. 1.1), there is a number $r_0 = r_0(G') > 0$ such that for all $x \in G'$ and $r < r_0$ Eq. (28) will be satisfied for the function $u(x)$.

We form the convolution

$$f_r(x) = \left(\frac{1}{\sigma_n r^{n+1}} \delta_{S_r} - \frac{1}{r^2} \delta \right) * \tilde{u} \tag{29}$$

where δ_{S_r} is a simple layer over the spherical surface S_r (cf. Sec. 5.6). Using formula (26) of Sec. 7.8, (a), we write the convolution (29) in integral form

$$f_r(x) = \frac{1}{\sigma_n r^{n+1}} \int_{S_r} \tilde{u}(x - y)\, dS_y - \frac{\tilde{u}(x)}{r^2}$$

From this, by virtue of (28), it follows that for all $r < r_0$, $f_r(x) = 0$ for $x \in G'$. On the other hand, if we use the limiting result (34) of Sec. 6.5 and the continuity of the convolution (cf. Sec. 7.5), from (29) we obtain

$$f_r \to \frac{1}{2n} \Delta\delta * \tilde{u} = \frac{1}{2n} \Delta\tilde{u}, \qquad r \to 0 \quad \text{in } \mathscr{D}'$$

therefore, $\Delta \tilde{u} = \Delta u = 0$ for $x \in G'$. From this, since $G' \subseteq G$ is arbitrary, we conclude that the function $u(x)$ is a generalized harmonic function and so is harmonic in the region G.

9. Analogy of Liouville's Theorem. The following theorem, which is analogous to Liouville's theorem for analytic functions, is valid for harmonic functions in the whole space R^n.

THEOREM. *If $u \in \mathscr{S}'$ satisfies Laplace's equation in the whole space R^n, then u is a polynomial.*

Proof. Applying the Fourier transform to the equation $\Delta u = 0$, we obtain [cf. Sec. 9.3, (b)] $-|\xi|^2 F[u](\xi) = 0$, from which it follows that $F[u] = 0$ for $\xi \neq 0$, that is, either $F[u] = 0$ or the support of $F[u]$ is the point $\{0\}$. By the theorem of Sec. 8.4, $F[u]$ can be represented in the form

$$F[u](\xi) = \sum_{|\alpha|=0}^{m} c_\alpha D^\alpha \delta(\xi)$$

from which it follows that u is a polynomial. The theorem is proved.

COROLLARY. *If the function u is harmonic in R^n and satisfies the inequality*

$$|u(x)| \leq C(1 + |x|)^m, \qquad x \in R^n, \qquad m \geq 0$$

then u is a (harmonic) polynomial of degree less than or equal to m.

10. Exercises. (a) Using the theorem of the mean value (cf. Sec. 21.3), prove the following modification of this theorem: If the function $u(x)$ is harmonic in the sphere U_R and continuous on $\bar{U}_R$, then

$$u(0) = \frac{n}{\sigma_n R^n} \int_{U_R} u(x)\, dx$$

(b) Using (a), prove Liouville's theorem: If the function $u(x)$ is harmonic in R^n and bounded above (or below), then $u(x) = $ const.

(c) Using statement (b) of Sec. 21.8, prove the following analogy of the Riemann–Schwartz symmetry principle: Let the boundary of the region G contain the open set Σ lying in the plane $x_n = 0$, let the function $u(x)$ be harmonic in G and become zero over Σ; then the odd continuation of the function $u(x)$ into the region $\tilde{G}$, which is symmetrical to G with respect to the plane $x_n = 0$, is a harmonic function in the region $G \cup \Sigma \cup \tilde{G}$ (Fig. 56).

§ 22. NEWTONIAN POTENTIAL

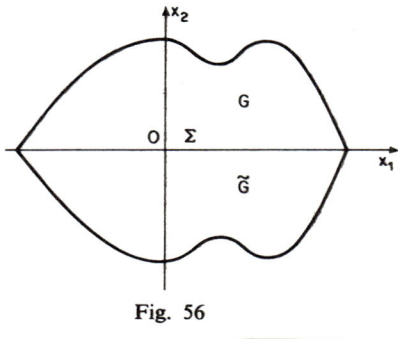

Fig. 56

§ 22. Newtonian Potential

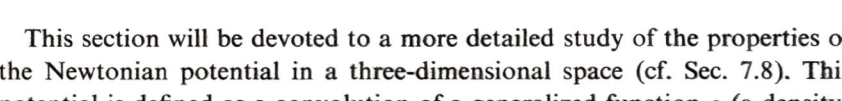

This section will be devoted to a more detailed study of the properties of the Newtonian potential in a three-dimensional space (cf. Sec. 7.8). This potential is defined as a convolution of a generalized function ϱ (a density) with the function $|x|^{-1}$,

$$V = \frac{1}{|x|} * \varrho = -4\pi \mathscr{E}_3 * \varrho \tag{1}$$

The potential V satisfies Poisson's equation

$$\Delta V = -4\pi\varrho \tag{2}$$

The foundations of the classical theory of potential were laid by A. M. Liapunov at the end of the last century.

1. Volume Potential. If ϱ is an (absolutely) integrable function over G and $\varrho(x) = 0$ for $x \in G_1 = R^3 \setminus \bar{G}$, then the Newtonian potential V, which is known as the *volume potential*, is expressed by the integral

$$V(x) = \int_G \frac{\varrho(y)}{|x-y|} \, dy \tag{3}$$

and is a locally integrable function in R^n [cf. Sec. 7.8, (c)].

If $\varrho \in C(\bar{G})$, then the volume potential V belongs to the class $C^1(R^3)$, is harmonic in G_1, and $V(\infty) = 0$.

In fact, since G is a bounded region and $\varrho \in C(\bar{G})$, then the integral in (3) converges uniformly with respect to x, defining a function $V(x)$ which is continuous in R^3 (cf. Lemma 1 of Sec. 15.4).

Moreover the potential $V(x)$ allows continuous single differentiation under the integral sign with respect to all the independent variables, and so $V \in C^1(R^3)$.

For $x \notin \bar{G}$ the potential $V(x)$ allows continuous differentiation under the integral sign in (3) an infinite number of times, so that $V \in C^\infty(G_1)$. From this and from Eq. (2) it follows that $\Delta V = \varrho = 0$ for $x \in G_1$; that is, the potential V is a harmonic function in the region G_1 (cf. lemma of Sec. 10.1).

It follows from the boundedness of the region G and from (3) that $V(x) \to 0$ as $|x| \to \infty$; that is, $V(\infty) = 0$.

If $\varrho \in C^1(G) \cap C(\bar{G})$, then $V \in C^2(G)$.

For our proof we shall take a subregion G' with a piecewise smooth boundary S', strictly lying in the region G. We shall divide the density ϱ into the sum of two terms, $\varrho = \varrho_1 + \varrho_2$, where $\varrho_1 = \varrho$ for $x \in R^3 \setminus \bar{G}'$ and $\varrho_1 = 0$ for $x \in \bar{G}'$. For this the potential V is divided into the sum of two volume potentials V_1 and V_2, $V = V_1 + V_2$, with the densities ϱ_1 and ϱ_2, respectively,

$$V_1(x) = \int_{G \setminus G'} \frac{\varrho_1(y)}{|x-y|}\, dy, \quad V_2(x) = \int_{G'} \frac{\varrho_2(y)}{|x-y|}\, dy$$

By what has been proved $V_1 \in C^\infty(G')$ and $V_2 \in C^1(R^3)$. If we differentiate the potential V_2 as a convolution, we shall obtain (cf. Sec. 7.5)

$$\operatorname{grad} V_2(x) = \operatorname{grad}\left(\frac{1}{|x|} * \varrho_2\right) = \frac{1}{|x|} * \operatorname{grad} \varrho_2 \qquad (4)$$

Since $\varrho_2 \in C^1(\bar{G}')$, then, by Eq. (15) of Sec. 6.5,

$$\operatorname{grad} \varrho_2 = \{\operatorname{grad} \varrho_2\} - \varrho \mathbf{n}' \delta_{S'}$$

Substituting this expression into (4) and using Eq. (3) for a volume potential and Eq. (32) of Sec. 7.8 for the potential of a simple layer, we shall obtain

$$\operatorname{grad} V_2(x) = \frac{1}{|x|} * \{\operatorname{grad} \varrho_2\} - \frac{1}{|x|} * \varrho \mathbf{n}' \delta_S$$

$$= \int_{G'} \frac{\operatorname{grad} \varrho(y)}{|x-y|}\, dy - \int_{S'} \frac{\varrho(y)\mathbf{n}'}{|x-y|}\, dS_y \qquad (5)$$

The first term on the right-hand side of (5), as a volume potential with a density $\operatorname{grad} \varrho \in C(\bar{G}')$, belongs to the class $C^1(R^3)$, and the second belongs

§ 22. NEWTONIAN POTENTIAL

to the class $C^\infty(G')$. Therefore, grad $V_2 \in C^1(G')$, that is, $V_2 \in C^2(G')$. But then $V = V_1 + V_2 \in C^2(G')$ and, since $G' \Subset G$ is arbitrary, $V \in C^2(G)$, as was to be shown.

2. Potentials of a Simple and a Double Layer. Let S be a bounded piecewise smooth two-sided* surface, let **n** be its normal, and let μ and ν be continuous functions over S. The Newtonian potentials

$$V^{(0)} = \frac{1}{|x|} * \mu \delta_S \quad \text{and} \quad V^{(1)} = -\frac{1}{|x|} * \frac{\partial}{\partial \mathbf{n}}(\nu \delta_S)$$

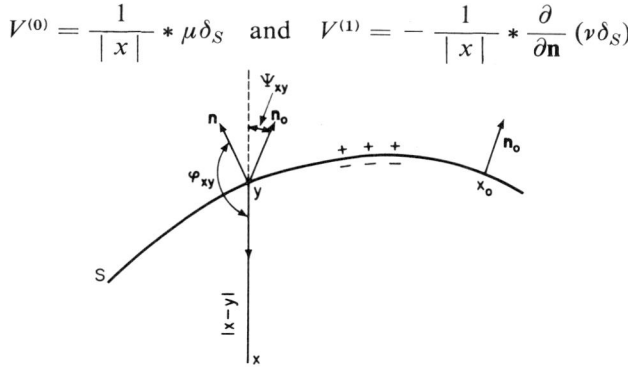

Fig. 57

which are said to be the *potentials of a simple* and a *double layer*, respectively, are expressed by the integrals

$$V^{(0)}(x) = \int_S \frac{\mu(y)}{|x-y|} dS_y \tag{6}$$

$$V^{(1)}(x) = \int_S \nu(y) \frac{\partial}{\partial \mathbf{n}_y} \frac{1}{|x-y|} dS_y \tag{7}$$

and are locally integrable functions in R^3 [cf. Sec. 7.8, (d)]. These potentials satisfy Poisson's equation

$$\Delta V^{(0)} = -4\pi \mu \delta_S, \quad \Delta V^{(1)} = 4\pi \frac{\partial}{\partial \mathbf{n}}(\nu \delta_S) \tag{8}$$

We shall fix the point x_0 on S and let $\mathbf{n}_0$ be its external normal with respect

---

* The side of the surface S with which the normal **n** is associated is considered positive, and the opposite side negative (Fig. 57).

to S. Differentiating Eq. (6) in the direction $\mathbf{n}_0$ with $x \notin S$ and using the equation

$$\frac{\partial}{\partial n_0} \frac{1}{|x-y|} = \sum_{i=1}^{3} \cos(\mathbf{n}_0 x_i) \frac{y_i - x_i}{|x-y|^3} = \frac{\cos \psi_{xy}}{|x-y|^2} \qquad (9)$$

where ψ_{xy} is the angle between the vector $y - x$ and the normal $\mathbf{n}_0$ (Fig. 57), we obtain an expression for the normal derivative of the simple layer potential

$$\frac{\partial V^{(0)}(x)}{\partial n_0} = \int_S \mu(y) \frac{\partial}{\partial n_0} \frac{1}{|x-y|} dS_y = \int_S \mu(y) \frac{\cos \psi_{xy}}{|x-y|^2} dS_y \qquad (10)$$
$$x \notin S$$

Analogously, by virtue of the equation

$$\frac{\partial}{\partial n_y} \frac{1}{|x-y|} = \sum_{i=1}^{3} \cos(\mathbf{n} x_i) \frac{x_i - y_i}{|x-y|^3} = \frac{\cos \varphi_{xy}}{|x-y|^2} \qquad (11)$$

where φ_{xy} is the angle between the vector $x - y$ and the normal $\mathbf{n}$ (Fig. 57), Eq. (7) for the double layer potential $V^{(1)}$ takes the form

$$V^{(1)}(x) = \int_S \nu(y) \frac{\cos \varphi_{xy}}{|x-y|^2} dS_y \qquad (12)$$

The potentials $V^{(0)}$ and $V^{(1)}$ are harmonic functions outside the surface S and $V^{(0)}(\infty) = 0$ and $V^{(1)}(\infty) = 0$; moreover, $V^{(0)} \in C(R^3)$.

These properties of the potentials $V^{(0)}$ and $V^{(1)}$ may be deduced from Eqs. (6) and (12) and from Eqs. (8) in the same way as for the volume potential (cf. Sec. 22.1).

LEMMA. *If $x \notin S$, then the potential of the double layer $V^{(1)}(x)$ with a density $\nu \equiv 1$ is equal to the solid angle $\omega_S(x)$ subtended by the surface S at the point x, that is,*

$$\int_S \frac{\cos \varphi_{xy}}{|x-y|^2} dS_y = \omega_S(x) \qquad (13)$$

Proof. Since $x \notin S$, then there is a sphere $U(x; r_0)$ which has no common points with S. We shall denote by σ the stereographic projection of the surface S from the center x onto the spherical surface $S(x; r_0)$, and let D be the region bounded by the surfaces S, σ, and the conical side surface Γ

§ 22. NEWTONIAN POTENTIAL

with vertex at the point x (Fig. 58). We shall consider that $D \neq \emptyset$. In the opposite case $\omega_S(x) = 0$ and $\cos \varphi_{xy} = 0$ over S, so that Eq. (13) will be trivially satisfied. Since the function $|x - y|^{-1}$ is harmonic for $y \neq x$,

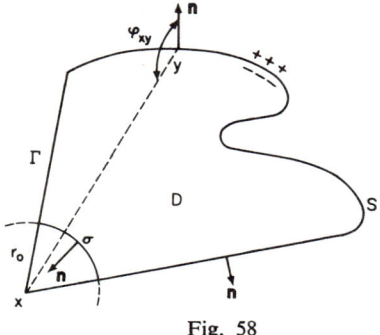

Fig. 58

then, if we apply formula (11) of Sec. 21.3 to the region D, we obtain

$$\int_S \frac{\partial}{\partial n_y} \frac{1}{|x-y|} dS_y + \int_\Gamma \frac{\partial}{\partial n_y} \frac{1}{|x-y|} dS_y$$

$$+ \int_\sigma \frac{\partial}{\partial n_y} \frac{1}{|x-y|} dS_y = 0 \qquad (14)$$

Allowing that, by virtue of (11),

$$\frac{\partial}{\partial n} \frac{1}{|x-y|} = \frac{\cos \varphi_{xy}}{|x-y|^2} = 0, \qquad y \in \Gamma$$

$$\frac{\partial}{\partial n} \frac{1}{|x-y|} = \frac{\cos \varphi_{xy}}{|x-y|^2} = \frac{1}{r_0^2}, \qquad y \in \sigma$$

from Eq. (14) we deduce

$$V^{(1)}(x) = \int_S \frac{\partial}{\partial n_y} \frac{1}{|x-y|} dS_y = -\frac{1}{r_0^2} \int_\sigma dS = \omega_S(x)$$

from which, by virtue of (12), formula (13) follows. The lemma is proved.

It follows from this lemma: *If the surface S is the boundary of the region G, then Gaussian formulas are valid*

$$\int_S \frac{\cos \varphi_{xy}}{|x-y|^2} dS_y = \begin{cases} -4\pi, & x \in G \\ 0, & x \in G_1 \end{cases} \qquad (15)$$

3. Physical Sense of Newtonian Potentials.

The potential $V = (1/|x|) * \varrho$ with an arbitrary density ϱ with compact support satisfies Poisson's equation $\Delta V = -4\pi\varrho$. Therefore, V is the Newtonian or Coulomb potential created by the masses or charges which are distributed in space with the density ϱ. Specifically, the continuous distribution of masses or charges creates a volume potential; if masses or charges are concentrated over a surface, they create a (Newtonian or Coulomb) potential of a simple layer; if there are dipoles concentrated over a surface, the Coulomb potential created by them is a potential of a double layer.

As an example we shall calculate the (Coulomb) potential $V^{(1)}(x;\mathbf{l})$ which is created by a dipole with a moment $+1$ at the point 0, oriented in the direction $\mathbf{l}, |\mathbf{l}| = 1$. This potential is created by the distribution [cf. Sec. 6.3, (b)]

$$\lim_{\varepsilon \to +0} \left[\frac{1}{\varepsilon} \delta(x - \mathbf{l}\varepsilon) - \frac{1}{\varepsilon} \delta(x) \right] = -\frac{\partial}{\partial \mathbf{l}} \delta(x)$$

(Fig. 59) and so

$$V^{(1)}(x, \mathbf{l}) = -\frac{1}{|x|} * \frac{\partial}{\partial \mathbf{l}} \delta = -\frac{\partial}{\partial \mathbf{l}} \left(\frac{1}{|x|} * \delta \right) = -\frac{\partial}{\partial \mathbf{l}} \frac{1}{|x|}$$

that is,

$$V^{(1)}(x;\mathbf{l}) = -\frac{\partial}{\partial \mathbf{l}} \frac{1}{|x|} = \frac{\cos \varphi}{|x|^2} \tag{16}$$

where φ is the angle between the vectors x and $\mathbf{l}$. Figure 60 shows the level surfaces of the potential $V^{(1)}(x;\mathbf{l})$ (equipotential surfaces).

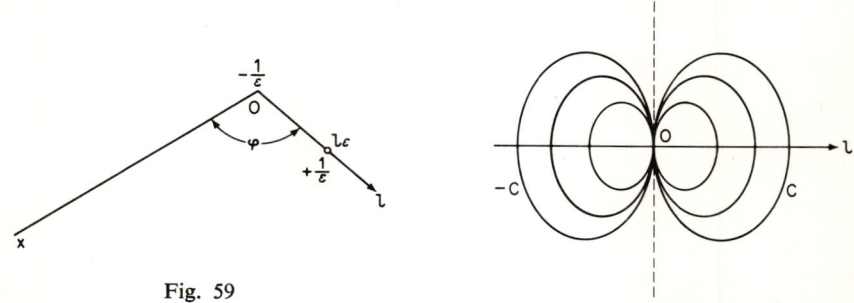

Fig. 59

Fig. 60

It follows from Eqs. (12) and (16) that the double layer potential is the "sum" of elementary potentials

$$v(y) V^{(1)}(x - y; \mathbf{n}) = v(y) \frac{\cos \varphi_{xy}}{|x - y|^2}$$

created by the dipoles over the surface S with a density of moment $v(y)$ and directed along the normal $\mathbf{n}$.

4. Liapunov's Surfaces. The proposition that S is a *Liapunov surface* involves other properties of the potentials of simple and double layers.

The closed bounded surface S is known as a *Liapunov surface* if it satisfies the following conditions:

(a) at each point of S there is a tangent plane;

(b) there is a number $r_0 > 0$ such that for any point $x \in S$ the set $S \cap U(x; r_0)$ is connected* and is intersected by straight lines parallel to the normal $\mathbf{n}_x$ at no more than one point;

(c) the normal $\mathbf{n}_x$ is Hölder continuous over S; that is, there are numbers $C > 0$ and $\alpha > 0$, $\alpha \leq 1$, such that

$$|\mathbf{n}_x - \mathbf{n}_y| \leq C |x - y|^\alpha, \qquad x, y \in S \tag{17}$$

It follows from this definition that Liapunov surfaces are contained in the class of surfaces with smoothness C^1; on the other hand, each bounded closed surface of the class C^2 is a Liapunov surface (for $\alpha = 1$).

LEMMA. *If S is a Liapunov surface, then there are constants a and b such that*

$$|\cos \varphi_{xy}| \leq a |x - y|^\alpha, \qquad x, y \in S \tag{18}$$

$$|\cos \varphi_{x'y} + \cos \psi_{x'y}| \leq b |x' - y|^\alpha, \qquad x, y \in S, \qquad x' \in \mathbf{n}_x \tag{19}$$

where $\psi_{x'y}$ is the angle between the vector $y - x'$ and the normal $\mathbf{n}_x$.

Proof. Obviously, it is sufficient to establish Eqs. (18) and (19) for small $|x - y|$ and $|x' - y|$. We shall choose a number $r_1 > 0$ such that $r_1 < r_0$ and $Cr_1^\alpha < \frac{1}{2}$, and we shall cover the (bounded) surface S with a finite number of neighborhoods $u_x = S \cap U(x; r_1)$. It is sufficient to prove Eqs. (18) and (19) for each such neighborhood.

By condition (a) there is a normal $\mathbf{n}_0$ to the surface S at the point x. We shall choose a local system of orthogonal coordinates with their origin at the point x, directing the y_3 axis along the normal $\mathbf{n}_0$; let $\mathbf{i}$ and $\mathbf{j}$ be the unit vectors along the positive directions of the y_1 and y_2 axes, respectively

---

* This means that the set $S \cap U(x; r_0)$ is a neighborhood of the point x on the surface S (cf. Sec. 1.1).

(Fig. 61). By condition (b) the piece u_x of the surface S is given by the equation $y_3 = f(y_1, y_2)$ for $f \in C^1(\bar{\sigma})$, where σ is the projection of u_x onto the (y_1, y_2) plane. Finally, from condition (c) we have

$$|\mathbf{n} - \mathbf{n}_0| \leq C |y|^\alpha < C r_1^\alpha < \tfrac{1}{2}, \qquad y \in u_x \tag{20}$$

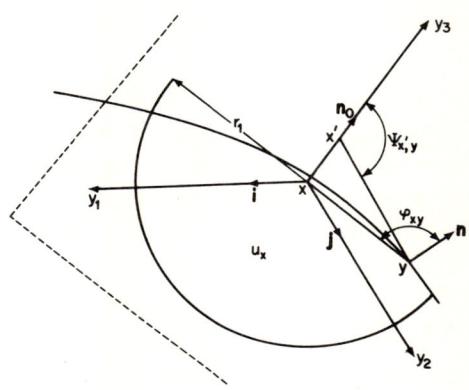

Fig. 61

The next equations follow from (20) for all $y \in u_x$:

$$\begin{aligned}
|(\mathbf{n}, \mathbf{i})| &= |(\mathbf{n} - \mathbf{n}_0, \mathbf{i}) + (\mathbf{n}_0, \mathbf{i})| \leq |\mathbf{n} - \mathbf{n}_0| \leq C |y|^\alpha \\
|(\mathbf{n}, \mathbf{j})| &= |(\mathbf{n} - \mathbf{n}_0, \mathbf{j}) + (\mathbf{n}_0, \mathbf{j})| \leq |\mathbf{n} - \mathbf{n}_0| \leq C |y|^\alpha \\
(\mathbf{n}, \mathbf{n}_0) &= (\mathbf{n} - \mathbf{n}_0, \mathbf{n}_0) + (\mathbf{n}_0, \mathbf{n}_0) \geq 1 - |\mathbf{n} - \mathbf{n}_0| \geq 1 - C |y|^\alpha > \tfrac{1}{2}
\end{aligned} \tag{21}$$

Allowing that the results

$$\frac{\partial f}{\partial y_1} = -\frac{(\mathbf{n}, \mathbf{i})}{(\mathbf{n}, \mathbf{n}_0)}, \qquad \frac{\partial f}{\partial y_2} = -\frac{(\mathbf{n}, \mathbf{j})}{(\mathbf{n}, \mathbf{n}_0)}$$

are valid over the surface u_x, from (21) we deduce the inequalities

$$\left|\frac{\partial f}{\partial y_1}\right| = \left|\frac{(\mathbf{n}, \mathbf{i})}{(\mathbf{n}, \mathbf{n}_0)}\right| \leq 2C |y|^\alpha, \qquad \left|\frac{\partial f}{\partial y_2}\right| \leq 2C |y|^\alpha, \qquad y \in u_x \tag{22}$$

Moreover, the estimate

$$|y_3| \leq k\varrho, \qquad \varrho^2 = y_1^2 + y_2^2 \tag{23}$$

is valid over the surface u_x for some $k > 0$. If we substitute result (23) into

§ 22. NEWTONIAN POTENTIAL

result (22), we obtain

$$\left|\frac{\partial f}{\partial y_1}\right| \leq 2C(\varrho^2 + y_3^2)^{\alpha/2} \leq d\varrho^\alpha, \qquad \left|\frac{\partial f}{\partial y_2}\right| \leq d\varrho^\alpha, \qquad (y_1, y_2) \in \bar{\sigma}$$

from which follows result

$$\left|\frac{\partial f}{\partial \varrho}\right| \leq |\operatorname{grad} f| = \left[\left(\frac{\partial f}{\partial y_1}\right)^2 + \left(\frac{\partial f}{\partial y_2}\right)^2\right]^{1/2} \leq \sqrt{2}\, d\varrho^\alpha \qquad (24)$$

Since $f = 0$ when $\varrho = 0$, from (24) we have

$$|f| = \left|\int_0^\varrho \frac{\partial f}{\partial \varrho}\, d\varrho\right| \leq \sqrt{2}\, d \int_0^\varrho \varrho^\alpha\, d\varrho = \frac{\sqrt{2}\, d}{\alpha + 1} \varrho^{\alpha + 1}$$

and so

$$|y_3| = |f| \leq \frac{\sqrt{2}\, d}{\alpha + 1} \varrho^{\alpha+1} \leq \frac{\sqrt{2}\, d}{\alpha + 1} |y|^{\alpha+1}, \qquad y \in u_x$$

Taking this result and result (20) into consideration, for all $y \in u_x$, we finally obtain inequality (18):

$$|\cos \varphi_{xy}| = \left|\left(\mathbf{n}, \frac{y}{|y|}\right)\right| = \left|\left(\mathbf{n} - \mathbf{n}_0, \frac{y}{|y|}\right) + \left(\mathbf{n}_0, \frac{y}{|y|}\right)\right|$$

$$\leq |\mathbf{n} - \mathbf{n}_0| + \frac{|y_3|}{|y|} \leq \left(C + \frac{\sqrt{2}\, d}{\alpha + 1}\right)|y|^\alpha = a|y|^\alpha$$

We shall now prove result (19) over u_x. Using Eqs. (9) and (11) and the Cauchy–Buniakowski inequality, for all $y \in u_x$ and $x' \in R^3$ we obtain the equation

$$|\cos \varphi_{x'y} + \cos \psi_{x'y}| = \left|\sum_{i=1}^3 \frac{y_i - x_i'}{|x' - y|}[\cos(\mathbf{n}_0 y_i) - \cos(\mathbf{n} y_i)]\right|$$

$$\leq \left\{\sum_{i=1}^3 [\cos(\mathbf{n}_0 y_i) - \cos(\mathbf{n} y_i)]^2\right\}^{1/2}$$

$$= \{(\mathbf{n}, \mathbf{i})^2 + (\mathbf{n}, \mathbf{j})^2 + [1 - (\mathbf{n}, \mathbf{n}_0)]^2\}^{1/2}$$

and so, by virtue of (21),

$$|\cos \varphi_{x'y} + \cos \psi_{x'y}| \leq \sqrt{3}\, C |y|^\alpha$$

From this, using inequality (23), for all $y \in u_x$ and $x' \in \mathbf{n}_0$ we obtain result (19):

$$|\cos \varphi_{x'y} + \cos \psi_{x'y}| \leq \sqrt{3}\, C(\varrho^2 + y_3^2)^{\alpha/2} \leq b\varrho^\alpha$$
$$\leq b[y_1^2 + y_2^2 + (y_3 - x_3')^2]^{\alpha/2} = b\, |\, x' - y\, |^\alpha$$

The lemma is proved.

5. Properties of Potentials of a Simple and a Double Layer over the Surface S. Supposing that the boundary S of the region G is a Liapunov surface, we shall establish some properties of the potentials $V^{(0)}$ and $V^{(1)}$ over S.

The following equations hold:

$$\int_S \frac{\cos \varphi_{xy}}{|\,x - y\,|^2}\, dS_y = \begin{cases} -4\pi, & x \in G \\ -2\pi, & x \in S \\ 0, & x \in G_1 \end{cases} \tag{25}$$

To prove Eqs. (25), by virtue of (15), it is sufficient to consider the case $x \in S$. If we reject from S the neighborhood u_x of the point x, we shall obtain

$$\int_S \frac{\cos \varphi_{xy}}{|\,x - y\,|^2}\, dS_y = \int_{S \setminus u_x} \frac{\cos \varphi_{xy}}{|\,x - y\,|^2}\, dS_y + \int_{u_x} \frac{\cos \varphi_{xy}}{|\,x - y\,|^2}\, dS_y \tag{26}$$

Since $x \notin S \setminus u_x$, then, by virtue of (13), the first integral on the right in (26) is the solid angle $\omega_{S \setminus u_x}(x)$, subtended by the surface $S \setminus u_x$ from the point x. Therefore, when u_x is contracted to the point x, this integral tends to -2π (Fig. 62). The second integral on the right in (26), by virtue of result (18), converges absolutely and therefore tends to zero as $u_x \to x$. Therefore, proceeding to the limit as $u_x \to x$ in (26), we shall obtain formula (25) for $x \in S$.

The potential of a double layer $V^{(1)}(x)$ is a continuous function over S.

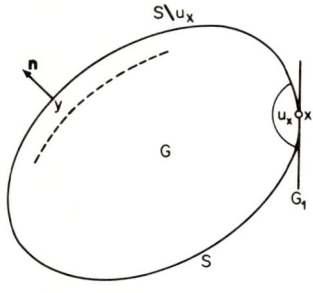

Fig. 62

§ 22. NEWTONIAN POTENTIAL

In fact, by virtue of inequality (18), which is valid over the Liapunov surface S, the potential $V^{(1)}(x)$, defined by formula (12), is an integral operator with the polar kernel

$$\frac{\cos \varphi_{xy}}{|x-y|^2}, \qquad x \in S, \qquad y \in S$$

and so, by Lemma 1 of Sec. 15.4, maps each function $v \in C(S)$ onto the function $V^{(1)} \in C(S)$.

We shall now prove that the *integral*

$$\int_S \mu(y) \frac{\cos \psi_{xy}}{|x-y|^2} dS_y \tag{27}$$

where ψ_{xy} *is the angle between the vector* $y - x$ *and the normal* $\mathbf{n}_x$, *is a continuous function of* x *over* S.

In fact, noting that

$$\psi_{xy} = \varphi_{yx}, \qquad x, y \in S \tag{28}$$

(Fig. 63), from (18) we deduce the result

$$|\cos \psi_{xy}| \le a |x-y|^\alpha, \qquad x, y \in S \tag{29}$$

from which, as for the potential $V^{(1)}$, it follows that the integral (27) is continuous over S.

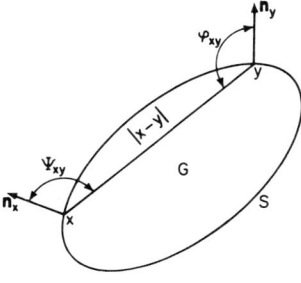

Fig. 63

In agreement with formula (10), we shall denote the integral (27) by $\partial V^{(0)}(x)/\partial \mathbf{n}$,

$$\frac{\partial V^{(0)}(x)}{\partial \mathbf{n}} = \int_S \mu(y) \frac{\cos \psi_{xy}}{|x-y|^2} dS_y = \int_S \mu(y) \frac{\partial}{\partial \mathbf{n}_x} \frac{1}{|x-y|} dS_y, \quad x \in S \tag{30}$$

The function $\partial V^{(0)}(x)/\partial \mathbf{n}$ is said to be the *direct value of the normal derivative of the simple layer potential over the surface S*; by what has been proved it is continuous over S.

We shall remark also that *the potential of the simple layer $V^{(0)}(x)$ is a continuous function over S*, since $V^{(0)} \in C(R^3)$ (cf. Sec. 22.2).

6. Discontinuity of the Potential of a Double Layer. We shall say that a Liapunov surface S satisfies condition (A) if there is a number $K > 0$ such that

$$\int_S \frac{|\cos \varphi_{xy}|}{|x - y|^2} \, dS_y \leq K, \qquad x \in R^3 \tag{A}$$

Condition (A) is automatically satisfied for convex Liapunov surfaces for $K = 4\pi$.

In fact, in this case $\cos \varphi_{xy} \leq 0$ for $x \in \bar{G}$, $y \in S$ (Fig. 63), and so, by virtue of (25),

$$\int_S \frac{|\cos \varphi_{xy}|}{|x - y|^2} \, dS_y = -\int_S \frac{\cos \varphi_{xy}}{|x - y|^2} \, dS_y = \begin{cases} 4\pi, & x \in G \\ 2\pi, & x \in S \end{cases}$$

Now let $x \in G_1$. Since each ray which emanates from the point x intersects S at no more than two points, then, by virtue of (13),

$$\int_S \frac{|\cos \varphi_{xy}|}{|x - y|^2} \, dS_y = 2 \int_{S_1} \frac{\cos \varphi_{xy}}{|x - y|^2} \, dS_y = 2\omega_{S_1}(x) \leq 4\pi$$

Figure 64 shows the surface S_1.

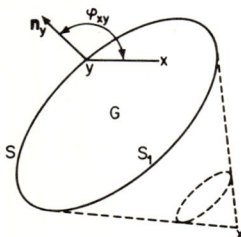

Fig. 64

Theorem. *If the Liapunov surface S satisfies condition (A) and $v \in C(S)$, then the double layer potential $V^{(1)}$ belongs to $C(\bar{G})$ and $C(\bar{G}_1)$ and its limiting values $V^{(1)}_+$ and $V^{(1)}_-$ over S from outside and inside S are expressed*

§ 22. NEWTONIAN POTENTIAL

by the formulas:

$$V_+^{(1)}(x) = 2\pi v(x) + V^{(1)}(x) = 2\pi v(x) + \int_S v(y) \frac{\cos \varphi_{xy}}{|x-y|^2} dS_y \qquad (31)$$

$$V_-^{(1)}(x) = -2\pi v(x) + V^{(1)}(x) = -2\pi v(x) + \int_S v(y) \frac{\cos \varphi_{xy}}{|x-y|^2} dS_y \qquad (31')$$

Proof. We shall introduce the function

$$W(x', x) = \int_S [v(y) - v(x)] \frac{\cos \varphi_{x'y}}{|x'-y|^2} dS_y, \qquad x' \in R^3, \qquad x \in S$$

The function $W(x', x)$ for $x' = x \in S$, by virtue of Eq. (25), is equal to

$$W(x, x) = \int_S v(y) \frac{\cos \varphi_{xy}}{|x-y|^2} dS_y + 2\pi v(x) = 2\pi v(x) + V^{(1)}(x) \qquad (32)$$

The function $W(x, x)$ is continuous over S by virtue of the continuity of the density v and the potential $V^{(1)}$ over S (cf. Sec. 22.5).

We shall prove that

$$W(x', x) \xrightarrow{x \in S} W(x, x), \qquad x' \to x \in S \qquad (33)$$

Let $\varepsilon > 0$. Since the function v is uniformly continuous over S, then there is a number $\delta = \delta_\varepsilon > 0$ such that for all $x \in S$ we have:

$$|v(y) - v(x)| < \frac{\varepsilon}{4K}, \qquad y \in u_x = S \cap U(x; \delta) \qquad (34)$$

where K is the number entering into condition (A).

We shall evaluate the difference

$$|W(x', x) - W(x, x)|$$

$$\leq \left(\int_{u_x} + \int_{S \setminus u_x} \right) |v(y) - v(x)| \left| \frac{\cos \varphi_{x'y}}{|x'-y|^2} - \frac{\cos \varphi_{xy}}{|x-y|^2} \right| dS_y \qquad (35)$$

By virtue of (34) and (A), the first integral on the right in (35) does not exceed $\varepsilon/2$,

$$\int_{u_x} |v(y) - v(x)| \left| \frac{\cos \varphi_{x'y}}{|x'-y|^2} - \frac{\cos \varphi_{xy}}{|x-y|^2} \right| dS_y$$

$$< \frac{\varepsilon}{4K} \int_S \left(\frac{|\cos \varphi_{x'y}|}{|x'-y|^2} + \frac{|\cos \varphi_{xy}|}{|x-y|^2} \right) dS_y \leq \frac{\varepsilon}{4K} 2K = \frac{\varepsilon}{2}$$

Moreover, the integrand in (35), which is a function of the variables (x, x', y), is uniformly continuous for $|x - x'| \leq \delta/2$, $x \in S$, $y \in S \setminus u_x$, and becomes zero for $x' = x$. Therefore there is a number $\delta' = \delta'_\varepsilon \leq \delta/2$, such that for all $x' \in u(x; \delta')$ the second integral on the right in (35) will be less than $\varepsilon/2$. Consequently,

$$|W(x', x) - W(x, x)| < \frac{\varepsilon}{2} + \frac{\varepsilon}{2} = \varepsilon, \qquad x' \in U(x; \delta'), \qquad x \in S$$

which proves the limiting result (32).

Considering $x' \in G_1$ and using Eq. (25), we may represent the potential $V^{(1)}(x')$ in the form

$$V^{(1)}(x') = \int_S [v(y) - v(x)] \frac{\cos \varphi_{x'y}}{|x' - y|^2} dS_y$$
$$+ v(x) \int_S \frac{\cos \varphi_{x'y}}{|x' - y|^2} dS_y = W(x', x) \qquad (36)$$

Proceeding to the limit in this equation as $x' \to x \in S$, $x' \in G_1$, and taking limiting result (33) into consideration, we obtain

$$V^{(1)}(x') \xrightarrow{x \in S} W(x, x) = V^{(1)}_+(x), \qquad x \in S$$

from which it follows that $V^{(1)} \in C(\bar{G}_1)$ and, by virtue of (32), Eq. (31) is valid.

The other case is considered analogously. The theorem is proved.

From Eqs. (31) and (31') follows the result

$$4\pi v(x) = V^{(1)}_+(x) - V^{(1)}_-(x), \qquad x \in S \qquad (37)$$

Note. Equations (31) and (31') are analogous to Sokhotski's formulas (10) and (10') of Sec. 5.7.

7. Discontinuity of the Normal Derivative of the Potential of a Simple Layer

THEOREM. *If the Liapunov surface S satisfies the condition* (A) *and $\mu \in C(S)$, then the simple layer potential $V^{(0)}$ has correct normal deriva-*

§ 22. NEWTONIAN POTENTIAL

tives $(\partial V^{(0)}/\partial \mathbf{n})_+$ and $(\partial V^{(0)}/\partial \mathbf{n})_-$ over S from outside and inside S, and

$$\left(\frac{\partial V^{(0)}}{\partial \mathbf{n}}\right)_+(x) = -2\pi\mu(x) + \frac{\partial V^{(0)}(x)}{\partial \mathbf{n}}$$

$$= -2\pi\mu(x) + \int_S \mu(y) \frac{\cos \psi_{xy}}{|x-y|^2} dS_y \qquad (38)$$

$$\left(\frac{\partial V^{(0)}}{\partial \mathbf{n}}\right)_-(x) = 2\pi\mu(x) + \frac{\partial V^{(0)}(x)}{\partial \mathbf{n}}$$

$$= 2\pi\mu(x) + \int_S \mu(y) \frac{\cos \psi_{xy}}{|x-y|^2} dS_y \qquad (38')$$

Proof. Let $V^{(1)}$ be the double layer potential over S with the density μ. We shall introduce the function

$$W_1(x', x) = \frac{\partial V^{(0)}(x')}{\partial \mathbf{n}_x} + V^{(1)}(x'), \qquad x' \notin S, \qquad x \in S$$

and we shall prove that as $x' \to x \in S$ for $x' \in \mathbf{n}_x$

$$W_1(x', x) \xrightarrow{x \in S} W_1(x, x) = \frac{\partial V^{(0)}(x)}{\partial \mathbf{n}} + V^{(1)}(x) \qquad (39)$$

By what has been proved (cf. Sec. 22.5) the function $W_1(x, x)$ is continuous over S.

Using formulas (10) and (12), we can represent the function W_1 in the form of the integral

$$W_1(x', x) = \int_S \mu(y) \frac{\cos \psi_{x'y} + \cos \varphi_{x'y}}{|x'-y|^2} dS_y$$

We take an $\varepsilon > 0$. Let us evaluate the difference

$$|W_1(x', x) - W_1(x, x)| \leq \left(\int_{u_x} + \int_{S \setminus u_x}\right) |\mu(y)| \left| \frac{\cos \psi_{x'y} + \cos \varphi_{x'y}}{|x'-y|^2} \right.$$

$$\left. - \frac{\cos \psi_{xy} + \cos \varphi_{xy}}{|x-y|^2} \right| dS_y, \quad u_x = S \cap U(x; \delta) \qquad (40)$$

By virtue of results (18), (19), and (29), the first integral on the right in

(40) does not exceed the (absolutely) convergent integral

$$\int_{u_x} |\mu(y)| \left(\frac{b}{|x'-y|^{2-\alpha}} + \frac{2a}{|x-y|^{2-\alpha}} \right) dS_y$$

and so may be made less than $\varepsilon/2$ for sufficiently small $\delta = \delta_\varepsilon$. Further, the integrand in (40), which is a function of the variables (x, x', y), is uniformly continuous for $|x - x'| \leq \delta/3$, $x \in S$, $y \in S \setminus u_x$, and becomes zero for $x' = x$. Therefore there will be a number $\delta' = \delta'_\varepsilon \leq \delta/3$ such that for all $x' \in U(x; \delta')$ the second integral on the right in (40) will be smaller than $\varepsilon/2$. Therefore,

$$|W_1(x', x) - W_1(x, x)| < \varepsilon, \qquad x' \in U(x; \delta'), \qquad x' \in \mathbf{n}_x, \qquad x \in S$$

which proves the limiting result (39).

By the theorem of Sec. 22.6, $V^{(1)} \in C(\bar{G})$ and

$$V_+^{(1)}(x) = 2\pi\mu(x) + V^{(1)}(x)$$

Therefore the limiting result (39) as $x' \to x \in S$, $x' \in \mathbf{n}_x$, takes the form

$$\frac{\partial V^{(0)}(x')}{\partial \mathbf{n}_x} \xrightarrow{x \in S} -V_+^{(1)}(x) + W_1(x, x) = -2\pi\mu(x) + \frac{\partial V^{(0)}(x)}{\partial \mathbf{n}_x}$$

from which we conclude that the correct normal derivative $(\partial V^{(0)}/\partial \mathbf{n})_+$ over S from the outside exists (cf. Sec. 21.1) and, because of formula (30), is expressed by Eqs. (38).

The other case is considered analogously. The theorem is proved.

From Eqs. (38) and (38') follows the result

$$4\pi\mu(x) = \left(\frac{\partial V^{(0)}}{\partial \mathbf{n}} \right)_- (x) - \left(\frac{\partial V^{(0)}}{\partial \mathbf{n}} \right)_+ (x), \qquad x \in S \qquad (41)$$

Note. It may be proved that if the density μ is Hölder continuous over S, then the potential $V^{(0)}$ belongs to the classes $C^1(\bar{G})$ and $C^1(\bar{G}_1)$ [cf., for example, S. L. Sobolev (*1*, Chap. XV)].

8. Exercises. (a) Show that the simple layer potential for the spherical surface S_R with a density $\mu = 1$ is equal to

$$V^{(0)}(x) = \begin{cases} \dfrac{4\pi R^2}{|x|}, & |x| \geq R \\ 4\pi R, & |x| \leq R \end{cases}$$

(b) Using (a), show that the volume potential for the sphere U_R with a density $\varrho = 1$ is equal to

$$V(x) = \begin{cases} \dfrac{4\pi R^3}{3|x|}, & |x| \geq R \\ 2\pi R^2 - \dfrac{2\pi}{3}|x|^2, & |x| \leq R \end{cases}$$

§ 23. Boundary Value Problems for Laplace and Poisson Equations in Space

1. Formulation of the Basic Boundary Value Problems. We shall study the following four boundary value problems of types I and II for the three-dimensional Laplace equation (cf. Sec. 4.4). We shall consider a region G such that $G_1 = R^3 \setminus \bar{G}$ is a region.

The interior Dirichlet problem: to find a function $u \in C(\bar{G})$, harmonic in the region G, which assumes prescribed (continuous) values u_0^- over S.

The exterior Dirichlet problem: to find a function $u \in C(\bar{G}_1)$, harmonic in the region G_1, which takes prescribed (continuous) values u_0^+ over S and which tends to zero at infinity.

The interior Neumann problem: to find a function $u \in C(\bar{G})$, harmonic in the region G, which has a prescribed (continuous) correct normal derivative u_1^- over S.

The exterior Neumann problem: to find a function $u \in C(\bar{G}_1)$, harmonic in the region G_1, which has a prescribed (continuous) correct normal derivative u_1^+ and which tends to zero at infinity.

Analogous boundary value problems are set for Poisson's equation

$$\Delta u = -f \tag{1}$$

and it is required that $u \in C^2(G) \cap C(\bar{G})$ for interior problems and $u \in C^2(G_1) \cap C(\bar{G}_1)$ with $u(\infty) = 0$ for exterior problems.

The substitution

$$u = v + V, \qquad V(x) = \frac{1}{4\pi} \int_G \frac{f(y)}{|x-y|}\, dy \tag{2}$$

reduces the interior boundary value problems for Poisson's equation to the corresponding interior boundary value problems for Laplace's equation if $f \in C^1(G) \cap C(\bar{G})$.

In fact, in this case the volume potential $V \in C^2(G) \cap C^1(\bar{G})$ and it satisfies Poisson's equation (1) (cf. Sec. 22.1). But then, by virtue of (2), the function v must satisfy Laplace's equation and the corresponding boundary condition.

The same may be done for exterior boundary value problems.

2. The Behavior of a Harmonic Function at Infinity. Let the point x lie outside the sphere U_R. We shall use the inversion mapping

$$x^* = \frac{R^2}{|x|^2} x, \qquad x = \frac{R^2}{|x^*|^2} x^* \tag{3}$$

The points x and x^* are said to be *symmetric with respect to the spherical surface* S_R. Symmetric points satisfy the equation

$$|x||x^*| = R^2 \tag{4}$$

and so the inversion mapping is a one-to-one mapping of the exterior of the sphere U_R onto $\bar{U}_R \setminus \{0\}$ (Fig. 65).

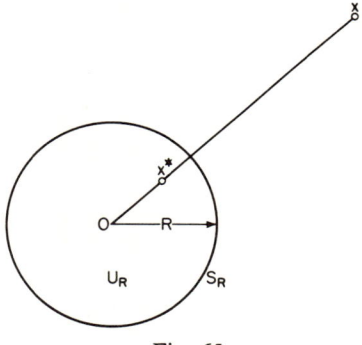

Fig. 65

Let the function $u(x)$ be harmonic outside the sphere $\bar{U}_R$. The function

$$u^*(x^*) = \frac{R}{|x^*|} u\left(\frac{R^2}{|x^*|^2} x^*\right) \tag{5}$$

is known as the *Kelvin transformation* of the function $u(x)$.

We shall prove that *the Kelvin transformation retains the harmonic property*; that is, the function $u^*(x^*)$ is harmonic in $U_R \setminus \{0\}$.

§ 23. LAPLACE AND POISSON EQUATIONS IN SPACE

For this we shall use spherical coordinates (cf. Sec. 3.2). Let $x = (r, \theta, \varphi)$ and $u(x) = \tilde{u}(r, \theta, \varphi)$. Then, by virtue of (3) and (4), $x^* = (\varrho, \theta, \varphi)$, $\varrho = R^2/r$ and, by virtue of (5),

$$u^*(x^*) = \tilde{u}^*(\varrho, \theta, \varphi) = \frac{R}{\varrho} \tilde{u}\left(\frac{R^2}{\varrho}, \theta, \varphi\right) \tag{6}$$

Therefore

$$\Delta u^*(x^*) = \frac{1}{\varrho^2} \frac{\partial}{\partial \varrho}\left(\varrho^2 \frac{\partial \tilde{u}^*}{\partial \varrho}\right) + \frac{1}{\varrho^2 \sin \theta} \frac{\partial}{\partial \theta}\left(\sin \theta \frac{\partial \tilde{u}^*}{\partial \theta}\right)$$
$$+ \frac{1}{\varrho^2 \sin^2 \theta} \frac{\partial^2 \tilde{u}^*}{\partial \varphi^2} = \frac{r^5}{R^5}\left[\frac{\partial^2 \tilde{u}}{\partial r^2} + \frac{2}{r}\frac{\partial \tilde{u}}{\partial r}\right.$$
$$\left. + \frac{1}{r^2 \sin \theta} \frac{\partial}{\partial \theta}\left(\sin \theta \frac{\partial \tilde{u}}{\partial \theta}\right) + \frac{1}{r^2 \sin^2 \theta} \frac{\partial^2 \tilde{u}}{\partial \varphi^2}\right] = \frac{r^5}{R^5} \Delta u(x)$$

that is,

$$\Delta u^*(x^*) = \frac{r^5}{R^5} \Delta u(x) \tag{7}$$

from which the required property follows.

THEOREM. *Let the function $u(x)$ be harmonic outside the sphere $\bar{U}_R$ and let $u(\infty) = 0$. Then as $|x| \to \infty$*

$$u(x) = O\left(\frac{1}{|x|}\right), \quad \text{grad } u(x) = O\left(\frac{1}{|x|^2}\right) \tag{8}$$

Proof. Performing the Kelvin transformation (5), we obtain a function $u^*(x^*)$ harmonic in $U_R \setminus \{0\}$ and, as $x^* \to 0$, satisfying the condition

$$u^*(x^*) = \frac{1}{|x^*|} o(1) = o\left(\frac{1}{|x^*|}\right) = o(|\mathscr{E}_3(x^*)|)$$

According to the theorem on the removal of singularities of harmonic functions (cf. Sec. 21.6), we conclude that the function $u^*(x^*)$ is harmonic in the sphere U_R. Performing the inverse Kelvin transformation, for the function u we obtain the equation

$$u(x) = \frac{R}{|x|} u^*\left(\frac{R^2}{|x|^2} x\right)$$

from which Eqs. (8) follow. The theorem is proved.

Kelvin's transformation allows us to change exterior boundary value problems into interior ones, and vice versa.

3. Uniqueness Theorems for the Solution of Boundary Value Problems.
We shall prove uniqueness theorems for the solution of the boundary value problems which were formulated in Sec. 23.1.

We shall say that the generalized function $u(x)$ *tends to zero at infinity* if it is continuous outside a sphere and if $u(\infty) = 0$.

THEOREM 1. *The solution of Poisson's equation is unique in the class of generalized functions which tend to zero at infinity.*

Proof. It is sufficient to establish that Laplace's equation has only a zero solution in the class of generalized functions which tend to zero at infinity. However, this follows from the analogy of Liouville's theorem (cf. Sec. 21.9).

THEOREM 2. *The solution of the interior or exterior Dirichlet problem is unique and depends continuously on the boundary value u_0^- or u_0^+, respectively, in the following sense: If $|u_0^{\mp} - \tilde{u}_0^{\mp}| \leq \varepsilon$ on S, then the corresponding solutions u and $\tilde{u}$ satisfy the equation*

$$|u(x) - \tilde{u}(x)| \leq \varepsilon, \qquad x \in \bar{G} \qquad (x \in \bar{G}_1) \qquad (9)$$

Proof. Applying inequalities (17) and (18) of Sec. 21.5 to the harmonic function $u - \tilde{u}$,

$$|u(x) - \tilde{u}(x)| \leq \max_{x \in S} |u_0^{\mp}(x) - \tilde{u}_0^{\mp}(x)|, \qquad x \in \bar{G} \qquad (x \in \bar{G}_1)$$

we obtain all the statements of the theorem.

We shall say that a Liapunov surface S is a *sufficiently smooth surface* if it satisfies condition (A) of Sec. 22.6 and if Green's formulas (7), (8) of Sec. 19.2 and (1) of Sec. 21.1 are valid for the functions u and v of the class $C^2(G) \cap C(\bar{G})$ which have a correct normal derivative over S and Δu, $\Delta v \in \mathscr{L}_2(G)$.

It may be proved that condition (A) is always satisfied for bounded closed surfaces of the class C^2, so that, by virtue of what was said in Sec. 21.2, such surfaces are sufficiently smooth.

§ 23. LAPLACE AND POISSON EQUATIONS IN SPACE

THEOREM 3. *If S is a sufficiently smooth surface, then the solution of the interior Neumann problem is defined uniquely as far as an arbitrary additive constant. A necessary condition of the solubility of this problem is the equality*

$$\int_S u_1^-(x)\, dS + \int_G f(x)\, dx = 0 \tag{10}$$

Proof. If u and $\tilde{u}$ are two solutions of the interior Neumann problem, then their difference $\eta \in C(\bar{G})$ is a harmonic function in G and has a zero correct normal derivative over S. Applying Green's formula (7) of Sec. 19.2 with $u = v = \eta$, we obtain

$$\int_G |\operatorname{grad} \eta|^2\, dx = \int_S \eta\, \frac{\partial \eta}{\partial \mathbf{n}}\, dS = 0$$

from which it follows that $\operatorname{grad} \eta = 0$ for $x \in G$, so that $\eta = u - \tilde{u} = \text{const.}$

The necessity of condition (10) for the solubility of the interior Neumann problem follows from formula (8) of Sec. 19.2 with $v \equiv 1$, according to which

$$\int_S u_1^-\, dS = \int_S \frac{\partial u}{\partial \mathbf{n}}\, dS = \int_G \Delta u\, dx = -\int_G f\, dx$$

if u is the solution of this problem. The theorem is proved.

THE PHYSICAL MEANING OF CONDITION (10). A steady state flux of heat (of an incompressible fluid, of an electric or magnetic field strength, cf. Sec. 2) through the closed surface S is equal to the total magnitude of all the sources (currents, etc.) outside S (conservation law).

THEOREM 4. *If S is a sufficiently smooth surface, then the solution of the exterior Neumann problem is unique.*

Proof. Let u and $\tilde{u}$ be two solutions of the exterior Neumann problem. Then their difference $\eta \in C(\bar{G})$ is a harmonic function in G_1, has a zero correct normal derivative over S, and tends to zero as $|x| \to \infty$. By the theorem of Sec. 23.2 the function η satisfies the inequalities

$$|\eta(x)| < \frac{c}{|x|}, \qquad |\operatorname{grad} \eta(x)| < \frac{c_1}{|x|^2}, \qquad |x| \to \infty \tag{11}$$

Applying Green's formula (7) of Sec. 19.2 with $u = v = \eta$ to the region

Q_R (Fig. 54), we obtain

$$\int_{Q_R} |\operatorname{grad} \eta|^2 \, dx = \int_S \eta \frac{\partial \eta}{\partial \mathbf{n}} \, dS + \int_{S_R} \eta \frac{\partial \eta}{\partial \mathbf{n}} \, dS = \int_{S_R} \eta \frac{\partial \eta}{\partial \mathbf{n}} \, dS \quad (12)$$

But from Eqs. (11) it follows that as $R \to \infty$

$$\left| \int_{S_R} \eta \frac{\partial \eta}{\partial \mathbf{n}} \, dS \right| < \int_{S_R} |\eta| \, |\operatorname{grad} \eta| \, dS < \frac{cc_1}{R^3} \int_{S_R} dS = 4\pi \frac{cc_1}{R}$$

Therefore, allowing R to tend to infinity in Eq. (12), we obtain

$$\int_{G_1} |\operatorname{grad} \eta|^2 \, dx = 0$$

from which it follows that $\operatorname{grad} \eta = 0$, that is, $\eta(x) = \text{const}$ for $x \in G_1$. Since $\eta \to 0$ as $|x| \to \infty$, then $\eta = u - \tilde{u} \equiv 0$ for $x \in G_1$. The theorem is proved.

4. Reduction of Boundary Value Problems to Integral Equations. We shall write out Green's formula (5) of Sec. 21.1 for $n = 3$

$$u(x) = \frac{1}{4\pi} \int_S \left[\frac{1}{|x-y|} \frac{\partial u(y)}{\partial \mathbf{n}_y} - u(y) \frac{\partial}{\partial \mathbf{n}_y} \frac{1}{|x-y|} \right] dS_y, \quad x \in G \quad (13)$$

Formula (13) is valid for functions $u \in C(\bar{G})$ which are harmonic in G and which have a correct normal derivative over S if S is a sufficiently smooth surface (cf. Sec. 23.3).

It follows from the uniqueness theorems for the Dirichlet and Neumann problems (cf. Sec. 23.3) that, generally speaking, there is no harmonic function u with arbitrarily prescribed values u and $\partial u / \partial \mathbf{n}$ over S. Therefore Green's formula (13) cannot be directly used to solve the boundary value problems formulated, as we did in solving Cauchy's problem (cf. Secs. 12.3, 14.3). This is the essential difference between a boundary value problem for elliptic equations and for the Cauchy problem.

If we use the theory of Newtonian potential, we may reduce the Dirichlet and Neumann problems for Laplace's equation to Fredholm integral equations with a polar kernel. Moreover, using the theory of integral equations, we shall prove that these boundary value problems are soluble.

Let S be a sufficiently smooth surface. We shall seek the solution of

§ 23. LAPLACE AND POISSON EQUATIONS IN SPACE

Dirichlet's problems (interior and exterior) in the form of the double layer potential

$$V^{(1)}(x) = \int_S \nu(y) \frac{\cos \varphi_{xy}}{|x-y|^2} dS_y$$

where ν is an unknown continuous density over S. The function $V^{(1)}$ is harmonic in G and G_1, and belongs to the classes $C(\bar{G})$, $C(\bar{G}_1)$ and $C(S)$ and $V^1(\infty) = 0$ (cf. Sec. 22.2 and 22.6). Therefore, in order that the potential $V^{(1)}$ give the solution of the interior or exterior Dirichlet problems, it is necessary and sufficient that the following respective equations be satisfied:

$$V^{(1)}_{\mp}(x) = u_0^{\mp}(x), \qquad x \in S \tag{14}$$

where $V^{(1)}_{\mp}$ are the limiting values of $V^{(1)}$ from inside and outside S. By the theorem for the discontinuity of the double layer potential (cf. Sec. 22.6), Eqs. (14) take the form

$$\mp 2\pi \nu(x) + \int_S \nu(y) \frac{\cos \varphi_{xy}}{|x-y|^2} dS_y = u_0^{\mp}(x), \qquad x \in S \tag{15}$$

Equations (15) are Fredholm integral equations with respect to the unknown density ν.

If we introduce a real parameter λ and the kernel

$$\mathscr{K}(x, y) = \frac{1}{2\pi} \frac{\cos \varphi_{xy}}{|x-y|^2} \tag{16}$$

we may write integral equations (15) in the single form

$$\nu(x) = \lambda \int_S \mathscr{K}(x, y) \nu(y) dS_y + f(x), \qquad x \in S \tag{17}$$

In this case, for the interior Dirichlet problem $\lambda = 1$ and $f = -(u_0^-/2\pi)$, and for the exterior Dirichlet problem $\lambda = -1$ and $f = u_0^+/2\pi$.

We shall seek the solution of Neumann's problems (interior and exterior) analogously in the form of the simple layer potential

$$V^{(0)}(x) = \int_S \frac{\mu(y)}{|x-y|} dS_y$$

where μ is an unknown continuous density over S. The function $V^{(0)}$ is harmonic in G and G_1, continuous in R^3, has correct normal derivatives

$(\partial V^{(0)}/\partial \mathbf{n})_{\mp}$ over S from inside and outside S, and tends to zero at infinity (cf. Secs. 22.2 and 22.7). Therefore, in order that the potential $V^{(0)}$ give the solution of the interior or exterior Neumann problems, it is necessary and sufficient that the following equations be satisfied, respectively,

$$\left(\frac{\partial V^{(0)}}{\partial \mathbf{n}}\right)_{\mp}(x) = u_1^{\mp}(x), \qquad x \in S \tag{18}$$

By the theorem for the discontinuity of the normal derivative of the simple layer potential (cf. Sec. 22.7), Eqs. (18) are equivalent to Fredholm's integral equations

$$\pm 2\pi\mu(x) + \int_S \mu(y) \frac{\cos \psi_{xy}}{|x-y|^2} dS_y = u_1^{\mp}(x), \qquad x \in S \tag{19}$$

with respect to the unknown density μ.

From the equation $\psi_{xy} = \varphi_{yx}$ for $x, y \in S$ (cf. Sec. 22.5), and from (16) it follows that the kernel of integral equations (19) is equal to $\mathscr{K}(y, x) = \mathscr{K}^*(x, y)$, so that Eqs. (15) and (19) are adjoint to each other. Introducing the parameter λ, we can write integral (19) in the single form:

$$\mu(x) = \lambda \int_S \mathscr{K}^*(x, y) \mu(y) \, dS_y + g(x), \qquad x \in S \tag{17*}$$

In this case, for the interior Neumann problem $\lambda = -1$ and $g = u_1^-/2\pi$, and for the exterior Neumann problem $\lambda = 1$ and $g = -(u_1^+/2\pi)$.

For a Liapunov surface S the function $\cos \varphi_{xy}$ is continuous over $S \times S$ and, by virtue of the lemma of Sec. 22.4, satisfies the equation

$$|\cos \varphi_{xy}| \leq a |x - y|^{\alpha}, \qquad \alpha > 0$$

Therefore, by virtue of (16), the kernel $\mathscr{K}(x, y)$ is continuous for $x \in S$, $y \in S$, $x \neq y$, and satisfies the inequality

$$|\mathscr{K}(x, y)| \leq \frac{a}{2\pi |x - y|^{2-\alpha}}$$

and so is a polar kernel (cf. Sec. 15.4). In this way, all the statements of Fredholm's theory (cf. note of Sec. 16.5) may be applied to the integral equation (17) and to Eq. (17*), which is adjoint to it.

§ 23. LAPLACE AND POISSON EQUATIONS IN SPACE

5. Investigation of Integral Equations of Potential Theory. We shall first prove that $\lambda = 1$ is not an eigenvalue of the kernel $\mathscr{K}^*(x, y)$. Assume the contrary, and let $\lambda = 1$ be an eigenvalue of this kernel and let μ^* be the eigenfunction corresponding to it

$$\mu^*(x) = \int_S \mathscr{K}^*(x, y)\mu^*(y) \, dS_y = \frac{1}{2\pi} \int_S \mu^*(y) \frac{\cos \psi_{xy}}{|x - y|^2} \, dS_y, \qquad x \in S \tag{20}$$

The eigenfunction $\mu^* \in C(S)$ (cf. Sec. 16.5). We shall construct the simple layer potential $V^{(0)}$ with a density μ^*. The function $V^{(0)}$ is harmonic outside S, continuous in R^3, and becomes zero at infinity (cf. Sec. 22.2). Moreover, by virtue of formula (38) of Sec. 22.7 and Eq. (20), its correct normal derivative over S from outside is equal to zero. From this, by Theorem 4 of Sec. 23.3 concerning the uniqueness of the solution of the exterior Neumann problem, we conclude that $V^{(0)}(x) \equiv 0$ for $x \in \bar{G}_1$ and, specifically, that $V^{(0)}|_S = 0$. But then, by Theorem 2 of Sec. 23.3 concerning the uniqueness of the solution of the interior Dirichlet problem, $V^{(0)}(x) \equiv 0$ for $x \in \bar{G}$. So $V^{(0)}(x) \equiv 0$ for $x \in R^3$. Therefore, using formula (41) of Sec. 22.7, we conclude that $\mu^*(x) \equiv 0$ for $x \in S$.

In this way, $\lambda = 1$ is not an eigenvalue of the kernel $\mathscr{K}^*(x, y)$. From this, by Fredholm's second theorem, $\lambda = 1$ is also not an eigenvalue of the kernel $\mathscr{K}(x, y)$. But then, by Fredholm's third and first theorems the integral equations (17) and (17*) for $\lambda = 1$ are uniquely soluble for any continuous f and g. Consequently, the following theorem is valid.

THEOREM 1. *The interior Dirichlet problem and the exterior Neumann problem are soluble for any continuous data u_0^- and u_1^+, and their solutions are the double and simple layer potentials, respectively.*

Now from formula (25) of Sec. 22.5,

$$-\frac{1}{2\pi} \int_S \frac{\cos \psi_{xy}}{|x - y|^2} \, dS_y = -\int_S \mathscr{K}(x, y) \, dS_y = 1, \qquad x \in S$$

it follows that $\lambda = -1$ is an eigenvalue of the kernel $\mathscr{K}(x, y)$, and $\nu \equiv 1$ is the eigenfunction corresponding to it. We shall prove that this is a simple eigenvalue. For this, by virtue of Fredholm's second theorem, it is sufficient to show that $\lambda = -1$ is a simple eigenvalue of the kernel $\mathscr{K}^*(x, y)$. Let

μ_0 be the corresponding eigenfunction,

$$\mu_0(x) = -\int_S \mathcal{K}^*(x, y)\mu_0(y)\, dS_y = -\frac{1}{2\pi}\int_S \frac{\cos \psi_{xy}}{|x-y|^2}\mu_0(y)\, dS_y \qquad (21)$$

The eigenfunction $\mu_0 \in C(S)$ (cf. Sec. 16.5).

We shall construct the simple layer potential with the density μ_0

$$V^{(0)}(x) = \int_S \frac{\mu_0(y)}{|x-y|}\, dS_y \qquad (22)$$

The function $V^{(0)}$ is harmonic outside S, continuous in R^3, and tends to zero as $|x| \to \infty$ (cf. Sec. 22.2). Moreover, by virtue of formula (38') of Sec. 22.7 and Eq. (21), its correct normal derivative over S from inside is equal to zero. From this, by Theorem 3 of Sec. 23.3 concerning the uniqueness of the solution of the interior Neumann problem, we conclude that $V^{(0)}(x) \equiv C = \text{const}$, $x \in \bar{G}$.

We shall prove that $C \neq 0$. On the contrary, let $V^{(0)}(x) \equiv 0$ for $x \in \bar{G}$ and, specifically, $V^{(0)}|_S = 0$. But then, by Theorem 2 of Sec. 23.3 concerning the uniqueness of the solution of the exterior Dirichlet problem, $V^{(0)}(x) \equiv 0$ for $x \in \bar{G}_1$. So $V^{(0)}(x) \equiv 0$ for $x \in R^3$. From this, using Eq. (41) of Sec. 22.7, we conclude that $\mu_0(x) \equiv 0$ for $x \in S$, which is impossible.

Let $\tilde{\mu}_0$ be another eigenfunction of the kernel $\mathcal{K}^*(x, y)$ corresponding to the eigenvalue $\lambda = -1$. By what has been proved, the simple layer potential $\tilde{V}^{(0)}$ with a density $\tilde{\mu}_0$ is equal to the constant $\tilde{C} \neq 0$ over $\bar{G}$. But then the simple layer potential $(\tilde{C}/C)V^{(0)} - \tilde{V}^{(0)}$ with the density $(\tilde{C}/C)\mu_0 - \tilde{\mu}_0$ is equal to zero over $\bar{G}$, from which it follows that this density is identically equal to zero over S; that is,

$$\tilde{\mu}_0(x) = \frac{\tilde{C}}{C}\mu_0(x), \qquad x \in S$$

Therefore $\lambda = -1$ is a simple eigenvalue of the kernel $\mathcal{K}^*(x, y)$, and therefore of the kernel $\mathcal{K}(x, y)$.

We shall normalize the eigenfunction μ_0 so that

$$V^{(0)}(x) = \int_S \frac{\mu_0(y)}{|x-y|}\, dS_y \equiv 1, \qquad x \in \bar{G} \qquad (23)$$

The simple layer potential $V^{(0)}$ with the density μ_0 is known as the **Robin potential**.

§ 23. LAPLACE AND POISSON EQUATIONS IN SPACE

PHYSICAL MEANING OF THE ROBIN POTENTIAL. This is a potential created by charges over a conducting surface S, while its density

$$\mu_0(x) = -\frac{1}{4\pi}\left(\frac{\partial V^{(0)}}{\partial \mathbf{n}}\right)_+(x)$$

is the charge density located on this surface.

We shall return to Eqs. (17) and (17*) for $\lambda = -1$. By Fredholm's third theorem the integral equation (17*) for $\lambda = -1$ is soluble when and only when the inhomogeneous term g is orthogonal to 1. So the following theorem holds.

THEOREM 2. *The interior Neumann problem is soluble for any continuous function u_1^- which satisfies the orthogonality condition*

$$\int_S u_1^-(x)\,dS = 0 \tag{24}$$

and its solution may be represented by the simple layer potential.

Moreover, for the solubility of Eq. (17) for $\lambda = -1$ it is necessary and sufficient that the inhomogeneous term f be orthogonal to μ_0. In this way, *the exterior Dirichlet problem has a solution, which is represented by the double layer potential, for any continuous function u_0^+ which is orthogonal to the density μ_0 of the Robin potential*

$$\int_S u_0^+(x)\mu_0(x)\,dS = 0 \tag{25}$$

The condition of solubility (25) arose because the solution of the exterior Dirichlet problem was sought in the form of a double layer potential since previously the solution was required to behave like $O(|x|^{-2})$ as $|x| \to \infty$. However, in formulating this problem it is required only that the solution should tend to zero as $|x| \to \infty$. In order to consider such solutions while at the same time eliminating condition (25), we proceed as follows.

We shall suppose that $0 \in G$. We shall seek the solution of the exterior Dirichlet problem in the form of a sum of the double layer potential $V^{(1)}$ with an unknown density ν over S and of the Newtonian potential $\alpha/|x|$ from a charge at the point $x = 0$ of unknown magnitude α,

$$u(x) = V^{(1)}(x) + \frac{\alpha}{|x|} = \int_S \nu(y)\frac{\cos \varphi_{xy}}{|x-y|^2}\,dS_y + \frac{\alpha}{|x|} \tag{26}$$

The corresponding integral equation (17) takes the form

$$v(x) = -\int_S \mathscr{K}(x,y) v(y)\, dS_y + \frac{u_0^+(x)}{2\pi} - \frac{\alpha}{2\pi |x|} \tag{27}$$

By what has been proved for the solubility of integral equation (27), it is necessary and sufficient that

$$\frac{1}{2\pi} \int_S \left[u_0^+(x) - \frac{\alpha}{|x|} \right] \mu_0(x)\, dS = 0 \tag{28}$$

Since $0 \in G$, then by virtue of (23),

$$\int_S \frac{\mu_0(y)}{|y|}\, dS = V^{(0)}(0) = 1$$

and therefore the condition of solubility (28) takes the form

$$\alpha = \int_S u_0^+(x) \mu_0(x)\, dS$$

In this way, the following theorem is seen to be true.

THEOREM 3. *The exterior Dirichlet problem is soluble for any continuous function u_0^+ and its solution may be represented in the form of a sum of the double layer potential and of the potential*

$$\frac{1}{|x|} \int_S u_0^+(x) \mu_0(x)\, dS$$

§ 24. The Green's Function for the Dirichlet Problem

1. Definition and Properties of the Green's Function. The function $\mathscr{G}(x,y)$ for $x \in \bar{G}$, $y \in G$ is known as the *Green's function of the interior Dirichlet problem* for the region G, if it satisfies the following properties:

(1) For each $y \in G$ it can be represented in the form

$$\mathscr{G}(x,y) = \frac{1}{4\pi |x-y|} + g(x,y) \tag{1}$$

where the function $g(x,y)$ is harmonic in G and continuous on $\bar{G}$ with respect to x.

§ 24. THE GREEN'S FUNCTION FOR THE DIRICHLET PROBLEM

(2) For each $y \in G$ it satisfies the boundary condition

$$\mathscr{G}(x, y)|_{x \in S} = 0 \qquad (2)$$

It follows from conditions (1) and (2) that the function $\mathscr{G}(x, y)$ is harmonic with respect to x in the region $G \setminus \{y\}$, is continuous in $\bar{G} \setminus \{y\}$, becomes zero over S, and tends to $+\infty$ as $x \to y$. It follows from this, by virtue of the maximum principle (cf. Sec. 21.4), that $\mathscr{G}(x, y) > 0$ for $x \in G$, $y \in G$. Moreover, the harmonic function $g(x, y)$ satisfies the boundary condition

$$g(x, y) = -\frac{1}{4\pi |x-y|}, \qquad x \in S, \qquad y \in G \qquad (3)$$

from which it follows that $g(x, y) < 0$ for $x \in S$, $y \in G$. But then, by virtue of the maximum principle, this inequality is also true in the region G; that is, $g(x, y) < 0$ for $x \in \bar{G}$, $y \in G$. So, by virtue of (1), the Green's function satisfies the inequalities

$$0 < \mathscr{G}(x, y) < \frac{1}{4\pi |x-y|}, \qquad x \in G, \qquad y \in G, \qquad x \neq y \qquad (4)$$

Since the solution of Dirichlet's problem is unique (cf. Sec. 23.3), it follows that the Green's function $\mathscr{G}(x, y)$ is unique, if it exists.

PHYSICAL MEANING OF THE GREEN'S FUNCTION. By definiton, the Green's function $\mathscr{G}(x, y)$ satisfies Poisson's equation in a generalized sense for each $y \in G$,

$$\Delta_x \mathscr{G}(x, y) = -\delta(x-y), \qquad x \in G$$

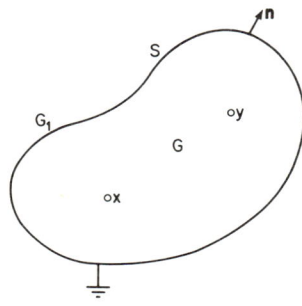

Fig. 66

and becomes zero over the boundary S. Therefore, the function $\mathscr{G}(x, y)$ may be interpreted as a Coulomb potential (cf. Sec. 22.3) generated inside the grounded conducting surface S by the charge $+(1/4\pi)$ at the point $y \in G$ (Fig. 66).

The function $g(x, y)$ is continuous with respect to the set of variables (x, y) in $\bar{G} \times G$.

Let $x_0 \in \bar{G}$, $y_0 \in G$, and $(x, y) \to (x_0, y_0)$ for $x \in \bar{G}$, $y \in G$. Using the continuity of the function $g(x, y)$ with respect to x, the maximum principle, and Eq. (3), we obtain

$$|g(x_0, y_0) - g(x, y)| \leq |g(x_0, y_0) - g(x, y_0)| + |g(x, y_0) - g(x, y)|$$
$$\leq |g(x_0, y_0) - g(x, y_0)|$$
$$+ \max_{x' \in S} \frac{1}{4\pi} \left| \frac{1}{|x' - y_0|} - \frac{1}{|x' - y|} \right| \to 0$$

which proves the continuity of the function g at the point (x_0, y_0).

THEOREM. *If S is a sufficiently smooth surface, then the Green's function $\mathscr{G}(x, y)$ exists, has a correct normal derivative $\partial \mathscr{G}(x, y)/\partial \mathbf{n}$ over S for all $y \in G$, and is symmetric:*

$$\mathscr{G}(x, y) = \mathscr{G}(y, x), \quad x \in G, \quad y \in G \quad (5)$$

Proof. It is sufficient to establish that there is a symmetric function $g(x, y)$ which has, for every $y \in G$, the following properties with respect to x: it is harmonic in G, continuous on $\bar{G}$, satisfies the boundary condition (3), and has a correct normal derivative over S.

We fix $y \in G$. The function $-(1/4\pi)|x - y|^{-1}$ for $x \in G_1$ is obviously the solution of the exterior Neumann problem with the boundary condition

$$u_1^+(x, y) = -\frac{1}{4\pi} \frac{\partial}{\partial \mathbf{n}_x} \frac{1}{|x - y|}, \quad x \in S \quad (6)$$

On the other hand, by Theorem 1 of Sec. 23.5, this solution is represented by the simple layer potential

$$V^{(0)}(x, y) = \int_S \frac{\mu(y', y)}{|x - y'|} dS_{y'}$$

with continuous density $\mu(y', y)$ with respect to $y' \in S$. Therefore, by virtue

§ 24. THE GREEN'S FUNCTION FOR THE DIRICHLET PROBLEM

of the uniqueness of the solution (cf. Sec. 23.3),

$$V^{(0)}(x, y) = -\frac{1}{4\pi |x - y|}, \qquad x \in G_1 \qquad (7)$$

The potential $V^{(0)}$ is harmonic in G and continuous in R^3 (cf. Sec. 22.2) and, by virtue of (7), satisfies the boundary condition (3). Therefore

$$g(x, y) = V^{(0)}(x, y) = \int_S \frac{\mu(y', y)}{|x - y'|} dS_{y'}, \qquad x \in G \qquad (8)$$

From this, by the theorem of Sec. 22.7, it follows that the function $g(x, y)$ has a correct normal derivative (from inside) over S and this derivative, by virtue of Eqs. (41) of Sec. 22.7 and (6) of Sec. 24.1, is equal to

$$\frac{\partial g(x, y)}{\partial \mathbf{n}_x} = 4\pi\mu(x, y) - \frac{1}{4\pi} \frac{\partial}{\partial \mathbf{n}_x} \frac{1}{|x - y|}, \qquad x \in S \qquad (9)$$

It remains to prove the symmetry of the function $g(x, y)$. Applying Green's formula (13) of Sec. 23.4 to the function $g(x, y)$ and using boundary conditions (3) and (9) and Eq. (8), for all $x \in G$ and $y \in G$ we obtain

$$g(x, y) = \frac{1}{4\pi} \int_S \left[\frac{1}{|x - y'|} \frac{\partial g(y', y)}{\partial \mathbf{n}_{y'}} - g(y', y) \frac{\partial}{\partial \mathbf{n}_{y'}} \frac{1}{|x - y'|} \right] dS_{y'}$$

$$= -\int_S g(y', x) \frac{\partial g(y', y)}{\partial \mathbf{n}_{y'}} dS_{y'}$$

$$+ \int_S g(y', y) \left[\frac{\partial g(y', x)}{\partial \mathbf{n}_{y'}} - 4\pi\mu(y', x) \right] dS_{y'}$$

$$= \int_S \left[g(y', y) \frac{\partial g(y', x)}{\partial \mathbf{n}_{y'}} - g(y', x) \frac{\partial g(y', y)}{\partial \mathbf{n}_{y'}} \right] dS_{y'}$$

$$- 4\pi \int_S g(y', y)\mu(y', x) dS_{y'}$$

$$= \int_S [g(y', y) \Delta g(y', x) - g(y', x) \Delta g(y', y)] dy'$$

$$+ \int_S \frac{\mu(y', x)}{|y - y'|} dS_{y'} = g(y, x)$$

The theorem is proved.

Other properties of the function $g(x, y)$ follow from its symmetry: *it is continuous with respect to (x, y) in $G \times \bar{G}$; for each $x \in G$ it is harmonic* with respect to y in G; it assumes the value $-(1/4\pi)|x-y|^{-1}$ when $y \in S$; and it has a correct normal derivative $\partial g(x, y)/\partial \mathbf{n}_y$ over S.

2. Examples of the Construction of a Green's Function (Reflection Method). The most effective way of constructing a Green's function for a region having a sufficiently wide symmetry group is by the reflection method, and we shall give a series of examples to illustrate this.

(a) A Sphere U_R. Let $y \in U_R$ for $y \neq 0$ and

$$y^* = y \frac{R^2}{|y|^2}, \qquad |y||y^*| = R^2 \tag{10}$$

be the symmetric point for y relative to the spherical surface S_R by an inversion mapping (cf. Sec. 23.2).

We shall seek a Green's function in the form

$$\mathscr{G}(x, y) = \frac{1}{4\pi |x-y|} - \frac{\alpha}{4\pi |x-y^*|} \tag{11}$$

where $-(\alpha/4\pi)$ is an unknown charge at the symmetric point y^*. The function

$$g(x, y) = -\frac{\alpha}{4\pi |x-y^*|}$$

is harmonic in U_R and belongs to the class $C^\infty(\bar{U}_R)$. We shall choose the magnitude α so that the function $\mathscr{G}(x, y)$ becomes zero over the boundary S_R. For this we note that when $|x| = R$ the triangles $(0, x, y^*)$ and $(0, x, y)$ are similar: one angle in them is common and the adjoining sides, by virtue of (10), are proportional (Fig. 67). Therefore when $|x| = R$ the result

$$\frac{R}{|y|} = \frac{|x - y^*|}{|x - y|}$$

is true and, therefore, by virtue of (11), it is necessary to put $\alpha = R/|y|$. So,

$$\begin{aligned}\mathscr{G}(x, y) &= \frac{1}{4\pi |x-y|} - \frac{R}{4\pi |y||x-y^*|} \\ &= \frac{1}{4\pi |x-y|} - \frac{R|y|}{4\pi |x|y|^2 - yR^2|}\end{aligned} \tag{12}$$

is the Green's function for a sphere. Formula (12) is also valid when $y = 0$,

$$\mathscr{G}(x, 0) = \frac{1}{4\pi |x|} - \frac{1}{4\pi R}.$$

§ 24. THE GREEN'S FUNCTION FOR THE DIRICHLET PROBLEM

(b) THE HALF-SPACE $x_3 > 0$. Let the point $y = (y_1, y_2, y_3)$ lie in this half-space, $y_3 > 0$. The point $\bar{y} = (y_1, y_2, -y_3)$ is said to be symmetric for the point y relative to the plane $x_3 = 0$ (Fig. 68). It is not difficult

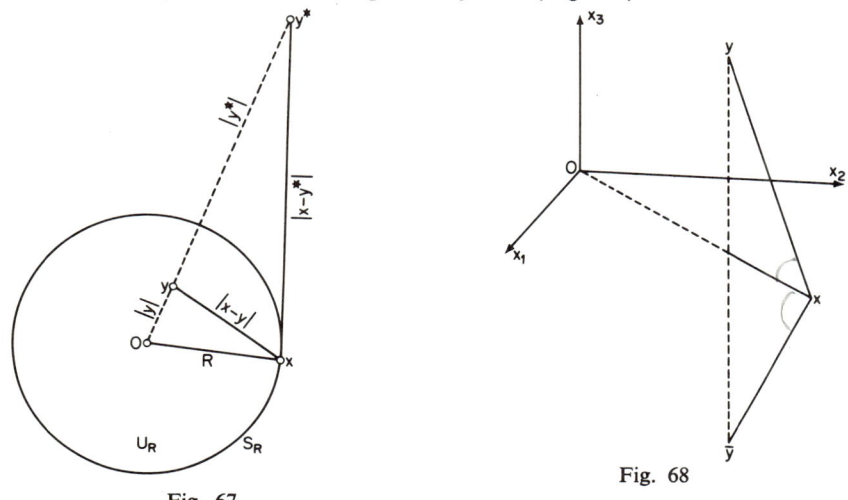

Fig. 67

Fig. 68

to see that the Green's function for the half-space $x_3 > 0$ is defined by the equation

$$\mathcal{G}(x, y) = \frac{1}{4\pi |x - y|} - \frac{1}{4\pi |x - \bar{y}|} \tag{13}$$

(c) THE HALF-SPHERE $|x| < R$ FOR $x_3 > 0$. Let the point y lie in this

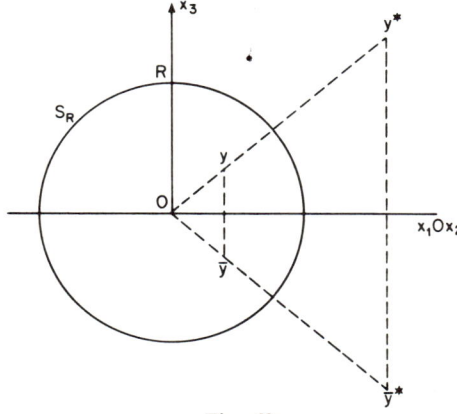

Fig. 69

half-sphere; y^* is a symmetric point for y relative to the spherical surface S_R; $\bar{y}$ and $\bar{y}^*$ are symmetric points for y and y^* relative to the plane $x_3 = 0$ (Fig. 69). The Green's function is expressed by the equation

$$\mathscr{G}(x, y) = \frac{1}{4\pi |x - y|} - \frac{R}{4\pi |y| |x - y^*|}$$
$$- \frac{1}{4\pi |x - \bar{y}|} + \frac{R}{4\pi |y| |x - \bar{y}^*|} \qquad (14)$$

(d) THE QUADRANT $x_2 > 0$, $x_3 > 0$. Let the point $y = (y_1, y_2, y_3)$ lie in the quadrant $y_2 > 0$, $y_3 > 0$; $\bar{y}$ and y' are points symmetric for y relative to the planes $x_3 = 0$ and $x_2 = 0$, respectively; $\bar{y}'$ is a point symmetric for $\bar{y}$ relative to the plane $x_3 = 0$ (Fig. 70).

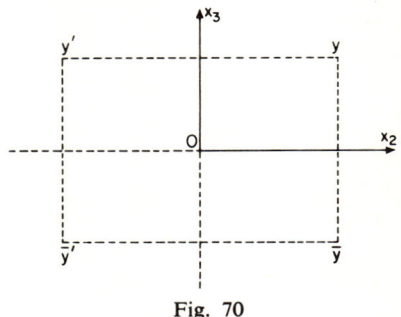

Fig. 70

The Green's function has the form

$$\mathscr{G}(x, y) = \frac{1}{4\pi |x - y|} - \frac{1}{4\pi |x - \bar{y}|}$$
$$- \frac{1}{4\pi |x - y'|} + \frac{1}{4\pi |x - \bar{y}'|} \qquad (15)$$

The Green's function is constructed analogously for a wedge of angle π/n, where n is an integer and $n \geq 3$.

3. Solution of Boundary Value Problems by Means of the Green's Function. In this subsection we shall suppose that S is a sufficiently smooth surface (cf. Sec. 23.3). We shall consider the interior Dirichlet problem for Poisson's

§ 24. THE GREEN'S FUNCTION FOR THE DIRICHLET PROBLEM

equation

$$\Delta u = -f(x), \qquad u|_S = u_0(x), \qquad u \in C^2(G) \cap C(\bar{G}) \qquad (16)$$

where $f \in C(G) \cap \mathscr{L}_2(G)$ and $u_0 \in C(S)$. As we established in Sec. 23.3, the solution of this problem is unique.

THEOREM. *If the solution $u(x)$ of problem* (16) *has a correct normal derivative over S, then it can be represented by the equation*

$$u(x) = -\int_S \frac{\partial \mathscr{G}(x, y)}{\partial \mathbf{n}_y} u_0(y) \, dS_y + \int_G \mathscr{G}(x, y) f(y) \, dy, \qquad x \in G \qquad (17)$$

Proof. By our conditions the solution $u \in C^2(G) \cap C(\bar{G})$ and has a correct normal derivative over S and $\Delta u = -f$ for $f \in \mathscr{L}_2(G)$. If we apply Green's formula (1) of Sec. 21.1 to the function $u(x)$ for $n = 3$ and use (16), we shall obtain

$$u(x) = \frac{1}{4\pi} \int_S \left[\frac{\partial u(y)}{\partial \mathbf{n}_y} \frac{1}{|x-y|} - u_0(y) \frac{\partial}{\partial \mathbf{n}_y} \frac{1}{|x-y|} \right] dS_y$$

$$+ \frac{1}{4\pi} \int_G \frac{f(y)}{|x-y|} \, dy, \qquad x \in G \qquad (18)$$

Moreover, for each $x \in S$ the function $g(x, y)$ is harmonic with respect to y in G, continuous with respect to y on $\bar{G}$, and has a correct normal derivative $\partial g(x, y)/\partial \mathbf{n}_y$ over S. If we apply Green's formula (8) of Sec. 19.2 to the functions $u(y)$ and $g(x, y)$, we deduce the equation

$$0 = \int_S \left[\frac{\partial u(y)}{\partial \mathbf{n}_y} g(x, y) - u_0(y) \frac{\partial g(x, y)}{\partial \mathbf{n}_y} \right] dS_y + \int_G f(y) g(x, y) \, dy, \qquad x \in G$$

If we add this equation to Eq. (18) and use (1) and (3), we obtain Eq. (17). The theorem is proved.

4. Poisson's Formula. We shall calculate the correct normal derivative of the Green's function for the sphere U_R over the spherical surface S_R.

Using expression (12) for this function, we obtain

$$\left.\frac{\partial \mathscr{G}(x, y)}{\partial \mathbf{n}_y}\right|_{S_R} = \frac{\partial}{\partial |y|} \left[\frac{1}{4\pi |x-y|} - \frac{R|y|}{4\pi |x|y|^2 - yR^2|} \right]\Bigg|_{|y|=R}$$

$$= \frac{1}{4\pi} \frac{\partial}{\partial \varrho} \left[\frac{1}{(|x|^2 + \varrho^2 - 2|x|\varrho \cos \gamma)^{1/2}} \right.$$

$$\left. - \frac{R}{(R^4 + |x|^2 \varrho^2 - 2R^2 |x| \varrho \cos \gamma)^{1/2}} \right]\Bigg|_{\varrho=R}$$

$$= \frac{|x|^2 - R^2}{4\pi R(R^2 + |x|^2 - 2R|x|\cos \gamma)^{3/2}}$$

$$= \frac{|x|^2 - R^2}{4\pi R |x-y|^3}\Bigg|_{S_R}$$

Therefore Eq. (17) for the sphere U_R when $f = 0$ takes the form

$$u(x) = \frac{1}{4\pi R} \int_{|y|=R} \frac{R^2 - |x|^2}{|x-y|^3} u_0(y) \, dS_y, \qquad |x| < R \qquad (19)$$

This is *Poisson's (integral) formula*. It is analogous to Cauchy's formula for analytic functions.

We shall prove that *Poisson's formula* (19) *gives the solution of the interior Dirichlet problem for the sphere* U_R,

$$\Delta u = 0, \qquad u|_{S_R} = u_0 \qquad (20)$$

for any function u_0 which is continuous over S_R.

In fact the solution $u(x)$ of this problem exists for any continuous function u_0 and is unique (cf. Sec. 23). In each smaller sphere U_ϱ, $\varrho < R$, the function $u(x)$ is the solution of the Dirichlet problem with the boundary data $u|_{S_R}$ and it belongs to the class $C^\infty(\bar{U}_\varrho)$. Therefore, by the theorem of the preceding subsection, this solution can be represented as Poisson's integral (19); that is,

$$u(x) = \frac{1}{4\pi \varrho} \int_{|y|=\varrho} \frac{\varrho^2 - |x|^2}{|x-y|^3} u(y) \, dS_y, \qquad |x| < \varrho$$

Proceeding to the limit as $\varrho \to R$ in this formula and using the continuity of $u(x)$ on $\bar{U}_R$ and the boundary condition (20), we obtain Eq. (19).

§ 24. THE GREEN'S FUNCTION FOR THE DIRICHLET PROBLEM

5. Reduction of a Boundary Value Problem to an Integral Equation.
In the region G we shall consider the boundary value problem for Poisson's equation

$$\Delta u = -f(x), \quad u|_S = 0, \quad u \in C^2(G) \cap C(\bar{G}) \tag{21}$$

We first prove the following lemma.

LEMMA. *If $f \in C(\bar{G})$, then the function*

$$\tilde{V}(x) = \int_G g(x, y) f(y) \, dy \tag{22}$$

is harmonic in the region G.

Proof. Since the function $g(x, y)$ is continuous with respect to (x, y) in $G \times \bar{G}$ and harmonic with respect to x in G (cf. Sec. 24.1), then $\tilde{V} \in C(G)$ and for any $\varphi \in \mathscr{D}(G)$ the following equalities are true:

$$\int \tilde{V}(x) \Delta \varphi(x) \, dx = \int [\int g(x, y) f(y) \, dy] \Delta \varphi(x) \, dx$$
$$= \int f(y) [\int g(x, y) \Delta \varphi(x) \, dx] \, dy$$
$$= \int f(y) [\int \Delta g(x, y) \varphi(x) \, dx] \, dy = 0$$

Therefore the function $\tilde{V}(x)$ is a generalized harmonic function and so is harmonic in the region G (cf. Sec. 21.7). The lemma is proved.

THEOREM. *If $f \in C^1(G) \cap C(\bar{G})$, then the (unique) solution of problem (21) is expressed by the equation*

$$u(x) = \int_G \mathscr{G}(x, y) f(y) \, dy \tag{23}$$

and has a correct normal derivative over S.

Proof. We shall prove that formula (23) gives the solution of problem (21). Using (1), we write (23) as the sum of two terms

$$u(x) = V(x) + \tilde{V}(x) \tag{24}$$

where V is a volume potential with the density $f/4\pi$ and the function $\tilde{V}$ is defined by Eq. (22).

By supposition $f \in C^1(G) \cap C(\bar{G})$. Therefore the volume potential $V \in C^2(G) \cap C^1(\bar{G})$ and it satisfies Poisson's equation (21) (cf. Sec. 22.1). By the lemma which has just been proved, the function $\tilde{V}(x)$ is harmonic in the region G. So, by virtue of (24), the function $u \in C^2(G)$ and in the region G it satisfies Poisson's equation (21).

We shall prove that $u \in C(\bar{G})$ and becomes zero over S. For this it is sufficient to show that

$$|u(x')| \xrightarrow{x \in S} 0, \quad x' \to x, \quad x' \in G \tag{25}$$

Let $\varepsilon > 0$. Then by virtue of Eq. (4), there is a subregion $G' \Subset G$ (Fig. 71), such that

$$\left| \int_{G \setminus G'} \mathscr{G}(x', y) f(y) \, dy \right| \leq \frac{1}{4\pi} \int_{G \setminus G'} \frac{|f(y)|}{|x' - y|} \, dy < \frac{\varepsilon}{2}, \quad x' \in \bar{G} \tag{26}$$

But the function $g(x', y)$ is uniformly continuous with respect to (x', y) over $\bar{G} \times \bar{G}'$ (cf. Sec. 24.1) and so, by virtue of (1), the Green's function $\mathscr{G}(x', y)$ is uniformly continuous with respect to (x', y) over $(G \setminus G'') \times \bar{G}'$,

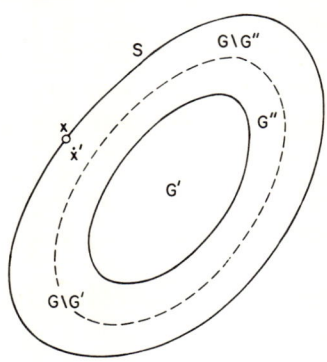

Fig. 71

where G'' is any subregion such that $G' \Subset G'' \Subset G$ (Fig. 71). Therefore, if we allow the function $\mathscr{G}(x', y)$ to become zero for $x' \in S$, $y \in \bar{G}'$, we conclude that there is a subregion $G'' \Subset G$, sufficiently close to G, such that

$$\left| \int_{G'} \mathscr{G}(x', y) f(y) \, dy \right| < \varepsilon/2, \quad x' \in G \setminus G''$$

§ 24. THE GREEN'S FUNCTION FOR THE DIRICHLET PROBLEM

from which, and from (26), follows the inequality

$$|u(x')| = \left|\int_G \mathcal{G}(x', y)f(y)\,dy\right| \le \left|\int_{G'} \mathcal{G}(x', y)f(y)\,dy\right|$$
$$+ \left|\int_{G\setminus G'} \mathcal{G}(x', y)f(y)\,dy\right| < \frac{\varepsilon}{2} + \frac{\varepsilon}{2} = \varepsilon$$

which is valid for all $x' \in G \setminus G''$. This proves the limiting result (25).

We shall prove that the function $u(x)$ has a correct normal derivative over S. Since $V \in C^1(R^3)$ (cf. Sec. 22.1), then, by virtue of (24), for our purpose it is sufficient to establish that the function $\tilde{V}(x)$ has a correct normal derivative over S. By what has been proved, $\tilde{V} \in C(\bar{G})$ is harmonic in G and satisfies the boundary condition $\tilde{V}|_S = -V|_S$. We shall construct the potential of a simple layer $V^{(0)}$ with a continuous density, solving the exterior Neumann problem with $u_1^+ = -(\partial V/\partial \mathbf{n})|_S$ (cf. Sec. 23.5). The volume potential $-V(x)$ is also the solution of this problem (cf. Sec. 22.1). Therefore, since the exterior Neumann problem has a unique solution (cf. Sec. 23.3), we conclude that $V^{(0)}(x) = -V(x)$ for $x \in G_1$. Specifically, $V^{(0)}|_S = -V|_S$. From this, by the uniqueness theorem for the solution of the interior Dirichlet problem (cf. Sec. 23.3), we obtain $\tilde{V}(x) = V^{(0)}(x)$ for $x \in G$. Since the potential of a simple layer $V^{(0)}$ has a correct normal derivative (from inside) over S (cf. Sec. 22.7), it follows that the function $\tilde{V}$ has the same properties. The theorem is proved.

We shall now establish that the *boundary value problem*

$$-\Delta u = \lambda u + f(x), \qquad u|_S = 0, \qquad u \in C^2(G) \cap C(\bar{G}) \qquad (27)$$

is equivalent to the integral equation

$$u(x) = \int_G \mathcal{G}(x, y)[\lambda u(y) + f(y)]\,dy, \qquad u \in C(\bar{G}) \qquad (28)$$

if $f \in C^1(G) \cap C(\bar{G})$.

In fact, let the function $u \in C(\bar{G})$ be the solution of the integral equation (28); that is, by virtue of (1),

$$u(x) = \frac{1}{4\pi} \int_G \frac{\lambda u(y) + f(y)}{|x-y|}\,dy + \int_G g(x, y)[\lambda u(y) + f(y)]\,dy \qquad (28')$$

The first term on the right in (28') is the volume potential and so belongs to the class $C^1(R^3)$ (cf. Sec. 22.1), while the second term is a harmonic function in the region G (cf. the lemma of this subsection). Therefore, $u \in C^1(G)$ and so $\lambda u + f \in C^1(G) \cap C(\bar{G})$. By the theorem of this subsection, the function $u(x)$ is the solution of the boundary value problem (27).

Conversely, if the function $u_1(x)$ is the solution of the boundary value problem (27), then it is the (unique) solution of the boundary value problem (21) subject to the change of f to $\lambda u_1 + f$. Since $\lambda u_1 + f \in C^1(G) \cap C(\bar{G})$, then, by the theorem of this subsection, this solution may be expressed by the integral (23) by changing f to $\lambda u_1 + f$; that is, the function u_1 satisfies the integral equation (28). This proves that problems (27) and (28) are equivalent.

6. Properties of Eigenvalues and Eigenfunctions. We shall consider the homogeneous boundary value problems involving the eigenvalues (the interior Dirichlet problem)

$$\Delta u + \lambda u = 0, \qquad u|_S = 0, \qquad u \in C^2(G) \cap C(\bar{G}) \tag{29}$$

It was shown in the preceding subsection that the problem (29) is equivalent to a problem involving the eigenvalues of the homogeneous integral equation

$$u(x) = \lambda \int_G \mathscr{G}(x, y) u(y) \, dy, \qquad u \in C(\bar{G}) \tag{30}$$

with a symmetric (and therefore Hermitian) kernel $\mathscr{G}(x, y)$ (cf. Sec. 24.1).

We shall prove that the kernel $\mathscr{G}(x, y)$ is weakly polar (cf. Sec. 15.4). For this the function $g(x, y)$, which was prescribed and is continuous over $(\bar{G} \times G) \cup (G \times \bar{G})$ (cf. Sec. 24.1), can be continued over $\bar{G} \times \bar{G}$ by putting, according to (3),

$$g(x, y) = -\frac{1}{4\pi |x - y|}, \qquad x \in S, \qquad y \in S$$

With this continuation the function $g(x, y)$ is continuous over $\bar{G} \times \bar{G}$, except at those points for which $x = y$ with $y \in S$. But then, by virtue of (1), the Green's function $\mathscr{G}(x, y)$ is continuous for $x \in \bar{G}$ and $y \in \bar{G}$ with $x \neq y$ and so, by virtue of (4), the kernel $\mathscr{G}(x, y)$ is weakly polar ($\alpha = 1$, $n = 3$).

Therefore all the propositions of the theory of integral equations with a symmetric weakly polar kernel which were proved in Secs. 17 and 18 are valid for Eq. (30). But the eigenvalues and eigenfunctions of the boundary

§ 24. THE GREEN'S FUNCTION FOR THE DIRICHLET PROBLEM

value problem (29) coincide with the eigenvalues and the corresponding eigenfunctions of the kernel $\mathscr{G}(x, y)$. So for the boundary value problem (29) it is possible to prove Theorem 1 of Sec. 19.4 completely, and also to establish some other properties concerning this problem.

THEOREM. *The set of eigenvalues $\{\lambda_k\}$ of the boundary value problem (29) is countable and has no finite limit points, and $\lambda_k > 0$; each eigenvalue λ_k has finite multiplicity. The smallest eigenvalue λ_1 is simple, while the eigenfunction corresponding to it $u_1(x) > 0$ for $x \in G$. The eigenfunctions $\{u_k\}$ may be chosen to be real and orthonormal; they have a correct normal derivative over S. Each function f belonging to $\mathscr{M}_\varDelta{}^*$ can be expanded into a regularly converging Fourier series involving the eigenfunctions $\{u_k\}$.*

Proof. Fredholm's theorems (cf. Sec. 16.5) prove that the set $\{\lambda_k\}$ has no finite limit points and that each eigenvalue λ_k has finite multiplicity. Since the kernel $\mathscr{G}(x, y)$ is real and Hermitian, the eigenfunctions $\{u_k\}$ may be chosen real and orthonormal (cf. Secs. 17.4 and 18.7).

The eigenfunction $u_k \in C^2(G) \cap C(\bar{G})$ is also the solution for $\lambda = \lambda_k$ of the boundary value problem (29) and of the integral equation (30). Therefore, by the theorem of Sec. 24.5, $u_k(x)$ has a correct normal derivative over S. From this, and from Green's formula (7) of Sec. 19.2, when $u = v = u_k$ it follows that

$$\lambda_k = \lambda_k(u_k, u_k) = -(\varDelta u_k, u_k) = \int_G |\operatorname{grad} u_k|^2 \, dx > 0$$

The fact that λ_1 is simple, together with the fact that $u_1(x)$ is positive in G, follow from Jentsch's theorem (cf. Sec. 18.7) since, by virtue of (4), the kernel $\mathscr{G}(x, y)$ is of positive type.

Let $f \in \mathscr{M}_\varDelta$. Then the function $f(x)$ is the (unique) solution of the boundary value problem

$$\varDelta f = -h, \quad f|_S = 0$$

where $h = -\varDelta f \in C(G) \cap \mathscr{L}_2(G)$. By the theorem of Sec. 24.3, the function $f(x)$ may be represented sourcewise by the kernel $\mathscr{G}(x, y)$ and so, according to the Hilbert–Schmidt theorem (cf. Sec. 18.6), it can be expanded into a regularly convergent Fourier series involving the eigenfunctions $\{u_k\}$. The theorem is proved.

* That is, $f \in C^2(G) \cap C^1(\bar{G})$, $\varDelta f \in \mathscr{L}_2(G)$, and $f|_S = 0$ (cf. Sec. 19.1).

In this way, Theorem 1 of Sec. 19.4 and its corollaries are true for the boundary value problem (29). Specifically, the system of eigenfunctions $\{u_k\}$ of this problem is complete in $\mathscr{L}_2(G)$.

Note. Using the note of Sec. 22.7, it is possible to prove that the eigenfunctions $u_k \in C^1(\bar{G})$; moreover $u_k \in C^\infty(G)$ (cf. Sec. 27).

7. Exercises. (a) Using Poisson's formula (19), prove Harnack's inequality

$$\frac{R(R-|x|)}{(R+|x|)^2} u(0) \leq u(x) \leq \frac{R(R+|x|)}{(R-|x|)^2} u(0), \qquad |x| < R$$

which is valid for any function $u(x) \geq 0$, which is harmonic in the sphere U_R and continuous on $\bar{U}_R$.

(b) Using Harnack's inequality, prove the theorem: Each increasing sequence of harmonic functions in the region G converges either to $+\infty$ (uniformly over each compactum $K \subset G$) or to a function harmonic in G.

(c) Prove the equation

$$\frac{1}{4\pi R} \int_{|y|=R} \frac{R^2-|x|^2}{|x-y|^3} dS_y = \begin{cases} 1, & |x| < R \\ -\dfrac{R}{|x|}, & |x| > R \end{cases}$$

(d) Using (c), prove that Poisson's integral

$$\frac{1}{4\pi R} \int_{|y|=R} \frac{|x|^2-R^2}{|x-y|^3} u_0(y)\, dS_y, \qquad |x| > R$$

gives the solution of the exterior Dirichlet problem for the sphere U_R.

(e) Show that the n-dimensional Poisson's integral

$$\frac{1}{\sigma_n R} \int_{|y|=R} \frac{R^2-|x|^2}{|x-y|^n} u_0(y)\, dS_y$$

gives the solution of the interior Dirichlet problem for the sphere $U_R \subset R^n$.

(f) Show that the solutions of the Dirichlet and Neumann problems for the half-space $x_3 > 0$ are given, respectively, by the equations

$$\frac{x_3}{2\pi} \int_{y_3=0} \frac{u_0(y)}{|x-y|^3} dS_y, \qquad \frac{1}{2\pi} \int_{y_3=0} \frac{u_1(y)}{|x-y|} dS_y$$

if

$$u_0(y_1, y_2) = O(|y|^{1-\varepsilon}), \qquad u_1(y_1, y_2) = O(|y|^{-1-\varepsilon})$$

as

$$|y| \to \infty; \qquad \varepsilon > 0.$$

§ 25. Spherical Functions

1. Definition of Spherical Functions. Each homogeneous harmonic polynomial of degree l considered over the unit spherical surface $S_1 \subset R^n$ is said to be a *spherical function* (*spherical harmonic*) of order $l = 0, 1, \ldots$. In this way, equation

$$Y_l(s) = u_l\left(\frac{x}{|x|}\right) = \frac{u_l(x)}{|x|^l}, \qquad s = \frac{x}{|x|} \tag{1}$$

establishes a one-to-one correspondence between the spherical functions $Y_l(s)$ for $s \in S_1$ of order l and the homogeneous harmonic polynomials $u_l(x)$ with $x \in R^n$.

The spherical functions Y_l and $Y_{l'}$ of different orders are orthogonal in $\mathscr{L}_2(S_1)$,

$$(Y_l, Y_{l'}) = \int_{S_1} Y_l(s) Y_{l'}(s)\, ds = 0, \qquad l \neq l' \tag{2}$$

In fact, if we apply Green's formula (8) of Sec. 19.2 for the sphere U_1 to the harmonic polynomials

$$u_l(x) = |x|^l \, Y_l\!\left(\frac{x}{|x|}\right), \qquad u_{l'}(x) = |x|^{l'} \, Y_{l'}\!\left(\frac{x}{|x|}\right)$$

we obtain

$$0 = \int_{S_1} \left[|x|^{l'} Y_{l'} \frac{\partial(|x|^l Y_l)}{\partial \mathbf{n}} - |x|^l Y_l \frac{\partial(|x|^{l'} Y_{l'})}{\partial \mathbf{n}} \right] ds$$

$$= (l - l') \int_{S_1} Y_l(s) Y_{l'}(s)\, ds$$

as was to be shown.

As an example we shall calculate all the spherical functions Y_l, $l = 0, 1, \ldots$, over the circumference S_1 ($n = 2$). It is convenient to do this in the polar coordinates (r, φ), $0 < r < \infty$, $0 \leq \varphi \leq 2\pi$. Applying Laplace's operator (cf. Sec. 3.2) to the harmonic polynomial

$$u_l(x) = r^l Y_l(\varphi) \tag{3}$$

for the spherical function Y_l we obtain the differential equation

$$Y_l'' + l^2 Y_l = 0$$

from which

$$Y_l(\varphi) = a_l \cos l\varphi + b_l \sin l\varphi, \qquad l = 0, 1, \ldots \qquad (4)$$

So spherical functions over the circumference are trigonometric functions. In this case, by virtue of (3) and (4), the formula

$$u_l(x) = r^l(a_l \cos l\varphi + b_l \sin l\varphi) = a_l \operatorname{Re} z^l + b_l \operatorname{Im} z^l, \qquad z = x_1 + ix_2$$

gives the general form of a homogeneous harmonic polynomial of order l in R^2.

Our problem is to calculate all the spherical functions Y_l, $l = 0, 1, \ldots$, over the spherical surface S_1 for $n = 3$.

2. The Differential Equation for Spherical Functions. We shall find the differential equation for spherical functions over the spherical surface S_1 for $n = 3$. It is convenient to do this in the spherical coordinates (r, θ, φ), $0 < r < \infty$, $0 \leq \theta \leq \pi$, $0 \leq \varphi \leq 2\pi$. Applying Laplace's operator (cf. Sec. 3.2) to the harmonic polynomial

$$u_l(x) = r^l Y_l(\theta, \varphi) \qquad (5)$$

for the spherical function Y_l we obtain the differential equation

$$\frac{1}{\sin \theta} \frac{\partial}{\partial \theta} \left(\sin \theta \frac{\partial Y_l}{\partial \theta} \right) + \frac{1}{\sin^2 \theta} \frac{\partial^2 Y_l}{\partial \varphi^2} + l(l+1) Y_l = 0 \qquad (6)$$

We shall seek the solution of Eq. (6) in the class of functions $C^\infty(S_1)$.

In order that the function Y_l be a spherical function of order l, it is necessary and sufficient that it belong to the class $C^\infty(S_1)$ and satisfy Eq. (6).

The necessity of these conditions has already been proved. We shall prove their sufficiency. Let the function $Y_l \in C^\infty(S_1)$ be the solution of Eq. (6). Then the function u_l, constructed according to formula (5), satisfies Laplace's equation in spherical coordinates and so is harmonic in $R^3 \setminus \{0\}$. Moreover, this function is bounded in the neighborhood of the point $x = 0$. According to the theorem concerning the removal of singularities of harmonic functions (cf. Sec. 21.6), the function u_l is harmonic in R^3. It is also homogeneous of degree l. From this, using the analogy of Liouville's theorem (cf. Sec. 21.9), we conclude that $u_l(x)$ is a homogeneous harmonic polynomial of order l. This also means that the function Y_l is a spherical function of order l.

§ 25. SPHERICAL FUNCTIONS

We shall find the solutions of Eq. (6) by separation of variables (cf. Sec. 19.5). According to the general scheme of this method, we seek the solution Y_l of Eq. (6) in the form of the product

$$Y_l(\theta, \varphi) = \mathscr{P}(\cos \theta)\Phi(\varphi) \tag{7}$$

Substituting this expression in Eq. (6), for the functions Φ and $\mathscr{P}$ we obtain the equations

$$\Phi'' + \nu\Phi = 0 \tag{8}$$

$$\frac{1}{\sin\theta} \frac{d}{d\theta}\left(\sin\theta \frac{d\mathscr{P}(\cos\theta)}{d\theta}\right) + \left[l(l+1) - \frac{\nu}{\sin^2\theta}\right]\mathscr{P}(\cos\theta) = 0 \tag{9}$$

where ν is an unknown parameter.

In order that the function (7) should be uniquely defined over the spherical surface S_1, it is necessary that Φ should be a 2π-periodic function. But Eq. (8) has such solutions only for $\nu = m^2$, $m = 0, 1, \ldots$, where

$$\Phi(\varphi) = e^{im\varphi} \tag{10}$$

In this way, the problem of finding spherical functions has been reduced to the solution of Eq. (9) for $\nu = m^2$, for $m = 0, 1, \ldots$. Making the change of variable $\mu = \cos\theta$ in this equation, for the function $\mathscr{P}(\mu)$ we obtain the equation

$$-[(1-\mu^2)\mathscr{P}']' + \frac{m^2}{1-\mu^2}\mathscr{P} = l(l+1)\mathscr{P} \tag{11}$$

The solutions of Eq. (11) at the points ± 1 must take finite values so that $|\mathscr{P}(\pm 1)| < \infty$.

3. Legendre Polynomials. For $m = 0$, Eq. (11) takes the form

$$[(1-\mu^2)\mathscr{P}']' + l(l+1)\mathscr{P} = 0 \tag{12}$$

We shall check that the polynomials

$$\mathscr{P}_l(\mu) = \frac{1}{2^l l!} \frac{d^l}{d\mu^l}(\mu^2 - 1)^l, \qquad l = 0, 1, \ldots \tag{13}$$

which are known as *Legendre polynomials*, satisfy Eq. (12).

In fact, putting $W_l = (\mu^2 - 1)^l$ and differentiating the identity

$$(\mu^2 - 1)W_l' - 2l\mu W_l = 0$$

$l + 1$ times, we obtain

$$(\mu^2 - 1)W_l^{(l+2)} + 2\mu W_l^{(l+1)} - l(l+1)W_l^{(l)} = 0$$

In this way, the function $W_l^{(l)}$, and therefore the polynomial $\mathscr{P}_l$, satisfy Eq. (12).

We shall display the first four Legendre polynomials:

$$\mathscr{P}_0(\mu) = 1, \quad \mathscr{P}_1(\mu) = \mu, \quad \mathscr{P}_2(\mu) = \tfrac{3}{2}\mu^2 - \tfrac{1}{2}, \quad \mathscr{P}_3(\mu) = \tfrac{5}{2}\mu^3 - \tfrac{3}{2}\mu$$

It follows directly from Eq. (13) that

$$\mathscr{P}_l(1) = 1 \tag{14}$$

The Legendre polynomial $\mathscr{P}_l$ is a unique linearly independent solution of Eq. (12) belonging to the class $C^2([-1, 1])$.

In fact, for each solution $\mathscr{P} \in C^2([-1, 1])$ of Eq. (12), the result

$$\mathscr{P}'(1) = \tfrac{1}{2}l(l+1)\mathscr{P}(1)$$

is valid. Therefore the Wronskian determinant for the solutions $\mathscr{P}_l$ and $\mathscr{P}$ becomes zero at the point $\mu = 1$, and so the solutions $\mathscr{P}_l$ and $\mathscr{P}$ are linearly dependent.

Legendre polynomials form an orthogonal system in $\mathscr{L}_2(-1, 1)$.

In fact, since the Legendre polynomial $\mathscr{P}_l(\mu)$ satisfies Eq. (11) for $m = 0$, then, by virtue of (10) and (7), $\mathscr{P}_l(\cos \theta) \in C^\infty(S_1)$ and satisfies Eq. (6). Therefore $\mathscr{P}_l(\cos \theta)$ is a spherical function of order l (cf. Sec. 25.2). However, spherical functions of different orders are orthogonal in $\mathscr{L}_2(S_1)$ (cf. Sec. 25.1). Therefore,

$$2\pi \int_{-1}^{1} \mathscr{P}_l(\mu) \mathscr{P}_{l'}(\mu) \, d\mu = \int_0^\pi \int_0^{2\pi} \mathscr{P}_l(\cos \theta) \mathscr{P}_{l'}(\cos \theta) \sin \theta \, d\theta \, d\varphi = 0, \quad l \neq l'$$

4. The Generating Function. Let $x = (r, \theta, \varphi)$ and $y = (0, 0, 1)$. We shall expand the function

$$\frac{1}{|x - y|} = \frac{1}{(1 - 2r \cos \theta + r^2)^{1/2}} = \frac{1}{[(1 - re^{i\theta})(1 - re^{-i\theta})]^{1/2}} \tag{15}$$

§ 25. SPHERICAL FUNCTIONS

into a series involving powers of r,

$$(1 - 2r\cos\theta + r^2)^{-1/2} = \sum_{l=0}^{\infty} a_l(\cos\theta) r^l \tag{16}$$

Series (16) converges for $|r| < 1$ and $\theta \in [0, \pi]$, and it can be differentiated term by term with respect to r and θ an infinite number of times, when the series obtained will converge uniformly with respect to (r, θ) in $[-r_0, r_0] \times [0, \pi]$ for any $r_0 < 1$. Applying Laplace's operator to successive terms of Eq. (16) and assuming that the function (15) is harmonic in the sphere $|x| < 1$, for all $r \in (0, 1)$ we obtain

$$0 = \sum_{l=0}^{\infty} \Delta[a_l(\cos\theta) r^l] = \sum_{l=0}^{\infty} r^{l-2} \left[\frac{1}{\sin\theta} \frac{d}{d\theta} \left(\sin\theta \frac{da_l}{d\theta} \right) + l(l+1) a_l \right]$$

It follows from this that each term in the last sum must be zero, and therefore the functions $\alpha_l(\mu)$ satisfy Eq. (12). Therefore $\alpha_l(\mu) = C_l \mathscr{P}_l(\mu)$ (cf. Sec. 25.3), and expansion (16) takes the form

$$(1 - 2r\cos\theta + r^2)^{-1/2} = \sum_{l=0}^{\infty} C_l \mathscr{P}_l(\cos\theta) r^l \tag{17}$$

To define the constants C_l we set $\theta = 0$ in (17) and use the equation $\mathscr{P}_l(1) = 1$. As a result we obtain

$$\frac{1}{1-r} = \sum_{l=0}^{\infty} r^l = \sum_{l=0}^{\infty} C_l r^l$$

from which it follows that $C_l = 1$.

So the expansion

$$(1 - 2r\mu + r^2)^{-1/2} = \sum_{l=0}^{\infty} \mathscr{P}_l(\mu) r^l, \quad |r| < 1 \tag{18}$$

must follow. The function $(1 - 2r\mu + r^2)^{-1/2}$ is known as the *generating function for Legendre polynomials*.

From formula (18) it is easy to obtain the recurrence relations between Legendre polynomials:

$$(l+1) \mathscr{P}_{l+1}(\mu) - (2l+1) \mu \mathscr{P}_l(\mu) + l \mathscr{P}_{l-1}(\mu) = 0 \tag{19}$$

$$(2l+1) \mathscr{P}_l(\mu) = \mathscr{P}'_{l+1}(\mu) - \mathscr{P}'_{l-1}(\mu) \tag{20}$$

For this, if we differentiate the identity (18) with respect to r and μ, and then multiply by $1 - 2r\mu + r^2$, we shall obtain the identities

$$(\mu - r) \sum_{l=0}^{\infty} \mathscr{P}_l(\mu) r^l = (1 - 2r\mu + r^2) \sum_{l=0}^{\infty} l \mathscr{P}_l'(\mu) r^{l-1}$$

$$r \sum_{l=0}^{\infty} \mathscr{P}_l(\mu) r^l = (1 - 2r\mu + r^2) \sum_{l=0}^{\infty} \mathscr{P}_l'(\mu) r^l$$

Comparing the coefficients for equal powers of r, we obtain the results (19) and

$$\mathscr{P}_l(\mu) = \mathscr{P}_{l-1}'(\mu) - 2\mu \mathscr{P}_l'(\mu) + \mathscr{P}_{l+1}'(\mu) \tag{21}$$

If we differentiate Eq. (19) we have

$$(l + 1)\mathscr{P}_{l+1}'(\mu) - (2l + 1)\mathscr{P}_l(\mu) - (2l + 1)\mu \mathscr{P}_l'(\mu) + l\mathscr{P}_{l-1}'(\mu) = 0$$

Excluding the product $\mu \mathscr{P}_l'(\mu)$ from this equation and from Eq. (21), we obtain Eq. (20).

We shall prove the result

$$\| \mathscr{P}_l \|^2 = \int_{-1}^{1} \mathscr{P}_l^2(\mu) \, d\mu = \frac{2}{2l + 1} \tag{22}$$

For this we express one of the factors $\mathscr{P}_l$ of the integrand in terms of $\mathscr{P}_{l-1}$ and $\mathscr{P}_{l-2}$ by using Eq. (19). Using the orthogonality of the polynomials $\mathscr{P}_l$ and $\mathscr{P}_{l-2}$, we obtain

$$\| \mathscr{P}_l \|^2 = \int_{-1}^{1} \mathscr{P}_l \mathscr{P}_l \, d\mu = \int_{-1}^{1} \mathscr{P}_l \left(\frac{2l - 1}{l} \mu \mathscr{P}_{l-1} - \frac{l - 1}{l} \mathscr{P}_{l-2} \right) d\mu$$

$$= \frac{2l - 1}{l} \int_{-1}^{1} \mu \mathscr{P}_l \mathscr{P}_{l-1} \, d\mu$$

Expressing the product $\mu \mathscr{P}_l$ in terms of Eq. (19) and using the orthogonality of the polynomials $\mathscr{P}_{l-1}$ and $\mathscr{P}_{l+1}$, we obtain

$$\| \mathscr{P}_l \|^2 = \frac{2l - 1}{l} \int_{-1}^{1} \mathscr{P}_{l-1} \left(\frac{l + 1}{2l + 1} \mathscr{P}_{l+1} + \frac{l}{2l + 1} \mathscr{P}_{l-1} \right) d\mu$$

$$= \frac{2l - 1}{2l + 1} \| \mathscr{P}_l \|^2$$

from which Eq. (22) follows

$$\| \mathscr{P}_l \|^2 = \frac{2l - 1}{2l + 1} \frac{2l - 3}{2l - 1} \cdots \frac{1}{3} \| \mathscr{P}_0 \|^2 = \frac{2}{2l + 1}$$

§ 25. SPHERICAL FUNCTIONS

The system of Legendre polynomials $\mathscr{P}_l$, $l = 0, 1, \ldots$, is complete in $\mathscr{L}_2(-1, 1)$.

This statement follows from the theorem of Sec. 1.7 and from Weierstrass's theorem (cf. Sec. 1.3), according to which a set of polynomials, and therefore a set of linear combinations of Legendre polynomials is dense in $C([-1, 1])$ and in $\mathscr{L}_2(-1, 1)$.

In this way, each function $f \in \mathscr{L}_2(-1, 1)$ can be expanded into a Fourier series involving Legendre polynomials

$$f(\mu) = \sum_{l=0}^{\infty} \frac{2l+1}{2} (f, \mathscr{P}_l) \mathscr{P}_l(\mu)$$

which is convergent in $\mathscr{L}_2(-1, 1)$ (cf. Sec. 1.7).

5. Associated Legendre Functions. We shall check that the functions

$$\mathscr{P}_l^m(\mu) = (1 - \mu^2)^{m/2} \mathscr{P}_l^{(m)}(\mu), \qquad l = 0, 1, \ldots, \qquad m = 0, 1, \ldots, l \quad (23)$$

which are known as *associated Legendre functions*, satisfy Eq. (11).

In fact, if in Eq. (11) we make the change of variable

$$\mathscr{P}(\mu) = (1 - \mu^2)^{m/2} z(\mu)$$

then for the function z we shall obtain the equation

$$(1 - \mu^2) z'' - 2\mu(m+1) z' + (l^2 + l - m^2 - m) z = 0 \quad (24)$$

On the other hand, if we differentiate Eq. (12) m times, we see that the derivative $\mathscr{P}_l^{(m)}$ satisfies Eq. (24). Consequently, the associated Legendre functions $\mathscr{P}_l^m$ satisfy Eq. (11).

Multiplying Eq. (24) by $(1 - \mu^2)^m$, we set $z = \mathscr{P}_l^{(m)}$ and write it in the form

$$[(1 - \mu^2)^{m+1} \mathscr{P}_l^{(m+1)}]' = -(l - m)(l + m + 1)(1 - \mu^2)^m \mathscr{P}_l^{(m)} \quad (25)$$

For each $m \geq 0$ the system of associated Legendre functions $\mathscr{P}_l^m$, $l = m, m+1, \ldots$, is orthogonal in $\mathscr{L}_2(-1, 1)$ and, moreover,

$$\|\mathscr{P}_l^m\|^2 = \frac{(l+m)!}{(l-m)!} \frac{2}{2l+1} \quad (26)$$

This statement is true when $m = 0$ because the Legendre polynomials $\mathscr{P}_l = \mathscr{P}_l^0$ (cf. Secs. 25.3 and 25.4). From this, if we use the definition of the functions $\mathscr{P}_l^m$, together with Eq. (25), replacing m by $m - 1$, we obtain

$$(\mathscr{P}_l^m, \mathscr{P}_{l'}^m) = \int_{-1}^{1} \mathscr{P}_l^m \mathscr{P}_{l'}^m \, d\mu = \int_{-1}^{1} (1 - \mu^2)^m \mathscr{P}_l^{(m)} \mathscr{P}_{l'}^{(m)} \, d\mu$$

$$= (1 - \mu^2)^m \mathscr{P}_l^{(m)} \mathscr{P}_{l'}^{(m-1)} \Big|_{-1}^{1} - \int_{-1}^{1} \mathscr{P}_{l'}^{(m-1)} [(1 - \mu^2)^m \mathscr{P}_l^{(m)}]' \, d\mu$$

$$= (l - m - 1)(l + m) \int_{-1}^{1} (1 - \mu^2)^{m-1} \mathscr{P}_l^{(m-1)} \mathscr{P}_{l'}^{(m-1)} \, d\mu$$

$$= (l + m)(l - m + 1)(\mathscr{P}_l^{m-1}, \mathscr{P}_{l'}^{m-1})$$

$$= (l + m)(l + m - 1)(l - m + 1)(l - m + 2)(\mathscr{P}_l^{m-2}, \mathscr{P}_{l'}^{m-2})$$

$$= \frac{(l + m)!}{(l - m)!} (\mathscr{P}_l, \mathscr{P}_{l'}) = \frac{(l + m)!}{(l - m)!} \frac{2}{2l + 1} \delta_{ll'} \qquad (27)$$

as was to be shown.

For each $m \geq 0$ the system of associated Legendre functions $\mathscr{P}_l^m$, $l = m, m + 1, \ldots,$ is complete in $\mathscr{L}_2(-1, 1)$.

In fact, let us take an arbitrary function f belonging to the class $\mathscr{D}(-1, 1)$ which is dense in $\mathscr{L}_2(-1, 1)$ (cf. Sec. 1.5). Then

$$\psi(\mu) = f(\mu)(1 - \mu^2)^{-m/2} \in \mathscr{D}(-1, 1)$$

According to Weierstrass's theorem (cf. Sec. 1.3), the function ψ may be approximated with arbitrary accuracy in $C([-1, 1])$ by polynomials, and therefore also by linear combinations of the derivatives $\mathscr{P}_l^{(m)}$, $l = m$, $m + 1, \ldots$. It follows that the function f may be approximated with arbitrary accuracy in $\mathscr{L}_2(-1, 1)$ by linear combinations of functions of the system $\mathscr{P}_l^m$, $l = m, m + 1, \ldots$, which, by virtue of the theorem of Sec. 1.7, also proves the completeness of this system.

6. Spherical Functions. By virtue of (7), (10), and (23), the following set of solutions of Eq. (6) has been obtained:

$$Y_l^m(\theta, \varphi) = \begin{cases} \mathscr{P}_l^m(\cos \theta) \cos m\varphi, & m = 0, 1, \ldots, l \\ \mathscr{P}_l^{|m|}(\cos \theta) \sin |m| \varphi, & m = -1, -2, \ldots, -l, \\ & l = 0, 1, \ldots \end{cases} \qquad (28)$$

§ 25. SPHERICAL FUNCTIONS

It is clear that these functions belong to the class $C^\infty(S_1)$. Therefore $Y_l^m(\theta, \varphi)$ are *spherical functions* (cf. Sec. 25.2).

The spherical functions Y_l^m, $m = 0 \pm 1, \ldots, \pm l$ of order l are linearly independent, and their linear combinations

$$Y_l(s) = \sum_{m=-l}^{l} a_l^{(m)} Y_l^m(s) \tag{29}$$

with arbitrary coefficients $a_l^{(m)}$, are also spherical functions of order l.

The spherical functions $\{Y_l^m\}$ *form an orthogonal and complete system in* $\mathscr{L}_2(S_1)$, *and*

$$\| Y_l^m \|^2 = 2\pi \frac{1 + \delta_{0m}}{2l + 1} \frac{(l + |m|)!}{(l - |m|)!} \tag{30}$$

In fact, the trigonometric system $\{e^{im\varphi}, m = 0, 1, \ldots\}$ is orthogonal and complete in $\mathscr{L}_2(0, 2\pi)$, and for each $m = 0, 1, \ldots$, the system of associated Legendre functions $\{\mathscr{P}_l^m(\mu), l = m, m + 1, \ldots\}$ is orthogonal and complete in $\mathscr{L}_2(-1, 1)$ (cf. Sec. 25.5). Therefore, by the lemma of Sec. 1.7, the system of functions

$$\{\mathscr{P}_l^m(\mu) \cos m\varphi, \mathscr{P}_l^{|m|}(\mu) \sin |m| \varphi, \quad l = 0, 1, \ldots, \quad m = 0, 1, \ldots, l\}$$

is orthogonal and complete in $\mathscr{L}_2[(-1, 1) \times (0, 2\pi)]$, and, consequently, the system of spherical functions $\{Y_l^m(\theta, \varphi)\}$ is orthogonal and complete in $\mathscr{L}_2(S_1)$. Formula (30) follows from (26):

$$\| Y_l^m \|^2 = \int_0^\pi \int_0^{2\pi} [Y_l^m(\theta, \varphi)]^2 \sin \theta \, d\theta \, d\varphi$$

$$= \int_{-1}^1 [\mathscr{P}_l^{|m|}(\mu)]^2 \, d\mu \int_0^{2\pi} \begin{Bmatrix} \cos^2 m\varphi \\ \sin^2 m\varphi \end{Bmatrix} d\varphi = 2\pi \frac{1 + \delta_{0m}}{2l + 1} \frac{(l + |m|)!}{(l - |m|)!}$$

The completeness of the orthogonal system of spherical functions $\{Y_l^m\}$ means that each function $f \in \mathscr{L}_2(S_1)$ can be expanded into a Fourier series in terms of these functions

$$f(s) = \sum_{l=0}^{\infty} \sum_{m=-l}^{l} a_l^{(m)} Y_l^m(s) = \sum_{l=0}^{\infty} Y_l(s) \tag{31}$$

which is convergent in $\mathscr{L}_2(S_1)$. In accordance with (30), the coefficients

$a_l^{(m)}$ of the series (31) are given by the formula

$$a_l^{(m)} = \frac{2l+1}{2\pi(1+\delta_{0m})} \frac{(l-|m|)!}{(l+|m|)!} \int_0^\pi \int_0^{2\pi} f(\theta,\varphi) Y_l^m(\theta,\varphi) \sin\theta \, d\theta \, d\varphi \qquad (32)$$

Let $Q_l(s)$ be an arbitrary spherical function of order l. Then $(Q_l, Y_{l'}) = 0$ for $l \neq l'$ (cf. Sec. 25.1) and in expansion (31) for the function Q_l there remains only one term Y_l, so that $Q_l = Y_l$. Thus we have proved that:

The spherical functions $\{Y_l^m\}$ exhaust all linearly independent spherical functions; Eq. (29) gives the general expression for a spherical function of order l.

Note. The spherical functions Y_l^m, $m = 0, \pm 1, \ldots, \pm l$, are eigenfunctions of Beltrami's operator

$$-\frac{1}{\sin\theta} \frac{\partial}{\partial\theta}\left(\sin\theta \frac{\partial Y}{\partial\theta}\right) - \frac{1}{\sin^2\theta} \frac{\partial^2 Y}{\partial\varphi^2}, \qquad Y \in C^\infty(S_1)$$

corresponding to the eigenvalue $\lambda = l(l+1)$ of multiplicity $2l+1$.

7. Laplace's Formula. Let $Y_l(s)$ be a spherical function of order l. If we apply Green's formula (13) of Sec. 23.4 for the sphere U_1 to the harmonic polynomial $r^l Y_l(s)$, then for $r < 1$ we obtain

$$r^l Y_l(s) = \frac{1}{4\pi} \int_{S_1} \left\{ \frac{\partial [|s'|^l Y_l(s')]}{\partial \mathbf{n}_{s'}} \frac{1}{|x-s'|} - Y_l(s') \frac{\partial}{\partial \mathbf{n}_{s'}} \frac{1}{|x-s'|} \right\} ds' \qquad (33)$$

But, by virtue of (18),

$$\frac{1}{|x-s'|}\bigg|_{S_1} = \frac{1}{[1-2r(s,s')+r^2]^{1/2}} = \sum_{k=0}^{\infty} \mathscr{P}_k((s,s')) r^k \qquad (34)$$

$$\frac{\partial}{\partial \mathbf{n}_{s'}} \frac{1}{|x-s'|}\bigg|_{S_1} = \frac{\partial}{\partial \varrho} \frac{1}{[\varrho^2 - 2r\varrho(s,s')+r^2]^{1/2}}\bigg|_{\varrho=1}$$

$$= \frac{\partial}{\partial \varrho} \sum_{k=0}^{\infty} \mathscr{P}_k((s,s')) \frac{r^k}{\varrho^{k+1}}\bigg|_{\varrho=1}$$

$$= -\sum_{k=0}^{\infty} (k+1) \mathscr{P}_k((s,s')) r^k \qquad (35)$$

§ 25. SPHERICAL FUNCTIONS

and the series (34) and (35) converge uniformly with respect to (s, s') for each $r < 1$ (cf. Sec. 25.4). Substituting Eqs. (34) and (35) in Eq. (33) and integrating term by term we obtain

$$r^l Y_l(s) = \frac{1}{4\pi} \int_{S_1} \left[l Y_l(s') \sum_{k=0}^{\infty} \mathscr{P}_k((s, s')) r^k \right. $$
$$\left. + Y_l(s') \sum_{k=0}^{\infty} (k+1) \mathscr{P}_k((s, s')) r^k \right] ds'$$
$$= \frac{1}{4\pi} \sum_{k=0}^{\infty} r^k \int_{S_1} (l + k + 1) Y_l(s') \mathscr{P}_k((s, s')) \, ds', \quad r < 1$$

From which, since r is arbitrary, we find the following important integral equation for spherical functions*

$$\int_{S_1} Y_l(s') \mathscr{P}_k((s, s')) \, ds' = \frac{4\pi}{2l+1} Y_l(s) \, \delta_{lk} \tag{36}$$

We shall replace s by s' in Eq. (31). If we multiply this equation by $\mathscr{P}_k((s, s'))$, integrate it term by term with respect to $s' \in S_1$, and use Eq. (36), we obtain the equation

$$Y_k(s) = \frac{2k+1}{4\pi} \int_{S_1} f(s') \mathscr{P}_k((s, s')) \, ds' \tag{37}$$

Equation (37) at once gives all the coefficients in the spherical function Y_k that occur in the expansion (31) of the arbitrary function $f \in \mathscr{L}_2(S)$. It is known as *Laplace's formula*.

8. Separation of Variables in Laplace's Equation. We shall construct solutions of Laplace's equation $\Delta u = 0$ in R^3 by separation of variables using the spherical coordinates (r, θ, φ). In these coordinates Laplace's equation has the form (cf. Sec. 3.2)

$$\Delta u = \frac{1}{r^2} \frac{\partial}{\partial r} \left(r^2 \frac{\partial \tilde{u}}{\partial r} \right) + \frac{1}{r^2 \sin \theta} \frac{\partial}{\partial \theta} \left(\sin \theta \frac{\partial \tilde{u}}{\partial \theta} \right) + \frac{1}{r^2 \sin^2 \theta} \frac{\partial^2 \tilde{u}}{\partial \varphi^2} = 0 \tag{38}$$

---
* An equivalent formula for the addition of Legendre polynomials has the form:
$$P_l((s, s')) = \sum_{m=-l}^{l} \frac{2}{(1+\delta_{0m})} \frac{(l-|m|)!}{(l+|m|)!} Y_l^m(s) Y_l^m(s')$$

where

$$\tilde{u}(r, \theta, \varphi) = u(r \sin \theta \cos \varphi, r \sin \theta \sin \varphi, r \cos \theta)$$

In accordance with the general scheme of the method of separation of variables (cf. Sec. 19.5), we seek the solution $\tilde{u}$ of Eq. (38) in the form of the product

$$\tilde{u}(r, \theta, \varphi) = \mathscr{R}(r) Y(\theta, \varphi) \tag{39}$$

Substituting this expression in Eq. (38), for the functions $\mathscr{R}$ and Y we obtain the equations

$$(r^2 \mathscr{R}')' - \mu \mathscr{R} = 0 \tag{40}$$

$$\frac{1}{\sin \theta} \frac{\partial}{\partial \theta} \left(\sin \theta \frac{\partial Y}{\partial \theta} \right) + \frac{1}{\sin^2 \theta} \frac{\partial^2 Y}{\partial \varphi^2} + \mu Y = 0 \tag{41}$$

where μ is an unknown parameter. Here $Y \in C^\infty(S_1)$.

When $\mu = l(l+1)$, $l = 0, 1, \ldots$, Eq. (41) has solutions of the class $C^\infty(S_1)$ and these solutions are the spherical functions Y_l^m, $m = 0, \pm 1, \ldots, \pm l$ (cf. Sec. 25.6). Equation (40) for $\lambda = l(l+1)$ has two linearly independent solutions, r^l and r^{-l-1}.

So, by virtue of (39), Laplace's equation has the following set of linearly independent solutions:

$$r^l Y_l(\theta, \varphi), \qquad r^{-l-1} Y_l(\theta, \varphi), \qquad l = 0, 1, \ldots \tag{42}$$

where $r^l Y_l$ is a harmonic polynomial of degree l and $r^{-l-1} Y_l$ is a harmonic function in $R^3 \setminus \{0\}$.

Example. We shall find the eigenvalues and the eigenfunctions of the boundary value problem

$$-\Delta u = \lambda u, \qquad u|_{S_R} = 0 \tag{43}$$

for the sphere U_R. The eigenfunctions of this problem in spherical coordinates will be sought in the form of the product (39). If we separate the variables, for the function $\mathscr{R}$ we obtain the boundary value problem

$$(r^2 \mathscr{R}')' + (\lambda r^2 - \mu) \mathscr{R} = 0, \qquad |\mathscr{R}(0)| < \infty, \qquad \mathscr{R}(R) = 0 \tag{44}$$

and for the function Y we obtain Eq. (41). When $\mu = l(l+1)$, $l = 0, 1, \ldots$, Eq. (41) has the solutions Y_l^m, $m = 0, \pm 1, \ldots, \pm l$ (cf. Sec. 25.6).

For $\mu = l(l+1)$, the solution of Eq. (44) which is bounded at the origin is the function

$$\mathcal{R}(r) = \frac{1}{\sqrt{r}} J_{l+1/2}(\sqrt{\lambda} r) \qquad (45)$$

To satisfy the boundary condition on the end $r = R$, it is necessary to put $\sqrt{\lambda} R = \mu_{lj}$ in (45), where μ_{lj} are the positive roots of the Bessel function $J_{l+1/2}$. So,

$$\lambda_{lj} = \frac{\mu_{lj}^2}{R^2}, \qquad u_{lmj}(x) = c_{lmj} \frac{1}{\sqrt{r}} J_{l+1/2}\left(\mu_{lj} \frac{r}{R}\right) Y_l^m(\theta, \varphi) \qquad (46)$$

$$l = 0, 1, \ldots, \qquad m = 0, \pm 1, \ldots, \pm l, \qquad j = 1, 2, \ldots$$

where, by virtue of (30),

$$c_{lmj} = \left\{ \int_0^R \int_0^\pi \int_0^{2\pi} J_{l+1/2}^2\left(\mu_{lj} \frac{r}{R}\right) [Y_l^m(\theta, \varphi)]^2 r \, dr \, d\theta \, d\varphi \right\}^{-1/2}$$

$$= \frac{1}{R | J'_{l+1/2}(\mu_{lj}) | } \left[\pi \frac{1 + \delta_{0m}}{2l + 1} \frac{(l + |m|)!}{(l - |m|)!} \right]^{1/2}$$

are the eigenvalues and eigenfunctions of the boundary value problem. As in Sec. 19.6, it can be established that the system of eigenfunctions (46) is complete in $\mathscr{L}_2(U_R)$, and therefore problem (43) has no other eigenvalues and eigenfunctions.

9. Solution of the Dirichlet and Neumann Problems for a Sphere. We shall use spherical functions to construct a solution of the Dirichlet and Neumann problems (interior and exterior) for the sphere U_R (cf. Sec. 23.1).

Let f be a given continuous function over the spherical surface S_R. Then $f(Rs)$ can be expanded into a Fourier series in terms of the spherical functions

$$f(Rs) = \sum_{l=0}^\infty Y_l(s), \qquad s \in S_1 \qquad (47)$$

where, by virtue of (37),

$$Y_l(s) = \frac{2l + 1}{4\pi} \int_{S_1} f(Rs') \, \mathscr{P}_l((s, s')) \, ds'$$

Series (47) converges in $\mathscr{L}_2(S_R)$ (cf. Sec. 25.6). We shall suppose that this series converges in $C(S_R)$.

Then
$$u(r, \theta, \varphi) = \sum_{l=0}^{\infty} \left(\frac{r}{R}\right)^l Y_l(\theta, \varphi), \qquad r < R \tag{48}$$

is the solution of the interior Dirichlet problem with $u_0^- = f$;

$$u(r, \theta, \varphi) = \sum_{l=1}^{\infty} \frac{R}{l} \left(\frac{r}{R}\right)^l Y_l(\theta, \varphi) + C, \qquad r < R \tag{49}$$

is the solution of the interior Neumann problem with $u_1^- = f$ subject to the condition that

$$Y_0 = \frac{1}{4\pi} \int_{S_1} f(Rs') \, ds' = \frac{1}{4\pi R^2} \int_{S_R} f(x) \, dS = 0 \tag{50}$$

$$u(r, \theta, \varphi) = \sum_{l=0}^{\infty} \left(\frac{R}{r}\right)^{l+1} Y_l(\theta, \varphi), \qquad r > R \tag{51}$$

is the solution of the exterior Dirichlet problem with $u_0^+ = f$;

$$u(r, \theta, \varphi) = -\sum_{l=0}^{\infty} \frac{R}{l+1} \left(\frac{R}{r}\right)^{l+1} Y_l(\theta, \varphi), \qquad r > R \tag{52}$$

is the solution of the exterior Neumann problem with $u_1^+ = f$.

In fact, the series (48) consists of harmonic polynomials and by supposition converges in $C(S_R)$. Therefore this series converges in $C(\bar{U}_R)$ (cf. Sec. 21.5) and defines a function u which is harmonic in U_R [cf. Sec. 21.8, (a)], continuous on $\bar{U}_R$, and takes, by virtue of (47), the value f over S_R. This means that series (48) gives the solution of the interior Dirichlet problem for the sphere U_R with $u_0^- = f$.

On the basis of Abel's test the series

$$\sum_{l=1}^{\infty} \frac{1}{l} Y_l(s)$$

converges together with the series (47) in $C(S_R)$. From this, repeating our previous arguments, we conclude that series (49) converges in $C(\bar{U}_R)$ and defines a function u which is harmonic in U_R and continuous on $\bar{U}_R$. Moreover, this series can be differentiated term by term with respect to r

$$\frac{\partial u(r, \theta, \varphi)}{\partial r} = \sum_{l=1}^{\infty} \left(\frac{r}{R}\right)^{l-1} Y_l(\theta, \varphi) \tag{53}$$

since the series (53), by virtue of Abel's test, converges in $C(\bar{U}_R)$. Finally, in accordance with (47), the sum of the series (53) over S_R coincides with f whenever the function f satisfies condition (50) for the solubility of the interior Neumann problem (cf. Sec. 23.5). This means that series (49) gives the solution of the interior Neumann problem for the sphere U_R with $u_1^- = f$ whenever the condition for solubility (50) is satisfied.

It can be proved analogously that series (51) and (52) give the solutions of the corresponding exterior boundary value problems.

§ 26. Boundary Value Problems for Laplace's Equation in a Plane

For a point (x, y) of the plane R^2 it is convenient to use the notation $z = x + iy$ or $\bar{z} = x - iy$. We shall consider that G is a bounded region in R^2 with a piecewise smooth boundary S.

The majority of results obtained in Secs. 22–24 for boundary value problems of three variables may be carried over to two-dimensional boundary value problems with a change of the fundamental solution $\mathscr{E}_3(x) = -(1/4\pi |x|)$ to the fundamental solution $\mathscr{E}_2(z) = (1/2\pi) \ln |z|$. There are, however, some differences in formulating and solving these problems which concern the peculiarity of the behavior of the fundamental solution $\mathscr{E}_2$ at infinity.

1. Behavior of a Harmonic Function at Infinity. As in Sec. 23.2, the inversion mapping with respect to the circumference S_R can be expressed by means of the equation

$$z^* = \frac{R^2}{|z|^2} z = \frac{R^2}{\bar{z}} \tag{1}$$

Let the function $u(x, y) \equiv u(z)$ be harmonic outside the circle U_R. Its Kelvin transformation (cf. Sec. 23.2) is defined by the formula

$$u^*(z^*) = u\left(\frac{R^2}{|z^*|^2} z^*\right) = u\left(\frac{R^2}{\bar{z}^*}\right) \tag{2}$$

The Kelvin transformation retains the harmonic property; that is, the function $u^*(z^*)$ is harmonic in $U_R \setminus \{0\}$.

In fact, in polar coordinates (cf. Sec. 3.2) the Kelvin transformation (2) takes the form

$$u^*(z^*) = \tilde{u}^*(\varrho, \varphi) = \tilde{u}\left(\frac{R^2}{\varrho}, \varphi\right)$$

and therefore

$$\Delta u^*(z^*) = \frac{1}{\varrho} \frac{\partial}{\partial \varrho}\left(\varrho \frac{\partial \tilde{u}^*}{\partial \varrho}\right) + \frac{1}{\varrho^2} \frac{\partial^2 \tilde{u}^*}{\partial \varphi^2}$$

$$= \frac{r^4}{R^4}\left[\frac{1}{r}\frac{\partial}{\partial r}\left(r\frac{\partial \tilde{u}}{\partial r}\right) + \frac{1}{r^2}\frac{\partial^2 \tilde{u}}{\partial \varphi^2}\right] = \frac{r^4}{R^4}\Delta u(z)$$

from which the required statement follows.

THEOREM. *Let the function $u(z)$ be harmonic in the region $G_1 = R^2 \setminus \bar{G}$, continuous on $\bar{G}_1$ and bounded. Then the following statements are valid:*

$$|u(z)| \leq \max_{z \in S} |u(z)|, \qquad z \in \bar{G}_1 \tag{3}$$

$$\lim u(z) = \alpha, \qquad |z| \to \infty \tag{4}$$

$$|\operatorname{grad} u(z)| = O\left(\frac{1}{|z|^2}\right), \qquad |z| \to \infty \tag{5}$$

Proof. Let the circle U_R lie outside the region G_1 and let $G_1^* \setminus \{0\}$ and S^* be the images of G_1 and S resulting from the inversion mapping (1) (Fig. 72). It is clear that S^* is the boundary of the region G_1^*. For this the function $u(z)$, in accordance with the Kelvin transformation (2), maps into

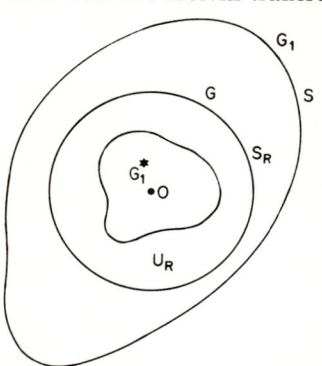

Fig. 72

the function $u^*(z^*)$ which is harmonic in $G_1^* \setminus \{0\}$, continuous in $\bar{G}_1^* \setminus \{0\}$, and bounded in the neighborhood of the point 0. According to the theorem concerning the removal of singularities of the harmonic functions (cf. Sec. 21.6), we conclude that the function $u^*(z^*)$ is harmonic in the region G_1^* and continuous on $\bar{G}_1^*$. Since G_1^* is a bounded region (it lies in the circle U_R), then by the maximum principle (cf. Sec. 21.5)

$$|u^*(z^*)| \leq \max_{z^* \in S^*} |u^*(z^*)| \tag{6}$$

Inverting the Kelvin transformation we obtain

$$u(z) = u^*\left(\frac{R^2}{|z|^2} z\right)$$

from which, and from (6), all the statements of the theorem follow.

2. Formulation and Uniqueness of the Solutions of the Basic Boundary Value Problems. The basic boundary value problems for Laplace's equation in a plane are formulated in the same way as the corresponding problems in a space (cf. Sec. 23.1), except that for exterior problems the solution need only be bounded as $|z| \to \infty$ (and not necessarily tending to zero). We assume that $G_1 = R^2 \setminus \bar{G}$ is a region.

A Liapunov line and a sufficiently smooth line are defined as for a space (cf. Secs. 22.4 and 23.3); in this case condition (A) has the form:

$$\int_S \frac{|\cos \varphi_{z\zeta}|}{|z - \zeta|} dS_\zeta \leq K, \qquad z \in R^2 \tag{A}$$

The following uniqueness theorems are valid for basic boundary value problems for Laplace's equation.

The solution of the interior or exterior Dirichlet problem is unique and depends continuously on the boundary data u_0^- and u_0^+, respectively.

If S is a sufficiently smooth line, then the solution of the interior or exterior Neumann problem is defined uniquely as far as an arbitrary additive constant, and

$$\int_S u_1^-(z) \, dS = 0 \quad \text{or} \quad \int_S u_1^+(z) \, dS = 0 \tag{7}$$

is a necessary condition for the solubility of the corresponding problem.

The proof of these statements resembles the proofs of Theorems 2–4 of Sec. 23.3, and the theorem of Sec. 26.1 is also used. The necessary solubility condition (7) for the exterior Neumann problem is the main difference. We shall prove the need for this condition. Let $u(z)$ be the solution of the exterior Neumann problem subject to the boundary condition u_1^+ over S. Since $u \in C(\bar{G}_1)$ is harmonic in G_1 and has a correct normal derivative over S equal to $-u_1^+$, then, applying Eq. (8) of Sec. 19.2 to the region Q_R when $v = 1$ (Fig. 54), we obtain

$$-\int_S u_1^+(z)\,dS + \int_{S_R} \frac{\partial u}{\partial \mathbf{n}}\,dS = 0$$

Proceeding to the limit as $R \to \infty$ in this equation and using Eq. (5), we obtain condition (7).

3. Logarithmic Potential. A logarithmic potential is defined as the convolution of a generalized function ϱ with the function $-\ln |z|$ (cf. Sec. 7.8):

$$V = -\ln |z| * \varrho = -2\pi \mathscr{E}_2 * \varrho \tag{8}$$

The logarithmic potential V satisfies Poisson's equation

$$\Delta V = -2\pi \varrho \tag{9}$$

Specific instances of logarithmic potentials are: the potential

$$V(z) = \int_G \varrho(\zeta) \ln \frac{1}{|z-\zeta|}\,d\xi\,d\eta, \qquad \zeta = \xi + i\eta \tag{10}$$

the simple layer potential

$$V^{(0)}(z) = \ln \frac{1}{|z|} * \mu \delta_S = \int_S \mu(\zeta) \ln \frac{1}{|z-\zeta|}\,dS_\zeta \tag{11}$$

and the double layer potential

$$V^{(1)}(z) = -\ln \frac{1}{|z|} * \frac{\partial}{\partial \mathbf{n}} (\nu \delta_S)$$

$$= \int_S \nu(\zeta) \frac{\partial}{\partial \mathbf{n}} \ln \frac{1}{|z-\zeta|}\,dS_\zeta = \int_S \nu(\zeta) \frac{\cos \varphi_{z\zeta}}{|z-\zeta|}\,dS_\zeta \tag{12}$$

These potentials have the following properties:

§ 26. LAPLACE'S EQUATION IN A PLANE

If $\varrho \in C(\bar{G})$, then the potential $V \in C^1(R^2)$ is harmonic in G_1 and when $|z| \to \infty$

$$V(z) = \ln\frac{1}{|z|} \int_G \varrho(\zeta)\, d\xi\, d\eta + O\left(\frac{1}{|z|}\right) \tag{13}$$

If, moreover, $\varrho \in C^1(G)$, then $V \in C^2(G)$.

If $\mu \in C(S)$, then the simple layer potential $V^{(0)} \in C(R^2)$, is harmonic outside S and as $|z| \to \infty$

$$V^{(0)}(z) = \ln\frac{1}{|z|} \int_S \mu(\zeta)\, dS + O\left(\frac{1}{|z|}\right) \tag{14}$$

If a Liapunov line S satisfies condition (A), then the potential $V^{(0)}(z)$ has the correct normal derivatives $(\partial V^{(0)}/\partial \mathbf{n})_+$ and $(\partial V^{(0)}/\partial \mathbf{n})_-$ over S from outside and inside S, respectively, and

$$\left(\frac{\partial V^{(0)}}{\partial \mathbf{n}}\right)_+(z) = -\pi\mu(z) + \frac{\partial V^{(0)}(z)}{\partial \mathbf{n}} = -\pi\mu(z) + \int_S \mu(\zeta) \frac{\cos \psi_{z\zeta}}{|z-\zeta|}\, dS_\zeta \tag{15}$$

$$\left(\frac{\partial V^{(0)}}{\partial \mathbf{n}}\right)_-(z) = \pi\mu(z) + \frac{\partial V^{(0)}(z)}{\partial \mathbf{n}} = \pi\mu(z) + \int_S \mu(\zeta) \frac{\cos \psi_{z\zeta}}{|z-\zeta|}\, dS_\zeta \tag{15'}$$

If $\nu \in C(S)$, then the double layer potential $V^{(1)}$ is harmonic outside S and

$$V^{(1)}(z) = O\left(\frac{1}{|z|}\right), \qquad |z| \to \infty \tag{16}$$

If S is a Liapunov line, then $V^{(1)} \in C(S)$ and

$$\int_S \frac{\cos \varphi_{z\zeta}}{|z-\zeta|}\, dS_\zeta = \begin{cases} -2\pi, & x \in G \\ -\pi, & x \in S \\ 0, & x \in G_1 \end{cases} \tag{17}$$

If a Liapunov line S satisfies condition (A), then the potential $V^{(1)}(z)$ belongs to the classes $C(\bar{G})$ and $C(\bar{G}_1)$ and its limiting values $V^{(1)}_+$ and $V^{(1)}_-$ over S from outside and inside S, respectively, are expressed by the equations

$$V^{(1)}_+(z) = \pi\nu(z) + V^{(1)}(z) = \pi\nu(z) + \int_S \nu(\zeta) \frac{\cos \varphi_{z\zeta}}{|z-\zeta|}\, dS_\zeta \tag{18}$$

$$V^{(1)}_-(z) = -\pi\nu(z) + V^{(1)}(z) = -\pi\nu(z) + \int_S \nu(\zeta) \frac{\cos \varphi_{z\zeta}}{|z-\zeta|}\, dS_\zeta \tag{18'}$$

Proof of these properties is analogous to the corresponding proof in a three-dimensional case (cf. Sec. 22). There is a difference only in proving Eqs. (13) and (14). We shall prove Eq. (13); Eq. (14) is proved analogously. Using (10), we have

$$V(z) - \ln \frac{1}{|z|} \int_G \varrho(\zeta) \, d\xi \, d\eta = \int_G \varrho(\zeta) \ln \frac{|z|}{|z - \zeta|} \, d\xi \, d\eta \quad (19)$$

Let $\bar{G}$ lie in the sphere U_R and let $|z| > 2R$. Then for all $\zeta \in \bar{G}$ we have the inequalities:

$$|z| - R \leq |z - \zeta| \leq |z| + R$$

$$-\frac{2R}{|z|} \leq \ln \frac{|z|}{|z| + R} \leq \ln \frac{|z|}{|z - \zeta|} \leq \ln \frac{|z|}{|z| - R} \leq \frac{2R}{|z|}$$

from which, and from (19), Eq. (13) follows:

$$\left| V(z) - \ln \frac{1}{|z|} \int_G \varrho(\zeta) \, d\xi \, d\eta \right| \leq \int_G |\varrho(\zeta)| \left| \ln \frac{|z|}{|z - \zeta|} \right| d\xi \, d\eta$$

$$\leq \frac{2R}{|z|} \int_G |\varrho(\zeta)| \, d\xi \, d\eta = \frac{C}{|z|}$$

PHYSICAL SENSE OF THE FUNDAMENTAL SOLUTION $\mathscr{E}_2(z)$. The electrostatic potential created by charges lying over the segment $|x_3| \leq N$ of the x_3 axis with linear density $-(1/4\pi)$, that is, by the distribution

$$\varrho_N(z, x_3) = -\frac{1}{4\pi} \delta(z) \cdot \theta(N - |x_3|)$$

is equal over the plane $x_3 = 0$ to*

$$V_N(z, 0) = -\frac{1}{\sqrt{|z|^2 + x_3^2}} * \frac{\delta(z)}{4\pi} \cdot \theta(N - |x_3|) + \frac{1}{2\pi} \ln(2N)$$

$$= -\frac{1}{4\pi} \int_{-N}^{N} \frac{dx_3}{\sqrt{|z|^2 + x_3^2}} + \frac{1}{2\pi} \ln(2N)$$

$$= -\frac{1}{2\pi} \ln(x_3 + \sqrt{|z|^2 + x_3^2}) \Big|_0^N + \frac{1}{2\pi} \ln(2N)$$

$$= \frac{1}{2\pi} \ln|z| + \frac{1}{2\pi} \ln \frac{2N}{N + \sqrt{|z|^2 + N^2}}$$

* We add to the convolution an appropriate constant $(1/2\pi) \ln(2N)$.

§ 26. LAPLACE'S EQUATION IN A PLANE

Proceeding to the limit as $N \to \infty$ in this equation, we obtain

$$\lim_{N \to \infty} V_N(z, 0) = \frac{1}{2\pi} \ln |z| = \mathscr{E}_2(z)$$

In this way, the fundamental solution $\mathscr{E}_2(z)$ is seen to be the electrostatic potential created by the charges which lie over the x_3 axis with a linear density $-(1/4\pi)$; that is, by the distribution $\varrho(z, x_3) = -(1/4\pi)\delta(z) \cdot 1(x_3)$ (cf. Sec. 10.4).

4. Solubility of the Boundary Value Problems Formulated. We shall suppose that the boundary S of the region G is a sufficiently smooth line and that $G_1 = R^2 \setminus \bar{G}$ is a region.

As in Sec. 23.4, the solution of the interior Dirichlet problem will be sought in the form of a double layer potential

$$V^{(1)}(z) = \int_S \nu(\zeta) \frac{\cos \varphi_{z\zeta}}{|z - \zeta|} dS_\zeta, \qquad \nu \in C(S) \qquad (20)$$

the solution of the exterior Dirichlet problem will be sought in the form of the sum of a double layer potential $V^{(1)}$ and an unknown constant α; the solution of the Neumann problems (interior or exterior) will be sought in the form of a simple layer potential

$$V^{(0)}(z) = \int_S \mu(\zeta) \ln \frac{1}{|z - \zeta|} dS_\zeta, \qquad \mu \in C(S) \qquad (21)$$

For the unknown densities ν and μ and the number α, by virtue of formulas (15), (15'), (18), and (18'), we obtain the integral equations

$$\nu(z) = \lambda \int_S \mathscr{K}(z, \zeta)\nu(\zeta) dS_\zeta + f(z), \qquad z \in S \qquad (22)$$

$$\mu(z) = \lambda \int_S \mathscr{K}^*(z, \zeta)\mu(\zeta) dS_\zeta + g(z), \qquad z \in S \qquad (22^*)$$

with the polar kernels (adjoint to each other)

$$\mathscr{K}(z, \zeta) = \frac{1}{\pi} \frac{\cos \varphi_{z\zeta}}{|z - \zeta|}, \qquad \mathscr{K}^*(z, \zeta) = \frac{1}{\pi} \frac{\cos \psi_{z\zeta}}{|z - \zeta|} \qquad (23)$$

For this, $\lambda = 1$ and $f = -(u_0^-/\pi)$ correspond to the interior, and $\lambda = -1$

and $f = (u_0^+ - \alpha)/\pi$ to the exterior Dirichlet problems; $\lambda = -1$ and $g = u_1^-/\pi$ to the interior, and $\lambda = 1$ and $g = -(u_1^+/\pi)$ to the exterior Neumann problems.

Let μ be the continuous solution of Eq. (22*) for $\lambda = 1$ and $g = -(u_1^+/\pi)$. Integrating this equation over the surface S and using (23) and (17), we obtain

$$\int_S \mu(z) \, dS = \frac{1}{\pi} \int_S \int_S \frac{\cos \varphi_{\zeta z}}{|\zeta - z|} \mu(\zeta) \, dS_\zeta \, dS_z - \frac{1}{\pi} \int_S u_1^+(z) \, dS$$

$$= \frac{1}{\pi} \int_S \mu(\zeta) \int_S \frac{\cos \varphi_{\zeta z}}{|\zeta - z|} \, dS_z \, dS_\zeta - \frac{1}{\pi} \int_S u_1^+(z) \, dS$$

$$= -\int_S \mu(\zeta) \, dS - \frac{1}{\pi} \int_S u_1^+(z) \, dS$$

that is,

$$\int_S \mu(z) \, dS = -\frac{1}{2\pi} \int_S u_1^+(z) \, dS \qquad (24)$$

We shall prove that $\lambda = 1$ is not an eigenvalue of the kernel $\mathcal{K}^*(z, \zeta)$. Conversely, let $\lambda = 1$ be an eigenvalue of this kernel and let μ^* be the eigenfunction corresponding to it

$$\mu^*(z) = \int_S \mathcal{K}^*(z, \zeta) \mu^*(\zeta) \, dS_\zeta = \frac{1}{\pi} \int_S \frac{\cos \psi_{z\zeta}}{|z - \zeta|} \mu^*(\zeta) \, dS_\zeta, \qquad z \in S \qquad (25)$$

Then $\mu^* \in C(S)$ and, in agreement with (24),

$$\int_S \mu^*(z) \, dS = 0 \qquad (26)$$

We now construct the simple layer potential $V^{(0)}$ with a density μ^*. The function $V^{(0)} \in C(R^2)$ is harmonic outside S and, by virtue of (26) and (14), $V^{(0)}(\infty) = 0$. Moreover, by virtue of (15) and (25), its correct normal derivative over S from outside is equal to zero. From this, using the uniqueness of the solution of an exterior Neumann problem and the interior Dirichlet problem (cf. Sec. 26.2), as in Sec. 23.5, we conclude that $V^{(0)}(z) \equiv 0$ for $z \in R^2$, and, consequently, $\mu^*(z) \equiv 0$ for $z \in S$, which is impossible.

According to Fredholm's theorems, Eqs. (22) and (22*) for $\lambda = 1$ are uniquely soluble in $C(S)$ for any continuous f and g. For this, for the solution μ of Eq. (22*) when $\lambda = 1$ and $g = -(u_1^+/\pi)$, result (24) is valid.

§ 26. LAPLACE'S EQUATION IN A PLANE

So if condition (7) is satisfied, then, by virtue of (14), $V^{(0)}(\infty) = 0$. So the following theorem is proved:

THEOREM 1. *The interior Dirichlet problem is soluble for any $u_0^- \in C(S)$. The exterior Neumann problem is soluble for any $u_1^+ \in C(S)$ which satisfies the solubility condition* (7).

It follows from formula (17) that $\lambda = -1$ is an eigenvalue of the kernel $\mathcal{K}(z, \zeta)$ and $v = 1$ is an eigenfunction corresponding to it. By Fredholm's second theorem $\lambda = -1$ is an eigenvalue of the adjoint kernel $\mathcal{K}^*(z, \zeta)$. Let μ_0 be the corresponding eigenfunction,

$$\mu_0(z) = -\int_S \mathcal{K}^*(z, \zeta) \mu_0(\zeta) \, dS_\zeta$$
$$= -\frac{1}{\pi} \int_S \frac{\cos \psi_{z\zeta}}{|z - \zeta|} \mu_0(\zeta) \, dS_\zeta, \qquad z \in S \qquad (27)$$

We know that $\mu_0 \in C(S)$ (cf. Sec. 16.5). We shall prove that

$$\int_S \mu_0(\zeta) \, dS = C \neq 0 \qquad (28)$$

On the other hand, let $C = 0$. We shall construct a simple layer potential with the density μ_0 (Robin's potential)

$$V^{(0)}(z) = \int_S \mu_0(\zeta) \ln \frac{1}{|z - \zeta|} \, dS_\zeta \qquad (29)$$

The function $V^{(0)} \in C(R^2)$ is harmonic outside S, and, by virtue of the condition that $C = 0$, $V^{(0)}(\infty) = 0$ (cf. Sec. 26.3). Moreover, it follows from (15′) and (27) that its correct normal derivative over S from inside is equal to zero. From this, since the solution of the interior Neumann problem is unique up to an additive constant, we conclude that $V^{(0)}(z) = \text{const} = C_1$ for $z \in \bar{G}$. But then, since the solution of the exterior Dirichlet problem is unique (cf. Sec. 26.2), $V^{(0)}(z) = C_1$ for $z \in \bar{G}_1$ and, consequently, $\mu_0(z) = 0$ for $z \in S$, which is impossible.

In this way we see $C \neq 0$. From this, reasoning as in Sec. 23.5, we conclude that $\lambda = -1$ is a simple eigenvalue of the kernel $\mathcal{K}^*(z, \zeta)$ and therefore of the kernel $\mathcal{K}(z, \zeta)$.

We shall normalize the eigenfunction μ_0 so that $C = 1$. By what has been proved the corresponding Robin's potential $V^{(0)}(z) = \text{const}$ for $z \in \bar{G}$.

By Fredholm's third theorem, integral equations (22) and (22*) for $\lambda = -1$ are soluble when, and only when, their inhomogeneous terms f and g are orthogonal to the eigenfunctions μ_0 and 1, respectively. For the exterior Dirichlet problem this condition takes the form

$$\frac{1}{\pi} \int_S [u_0^+(z) - \alpha]\mu_0(z)\, dS = 0$$

that is, by virtue of (28) (as $C = 1$), it can always be satisfied by a suitable choice of constant α

$$\alpha = \int_S u_0^+(z)\mu_0(z)\, dS \tag{30}$$

So the following theorem is true:

THEOREM 2. *The exterior Dirichlet problem is soluble for any $u_0^+ \in C(S)$. The interior Neumann problem is soluble for any $u_1^- \in C(S)$ which satisfies the solubility condition* (7).

5. Solution of Boundary Value Problems for a Circle. It is easy to solve integral equations (22) and (22*) for the circle comprising the circumference S_R, and this makes it possible to construct solutions of boundary value problems for the circle U_R in explicit form.

In fact, by virtue of the result (cf. Fig. 73)

$$|z|^2 = |z - \zeta|^2 + R^2 + 2R|z - \zeta|\cos\varphi_{z\zeta}, \qquad |\zeta| = R \tag{31}$$

we obtain

$$\mathscr{K}(z, \zeta) = \frac{1}{\pi} \frac{\cos \varphi_{z\zeta}}{|z - \zeta|} = -\frac{1}{2\pi R} = \mathscr{K}^*(z, \zeta), \qquad z, \zeta \in S_R$$

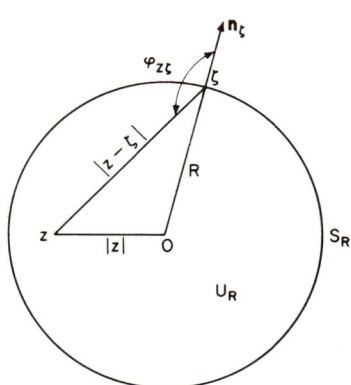

Fig. 73

§ 26. LAPLACE'S EQUATION IN A PLANE

Therefore the integral equations (22) and (22*) take the single form:

$$v(z) = -\frac{\lambda}{2\pi R} \int_{|\zeta|=R} v(\zeta) \, dS + f(z), \qquad |z| = R \tag{32}$$

Solving this equation (cf. Sec. 16.1), we obtain

$$v(z) = -\frac{1}{\pi} u_0^-(z) + \frac{1}{4\pi^2 R} \int_{|\zeta|=R} u_0^-(\zeta) \, dS \tag{33}$$

for $\lambda = 1$ and $f = -(u_0^-/\pi)$ (interior Dirichlet problem);

$$v(z) = \frac{1}{\pi} u_0^+(z) - \frac{1}{2\pi^2 R} \int_{|\zeta|=R} u_0^+(\zeta) \, dS$$

$$\alpha = \frac{1}{2\pi R} \int_{|\zeta|=R} u_0^+(\zeta) \, dS \tag{34}$$

for $\lambda = -1$ and $f = u_0^+/\pi - \alpha/\pi$ (exterior Dirichlet problem);

$$\mu(z) = \frac{1}{\pi} u_1^-(z), \qquad \text{if} \int_{|\zeta|=R} u_1^-(\zeta) \, dS = 0 \tag{35}$$

for $\lambda = -1$ and $f = u_1^-/\pi$ (interior Neumann problem);

$$\mu(z) = -\frac{1}{\pi} u_1^+(z), \qquad \text{if} \int_{|\zeta|=R} u_1^+(\zeta) \, dS = 0 \tag{36}$$

for $\lambda = 1$ and $f = (1/\pi) u_1^+$ (exterior Neumann problem).

If we substitute expression (33) for the density $v(z)$ in the double layer potential (20) and use Eqs. (17) and (31), we obtain the solution of the interior Dirichlet problem:

$$u(z) = \int_{|\zeta|=R} \left[-\frac{1}{\pi} u_0^-(\zeta) + \frac{1}{4\pi^2 R} \int_{|\zeta'|=R} u_0^-(\zeta') \, dS \right] \frac{\cos \varphi_{z\zeta}}{|z-\zeta|} \, dS_\zeta$$

$$= \frac{1}{2\pi R} \int_{|\zeta|=R} u_0^-(\zeta) \left(-\frac{2R \cos \varphi_{z\zeta}}{|z-\zeta|} + \frac{1}{2\pi} \int_{|\zeta'|=R} \frac{\cos \varphi_{z\zeta'}}{|z-\zeta'|} \, dS_{\zeta'} \right) dS$$

$$= \frac{1}{2\pi R} \int_{|\zeta|=R} u_0^-(\zeta) \frac{-2R|z-\zeta|\cos \varphi_{z\zeta} - |z-\zeta|^2}{|z-\zeta|^2} \, dS_\zeta$$

$$= \frac{1}{2\pi R} \int_{|\zeta|=R} u_0^-(\zeta) \frac{R^2 - |z|^2}{|z-\zeta|^2} \, dS_\zeta$$

that is, this solution is represented by Poisson's integral formula

$$u(z) = \frac{1}{2\pi R} \int_{|\zeta|=R} u_0^-(\zeta) \frac{R^2 - |z|^2}{|z - \zeta|^2} dS_\zeta, \qquad |z| < R \qquad (37)$$

Analogously, if we substitute expressions (34) for the density $v(z)$ and the constant α in the sum $V^{(1)}(z) + \alpha$, we shall obtain the solution of the exterior Dirichlet problem:

$$u(z) = \frac{1}{2\pi R} \int_{|\zeta|=R} u_0^+(\zeta) \frac{|z|^2 - R^2}{|z - \zeta|^2} dS_\zeta, \qquad |z| > R \qquad (38)$$

Finally, if we substitute expressions (35) and (36) for the density $\mu(z)$ in the simple layer potential (21), we shall obtain the solutions of the interior and exterior Neumann problems, respectively:

$$u(z) = \frac{1}{\pi} \int_{|\zeta|=R} u_1^-(\zeta) \ln \frac{1}{|z - \zeta|} dS_\zeta + C, \qquad |z| < R \qquad (39)$$

$$u(z) = \frac{1}{\pi} \int_{|\zeta|=R} u_1^+(\zeta) \ln |z - \zeta| \, dS_\zeta + C, \qquad |z| > R \qquad (40)$$

6. The Green's Function for the Dirichlet Problem. *The Green's function for the interior Dirichlet problem* for the region G is the function $\mathscr{G}(z, \zeta)$ which has the following properties (cf. Sec. 24.1): for each $\zeta \in G$ it can be written in the form

$$\mathscr{G}(z, \zeta) = \frac{1}{2\pi} \ln \frac{1}{|z - \zeta|} + g(z, \zeta) \qquad (41)$$

where the function $g(z, \zeta)$ is harmonic in G and continuous on $\bar{G}$ with respect to z, and satisfies the boundary condition

$$\mathscr{G}(z, \zeta)|_{z \in S} = 0 \qquad (42)$$

It follows from the maximum principle that the Green's function satisfies the inequality

$$0 < \mathscr{G}(z, \zeta) < \frac{1}{2\pi} \ln \frac{D}{|z - \zeta|}, \qquad z \in G, \qquad \zeta \in G, \qquad z \neq \zeta \qquad (43)$$

where D is the diameter of the region G. The Green's function $\mathscr{G}(z, \zeta)$ possesses all the other properties set out in Sec. 24.1.

§ 26. LAPLACE'S EQUATION IN A PLANE

Now let G be a simply-connected region bounded by a piecewise smooth curve S, and let $w = w(z)$ be a function giving a conformal mapping of the region G onto the unit circle $|w| < 1$. Then the function

$$\omega(z, \zeta) = \frac{w(z) - w(\zeta)}{1 - \overline{w(\zeta)}w(z)} \tag{44}$$

gives a conformal mapping of the region G onto the unit circle $|\omega| < 1$, and the point $\zeta \in G$ maps into the point 0. Therefore, for every $\zeta \in G$, this function can be represented in the form

$$\omega(z, \zeta) = (z - \zeta)\psi(z, \zeta) \tag{45}$$

where the function $\psi(z, \zeta)$ is analytic in the region G, $\psi(z, \zeta) \neq 0$ for $z \in \bar{G}$; and $\psi \in C(\bar{G})$ [cf., for instance, M. A. Evgrafov (*1*, Chaps. V and IX)].

We shall check that *the function*

$$\mathscr{G}(z, \zeta) = -\frac{1}{2\pi}\ln|\omega(z, \zeta)| = -\frac{1}{2\pi}\operatorname{Re}\ln\omega(z, \zeta) \tag{46}$$

is the Green's function of the Dirichlet problem for the region G.

In fact, it follows from (45) that function (46) appears in the form (41), and the function

$$g(z, \zeta) = -\frac{1}{2\pi}\ln|\psi(z, \zeta)| = -\frac{1}{2\pi}\operatorname{Re}\ln\psi(z, \zeta)$$

the real part of the analytic function $\ln\psi(z, \zeta)$ with $\psi(z, \zeta) \neq 0$, is harmonic in G and continuous on $\bar{G}$. Moreover, by virtue of the equation $|\omega(z, \zeta)| = 1$ for $z \in S$, the function (46) also satisfies condition (42).

As an example we shall construct the Green's function for the circle $|z| < R$. The function

$$\omega(z, \zeta) = \frac{R(z - \zeta)}{R^2 - z\bar{\zeta}}$$

maps the circle $|z| < R$ onto the unit circle $|\omega| < 1$, and the point ζ maps into the point 0. Therefore, by virtue of (46),

$$\mathscr{G}(z, \zeta) = \frac{1}{2\pi}\ln\left|\frac{R^2 - z\bar{\zeta}}{R(z - \zeta)}\right| = \frac{1}{2\pi}\operatorname{Re}\ln\frac{R^2 - z\bar{\zeta}}{R(z - \zeta)} \tag{47}$$

is the Green's function for the circle $|z| < R$.

7. Solution of the Dirichlet Problem for a Simply-Connected Region.

The method of conformal mappings allows us to obtain a representation of the solution of the Dirichlet problem for any simply-connected region. This representation is a generalization of Poisson's formula.

First, using the equation

$$\frac{R^2 - |z|^2}{|z - \zeta|^2} = \operatorname{Re} \frac{\zeta + z}{\zeta - z}, \qquad |\zeta| = R$$

we shall represent Poisson's formula (37) in the form

$$u(z) = \operatorname{Re} \frac{1}{2\pi i} \int_{|\zeta|=R} u_0(\zeta) \frac{\zeta + z}{\zeta - z} \frac{d\zeta}{\zeta} \tag{48}$$

Let the function u_0 be continuous over the boundary S of the simply-connected region G. Moreover, let the function $z = z(w)$ conformally map the circle $|w| < 1$ onto the region G and let $w = w(z)$ be the inverse mapping. Then $z \in C(\bar{U}_1)$; we shall suppose that $w \in C^1(\bar{G})$. Under this mapping the function $u_0(z)$ maps into the function $u_0[z(w)]$, which is continuous over the circumference $|w| = 1$. Using formula (48) we may construct the solution of the Dirichlet problem for the circular region $|w| < 1$ with the boundary condition $u_0[z(w)]$:

$$U(w) = \operatorname{Re} \frac{1}{2\pi i} \int_{|\omega|=1} u_0[z(\omega)] \frac{\omega + w}{\omega - w} \frac{d\omega}{\omega}$$

Changing to the old variables z and ζ in this formula with $w = w(z)$ and $\omega = w(\zeta)$, we obtain the required solution of the Dirichlet problem for the region G with the boundary condition u_0:

$$u(z) = \operatorname{Re} \frac{1}{2\pi i} \int_S u_0(\zeta) \frac{w(\zeta) + w(z)}{w(\zeta) - w(z)} \frac{w'(\zeta)}{w(\zeta)} d\zeta, \qquad z \in G \tag{49}$$

Note. As we know, the real and imaginary parts of an analytic function are harmonic functions. Conversely, if the function $u(z)$ is harmonic, then, if we construct the conjugate function

$$v(z) = \int^z \left[-\frac{\partial u(\zeta)}{\partial \eta} d\xi + \frac{\partial u(\zeta)}{\partial \xi} d\eta \right] + C, \qquad \zeta = \xi + i\eta$$

we shall obtain the analytic function $f(z) = u(z) + iv(z)$, of which the real part is the function $u(z)$. This makes it possible to use the method of the theory of analytic functions when solving and investigating boundary value problems for harmonic functions in a plane [cf. M. A. Lavrentiev and B. V. Shabat (1)].

8. Exercises. (a) Using formula (35) of Sec. 6.5 and the fundamental solution $1/\pi z$ of the Cauchy–Riemann operator (cf. Sec. 10.10), deduce the analog of Green's formula: If $u \in C^1(G) \cap C(\bar{G})$, then the result

$$u(z) = \frac{1}{\pi} \int_G \frac{1}{z-\zeta} \frac{\partial u(\zeta)}{\partial \bar{\zeta}} d\xi\, d\eta + \frac{1}{2\pi i} \int_S \frac{u(\zeta)}{\zeta - z} d\zeta, \qquad z \in G$$

is true.

(b) Using (a), prove: If $u \in C^1(G) \cap C(\bar{G})$ and satisfies the Cauchy–Riemann condition, $\partial u/\partial \bar{z} = 0$ for $z \in G$, then $u(z)$ is an analytic function in the region G and Cauchy's formula is valid:

$$u(z) = \frac{1}{2\pi i} \int_S \frac{u(\zeta)}{\zeta - z} d\zeta, \qquad z \in G$$

(c) Using the method of Sec. 21.9, prove Liouville's theorem for analytic functions.

(d) Show that the simple layer potential for the circle $|z| = R$ having the density $\mu = 1$ is equal to

$$V^{(0)}(z) = -\int_{|\zeta|=R} \ln|z-\zeta|\, dS_\zeta = \begin{cases} -2\pi R \ln R, & |z| \le R \\ -2\pi R \ln|z|, & |z| \ge R \end{cases}$$

(e) Using (d), show that the logarithmic potential for the circular region $|z| < R$ with density $\varrho = 1$ is equal to

$$V(z) = \begin{cases} -\pi R^2 \left(\ln R - \dfrac{1}{2}\right) - \dfrac{\pi}{2}|z|^2, & |z| \le R \\ -\pi R^2 \ln|z|, & |z| \ge R \end{cases}$$

(f) Show that the following functions are Green's functions of the Dirichlet problem:

$\dfrac{1}{2\pi} \ln \left|\dfrac{z-\bar{\zeta}}{z-\zeta}\right|$ for the half plane $y > 0$

$\dfrac{1}{2\pi} \ln \left|\dfrac{(z-\zeta^*)(z-\bar{\zeta})}{(z-\zeta)(z-\bar{\zeta}^*)}\right|$ for the semicircle $|z| < R$, $y > 0$

$\dfrac{1}{2\pi} \ln \left|\dfrac{z^2-\bar{\zeta}^2}{z^2-\zeta^2}\right|$ for the quarter plane $x > 0$, $y > 0$

$\dfrac{1}{2\pi} \ln \left|\dfrac{e^{z-\bar{\zeta}}-1}{e^{z-\zeta}-1}\right|$ for the strip $0 < y < \pi$

(g) Show that the solutions of the Dirichlet and Neumann problems for the half-plane $y > 0$ are given, respectively, by the equations

$$\frac{y}{\pi} \int_{-\infty}^{\infty} u_0(\xi) \frac{d\xi}{(x-\xi)^2 + y^2}, \qquad -\frac{1}{2\pi} \int_{-\infty}^{\infty} u_1(\xi) \ln[(x-\xi)^2 + y^2] \, d\xi$$

if

$$u_0(\xi) = O(|\xi|^{1-\varepsilon}), \qquad u_1(\xi) = O(|\xi|^{-1-\varepsilon})$$

as

$$|\xi| \to \infty; \qquad \varepsilon > 0.$$

§ 27. Helmholtz's Equation

The equation

$$\Delta u + k^2 u = -f(x) \tag{1}$$

is known as Helmholtz's equation (cf. Sec. 2.3). For $k = 0$ it coincides with Poisson's equation. The theory of Helmholtz's equation is close to that of Poisson's equation, but there are a few peculiarities concerning the uniqueness of the solution (for $k^2 > 0$).

We shall consider Eq. (1) in a three-dimensional space so that $n = 3$. The corresponding fundamental solutions are expressed by the equations (cf. Sec. 10.9)

$$\mathscr{E}(x) = -\frac{e^{ik|x|}}{4\pi |x|}, \qquad \bar{\mathscr{E}}(x) = -\frac{e^{-ik|x|}}{4\pi |x|}$$

In future we suppose that $k > 0$.

1. Sommerfeld Radiation Conditions. As we showed in Sec. 23.3, the solution of Poisson's equation in a whole space is unique in the class of (generalized) functions and tends to zero at infinity. But this statement is not true for Helmholtz's equation since the corresponding homogeneous equation

$$\Delta u + k^2 u = 0 \tag{2}$$

has a nonzero solution in R^3

$$\operatorname{Im} \mathscr{E}(x) = -\frac{\sin k |x|}{4\pi |x|}$$

but tending to zero as $|x| \to \infty$ [like $O(|x|^{-1})$].

In order to isolate the class of unique solutions for Helmholtz's equation

§ 27. HELMHOLTZ'S EQUATION

in nonbounded regions which are the exterior of bounded regions, there must be additional restrictions on the behavior of the solution at infinity. These restrictions are the so-called Sommerfeld radiation conditions:

$$u(x) = O(|x|^{-1}), \quad \frac{\partial u(x)}{\partial |x|} - iku(x) = o(|x|^{-1}), \quad |x| \to \infty \tag{3}$$

or

$$u(x) = O(|x|^{-1}), \quad \frac{\partial u(x)}{\partial |x|} + iku(x) = o(|x|^{-1}), \quad |x| \to \infty \tag{$\bar{3}$}$$

Later (cf. Sec. 27.5) we shall explain the physical meaning of the radiation conditions: Conditions (3) correspond to divergent waves (outgoing to infinity), while conditions ($\bar{3}$) correspond to convergent waves (incoming from infinity). It is easy to check that the fundamental solutions $\mathscr{E}(x)$ and $\bar{\mathscr{E}}(x)$ satisfy the radiation conditions (3) and ($\bar{3}$), respectively. We note that for harmonic functions ($k = 0$), the radiation conditions follow from only one requirement: $u(x) = o(1)$ as $|x| \to \infty$ (cf. Sec. 23.2). On the other hand, it can be shown* that for $k > 0$ each solution of the homogeneous Helmholtz equation which satisfies the second of the radiation conditions (3) or ($\bar{3}$) also satisfies the first condition: $u(x) = O(|x|^{-1})$.

2. The Homogeneous Helmholtz Equation. The solutions of the homogeneous Helmholtz equation (2) have properties which are analogous to those of harmonic functions. We shall note some of these.

(a) *If the function $u \in C(G)$ satisfies Eq. (2) in the generalized sense in the region G, then $u \in C^\infty(G)$.* This statement is proved in the same way as for harmonic functions (cf. Sec. 21.7).

(b) Let the boundary S of the region G be a sufficiently smooth surface (in the sense of Sec. 23.3). *If the function $u \in C(\bar{G})$ satisfies Eq. (2) in the region G and has a correct normal derivative over S, then the following equations are true:*

$$u(x) = \frac{1}{4\pi} \int_S \left[\frac{\exp(ik|x-y|)}{|x-y|} \frac{\partial u(y)}{\partial n} - u(y) \frac{\partial}{\partial n_y} \frac{\exp(ik|x-y|)}{|x-y|} \right] dS_y \tag{4}$$

$$u(x) = \frac{1}{4\pi} \int_S \left[\frac{\exp(-ik|x-y|)}{|x-y|} \frac{\partial u(y)}{\partial n} - u(y) \frac{\partial}{\partial n_y} \frac{\exp(-ik|x-y|)}{|x-y|} \right] dS_y \tag{$\bar{4}$}$$

---
* Compare I. N. Vekua (2).

Proof of these equations is analogous to the proof of Green's formula (5) of Sec. 21.1 for harmonic functions.

(c) *If the generalized function u belonging to $\mathscr{S}'$ satisfies the homogeneous Helmholtz equation in a whole space, then $u \in \theta_M$.* In fact, applying the Fourier transform to Eq. (2), we obtain

$$(-|\xi|^2 + k^2)F[u] = 0$$

from which it follows that $F[u] = 0$ when $|\xi| \neq k$; that is, $F[u]$ is a generalized function with compact support. But then, by the theorem of Sec. 9.4,

$$u \in F^{-1}[F[u]] \in \theta_M$$

(d) We shall say that the generalized function u *satisfies the radiation conditions* (3) or ($\bar{3}$) if there is a number $R = R(u) > 0$ such that $u \in C^1(R^3 \setminus U_R)$ and satisfies the conditions (3) or ($\bar{3}$).

If the generalized function u satisfies the homogeneous Helmholtz equation in R^3 and the radiation conditions (3) *or* ($\bar{3}$), *then $u(x) \equiv 0$ for $x \in R^3$.*

In fact, let the solution u of Eq. (2) satisfy conditions (3). Then $u \in \mathscr{S}'$ and, by virtue of (c), $u \in C^\infty(R^3)$. If we apply formula (4) to the sphere U_R of arbitrary radius R, for $|x| < R$ we obtain

$$u(x) = \frac{1}{4\pi} \int_{S_R} \left[\frac{\exp(ik|x-y|)}{|x-y|} \frac{\partial u(y)}{\partial |y|} - u(y) \frac{\partial}{\partial |y|} \frac{\exp(ik|x-y|)}{|x-y|} \right] dS_y$$

$$= \frac{1}{4\pi} \int_{S_R} \frac{\exp(ik|x-y|)}{|x-y|} \left[\frac{\partial u(y)}{\partial |y|} - iku(y) \right.$$

$$\left. + iku(y)\left(1 - \frac{R - |x|\cos\gamma}{|x-y|}\right) + u(y)\frac{R - |x|\cos\gamma}{|x-y|^2} \right] dS_y$$

Taking conditions (3) into account

$$|u(y)| < \frac{C}{R}, \qquad \left|\frac{\partial u(y)}{\partial |y|} - iku(y)\right| < \frac{\eta(R)}{R}, \qquad |y| = R$$

where $\eta(R) \to 0$ as $R \to \infty$, and using the inequalities

$$R - |x| \leq |y - x| \leq R + |x|, \qquad |x| < R, \qquad |y| = R$$

§ 27. HELMHOLTZ'S EQUATION

we shall evaluate the last integral for large R:

$$|u(x)| \leq \frac{1}{4\pi} \int_{S_R} \frac{1}{R - |x|} \left[\frac{\eta(R)}{R} \right.$$
$$\left. + \frac{kC}{R(R - |x|)} (||x - y| - R| + |x|) + \frac{C(R + |x|)}{R(R - |x|)^2} \right] dS_y$$
$$\leq \frac{R}{R - |x|} \left[\eta(R) + \frac{2kC|x|}{R - |x|} + C \frac{R + |x|}{(R - |x|)^2} \right], \qquad |x| < R$$

Letting R tend to infinity in the right-hand side of this inequality, we conclude that $u(x) = 0$, as was to be shown.

The case of conditions $(\bar{3})$ may be considered analogously.

Note. A more general statement* is true: If the boundary S of the region G is a sufficiently smooth surface, the function $u \in C(\bar{G}_1)$ satisfies the homogeneous Helmholtz equation in the region $G_1 = R^3 \setminus \bar{G}$, has a correct normal derivative over S, and satisfies the radiation conditions (3) or $(\bar{3})$ and the boundary condition $u|_S = 0$ or $(\partial u/\partial \mathbf{n})|_S = 0$, then $u(x) \equiv 0$ for $x \in G_1$.

3. Potentials. Let ϱ be a generalized function. The convolutions

$$V = \frac{e^{ik|x|}}{|x|} * \varrho = -4\pi \mathscr{E} * \varrho, \qquad \bar{V} = \frac{e^{-ik|x|}}{|x|} * \varrho = -4\pi \bar{\mathscr{E}} * \varrho$$

are analogous to Newtonian potentials. If ϱ has compact support then these potentials belong to $\mathscr{S}'$ (cf. Sec. 8.6) and satisfy Helmholtz's equation (cf. Sec. 10.3):

$$\Delta u + k^2 u = -4\pi \varrho$$

In this way, the solutions of Helmholtz's equation (1) exist in $\mathscr{S}'$ for any generalized function f with compact support and are represented by the potentials

$$u = -\mathscr{E} * f, \qquad \bar{u} = -\bar{\mathscr{E}} * f \tag{5}$$

For this the solution is unique in the class of generalized functions which satisfy the radiation conditions (3) *or* $(\bar{3})$ [cf. Sec. 27.2, (d)].

* Compare I. N. Vekua (2) and V. I. Smirnov (2, Chap. IV, Sec. 2).

366 **5. BOUNDARY VALUE PROBLEMS FOR ELLIPTIC EQUATIONS**

If $\varrho \in C(\bar{G})$ and $\varrho(x) = 0$ for $x \in G_1 = R^3 \setminus \bar{G}$, then the potentials V and $\bar{V}$ are expressed by the integrals

$$V(x) = \int_G \frac{\exp(ik|x-y|)}{|x-y|} \varrho(y)\, dy, \quad \bar{V}(x) = \int_G \frac{\exp(-ik|x-y|)}{|x-y|} \varrho(y)\, dy$$

These potentials belong to the class $C^1(R^3) \cap C^\infty(G_1)$, and in the region G_1 satisfy the homogeneous equation (2) and the conditions (3) and ($\bar{3}$), respectively.

This statement is proved in the same way as for a volume Newtonian potential (cf. Sec. 22.1). Only the radiation conditions need be checked. We shall suppose that $G \subset U_R$ for $|x| > R$ and, consequently,

$$|x| - R \leq |x| - |y| \leq |x - y| \leq |x| + |y| \leq |x| + R \qquad (6)$$

for all $y \in G$. Therefore

$$|V(x)| \leq \int_G \frac{|\varrho(y)|}{|x-y|}\, dy \leq \frac{1}{|x|-R} \int_G |\varrho(y)|\, dy = O(|x|^{-1})$$

and the first of conditions (3) is established. To establish the second of conditions (3), we shall isolate the dependence of the potential $V(x)$ on $|x|$, putting $x = |x|\, s$ with $|s| = 1$:

$$V(|x|s) = \int_G \varrho(y) \frac{\exp[ik(|x|^2 + |y|^2 - 2|x||y|\cos\gamma)^{1/2}]}{(|x|^2 + |y|^2 - 2|x||y|\cos\gamma)^{1/2}}\, dy$$

$$\cos\gamma = \left(s, \frac{y}{|y|}\right)$$

Then

$$\frac{\partial V(x)}{\partial |x|} - ikV(x) = \int_G \varrho(y) \frac{\exp(ik|x-y|)}{|x-y|^2} \left[\frac{|x|-|y|\cos\gamma}{|x-y|}\right.$$

$$\left. + ik(|x| - |y|\cos\gamma - |x-y|)\right] dy$$

and so, by virtue of inequalities (6),

$$\left|\frac{\partial V(x)}{\partial |x|} - ikV(x)\right|$$

$$\leq \int_G \frac{|\varrho(y)|}{|x-y|^2} \left(\frac{|x|+|y|}{|x-y|} + k|y| + k||x|-|x-y||\right) dy$$

$$\leq \frac{1}{(|x|-R)^2} \left(\frac{|x|+R}{|x|-R} + 2kR\right) \int_G |\varrho(y)|\, dy = o(|x|^{-1})$$

§ 27. HELMHOLTZ'S EQUATION

The potential $\bar{V}$ is considered analogously.

If $\varrho = \mu \delta_S$ or $\varrho = -(\partial/\partial \mathbf{n})(\nu \delta_S)$, where μ and $\nu \in C(S)$, then the corresponding potentials V and $\bar{V}$ are analogous to surface (simple and double layer) potentials and are expressed by the integrals:

$$V^{(0)}(x) = \int_S \frac{\exp(ik|x-y|)}{|x-y|} \mu(y) \, dS_y$$

$$\bar{V}^{(0)}(x) = \int_S \frac{\exp(-ik|x-y|)}{|x-y|} \mu(y) \, dS_y$$

$$V^{(1)}(x) = \int_S \nu(y) \frac{\partial}{\partial \mathbf{n}_y} \frac{\exp(ik|x-y|)}{|x-y|} \, dS_y$$

$$\bar{V}^{(1)}(x) = \int_S \nu(y) \frac{\partial}{\partial \mathbf{n}_y} \frac{\exp(-ik|x-y|)}{|x-y|} \, dS_y$$

The properties of the potentials $V^{(0)}$, $\bar{V}^{(0)}$, $V^{(1)}$, and $\bar{V}^{(1)}$ are analogous to the properties of the corresponding Newtonian potentials (cf. Sec. 22). Outside the surface S these potentials are infinitely differentiable and satisfy the homogeneous equation (2) and the radiation conditions (3) or ($\bar{3}$), respectively, for this $V^{(0)}$ and $\bar{V}^{(0)} \in C(R^3)$. If S is a Liapunov surface and satisfies condition (A) of Sec. 22.6, then the potentials $V^{(0)}$ and $\bar{V}^{(0)}$ have a correct normal derivative over S from outside and inside S and these derivatives are, respectively, equal to

$$\left(\frac{\partial V^{(0)}}{\partial \mathbf{n}}\right)_{\pm}(x) = \mp 2\pi \mu(x) + \int_S \mu(y) \frac{\partial}{\partial \mathbf{n}_x} \frac{\exp(ik|x-y|)}{|x-y|} \, dS_y \quad (7)$$

$$\left(\frac{\partial \bar{V}^{(0)}}{\partial \mathbf{n}}\right)_{\pm}(x) = \mp 2\pi \mu(x) + \int_S \mu(y) \frac{\partial}{\partial \mathbf{n}_x} \frac{\exp(-ik|x-y|)}{|x-y|} \, dS_y \quad (\bar{7})$$

the double layer potentials $V^{(1)}$ and $\bar{V}^{(1)}$ belong to the class $C(\bar{G}) \cap C(\bar{G}_1)$ $\cap\, C(S)$ and their limiting values over S from outside and inside S are, respectively, equal to

$$V^{(1)}_{\pm}(x) = \pm 2\pi \nu(x) + \int_S \nu(y) \frac{\partial}{\partial \mathbf{n}_y} \frac{\exp(ik|x-y|)}{|x-y|} \, dS_y \quad (8)$$

$$\bar{V}^{(1)}_{\pm}(x) = \pm 2\pi \nu(x) + \int_S \nu(y) \frac{\partial}{\partial \mathbf{n}_y} \frac{\exp(-ik|x-y|)}{|x-y|} \, dS_y \quad (\bar{8})$$

4. The Limiting Absorption Principle. We shall add the term $i\varepsilon u$ to the left-hand side of Helmholtz's equation:

$$\Delta u_\varepsilon + (k^2 + i\varepsilon)u_\varepsilon = -f(x) \tag{9}$$

When $\varepsilon \neq 0$, for any generalized function f with compact support Eq. (9) has a unique solution in the class $\mathscr{S}'$, and this solution is expressed by the convolution

$$u_\varepsilon = \frac{1}{4\pi |x|} \exp(i\sqrt{k^2 + i\varepsilon}\,|x|) * f \tag{10}$$

In fact, the convolution (10) exists in $\mathscr{S}'$ (cf. Sec. 8.6) and satisfies Eq. (9), since the function

$$-\frac{1}{4\pi |x|} \exp(i\sqrt{k^2 + i\varepsilon}\,|x|)$$

is the corresponding fundamental solution (cf. Sec. 10.9). The uniqueness of the solution of Eq. (9) in the class $\mathscr{S}'$ follows from the uniqueness of the solution of the homogeneous equation

$$\Delta u + (k^2 + i\varepsilon)u = 0$$

that is, of the equation

$$(-|\xi|^2 + k^2 + i\varepsilon)F[u] = 0$$

Let $f \in C(\bar{G})$, $f(x) = 0$ for $x \in G_1$. In this case solution (10) of Eq. (9) can be written in the form of the integral

$$u_\varepsilon(x) = \frac{1}{4\pi} \int_G \frac{\exp(i\sqrt{k^2 + i\varepsilon}\,|x-y|)}{|x-y|} f(y)\,dy \tag{11}$$

Proceeding to the limit in formula (11) as $\varepsilon \to +0$ or -0, and putting $\sqrt{k^2 \pm i0} = \pm k$, we obtain, by virtue of (5), the solution u or $\bar{u}$ of Eq. (1) that satisfies conditions (3) or ($\bar{3}$), respectively,

$$\lim_{\varepsilon \to +0} u_\varepsilon(x) = \frac{1}{4\pi} \int_G \frac{\exp(ik\,|x-y|)}{|x-y|} f(y)\,dy = u(x)$$

$$\lim_{\varepsilon \to -0} u_\varepsilon(x) = \frac{1}{4\pi} \int_G \frac{\exp(-ik\,|x-y|)}{|x-y|} f(y)\,dy = \bar{u}(x)$$

§ 27. HELMHOLTZ'S EQUATION

In this way, the following statement, which is known as the *limiting absorption principle*, is established: *The solution of Eq.* (1) *which satisfies conditions* (3) *or* ($\bar{3}$) *is the limit uniformly in x of the unique solution of Eq.* (9) *as* $\varepsilon \to \pm 0$, *respectively*.

The limiting absorption principle allows us to isolate the unique solution of Helmholtz's equation without troubling about its behavior at infinity; in this case the solution obtained will automatically satisfy the radiation conditions (3) or ($\bar{3}$).

5. The Limiting Amplitude Principle. This principle is that *the solutions u and $\bar{u}$ of Eq.* (1) *which satisfy conditions* (3) *or* ($\bar{3}$) *are the respective limits*

$$u(x) = \lim_{t\to\infty} e^{ikt} v_+(x, t) \tag{12}$$

$$\bar{u}(x) = \lim_{t\to\infty} e^{-ikt} v_-(x, t) = \lim_{t\to-\infty} e^{ikt} v_-(x, -t) \tag{$\overline{12}$}$$

where $v_\pm(x, t)$ is the solution of the (generalized) Cauchy problem for a wave equation with the right-hand side $\theta(t) e^{\mp ikt} f(x)$ and with zero initial data:

$$\Box v_\pm = \theta(t) e^{\mp ikt} f(x), \qquad v_\pm(x, t) = 0, \qquad t < 0 \tag{13}$$

In fact, let the function $f \in C(\bar{G})$, $f(x) = 0$ for $x \in G_1$. Then the unique (generalized) solution v_+ of the Cauchy problem (13) is expressed by the retarded potential (cf. Sec. 12.3):

$$v_+(x, t) = \frac{e^{-ikt}}{4\pi} \int_{U(x,t)} \frac{\exp(ik|x-y|)}{|x-y|} f(y)\, dy \tag{14}$$

Let $G \subset U_R$. Then $|x - y| \leq |x| + R$, and therefore $|x - y| \leq t$ for all $y \in G$ if $t \geq |x| + R$ (Fig. 74). Therefore formula (14) in the region $t \geq |x| + R$ takes the form

$$v_+(x, t) = \frac{e^{-ikt}}{4\pi} \int_G \frac{\exp(ik|x-y|)}{|x-y|} f(y)\, dy$$

that is, by virtue of (5),

$$v_+(x, t) = e^{-ikt} u(x)$$

from which limiting result (12) for $u(x)$ follows. The solution $\bar{u}(x)$ is considered analogously.

In this way, the solution of Helmholtz's equation (1) which satisfies the radiation conditions (3) or (3̄) may be considered as the amplitude of a steady state oscillation, obtained by a limiting process from transient oscillations caused by a periodic exterior perturbation with the frequency k and the amplitude $f(x)$. In this case the limiting amplitude $u(x)$ corresponds to a divergent wave, and the amplitude $\bar{u}(x)$ to a convergent wave.

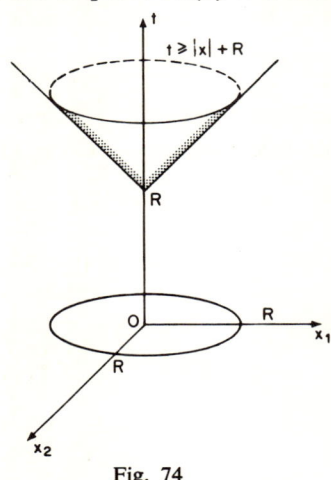

Fig. 74

6. Boundary Value Problems for Helmholtz's Equation. Let the boundary S of the region G be a sufficiently smooth surface, while $G_1 = R^3 \setminus \bar{G}$ is a region. The Dirichlet and Neumann problems (interior and exterior) for Helmholtz's equation are formulated as for Poisson's equation (cf. Sec. 23.1). In the case of the exterior problems, it is required that the solution should satisfy radiation conditions (3) or (3̄) at infinity.

If $\lambda = k^2$ is not an eigenvalue of the interior Dirichlet or Neumann problem for Laplace's equation, then the solution of the corresponding interior boundary value problem for Helmholtz's equation is unique.

We note that the set of exceptional values k, for which the uniqueness of solution of the interior boundary value problems is violated, is countable (cf. Secs. 19.4 and 24.6).

The solutions of the boundary value problems which have been formulated for the homogeneous Helmholtz equation are constructed by the methods of the theory of the potential in the same way that was used in Sec. 23 for Laplace's equation.

§ 27. HELMHOLTZ'S EQUATION

The solution of the Dirichlet problem [interior and exterior, satisfying conditions (3)] is sought in the form of a double layer potential $V^{(1)}$ with unknown density $v \in C(S)$. By virtue of (8) the function v must satisfy the integral equation

$$v(x) = \lambda \int_S \mathcal{H}(x, y) v(y) \, dS_y + f(x), \qquad x \in S \tag{15}$$

where

$$\mathcal{H}(x, y) = \frac{1}{2\pi} \frac{\partial}{\partial \mathbf{n}_y} \frac{\exp(ik |x - y|)}{|x - y|}$$

$$= (1 - ik |x - y|) \frac{\cos \varphi_{xy}}{|x - y|^2} \exp(ik |x - y|)$$

For this, $\lambda = 1$ and $f = -(u_0^-/2\pi)$ correspond to the interior Dirichlet problem, and $\lambda = -1$ and $f = u_0^+/2\pi$ to the exterior one.

The solution of the Neumann problem [interior and exterior, satisfying conditions (3̄)] is sought in the form of a simple layer potential $V^{(0)}$ with an unknown density $\mu \in C(S)$. By virtue of (7̄) the function μ must satisfy the integral equation adjoint to Eq. (15),

$$\mu(x) = \lambda \int_S \mathcal{H}^*(x, y) \mu(y) \, dS_y + g(x), \qquad x \in S \tag{15*}$$

and $\lambda = -1$ and $g = u_1^-/2\pi$ correspond to the interior Neumann problem, and $\lambda = 1$ and $g = -(u_1^+/2\pi)$ to the exterior one.

Applying Fredholm's theorems to integral equations (15) and (15*), and using the uniqueness theorem (cf. note of Sec. 27.2) as for Laplace's equation (cf. Sec. 23.5), we obtain the following theorem.

THEOREM. *If $\lambda = k^2$ is not an eigenvalue of the interior Dirichlet and Neumann problems for Laplace's equation, then the boundary value problems (interior and exterior) for the homogeneous Helmholtz equation are uniquely soluble in the form of the corresponding potentials for any continuous boundary values.**

Note. Scattering problems may be reduced to boundary value problems for Helmholtz's equation. For instance, let there be an incident plane wave $\exp[ik(a, x)]$ with $|a| = 1$, which is subjected to change because of some

---

* The solubility of the exterior boundary value problems occurs for all values of the parameter k^2 [cf. I. N. Vekua (2)].

obstruction which generates a scattered wave $u_1(x)$. The obstruction may be described by the boundary condition $u|_S = 0$ or $(\partial u/\partial \mathbf{n})|_S = 0$. Then the scattered wave u_1 must satisfy the boundary condition

$$u_1|_S = -\exp[ik(a,x)]|_S \quad \text{or} \quad \frac{\partial u_1}{\partial \mathbf{n}}\bigg|_S = -\frac{\partial}{\partial \mathbf{n}}\exp[ik(a,x)]\bigg|_S$$

and the radiation conditions (3); that is, it must be a divergent wave.

7. Exercises. (a) Let ϱ be a generalized function with compact support. Prove that the potentials V and $\bar{V}$ with the density ϱ satisfy the radiation conditions (3) and $(\bar{3})$; respectively.

(b) Let the function $u(x)$ of the class $C(\bar{U}_R)$ satisfy the homogeneous Helmholtz equation in the sphere U_R. Prove the analog of the theorem of the arithmetic mean:

$$u(0) = \frac{kR}{4\pi \sin kR} \int_{S_1} u(Rs)\, ds$$

(c) Let k^2 lie in the plane of the complex variable z with the cut: $\operatorname{Im} z = 0$, $\operatorname{Re} z \geq 0$. Prove that the solution of Helmholtz's equation is unique in the class $\mathscr{S}'$.

(d) Show that in the n-dimensional case the Sommerfeld radiation conditions

$$u(x) = O(|x|^{(1-n)/2}), \quad \frac{\partial u(x)}{\partial |x|} \mp k u(x) = o(|x|^{(1-n)/2}), \quad |x| \to \infty$$

guarantee the uniqueness of the solution of exterior boundary value problems for Helmholtz's equation.

(e) Prove the following positive maximum principle: The solution of the homogeneous Helmholtz equation for $k^2 < 0$ cannot assume a positive maximum value at interior points of the region where this solution is defined.

(f) Using (e), prove the uniqueness of the solution of boundary value problems for Helmholtz's equation when $k^2 < 0$ (for exterior problems, in the class of functions tending to zero at infinity) and its continuous dependence on the data of the problem.

(g) Construct the theory of the potential for Helmholtz's equation when $k^2 < 0$.

(h) Prove: If $f \in \mathscr{D}'(R^n)$ satisfies the homogeneous Helmholtz equation in the region G, then $f \in C^\infty(G)$; k^2 is any (complex) number.

CHAPTER
6

The Mixed Problem

In this chapter we shall consider the mixed problem for equations of hyperbolic and parabolic types.

§ 28. Fourier's Method

Fourier's method (the separation of variables) is one of the most effective ways of solving many-dimensional boundary value problems. In Sec. 19 this method was applied to boundary value problems involving eigenvalues. In this section Fourier's method is formally applied to the solution of boundary value problems for equations of different types. The following sections of this chapter deal with foundations of Fourier's method applied to mixed problems for equations of hyperbolic and parabolic types.

Let the operator L be defined by the differential expression

$$Lu = -\operatorname{div}(p \operatorname{grad} u) + qu, \qquad x \in G$$

and the boundary condition

$$\alpha u + \beta \left.\frac{\partial u}{\partial \mathbf{n}}\right|_S = 0$$

and the functions $p(x)$, $q(x)$, $\alpha(x)$, and $\beta(x)$ satisfy conditions (3) of Sec. 19.1.

We assume that the eigenvalues $\{\lambda_k\}$ of the operator L are positive, $0 < \lambda_1 \leq \lambda_2 \leq \ldots$, while the corresponding eigenfunctions $\{X_k\}$,

$$LX_k = \lambda_k \varrho X_k, \qquad X_k \in \mathcal{M}_L, \qquad k = 1, 2, \ldots$$

are real and form a complete orthonormal system in the space $\mathcal{L}_2(G; \varrho)$ with the scalar product $(f, g)_\varrho$ involving the weight function $\varrho(x) > 0$, $x \in \bar{G}$, $\varrho \in C(\bar{G})$. (Sufficient conditions under which these assumptions are realized are given in Sec. 19.4.)

1. The Homogeneous Hyperbolic Equation. We shall consider the mixed problem for the homogeneous equation of hyperbolic type (cf. Sec. 4.5) in the infinite cylinder $\Pi_\infty = G \times (0, \infty)$

$$\varrho \frac{\partial^2 u}{\partial t^2} = -Lu \tag{1}$$

$$u\big|_{t=+0} = u_0(x), \qquad \frac{\partial u}{\partial t}\bigg|_{t=+0} = u_1(x), \qquad x \in \bar{G} \tag{2}$$

$$\alpha u + \beta \frac{\partial u}{\partial \mathbf{n}}\bigg|_S = 0, \qquad t \geq 0 \tag{3}$$

Fourier's method is essentially as follows: We shall construct a sufficient number of solutions of Eq. (1) which may be represented by the product

$$T(t)X(x) \tag{4}$$

and which satisfy the boundary condition (3); out of these solutions we shall construct a linear combination which satisfies the initial conditions (2); under certain conditions it is natural to expect that the linear combination obtained will satisfy Eq. (1) and the boundary condition (3); that is, will be the solution of the problem (1)-(2)-(3).

So we shall seek the solution of Eq. (1) in the form of the product (4), and we shall require that the function $X(x)$ should satisfy the boundary condition (3). Substituting expression (4) in Eq. (1) and dividing it by $\varrho T X$, we obtain

$$\frac{T''(t)}{T(t)} = -\frac{LX(x)}{\varrho(x)X(x)} \tag{5}$$

The left-hand side of Eq. (5) does not depend on x, nor the right-hand

§ 28. FOURIER'S METHOD

side on t. Therefore, each of these magnitudes does not depend either on x or on t, that is, it is a constant. If we denote this constant by $-\lambda$ (cf. Sec. 19.5), from Eq. (5) for the unknown functions T and X and the parameter λ we obtain the equations

$$LX = \lambda \varrho X \tag{6}$$

$$T'' + \lambda T = 0 \tag{7}$$

Therefore Eq. (1) has split into the two equations (6) and (7) with a smaller number of independent variables; that is, the variables have been separated.

The solutions $X(x)$ of Eq. (6) must satisfy the boundary condition (3). Therefore we may take the eigenfunctions X_k and the eigenvalues λ_k of the operator L as X and λ, respectively. The general solution of Eq. (6)/(7) for $\lambda = \lambda_k > 0$ has the form

$$T_k(t) = a_k \cos \sqrt{\lambda_k}\, t + b_k \sin \sqrt{\lambda_k}\, t \tag{8}$$

where a_k and b_k are arbitrary constants.

In this way, by virtue of (4) and (8), a countable number of particular (linearly independent) solutions of Eq. (1) have been constructed

$$T_k(t)X_k(x) = (a_k \cos \sqrt{\lambda_k}\, t + b_k \sin \sqrt{\lambda_k}\, t)X_k(x), \quad k = 1, 2, \ldots \tag{9}$$

which satisfy the boundary condition (3) and which contain the arbitrary constants a_k and b_k. Naturally, each finite sum of solutions (9) will again satisfy Eq. (1) and the boundary condition (3).

We shall construct the formal series

$$\sum_{k=1}^{\infty} T_k(t)X_k(x) = \sum_{k=1}^{\infty} (a_k \cos \sqrt{\lambda_k}\, t + b_k \sin \sqrt{\lambda_k}\, t)X_k(x) \tag{10}$$

We choose the coefficients a_k and b_k such that the series (10) formally satisfies the initial conditions (2):

$$\sum_{k=1}^{\infty} a_k X_k(x) = u_0(x), \quad \sum_{k=1}^{\infty} \sqrt{\lambda_k}\, b_k X_k(x) = u_1(x)$$

that is, since the orthonormal system $\{X_k\}$ is complete in $\mathscr{L}_2(G; \varrho)$,

$$a_k = (u_0, X_k)_\varrho = \int_G \varrho u_0 X_k\, dx, \quad b_k = \frac{1}{\sqrt{\lambda_k}} (u_1, X_k)_\varrho \tag{11}$$

So, for the solution $u(x, t)$ of the mixed problem (1)-(2)-(3) we have obtained a formal expansion involving the eigenfunctions $\{X_k\}$ of the operator L,

$$u(x, t) = \sum_{k=1}^{\infty} (a_k \cos \sqrt{\lambda_k}\, t + b_k \sin \sqrt{\lambda_k}\, t) X_k(x) \qquad (12)$$

We shall call this series the *formal solution* of the mixed problem (1)-(2)-(3); the k-th term of series (12), equal to

$$T_k(t) X_k(x) = N_k X_k(x) \sin(\sqrt{\lambda_k}\, t + \alpha_k)$$

where

$$N_k = \sqrt{a_k^2 + b_k^2}, \qquad \sin \alpha_k = \frac{a_k}{N_k}, \qquad \cos \alpha_k = \frac{b_k}{N_k}$$

is said to be the *harmonic component with fundamental frequency* $\sqrt{\lambda_k}$ and amplitude $N_k X_k(x)$.

2. The Inhomogeneous Hyperbolic Equation. We shall set out another, more general, variant of Fourier's method, which is used to construct a formal solution of the mixed problem for the inhomogeneous equation of hyperbolic type

$$\varrho \frac{\partial^2 u}{\partial t^2} = -Lu + F(x, t) \qquad (13)$$

For each $t > 0$ we shall expand the solution $u(x, t)$ of the problem (13)-(2)-(3) into a Fourier series involving the eigenfunctions $\{X_k\}$ of the operator L,

$$u(x, t) = \sum_{k=1}^{\infty} T_k(t) X_k(x), \qquad T_k(t) = (u, X_k)_\varrho \qquad (14)$$

By virtue of (2), (14), and (11), the unknown functions $T_k(t)$ must satisfy the initial conditions:

$$T_k(0) = \int_G \varrho(x) u(x, 0) X_k(x)\, dx = (u_0, X_k)_\varrho = a_k$$

$$T'_k(0) = \int_G \varrho(x) \frac{\partial u(x, 0)}{\partial t} X_k(x)\, dx = (u_1, X_k)_\varrho = \sqrt{\lambda_k}\, b_k \qquad (15)$$

We shall construct a differential equation for the functions T_k. If we

§ 28. FOURIER'S METHOD

multiply Eq. (13) by X_k, integrate over G, and carry out the formal calculations, we obtain

$$\int_G \varrho \frac{\partial^2 u}{\partial t^2} X_k \, dx = \frac{d^2}{dt^2} \int_G \varrho u X_k \, dx = \frac{d^2}{dt^2} (u, X_k)_\varrho$$
$$= -(Lu, X_k) + (F, X_k)$$
$$= -(u, LX_k) + (F, X_k) = -\lambda_k (u, X_k)_\varrho + (F, X_k)$$

that is, by virtue of (14), the functions T_k satisfy the equation

$$T_k'' + \lambda_k T_k = c_k(t), \qquad k = 1, 2, \ldots \tag{16}$$

where

$$c_k(t) = (F, X_k) = \int_G F(x, t) X_k(x) \, dx \tag{17}$$

Solving the Cauchy problem for Eq. (16) with initial conditions (15), we have (cf. Sec. 12.1)

$$T_k(t) = a_k \cos \sqrt{\lambda_k} \, t + b_k \sin \sqrt{\lambda_k} \, t$$
$$+ \frac{1}{\sqrt{\lambda_k}} \int_0^t c_k(\tau) \sin \sqrt{\lambda_k} \, (t - \tau) \, d\tau \tag{18}$$

If we substitute expression (18) into the series (14), we obtain the formal solution of the mixed problem (13)-(2)-(3)

$$u(x, t) = \sum_{k=1}^\infty \left[a_k \cos \sqrt{\lambda_k} \, t + b_k \sin \sqrt{\lambda_k} \, t \right.$$
$$\left. + \frac{1}{\sqrt{\lambda_k}} \int_0^t c_k(\tau) \sin \sqrt{\lambda_k} \, (t - \tau) \, d\tau \right] X_k(x) \tag{19}$$

We note that the first two terms in the series (19), by virtue of (12), give a formal solution of the mixed problem for $F = 0$; the third term is the solution of this problem for $u_0 = u_1 = 0$.

Let $u_0 = u_1 = 0$ and

$$F(x, t) = C \sin \omega t \varrho(x) X_i(x) \tag{20}$$

Then

$$a_k = b_k = 0, \qquad c_k(t) = C \sin \omega t (X_i, X_k)_\varrho = C \delta_{ik} \sin \omega t$$

and therefore, by virtue of (18),

$$T_k(t) = \frac{C\delta_{ik}}{\sqrt{\lambda_k}} \int_0^t \sin \omega\tau \sin \sqrt{\lambda_k}(t-\tau)\, d\tau$$

$$= \frac{C\delta_{ik}}{\omega^2 - \lambda_i}\left(\frac{\omega}{\sqrt{\lambda_i}} \sin \sqrt{\lambda_i}\, t - \sin \omega t\right)$$

Therefore the formal series (19) is reduced to the single term

$$u(x, t) = \frac{C}{\omega^2 - \lambda_i}\left(\frac{\omega}{\sqrt{\lambda_i}} \sin \sqrt{\lambda_i}\, t - \sin \omega t\right) X_i(x) \qquad (21)$$

which is the real solution of the problem. For $\omega \to \sqrt{\lambda_i}$ solution (21) takes the form

$$u(x, t) = \frac{C}{2\sqrt{\lambda_i}}\left(\frac{\sin \sqrt{\lambda_i}\, t}{\sqrt{\lambda_i}} - t \cos \sqrt{\lambda_i}\, t\right) X_i(x) \qquad (22)$$

It follows from Eq. (22) that the effect of the periodic external perturbation (20) with a frequency equal to one of the fundamental frequencies $\sqrt{\lambda_i}$ is to cause the amplitude of oscillations to increase without limit for $t \to \infty$; that is, *resonance* is said to take place.

3. The Parabolic Equation. We shall consider the mixed problem for an equation of parabolic type (cf. Sec. 4.5)

$$\varrho \frac{\partial u}{\partial t} = -Lu + F(x, t) \qquad (23)$$

$$u\big|_{t=+0} = u_0(x), \qquad x \in \bar{G} \qquad (24)$$

$$\alpha u + \beta \frac{\partial u}{\partial \mathbf{n}}\bigg|_S = 0, \qquad t \geq 0 \qquad (25)$$

in the cylinder $\Pi_\infty = G \times (0, \infty)$.

We shall use Fourier's method in the form given in the preceding subsection to construct a formal solution of the mixed problem (23)-(24)-(25). In accordance with this method, the solution $u(x, t)$ will be sought in the

§ 28. FOURIER'S METHOD

form of series (14). For the functions $T_k(t)$ we obtain the Cauchy problem:

$$T'_k + \lambda_k T = c_k(t), \quad T_k(0) = a_k, \quad k = 1, 2, \ldots \quad (26)$$

where $c_k(t)$ and a_k are defined by Eqs. (17) and (15), respectively. Solving the Cauchy problem (26) we obtain (cf. Sec. 12.1)

$$T_k(t) = a_k e^{-\lambda_k l} + \int_0^t c_k(\tau) \exp[-\lambda_k(t-\tau)] \, d\tau \quad (27)$$

and so the formal solution of the mixed problem (23)-(24)-(25) is expressed by the series

$$u(x, t) = \sum_{k=1}^{\infty} \left[a_k \exp(-\lambda_k t) + \int_0^t c_k(\tau) \exp[-\lambda_k(t-\tau)] \, d\tau \right] X_k(x) \quad (28)$$

4. Schrödinger's Equation. The mixed problem for Schrödinger's equation (cf. Sec. 2.7)

$$i\hbar \frac{\partial \psi}{\partial t} = -\frac{\hbar^2}{2m} \Delta \psi + V(x)\psi \quad (29)$$

$$\psi|_{t=+0} = \psi_0(x), \quad x \in \bar{G} \quad (30)$$

$$\alpha \psi + \beta \frac{\partial \psi}{\partial \mathbf{n}}\bigg|_S = 0, \quad t \geq 0 \quad (31)$$

is considered in the same way as the mixed problem (23)-(24)-(25). For the functions T_k we have the following Cauchy problem:

$$i\hbar T'_k - \lambda_k T_k = 0, \quad T_k(0) = a_k = (\psi_0, X_k) \quad (32)$$

from which

$$T_k(t) = a_k \exp\left(-\frac{i}{\hbar} \lambda_k t\right) \quad (33)$$

and therefore the formal solution of the mixed problem (29)-(30)-(31) is expressed by the series

$$\psi(x, t) = \sum_{k=1}^{\infty} a_k \exp\left(-\frac{i}{\hbar} \lambda_k t\right) X_k(x) \quad (34)$$

where X_k are the eigenfunctions of the operator L for $p = \hbar^2/2m$ and $q = V$.

5. The Elliptic Equation.

We shall consider the boundary value problem for the equation of elliptic type

$$\varrho \frac{\partial^2 u}{\partial t^2} = Lu + F(x; t) \tag{35}$$

$$u\big|_{t=0} = u_0(x), \qquad u\big|_{t=l} = u_l(x), \qquad x \in \bar{G} \tag{36}$$

$$\alpha u + \beta \frac{\partial u}{\partial \mathbf{n}}\bigg|_S = 0, \qquad 0 \le t \le l \tag{37}$$

in the finite cylinder $\varPi_l = G \times (0, l)$.

The formal solution of this problem will be sought in the form of series (14). The unknown functions $T_k(t)$ must satisfy equation

$$T_k'' - \lambda_k T_k = c_k(t), \qquad k = 1, 2, \ldots \tag{38}$$

and the boundary conditions

$$T_k(0) = (u_0, X_k)_\varrho = a_k, \qquad T_k(l) = (u_l, X_k)_\varrho = b_k \tag{39}$$

the functions $c_k(t)$ are defined by Eq. (17).

We shall construct the solution of the boundary value problem (38)-(39). The function

$$v_k(t) = T_k(t) - a_k \frac{\sinh \sqrt{\lambda_k}\,(l-t)}{\sinh \sqrt{\lambda_k}\,l} - b_k \frac{\sinh \sqrt{\lambda_k}\,t}{\sinh \sqrt{\lambda_k}\,l}$$

satisfies Eq. (38) and the boundary conditions $v_k(0) = v_k(l) = 0$. Therefore this function is expressible by the equation (cf. Sec. 20.2)

$$v_k(t) = -\int_0^l \mathcal{G}_k(t, \tau) c_k(\tau)\, d\tau \tag{40}$$

where

$$\mathcal{G}_k(t, \tau) = \frac{1}{\sqrt{\lambda_k}\, \sinh \sqrt{\lambda_k}\, l} \begin{cases} \sinh \sqrt{\lambda_k}\, t \, \sinh \sqrt{\lambda_k}\,(l-\tau), & 0 \le t \le \tau \\ \sinh \sqrt{\lambda_k}\,(l-t) \sinh \sqrt{\lambda_k}\, \tau, & \tau \le t \le l \end{cases}$$

is the Green's function for the boundary value problem (cf. Sec. 20.1)

$$-v'' + \lambda_k v = -c_k(t), \qquad v(0) = v(l) = 0$$

§ 28. FOURIER'S METHOD

Consequently,

$$T_k(t) = a_k \frac{\sinh \sqrt{\lambda_k}\,(l-t)}{\sinh \sqrt{\lambda_k}\,l} + b_k \frac{\sinh \sqrt{\lambda_k}\,t}{\sinh \sqrt{\lambda_k}\,l} - \int_0^l \mathcal{G}_k(t,\tau) c_k(\tau)\,d\tau \quad (41)$$

In this way, the formal solution of the boundary value problem (35)-(36)-(37) is expressed by the series

$$u(x,t) = \sum_{k=1}^{\infty} \left[a_k \frac{\sinh \sqrt{\lambda_k}\,(l-t)}{\sinh \sqrt{\lambda_k}\,l} + b_k \frac{\sinh \sqrt{\lambda_k}\,t}{\sinh \sqrt{\lambda_k}\,l} \right.$$
$$\left. - \int_0^l \mathcal{G}_k(t,\tau) c_k(\tau)\,d\tau \right] X_k(x) \quad (42)$$

6. Examples

(a) OSCILLATION OF A FIXED STRING. This problem is reducible to the solution of the mixed problem in the half-strip $(0, l) \times (0, \infty)$ for the one-dimensional wave equation (cf. Sec. 2.1)

$$\Box_a u = 0, \quad u\big|_{t=+0} = u_0(x), \quad \frac{\partial u}{\partial t}\bigg|_{t=+0} = u_1(x)$$
$$u\big|_{x=0} = u\big|_{x=l} = 0 \quad (43)$$

The corresponding problem involving eigenvalues is the Sturm–Liouville problem:

$$-a^2 X'' = \lambda X, \quad X(0) = X(l) = 0$$

Therefore (cf. Sec. 19.6)

$$\lambda_k = \left(\frac{k\pi a}{l}\right)^2, \quad X_k(x) = \sqrt{\frac{2}{l}} \sin \frac{k\pi x}{l}, \quad k = 1, 2, \ldots$$

and the formal solution of problem (43) is expressed by the series

$$u(x,t) = \frac{2}{l} \sum_{k=1}^{\infty} \left(a_k \cos \frac{k\pi a}{l} t + b_k \sin \frac{k\pi a}{l} t \right) \sin \frac{k\pi x}{l} \quad (44)$$

where

$$a_k = \int_0^l u_0(x) \sin \frac{k\pi x}{l}\,dx, \quad b_k = \frac{l}{k\pi a} \int_0^l u_1(x) \sin \frac{k\pi x}{l}\,dx$$

Each harmonic component

$$T_k(t)X_k(x) = N_k\sqrt{\frac{2}{l}}\sin\frac{k\pi x}{l}\sin\left(\frac{k\pi a}{l}t + \alpha_k\right), \qquad k = 1, 2, \ldots$$

comprises a *standing wave* with the fundamental frequency $k\pi a/l$ and the amplitude

$$N_k\sqrt{\frac{2}{l}}\sin\frac{k\pi x}{l}$$

The zero $(n/k)l$, $n = 0, 1, \ldots, k$, of the amplitude are known as *nodes*, while its extrema occurring at the points $[(n + 0.5)/k]l$, $n = 0, 1, \ldots, k - 1$, are the *antinodes* of this standing wave (Fig. 75).

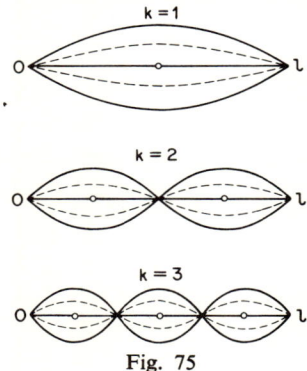

Fig. 75

The harmonic component T_1X_1 with the smallest fundamental frequency $\sqrt{\lambda_1} = \pi a/l$ is known as the *basic tone*; the other harmonic components $T_2X_2, T_3X_3, \ldots$, with the fundamental frequencies

$$\sqrt{\lambda_2} = \frac{2\pi a}{l}, \qquad \sqrt{\lambda_3} = \frac{3\pi a}{l}, \qquad \ldots$$

form a series of sequential *overtones*. The solution (44) is formed from the separate tones (the basic tone and the overtones), and their combined effect creates the *timbre of the sound* emitted by a fixed string.

(b) HEAT DIFFUSION IN A BOUNDED ROD. We shall consider the mixed problem for a one-dimensional heat conduction equation:

$$\frac{\partial u}{\partial t} = a^2\frac{\partial^2 u}{\partial x^2}, \qquad u\big|_{t=+0} = u_0(x) \qquad u\big|_{x=0} = u\big|_{x=l} = 0 \qquad (45)$$

§ 28. FOURIER'S METHOD

The formal solution of problem (45) is expressed by the series

$$u(x, t) = \frac{2}{l} \sum_{k=1}^{\infty} a_k \exp\left(-\frac{k^2\pi^2 a^2}{l^2} t\right) \sin \frac{k\pi x}{l} \qquad (46)$$

(c) OSCILLATIONS OF A FIXED MEMBRANE. The problem is reducible to the solution of a mixed problem for the two-dimensional wave equation

$$\Box_a u = 0, \qquad u\big|_{t=+0} = u_0(x), \qquad \frac{\partial u}{\partial t}\bigg|_{t=+0} = u_1(x), \qquad u\big|_S = 0 \qquad (47)$$

The corresponding problem involving eigenvalues takes the form

$$-a^2 \Delta X = \lambda X, \qquad X\big|_S = 0$$

For the rectangle $(0, l) \times (0, m)$ (cf. Sec. 19.6)

$$\lambda_{kj} = \pi^2 a^2\left(\frac{k^2}{l^2} + \frac{j^2}{m^2}\right), \qquad X_{kj}(x, y) = \frac{2}{\sqrt{lm}} \sin \frac{k\pi x}{l} \sin \frac{j\pi y}{m}$$

$$k, j = 1, 2, \ldots$$

and the formal solution of problem (47) is expressed by the double series

$$u(x, y, t) = \frac{4}{lm} \sum_{k,j=1}^{\infty} \left[a_{kj} \cos \pi a\left(\frac{k^2}{l^2} + \frac{j^2}{m^2}\right)^{1/2} t \right.$$

$$\left. + b_{kj} \sin \pi a\left(\frac{k^2}{l^2} + \frac{j^2}{m^2}\right)^{1/2} t\right] \sin \frac{k\pi x}{l} \sin \frac{j\pi y}{m} \qquad (48)$$

where

$$a_{kj} = \int_0^l \int_0^m u_0(x, y) \sin \frac{k\pi x}{l} \sin \frac{j\pi y}{m} dx\, dy$$

$$b_{kj} = \frac{lm}{\pi a \sqrt{k^2 m^2 + j^2 l^2}} \int_0^l \int_0^m u_1(x, y) \sin \frac{k\pi x}{l} \sin \frac{j\pi y}{m} dx\, dy$$

For the circle U_R (cf. Sec. 19.6)

$$\lambda_{kj} = \mu_{kj}^2 \frac{a^2}{R^2}, \qquad X_{kj}(x, y) = \frac{1}{R\sqrt{\pi} \,|J_k'(\mu_{kj})|} J_k\left(\mu_{kj} \frac{r}{R}\right) e^{ik\varphi}$$

$$k = 0, 1, \ldots \qquad j = 1, 2, \ldots$$

where μ_{kj} are the positive roots of the equation $J_k(\mu) = 0$. The formal solution of problem (47) is expressed by the double series

$$u(x, y, t) = \frac{1}{\pi R^2} \sum_{k=0}^{\infty} \sum_{j=1}^{\infty} \left(a_{kj} \cos \frac{\mu_{kj} a}{R} t \right.$$
$$\left. + b_{kj} \sin \frac{\mu_{kj} a}{R} t \right) \frac{J_k[\mu_{kj}(r/R)]}{[J'_k(\mu_{kj})]^2} e^{ik\varphi} \quad (49)$$

where

$$a_{kj} = \int_0^R \int_0^{2\pi} u_0(x, y) J_k\left(\mu_{kj} \frac{r}{R}\right) e^{-ik\varphi} r \, dr \, d\varphi$$

$$b_{kj} = \frac{R}{a\mu_{kj}} \int_0^R \int_0^{2\pi} u_1(x, y) J_k\left(\mu_{kj} \frac{r}{R}\right) e^{-ik\varphi} r \, dr \, d\varphi$$

(d) The formal solution of the mixed problem for the two-dimensional heat conduction equation,

$$\frac{\partial u}{\partial t} = a^2 \Delta u, \quad u|_{t=+0} = u_0(x), \quad u|_S = 0 \quad (50)$$

has the form: for the rectangle $(0, l) \times (0, m)$

$$u(x, y, t) = \frac{4}{lm} \sum_{k,j=1}^{\infty} a_{kj} \exp\left[-\pi^2 a^2 \left(\frac{k^2}{l^2} + \frac{j^2}{m^2}\right) t\right] \sin \frac{k\pi x}{l} \sin \frac{j\pi y}{m} \quad (51)$$

for the circle U_R

$$u(x, y, t) = \frac{1}{\pi R^2} \sum_{k=0}^{\infty} \sum_{j=1}^{\infty} a_{kj} \exp\left(-\frac{\mu_{kj}^2 a^2}{R^2} t\right) \frac{J_k[\mu_{kj}(r/R)]}{[J'_k(\mu_{kj})]^2} e^{ik\varphi} \quad (52)$$

(e) OSCILLATION OF A SPHERICAL VOLUME. We shall consider the mixed problem (47) for the three-dimensional wave equation in the cylinder $U_R \times (0, \infty)$. The corresponding eigenvalues and eigenfunctions were calculated in Sec. 25.8:

$$\lambda_{lj} = \mu_{lj}^2 \frac{a^2}{R^2}$$

$$X_{ljm}(x) = \frac{J_{l+1/2}\left(\mu_{lj} \frac{r}{R}\right) Y_l^m(\theta, \varphi)}{\sqrt{rR} |J'_{l+1/2}(\mu_{lj})| \left[\pi \frac{1 + \delta_{0m}}{2l + 1} \frac{(l + |m|)!}{(l - |m|)!}\right]^{1/2}}$$

$$m = 0, \pm 1, \ldots, \pm l, \quad j = 1, 2, \ldots, \quad l = 0, 1, \ldots$$

§ 28. FOURIER'S METHOD

where μ_{lj} are the positive roots of the equation $J_{l+1/2}(\mu) = 0$. The formal solution of the problem is expressed by the series

$$u(x, t) = \frac{1}{\pi R^2 \sqrt{r}} \sum_{l=0}^{\infty} \sum_{j=1}^{\infty} \sum_{m=-l}^{l} \left(a_{ljm} \cos \frac{\mu_{lj}a}{R} t \right.$$
$$\left. + b_{ljm} \sin \frac{\mu_{lj}a}{R} t\right) \frac{2l+1}{1+\delta_{0m}} \frac{(l-|m|)!}{(l+|m|)!} \frac{1}{[J'_{l+1/2}(\mu_{lj})]^2}$$
$$\times J_{l+1/2}\left(\mu_{lj} \frac{r}{R}\right) Y_l^m(\theta, \varphi) \tag{53}$$

where

$$a_{ljm} = \int_0^R \int_0^\pi \int_0^{2\pi} u_0(x) J_{l+1/2}\left(\mu_{lj} \frac{r}{R}\right) \overline{Y_l^m(\theta, \varphi)} r^{3/2} \, dr \sin\theta \, d\theta \, d\varphi$$

$$b_{ljm} = \frac{R}{a\mu_{lj}} \int_0^R \int_0^\pi \int_0^{2\pi} u_1(x) J_{l+1/2}\left(\mu_{lj} \frac{r}{R}\right) \overline{Y_l^m(\theta, \varphi)} r^{3/2} \, dr \sin\theta \, d\theta \, d\varphi$$

(f) The formal solution of the mixed problem (50) for the three-dimensional heat conduction equation in the cylinder $U_R \times (0, \infty)$ is expressed by the series

$$u(x, t) = \frac{1}{\pi R^2 \sqrt{r}} \sum_{l=0}^{\infty} \sum_{j=1}^{\infty} \sum_{m=-l}^{l} a_{ljm} \exp\left(-\mu_{lj}^2 \frac{a^2}{R^2} t\right)$$
$$\times \frac{2l+1}{1+\delta_{0m}} \frac{(l-|m|)!}{(l+|m|)!} \frac{J_{l+1/2}[\mu_{lj}(r/R)] Y_l^m(\theta, \varphi)}{[J'_{l+1/2}(\mu_{lj})]^2} \tag{54}$$

(g) We shall consider in the cylinder $U_R \times (0, \infty)$ the mixed problem for Schrödinger's equation

$$i\hbar \frac{\partial \psi}{\partial t} = -\frac{\hbar^2}{2m_0} \Delta\psi + V(|x|)\psi, \quad \psi|_{t=+0} = \psi_0(x), \quad \psi|_{S_R} = 0 \tag{55}$$

with the potential V depending only on $|x|$. The corresponding problem involving eigenvalues takes the form

$$-\frac{\hbar^2}{2m_0} \Delta X + V(|x|)X = \lambda X, \quad X|_{S_R} = 0$$

or, in spherical coordinates,

$$-\frac{\hbar^2}{2m_0}\left[\frac{1}{r^2}\frac{\partial}{\partial r}\left(r^2 \frac{\partial X}{\partial r}\right) + \frac{1}{r^2 \sin^2\theta} \frac{\partial}{\partial \theta}\left(\sin\theta \frac{\partial X}{\partial \theta}\right)\right.$$
$$\left. + \frac{1}{r^2 \sin\theta} \frac{\partial^2 X}{\partial \varphi^2}\right] + V(r)X = \lambda X$$
$$|X(0, \theta, \varphi)| < \infty, \quad X(R, \theta, \varphi) = 0 \tag{56}$$

The eigenvalues and eigenfunctions of the boundary value problem (56) are defined by the method of separation of variables. Putting

$$u = \frac{\mathscr{R}(r)}{r} Y(\theta, \varphi)$$

and proceding as in Sec. 25.8, we shall obtain

$$\lambda_{lj}, X_{ljm}(x) = \left[\frac{2l+1}{\pi(1+\delta_{0m})} \frac{(l-|m|)!}{(l+|m|)!}\right]^{1/2} \frac{\mathscr{R}_{lj}(r)}{r} Y_l^m(\theta, \varphi)$$

$$m = 0, \pm 1, \ldots, \pm l, \qquad j = 1, 2, \ldots, \qquad l = 0, 1, \ldots$$

where λ_{lj} and $\mathscr{R}_{lj}(r)$, $j = 1, 2, \ldots$ are the eigenvalues and the eigenfunctions of the one-dimensional boundary value problem

$$-\mathscr{R}'' + \frac{l(l+1)}{r^2}\mathscr{R} + \frac{2m_0}{\hbar^2}(V-\lambda)\mathscr{R} = 0, \qquad \mathscr{R}(0) = \mathscr{R}(R) = 0$$

The formal solution of problem (55) is expressed by the series

$$u(x, t) = \frac{1}{\pi r} \sum_{l=0}^{\infty} \sum_{j=1}^{\infty} \sum_{m=-l}^{l} a_{ljm} \exp\left(-\frac{i}{\hbar}\lambda_{lj}t\right) \frac{2l+1}{1+\delta_{0m}} \frac{(l-|m|)!}{(l+|m|)!}$$
$$\times \mathscr{R}_{lj}(r) Y_l^m(\theta, \varphi) \qquad (57)$$

where

$$a_{ljm} = \int_0^R \int_0^\pi \int_0^{2\pi} u_0(x)\mathscr{R}_{lj}(r) Y_l^m(\theta, \varphi) r \, dr \sin\theta \, d\theta \, d\varphi$$

The eigenvalues λ_{lj} define the energy levels of the quantum particle; the indices l and m are said to be the *orbital* (azimuth) and *magnetic quantum numbers*, respectively.

(h) The formal solution of the Dirichlet problem

$$\Delta u = 0, \quad u|_{x=0} = u|_{x=a} = 0, \quad u|_{y=0} = u_0(x), \quad u|_{y=l} = u_l(x) \quad (58)$$

in the rectangle $(0, a) \times (0, l)$ is expressed by the series

$$u(x, y) = \frac{2}{a} \sum_{k=1}^{\infty} \left(a_k \sinh k\pi \frac{l-y}{a} + b_k \sinh \frac{k\pi y}{a}\right) \frac{\sin(k\pi x/a)}{\sinh k\pi l/a} \quad (59)$$

where

$$a_k = \int_0^l u_0(x) \sin \frac{k\pi x}{a} dx, \qquad b_k = \int_0^l u_l(x) \sin \frac{k\pi x}{a} dx$$

§ 28. FOURIER'S METHOD

(i) We shall consider the Dirichlet problem in the three-dimensional cylinder $U_R \times (0, h)$:

$$\Delta u = 0, \quad u|_{S_R} = u_0(z), \quad u|_{z=0} = u|_{z=h} = 0 \tag{60}$$

In cylindrical coordinates, for the function $\tilde{u}(r, z) = u(x, y, z)$ this problem takes the form*

$$\frac{1}{r}\frac{\partial}{\partial r}\left(r \frac{\partial \tilde{u}}{\partial r}\right) + \frac{\partial^2 \tilde{u}}{\partial z^2} = 0$$

$$\tilde{u}(R, z) = u_0(z), \quad \tilde{u}(r, 0) = \tilde{u}(r, h) = 0 \tag{61}$$

If we solve the boundary value problem (61) by the method of separation of variables, $\tilde{u}(r, z) = \mathscr{R}(r)Z(z)$, for the functions $\mathscr{R}$ and Z we shall obtain the boundary value problems

$$Z'' + \lambda Z = 0, \quad Z(0) = Z(h) = 0$$

$$\mathscr{R}'' + \frac{\mathscr{R}'}{r} - \lambda \mathscr{R} = 0, \quad |\mathscr{R}(0)| < \infty$$

The solutions of these boundary value problems are easily found to be

$$\lambda_k = \frac{k^2 \pi^2}{h^2}, \quad Z_k(z) = \sqrt{\frac{2}{h}} \sin \frac{k\pi z}{h}, \quad \mathscr{R}_k(r) = c_k I_0\left(k\pi \frac{r}{h}\right)$$

where I_0 is the modified Bessel function with an imaginary argument $I_0(x) = J_0(ix)$.

The formal solution of problem (61), and therefore of problem (60), is expressed by the series

$$u(x) = \tilde{u}(r, z) = \frac{2}{h} \sum_{k=1}^{\infty} a_k \frac{I_0[k\pi(r/h)]}{I_0[k\pi(R/h)]} \sin \frac{k\pi z}{h} \tag{62}$$

where

$$a_k = \int_0^h u_0(z) \sin \frac{k\pi z}{h} dz$$

7. Exercises. (a) Prove: If $u_0''' \in \mathscr{L}_2(0, l)$, $u_0(0) = u_0(l) = u_0''(0) = u_0''(l) = 0$, $u_1'' \in \mathscr{L}_2(0, l)$, $u_1(0) = u_1(l) = 0$, then the series (44) represents the classical solution of problem (43).

(b) Prove: If $u_0' \in \mathscr{L}_2(0, l)$, $u_0(0) = u_0(l) = 0$, then the series (46) represents the classical solution of problem (45).

---

* The solution $\tilde{u}(r, z)$ evidently does not depend on φ.

(c) Prove: If u_0 and $u_l \in C^2(\bar{G})$, $u_0|_S = u_l|_S = 0$, then the series

$$u(x, y) = \sum_{k=1}^{\infty} \left[(u_0, X_k) \frac{\sinh \sqrt{\lambda_k}(l-y)}{\sinh \sqrt{\lambda_k}\, l} + (u_l, X_k) \frac{\sinh \sqrt{\lambda_k}\, y}{\sinh \sqrt{\lambda_k}\, l} \right] X_k(x)$$

gives the solution of the following Dirichlet problem for Laplace's equation in the cylinder $G \times (0, l)$:

$$\frac{\partial^2 u}{\partial y^2} + \Delta_x u = 0, \qquad u|_{y=0} = u_0(x), \qquad u|_{y=l} = u_l(x), \qquad u|_S = 0$$

§ 29. The Mixed Problem for an Equation of Hyperbolic Type

In this section we shall consider the mixed problem for an equation of hyperbolic type (cf. Sec. 4.5):

$$\varrho \frac{\partial^2 u}{\partial t^2} = \operatorname{div}(p \operatorname{grad} u) - qu + F(x, t) \equiv -Lu + F(x, t)$$
$$(x, t) \in \mathit{\Pi}_\infty = G \times (0, \infty) \tag{1}$$

$$u|_{t=+0} = u_0(x), \qquad \left.\frac{\partial u}{\partial t}\right|_{t=+0} = u_1(x), \qquad x \in \bar{G} \tag{2}$$

$$\left.\alpha u + \beta \frac{\partial u}{\partial \mathbf{n}}\right|_S = 0, \qquad t \geq 0 \tag{3}$$

We shall assume that the functions ϱ, p, q, α, and β do not depend on t, and that they satisfy other conditions of Sec. 28; G is a bounded region and its boundary S is a piecewise smooth surface, S_0 is that part of S for which $\alpha(x) > 0$ and $\beta(x) > 0$ simultaneously.

1. The Classical Solution. The Energy Integral. The function $u(x, t)$ of the class $C^2(\mathit{\Pi}_\infty) \cap C^1(\bar{\mathit{\Pi}}_\infty)$ which satisfies Eq. (1) in the cylinder $\mathit{\Pi}_\infty$, initial conditions (2) on the lower base, and boundary condition (3) on the side surface of this cylinder is said to be the classical solution of the mixed problem (1)-(2)-(3).

The smoothness conditions

$$F \in C(\mathit{\Pi}_\infty), \qquad u_0 \in C^1(\bar{G}), \qquad u_1 \in C(\bar{G})$$

and the consistency condition

$$\left.\alpha u_0 + \beta \frac{\partial u_0}{\partial \mathbf{n}}\right|_S = 0$$

§ 29. EQUATION OF HYPERBOLIC TYPE

are necessary conditions for the existence of the classical solution of the problem (1)-(2)-(3).

The method of energy integrals is most effective when studying boundary value problems for hyperbolic equations. Let $u(x, t)$ be the classical solution of the problem (1)-(2)-(3). The magnitude

$$J^2(t) = \frac{1}{2} \int_G \left[\varrho \left(\frac{\partial u}{\partial t} \right)^2 + p \mid \text{grad } u \mid^2 + qu^2 \right] dx + \frac{1}{2} \int_{S_0} p \frac{\alpha}{\beta} u^2 \, dS$$

which is the sum of the kinetic and potential energy of an oscillating system at the instant of time t, is known as the *energy integral*.

Let $u(x, t)$ be the classical solution of the problem (1)-(2)-(3) and $F \in C(\bar{\Pi}_\infty)$. Then the result

$$J^2(t) = J^2(0) + \int_0^t \int_G F(x, \tau) \frac{\partial u(x, \tau)}{\partial \tau} d\tau, \qquad t \geq 0 \qquad (4)$$

is true, where

$$J^2(0) = \frac{1}{2} \int_G (\varrho u_1^2 + p \mid \text{grad } u_0 \mid^2 + q u_0^2) \, dx + \frac{1}{2} \int_{S_0} p \frac{\alpha}{\beta} u_0^2 \, dS$$

For our proof we shall take the arbitrary number $\varepsilon > 0$ and the arbitrary region $G' \Subset G$ with the piecewise smooth surface S' (Fig. 76). Multiplying Eq. (1) by $\partial u/\partial t$, integrating over the cylinder $G' \times (\varepsilon, T)$, and using Green's first formula (cf. Sec. 19.2), we obtain

$$\int_{G' \times (\varepsilon, T)} F \frac{\partial u}{\partial t} \, dx \, dt = \int_{G' \times (\varepsilon, T)} \frac{\partial u}{\partial t} \left(\varrho \frac{\partial^2 u}{\partial t^2} + Lu \right) dx \, dt$$

$$= \int_{G'} \varrho \int_\varepsilon^T \frac{\partial u}{\partial t} \frac{\partial^2 u}{\partial t^2} \, dt \, dx + \int_\varepsilon^T \int_{G'} \frac{\partial u}{\partial t} Lu \, dx \, dt$$

$$= \frac{1}{2} \int_{G'} \varrho \left(\frac{\partial u}{\partial t} \right)^2 \Big|_\varepsilon^T dx + \int_\varepsilon^T \left[\int_{G'} p \left(\text{grad } u, \text{grad } \frac{\partial u}{\partial t} \right) dx \right.$$

$$\left. - \int_{S'} p \frac{\partial u}{\partial t} \frac{\partial u}{\partial n} \, dS' + \int_{G'} qu \frac{\partial u}{\partial t} \, dx \right] dt$$

$$= \frac{1}{2} \int_{G'} \left[\varrho \left(\frac{\partial u}{\partial t} \right)^2 + p \mid \text{grad } u \mid^2 + qu^2 \right] \Big|_\varepsilon^T dx$$

$$- \int_\varepsilon^T \int_{S'} p \frac{\partial u}{\partial t} \frac{\partial u}{\partial n} \, dS' \, dt$$

Proceeding to the limit as $\varepsilon \to +0$ and $G' \to G$ and using the fact that $u \in C^1(\bar{\Pi}_T)$ and $F \in C(\bar{\Pi}_T)$, we obtain the equation

$$\frac{1}{2}\int_G \left[\varrho\left(\frac{\partial u}{\partial t}\right)^2 + p\,|\operatorname{grad} u|^2 + qu^2\right]\Big|_0^T dx - \int_0^T\int_S p\,\frac{\partial u}{\partial t}\,\frac{\partial u}{\partial \mathbf{n}}\,dS\,dt$$
$$= \int_{\Pi_T} F\,\frac{\partial u}{\partial t}\,dx\,dt \tag{5}$$

From boundary condition (3) follows the result over S: $\partial u/\partial \mathbf{n} = -(\alpha/\beta)u$ if $\beta > 0$; $u = 0$ if $\beta = 0$. Therefore

$$-\int_0^T\int_S p\,\frac{\partial u}{\partial t}\,\frac{\partial u}{\partial \mathbf{n}}\,dS\,dt = \int_0^T\int_{S_0} p\,\frac{\alpha}{\beta}\,u\,\frac{\partial u}{\partial t}\,dS\,dt = \frac{1}{2}\int_{S_0} p\,\frac{\alpha}{\beta}\,u^2\Big|_0^T dS$$

from which and from (5), replacing T by t, we obtain formula (4). The theorem is proved.

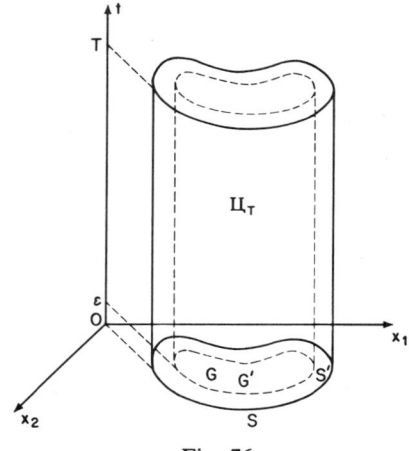

Fig. 76

COROLLARY. *For $F = 0$ Eq. (4) takes the form*

$$J^2(t) = J^2(0), \qquad t \geq 0 \tag{6}$$

THE PHYSICAL MEANING OF EQ. (6). The complete energy of an oscillating system does not change with time when exterior perturbations are absent (the law of energy conservation).

§ 29. EQUATION OF HYPERBOLIC TYPE

2. Uniqueness and Continuous Dependence of the Classical Solution. We shall use the method of energy integrals to prove that the classical solution of the mixed problem (1)-(2)-(3) is unique and continuously dependent.

Differentiating Eq. (4) with respect to t, we obtain

$$2J(t)J'(t) = \int_G F(x,t) \frac{\partial u(x,t)}{\partial t} dx, \qquad t \geq 0 \qquad (7)$$

If we apply the Cauchy–Buniakowsky inequality to the right-hand side of Eq. (7), we deduce the inequality

$$2JJ' \leq \|F\| \left\| \frac{\partial u}{\partial t} \right\| \qquad (8)$$

Assuming now that $\varrho(x) > 0$, $\varrho \in C(\bar{G})$, and therefore $\varrho(x) \geq \varrho_0$ for a certain $\varrho_0 > 0$, we obtain the sequence of inequalities

$$\left\| \frac{\partial u}{\partial t} \right\|^2 \leq \frac{1}{\varrho_0} \int_G \varrho \left(\frac{\partial u}{\partial t} \right)^2 dx \leq \frac{2}{\varrho_0} J^2(t)$$

that is,

$$\left\| \frac{\partial u}{\partial t} \right\| \leq \sqrt{\frac{2}{\varrho_0}} J(t) \qquad (9)$$

We may prove the validity of the result

$$\| |\operatorname{grad} u| \| \leq \sqrt{\frac{2}{p_0}} J(t) \qquad (10)$$

analogously, where $p_0 = \min p(x)$, $x \in \bar{G}$, $p_0 > 0$.

Substituting inequality (9) into inequality (8) and simplifying, we obtain the inequality

$$J'(t) \leq \frac{1}{\sqrt{2\varrho_0}} \|F\|, \qquad t \geq 0$$

Integrating, we obtain the differential inequality for J,

$$J(t) \leq J(0) + \frac{1}{\sqrt{2\varrho_0}} \int_0^t \|F\| \, d\tau \qquad (11)$$

From (9), (10), and (11) we deduce the results

$$\left\| \frac{\partial u}{\partial t} \right\| \leq \sqrt{\frac{2}{\varrho_0}} J(0) + \frac{1}{\varrho_0} \int_0^t \| F \| \, d\tau, \qquad t \geq 0 \qquad (12)$$

$$\| | \operatorname{grad} u | \| \leq \sqrt{\frac{2}{p_0}} J(0) + \frac{1}{\sqrt{\varrho_0 p_0}} \int_0^t \| F \| \, d\tau, \qquad t \geq 0 \qquad (13)$$

We shall now evaluate the function $\| u \|$. Differentiating the equation

$$\| u \|^2 = \int_G u^2(x, t) \, dx$$

with respect to t, using the Cauchy–Buniakowsky inequality and allowing inequality (12), we obtain

$$2 \| u \| \, \| u \|' = 2 \int_G u \frac{\partial u}{\partial t} \, dx \leq 2 \| u \| \left\| \frac{\partial u}{\partial t} \right\|$$

$$\leq 2 \| u \| \left[\sqrt{\frac{2}{\varrho_0}} J(0) + \frac{1}{\varrho_0} \int_0^t \| F \| \, d\tau \right]$$

that is, after dividing by $2 \| u \|$,

$$\| u \|' \leq \sqrt{\frac{2}{\varrho_0}} J(0) + \frac{1}{\varrho_0} \int_0^t \| F \| \, d\tau, \qquad t \geq 0$$

Integrating this differential inequality, we have

$$\| u \| \leq \| u \|_0 + \sqrt{\frac{2}{\varrho_0}} J(0)t + \frac{1}{\varrho_0} \int_0^t \int_0^{t'} \| F \| \, d\tau \, dt'$$

where $\| u \|_0$ is the value of the function $\| u \|$ at the point $t = 0$; that is,

$$\| u \|_0^2 = \int_G u^2(x, 0) \, dx = \int_G u_0^2(x) \, dx = \| u_0 \|^2$$

Changing the order of integration in the last integral, we obtain the required result

$$\| u \| \leq \| u_0 \| + \sqrt{\frac{2}{\varrho_0}} J(0)t + \frac{1}{\varrho_0} \int_0^t (t - \tau) \| F \| \, d\tau, \qquad t \geq 0 \qquad (14)$$

Now, using the results (12) through (14), we shall prove the following theorem.

§ 29. EQUATION OF HYPERBOLIC TYPE

THEOREM. *The classical solution of the problem* (1)-(2)-(3) *is unique and depends continuously on* u_0, u_1, *and* F *in the sense that if* $F \in C(\bar{\Pi}_T)$, $\tilde{F} \in C(\bar{\Pi}_T)$, *and*

$$\| F - \tilde{F} \| \leq \varepsilon, \qquad 0 \leq t \leq T; \qquad \| u_0 - \tilde{u}_0 \|_C \leq \varepsilon_0$$

$$\| | \operatorname{grad} u_0 - \operatorname{grad} \tilde{u}_0 | \| \leq \varepsilon_0', \qquad \| u_1 - \tilde{u}_1 \| \leq \varepsilon_1 \qquad (15)$$

then the corresponding (classical) solutions $u(x, t)$ *and* $\tilde{u}(x, t)$ *satisfy the inequalities for* $0 \leq t \leq T$:

$$\| u - \tilde{u} \| \leq C\left(\varepsilon_0 + T\varepsilon_0 + T\varepsilon_0' + T\varepsilon_1 + \frac{T^2}{2}\varepsilon\right) \qquad (16)$$

$$\| | \operatorname{grad} u - \operatorname{grad} \tilde{u} | \| \leq C(\varepsilon_0 + \varepsilon_0' + \varepsilon_1 + T\varepsilon) \qquad (17)$$

$$\left\| \frac{\partial u}{\partial t} - \frac{\partial \tilde{u}}{\partial t} \right\| \leq C(\varepsilon_0 + \varepsilon_0' + \varepsilon_1 + T\varepsilon) \qquad (18)$$

and the number C *does not depend on* u_0, u_1, F, t, *or* T.

Proof. To prove uniqueness, it is sufficient to establish that the classical solution $u(x, t)$ of the homogeneous problem (1)-(2)-(3) (for $u_0 = u_1 = 0$ and $F = 0$) is unique, that is, $u(x, t) = 0$, $(x, t) \in \Pi_\infty$ (cf. Sec. 1.9). But this follows from inequality (14), since $\| u_0 \| = 0$, $J(0) = 0$, and $F = 0$.

To prove continuous dependence we shall form the difference $\eta = u - \tilde{u}$. The function η is the classical solution of the problem (1)-(2)-(3) with the replacement of F, u_0, and u_1 by $F - \tilde{F}$, $u_0 - \tilde{u}_0$, and $u_1 - \tilde{u}_1$, respectively. Using inequalities (15) for the solution η we may evaluate the magnitude of the energy integral $\tilde{J}^2(0)$,

$$2\tilde{J}^2(0) = \int_G [\varrho(u_1 - \tilde{u}_1)^2 + p | \operatorname{grad} u_0 - \operatorname{grad} \tilde{u}_0 |^2 + q(u_0 - \tilde{u}_0)^2] \, dx$$

$$+ \int_{S_0} p \frac{\alpha}{\beta} (u_0 - \tilde{u}_0)^2 \, dS \leq \max_{x \in \bar{G}} \varrho(x) \varepsilon_1^2 + \max_{x \in \bar{G}} p(x)(\varepsilon_0')^2$$

$$+ \left[V \max_{x \in \bar{G}} q(x) + \sigma \max_{x \in S_0} p \frac{\alpha}{\beta}(x) \right] \varepsilon_0^2 \leq C_1^2(\varepsilon_0^2 + \varepsilon_0'^2 + \varepsilon_1^2)$$

$$\leq C_1^2(\varepsilon_0 + \varepsilon_0' + \varepsilon_1)^2$$

where V is the volume of the region G, σ is the area of the piece S_0, and C_1^2 is a number greater than the numbers $\max \varrho$, $\max p$, and $V \max q + \sigma$

max $p(\alpha/\beta)$. In this way we obtain the result

$$\sqrt{2}\,\tilde{J}(0) \leq C_1(\varepsilon_0 + \varepsilon_0' + \varepsilon_1) \tag{19}$$

If we now apply inequality (14) to the solution η,

$$\|\eta\| \leq \|u_0 - \tilde{u}_0\| + \sqrt{\frac{2}{\varrho_0}}\,\tilde{J}(0)t + \frac{1}{\varrho_0}\int_0^t (t-\tau)\,\|F - \tilde{F}\|\,d\tau$$

and use inequalities (15) and (19), we obtain for all $t \in [0, T]$ the result (16)

$$\|\eta\| \leq \left[\int_G (u_0 - \tilde{u}_0)^2\,dx\right]^{1/2} + \frac{1}{\sqrt{\varrho_0}}\,C_1(\varepsilon_0 + \varepsilon_0' + \varepsilon_1)t$$
$$+ \frac{\varepsilon}{\varrho_0}\int_0^t (t-\tau)\,d\tau \leq \varepsilon_0\sqrt{V} + \frac{T}{\sqrt{\varrho_0}}\,C_1(\varepsilon_0 + \varepsilon_0' + \varepsilon_1)$$
$$+ \frac{\varepsilon}{2\varrho_0}T^2 \leq C\left(\varepsilon_0 + \varepsilon_0 T + \varepsilon_0' T + \varepsilon_1 T + \frac{\varepsilon}{2}T^2\right)$$

for a suitable choice of constant C.

Inequalities (17) and (18) are established analogously, by means of inequalities (12), (13), and (19). The theorem is proved.

Proof that the classical solution of the problem (1)-(2)-(3) exists presents considerable difficulty. To avoid these difficulties the concept of a generalized solution of the problem is introduced, as for the Cauchy problem, since the existence of a generalized solution is established much more simply. Before doing this we shall study in more detail the functions belonging to $\mathscr{L}_2(G)$ which depend on a parameter.

3. Functions Continuous in $\mathscr{L}_2(G)$. For every $t \in [a, b]$ let the function $u(x, t)$ belong to $\mathscr{L}_2(G)$. The function $u(x, t)$ is said to be *continuous in* $\mathscr{L}_2(G)$ *with respect to the variable* t *on* $[a, b]$ if for any $t \in [a, b]$

$$u(x, t') \to u(x, t), \qquad t' \to t \quad \text{in} \quad \mathscr{L}_2(G)$$

It follows from this definition: If the function $u(x, t)$ is continuous in $\mathscr{L}_2(G)$ with respect to t on $[a, b]$, then the norm $\|u(x, t)\|$ is continuous with respect to t on $[a, b]$; for any $f \in \mathscr{L}_2(G)$ the scalar product $(u(x, t), f)$ is continuous with respect to t in $[a, b]$; $u \in \mathscr{L}_2(G \times (a, b))$.

§ 29. EQUATION OF HYPERBOLIC TYPE

In fact, the continuity of $\|u\|$ follows from the inequality

$$|\,\|u(x,t')\| - \|u(x,t)\|\,| \leq \|u(x,t') - u(x,t)\|$$

which is a consequence of Minkowski's inequality. The continuity of (u, f) follows from the Cauchy–Buniakowsky inequality

$$|(u(x,t'),f) - (u(x,t),f)| \leq \|u(x,t') - u(x,t)\|\,\|f\|$$

The fact that u belongs to $\mathscr{L}_2(G \times (a,b))$ follows from the finiteness of (a,b), the continuity of $\|u\|$, and the equation

$$\int_a^b \int_G |u(x,t)|^2 \, dx \, dt = \int_a^b \|u(x,t)\|^2 \, dt$$

The sequence of functions $u_k(x,t)$, $k = 1, 2, \ldots$, is said to *converge to the function* $u(x,t)$ in $\mathscr{L}_2(G)$ *uniformly with respect to* t on $[a,b]$ if

$$\|u_k(x,t) - u(x,t)\| \xrightarrow{t \in [a,b]} 0, \qquad k \to \infty$$

for this we shall write

$$u_k \xrightarrow{t \in [a,b]} u, \qquad k \to \infty \quad \text{in} \quad \mathscr{L}_2(G)$$

It follows from this definition that

$$u_k \to u, \qquad k \to \infty \quad \text{in} \quad \mathscr{L}_2(G \times (a,b))$$

$$\|u_k(x,t)\| \xrightarrow{t \in [a,b]} \|u(x,t)\|, \qquad k \to \infty$$

LEMMA 1. *If the sequence* $u_k(x,t)$, $k = 1, 2, \ldots$, *of functions which are continuous in* $\mathscr{L}_2(G)$ *with respect to* t *on* $[a,b]$, *converge to the function* $u(x,t)$ *in* $\mathscr{L}_2(G)$ *uniformly with respect to* t *on* $[a,b]$ *then* $u(x,t)$ *is a continuous function in* $\mathscr{L}_2(G)$ *with respect to* t *on* $[a,b]$.

Proof. We take an arbitrary $\varepsilon > 0$. There is a number $m = m_\varepsilon$ such that

$$\|u_m(x,t) - u(x,t)\| < \varepsilon/3, \qquad t \in [a,b]$$

By the stated conditions the function $u_m(x,t)$ is continuous in $\mathscr{L}_2(G)$ with respect to $t \in [a, b]$. Therefore there is a number $\delta = \delta_\varepsilon$ such that

$$\| u_m(x, t') - u_m(x, t) \| < \varepsilon/3, \qquad |t' - t| < \delta, \qquad t', t \in [a, b]$$

Consequently, using Minkowski's inequality, we obtain

$$\| u(x, t') - u(x, t) \| \leq \| u(x, t') - u_m(x, t') \| + \| u_m(x, t') - u_m(x, t) \| + \| u_m(x, t) - u(x, t) \|$$
$$< \frac{\varepsilon}{3} + \frac{\varepsilon}{3} + \frac{\varepsilon}{3} = \varepsilon$$

for all $|t - t'| < \delta$, $t', t \in [a, b]$. The lemma is proved.

The sequence of functions $u_k(x, t)$, $k = 1, 2, \ldots$, is said to *converge in itself* in $\mathscr{L}_2(G)$ uniformly with respect to t on $[a, b]$ if

$$u_k - u_p \xrightarrow{t \in [a,b]} 0, k, \quad p \to \infty \quad \text{in} \quad \mathscr{L}_2(G)$$

LEMMA 2. *If the sequence of functions $u_k(x, t)$, $k = 1, 2, \ldots$, converges in itself in $\mathscr{L}_2(G)$ uniformly with respect to t on $[a, b]$, then there is a function $u(x, t)$, continuous in $\mathscr{L}_2(G)$ with respect to t on $[a, b]$, such that*

$$u_k \xrightarrow{t \in [a,b]} u, \quad k \to \infty \quad \text{in} \quad \mathscr{L}_2(G)$$

Proof. By the Riesz–Fischer theorem (cf. Sec. 1.5), for each $t \in [a, b]$ there is a function $u(x, t) \in \mathscr{L}_2(G)$ such that

$$u_k \to u, \quad k \to \infty \quad \text{in} \quad \mathscr{L}_2(G) \tag{20}$$

Moreover it is possible to choose a subsequence $u_{k_i}(x, t)$, $i = 1, 2, \ldots$, such that

$$\| u_{k_{i+1}}(x, t) - u_{k_i}(x, t) \| < \frac{1}{2^i}, \qquad t \in [a, b] \tag{21}$$

But, by virtue of (20), for each $t \in [a, b]$

$$u = \lim_{p \to \infty} u_{k_p} = u_{k_i} + (u_{k_{i+1}} - u_{k_i}) + (u_{k_{i+2}} - u_{k_{i+1}}) + \ldots$$

and so, by virtue of (21),

$$\| u - u_{k_i} \| \leq \| u_{k_{i+1}} - u_{k_i} \| + \| u_{k_{i+2}} - u_{k_{i+1}} \| + \ldots$$
$$< \frac{1}{2^i} + \frac{1}{2^{i+1}} + \ldots = \frac{1}{2^{i-1}}, \qquad i = 1, 2, \ldots$$

§ 29. EQUATION OF HYPERBOLIC TYPE

from which it follows that the subsequence $\{u_{k_i}\}$ converges to u in $\mathscr{L}_2(G)$ uniformly with respect to $t \in [a, b]$. And then from the inequality

$$\| u_k - u \| \leq \| u_k - u_{k_i} \| + \| u_{k_i} - u \|$$

we conclude that the sequence $\{u_k\}$ converges to the function u in $\mathscr{L}_2(G)$ uniformly with respect to t on $[a, b]$. By Lemma 1 of this subsection, the function $u(x, t)$ is continuous in $\mathscr{L}_2(G)$ with respect to t on $[a, b]$. The lemma is proved.

4. The Generalized Solution. Let there be sequences of functions $F_k \in C(\bar{\Pi}_\infty)$, $u_{k0} \in C^1(\bar{G})$, and $u_{k1} \in C(\bar{G})$, $k = 1, 2, \ldots$, such that for $k \to \infty$

$$F_k \xrightarrow{t \in [0,T]} F \quad \text{in} \quad \mathscr{L}_2(G) \quad \text{for any} \quad T > 0$$

$$u_{k0} \to u_0 \quad \text{in} \quad C(\bar{G}), \quad \text{grad } u_{k0} \to \text{grad } u_0 \quad \text{in} \quad \mathscr{L}_2(G) \quad (22)$$

$$u_{k1} \to u_1 \quad \text{in} \quad \mathscr{L}_2(G)$$

and for each $k = 1, 2, \ldots$, there is a classical solution $u_k(x, t)$ of the mixed problem

$$\varrho \frac{\partial^2 u_k}{\partial t^2} = -L u_k + F_k(x, t) \tag{1'}$$

$$u_k \big|_{t=+0} = u_{k0}(x), \quad \frac{\partial u_k}{\partial t}\bigg|_{t=+0} = u_{k1}(x) \tag{2'}$$

$$\alpha u_k + \beta \frac{\partial u_k}{\partial \mathbf{n}}\bigg|_S = 0 \tag{3'}$$

We shall prove that *there is a function $u(x, t)$, continuous in $\mathscr{L}_2(G)$ with respect to t on $[0, \infty)$, and such that for any $T > 0$*

$$u_k \xrightarrow{t \in [0,T]} u, \quad k \to \infty \quad \text{in} \quad \mathscr{L}_2(G) \tag{23}$$

The function $u(x, t)$ is said to be the *generalized solution* of the problem (1)-(2)-(3).

In fact, if we apply inequality (16) of the theorem of Sec. 29.2 to the difference $u_k - u_p$, for all $t \in [0, T]$ and $T > 0$ we obtain

$$\| u_k - u_p \| \leq C[(1 + T) \| u_{k0} - u_{p0} \|_C + T \| | \text{grad } u_{k0} - \text{grad } u_{p0} \| |$$
$$+ T \| u_{k1} - u_{p1} \| + \frac{T^2}{2} \max_{0 \leq t \leq T} \| F_k - F_p \|]$$

from which, by virtue of (22), it follows that the sequence $\{u_k\}$ converges in itself in $\mathscr{L}_2(G)$ uniformly with respect to t in $[0, T]$. By Lemma 2 of the preceding subsection there is a function $u(x, t)$, continuous in $\mathscr{L}_2(G)$ with respect to t on $[0, \infty)$, such that for any $T > 0$ limit result (23) is valid.

It follows from the definition of the generalized solution that each classical solution of the problem (1)-(2)-(3) is also its generalized solution and for the existence of the generalized solution the following conditions must be satisfied: F is continuous in $\mathscr{L}_2(G)$ with respect to t on $[0, \infty)$, $u_0 \in C(\bar{G})$, grad $u_0 \in \mathscr{L}_2(G)$, and $u_1 \in \mathscr{L}_2(G)$.

We shall now establish other properties of generalized solutions.

(a) *The generalized solution $u(x, t)$ of the problem (1)-(2)-(3) satisfies Eq. (1) in the generalized sense; that is, for any $\varphi \in \mathscr{D}(\varPi_\infty)$ the integral equality*

$$\int u(x, t)\left(\varrho \frac{\partial^2 \varphi}{\partial t^2} + L\varphi\right) dx\, dt = \int F(x, t)\varphi\, dx\, dt \qquad (24)$$

is satisfied.

In fact, let $\varphi \in \mathscr{D}(\varPi_\infty)$; then supp $\varphi \in \varPi_T$ for some $T > 0$. Multiplying Eq. (1′) by the function φ and integrating over the cylinder $\varPi_T$, we obtain

$$\int_{\varPi_T} \left[\varrho \frac{\partial^2 u_k}{\partial t^2} - \operatorname{div}(p \operatorname{grad} u_k) + qu_k\right] \varphi\, dx\, dt = \int_{\varPi_T} F_k \varphi\, dx\, dt$$

In the integral displayed in the left-hand side of this equation, by integrating by parts we may transfer the operation of the differentiation to the test function φ. Since φ becomes zero in the neighborhood of the boundary of the cylinder $\varPi_T$, the terms outside the integral disappear, and as a result we obtain

$$\int_{\varPi_T} u_k \left[\varrho \frac{\partial^2 \varphi}{\partial t^2} - \operatorname{div}(p \operatorname{grad} \varphi) + q\varphi\right] dx\, dt = \int_{\varPi_T} F_k \varphi\, dx\, dt, \quad k = 1, 2, \ldots$$

Allowing now that, by virtue of (23) and (22),

$$u_k \to u \quad \text{and} \quad F_k \to F, \quad k \to \infty \quad \text{in} \quad \mathscr{L}_2(\varPi_T)$$

and in the last equation proceeding to the limit as $k \to \infty$, we obtain integral equality (24).

(b) *The generalized solution $u(x, t)$ has the first (generalized) derivatives*

§ 29. EQUATION OF HYPERBOLIC TYPE

$\partial u/\partial t$, grad u, *continuous in* $\mathscr{L}_2(G)$ *with respect to* t *on* $[0, \infty)$, *and for all* $T > 0$

$$\frac{\partial u_k}{\partial t} \xrightarrow{t \in [0,T]} \frac{\partial u}{\partial t}, \quad \text{grad } u_k \xrightarrow{t \in [0,T]} \text{grad } u, \quad k \to \infty \quad \text{in} \quad \mathscr{L}_2(G) \qquad (25)$$

In fact, if we apply inequality (18) of the theorem of Sec. 29.2 to the difference $u_k - u_p$, for all $t \in [0, T]$ and $T > 0$ we obtain

$$\left\| \frac{\partial u_k}{\partial t} - \frac{\partial u_p}{\partial t} \right\| \leq C(\| u_{k0} - u_{p0} \|_C + \| |\text{grad } u_{k0} - \text{grad } u_{p0}| \| \\ + \| u_{k1} - u_{p1} \| + T \max_{0 \leq t \leq T} \| F_k - F_p \|)$$

from which, by virtue of (22), it follows that the sequence of derivatives $\partial u_k/\partial t$, $k = 1, 2, \ldots$, converges in itself in $\mathscr{L}_2(G)$ uniformly with respect to t on $[0, T]$ for all $T > 0$. By Lemma 2 of Sec. 29.3 there is a function $\tilde{u}(x, t)$, continuous in $\mathscr{L}_2(G)$ with respect to t on $[0, \infty)$, such that for all $T > 0$

$$\frac{\partial u_k}{\partial t} \xrightarrow{t \in [0,T]} \tilde{u}, \quad k \to \infty \quad \text{in} \quad \mathscr{L}_2(G) \qquad (26)$$

On the other hand, it follows from (23) that $u_k \to u$ as $k \to \infty$ in $\mathscr{D}'$ (the functions u_k and u are assumed to be continued as zero outside the cylinder $\bar{\varPi}_\infty$). From this, using the continuity in $\mathscr{D}'$ of the operation of generalized differentiation, [cf. Sec. 6.2, (e)], we conclude that

$$\frac{\partial u_k}{\partial t} \to \frac{\partial u}{\partial t}, \quad k \to \infty \quad \text{in} \quad \mathscr{D}'$$

Applying the result obtained to the test functions φ belonging to $\mathscr{D}(\varPi_\infty)$ and using result (26), we obtain

$$\int \tilde{u} \varphi \, dx \, dt \leftarrow \int \frac{\partial u_k}{\partial t} \varphi \, dx \, dt \to \left(\frac{\partial u}{\partial t}, \varphi \right), \quad k \to \infty$$

from which follows the equation (cf. Sec. 5.4)

$$\frac{\partial u}{\partial t} = \tilde{u}, \quad (x, t) \in \varPi_\infty$$

In this way, by virtue of (26), the first limiting result (25) has been proved. The second result of (25) is proved analogously.

(c) *The generalized solution $u(x, t)$ satisfies the initial conditions* (2) *in the $\mathscr{L}_2(G)$ sense; that is,*

$$\| u(x, t) - u_0(x) \| \to 0, \qquad \left\| \frac{\partial u(x, t)}{\partial t} - u_1(x) \right\| \to 0, \qquad t \to \infty \quad (27)$$

To prove this we shall proceed to the limit as $k \to \infty$ in the inequality

$$\| u(x, 0) - u_0(x) \| \le \| u(x, 0) - u_k(x, 0) \| + \| u_{k0}(x) - u_0(x) \|$$

using the limiting results $u_k(x, 0) \to u(x, 0)$ and $u_{k0}(x) \to u_0(x)$ in $\mathscr{L}_2(G)$. As a result we obtain $u(x, 0) = u_0(x)$. From this, since the function $u(x, t)$ is continuous in $\mathscr{L}_2(G)$ with respect to $t \in [0, \infty)$, we see that the first limiting result (27) is true. Analogously, using property (b), we obtain the second result of (27).

The sense in which the generalized solution $u(x, t)$ satisfies the boundary condition (3) is a question subject to further explanation and more precise definition.

5. Uniqueness and Continuous Dependence of the Generalized Solution.

We shall prove that *the results* (12), (13), *and* (14) *remain valid for the generalized solution $u(x, t)$ of the problem* (1)-(2)-(3).

In fact, let $u_k(x, t)$, $k = 1, 2, \ldots$, be a sequence of classical solutions of the problem (1')-(2')-(3'), converging to the generalized solution $u(x, t)$ in the sense of (23). Applying inequality (14) to the solutions u_k, we obtain

$$\| u_k \| \le \| u_{k0} \| + \sqrt{\frac{2}{\varrho_0} J_k(0)} \, t + \frac{1}{\varrho_0} \int_0^t (t - \tau) \| F_k \| \, d\tau, \qquad t \ge 0 \quad (28)$$

where

$$J_k^2(0) = \frac{1}{2} \int_G (\varrho u_{k1}^2 + p \, | \operatorname{grad} u_{k0} |^2 + q u_{k0}^2) \, dx + \frac{1}{2} \int_{S_0} p \frac{\alpha}{\beta} u_{k0}^2 \, dS \quad (29)$$

Using the fact that, by virtue of (23) and (22) (cf. Sec. 29.4),

$$\| u_k \| \xrightarrow{t \in [0, T]} \| u \|, \qquad \| F_k \| \xrightarrow{t \in [0, T]} \| F \|, \qquad T > 0 \quad \text{is arbitrary}$$

$$\| u_{k0} - u_0 \|_G \to 0, \qquad \| \, | \operatorname{grad} u_{k0} | \, \| \to \| \, | \operatorname{grad} u_0 | \, \|$$

$$\| u_{k1} \| \to \| u_1 \|, \qquad k \to \infty$$

and proceeding to the limit in (28) and (29), we see that result (14) is valid.

§ 29. EQUATION OF HYPERBOLIC TYPE

Results (12) and (13) are established analogously if the limiting results (25) are used.

From results (12) through (14), as for the classical solution, it follows that *the generalized solution of the problem* (1)-(2)-(3) *is unique and depends continuously on* u_0, u_1, *and F in the sense of the theorem of Sec.* 29.2.

6. Existence of the Generalized Solution. In Sec. 28.2 we constructed a formal solution of the problem (1)-(2)-(3) in the form of a Fourier series involving the eigenfunctions $\{X_j\}$ of the operator L,

$$u(x, t) = \sum_{j=1}^{\infty} T_j(t) X_j(x) \tag{30}$$

where

$$T_j(t) = a_j \cos \sqrt{\lambda_j} t + b_j \sin \sqrt{\lambda_j} t + \frac{1}{\sqrt{\lambda_j}} \int_0^t c_j(\tau) \sin \sqrt{\lambda_j}(t-\tau) \, d\tau \tag{31}$$

$$a_j = (u_0, X_j)_\varrho, \quad b_j = \frac{1}{\sqrt{\lambda_j}} (u_1, X_j)_\varrho, \quad c_j(t) = (F, X_j) \tag{32}$$

Let the boundary S of the region G and the coefficients p, q, α, and β be such that Theorem 1 of Sec. 19.4 is valid. We shall suppose, moreover, that $u_0 \in \mathcal{M}_L$, $u_1 \in \mathcal{L}_2(G)$ and F is continuous in $\mathcal{L}_2(G)$ with respect to t on $[0, \infty)$. We shall prove that in these conditions the series (30), representing the formal solution of problem (1)-(2)-(3), converges in $\mathcal{L}_2(G)$ uniformly with respect to t on $[0, T]$ for all $T > 0$ and defines the generalized solution $u(x, t)$ of this problem.

In fact, using the expansion theorems 1, 2, and 3 of Sec. 19.4 (cf. note), we shall represent the functions u_0, grad u_0, u_1, and F/ϱ as Fourier series involving the eigenfunctions $\{X_j\}$ of the operator L,

$$u_0(x) = \sum_{j=1}^{\infty} a_j X_j(x) \tag{33}$$

$$\text{grad } u_0(x) = \sum_{j=1}^{\infty} a_j \text{ grad } X_j(x), \quad u_1(x) = \sum_{j=1}^{\infty} \sqrt{\lambda_j} b_j X_j(x) \tag{34}$$

$$F(x, t) = \varrho(x) \sum_{j=1}^{\infty} c_j(t) X_j(x) \tag{35}$$

where a_j, b_j, and $c_j(t)$ are defined by formulas (32), and the functions

$c_j(t)$ are continuous on $[0, \infty)$ (cf. Sec. 29.3). On account of this the series (33) converges in $C(\bar{G})$ and the series (34) converge in $\mathscr{L}_2(G)$.

We shall prove that the series (35) converges in $\mathscr{L}_2(G)$ uniformly with respect to t on $[0, T]$ for any $T > 0$. In fact, for each $t \in [0, \infty)$, the following Parseval equation (cf. Sec. 1.6) is valid for the function $F(x, t)/\varrho(x)$:

$$\sum_{j=1}^{\infty} c_j^2(t) = \left\| \frac{F(x, t)}{\varrho(x)} \right\|_\varrho^2 = \int_G \frac{F^2(x, t)}{\varrho(x)} dx \tag{36}$$

Each term of the series (36) is a nonnegative continuous function $c_j^2(t)$ and this series converges to a continuous function (cf. Sec. 29.3). By Dini's lemma (cf. Sec. 1.3), the series (36) converges uniformly on any finite interval $[0, T]$. From this, evaluating the remainder of the series (35) in $\mathscr{L}_2(G)$,

$$\left\| \varrho(x) \sum_{j=k}^{\infty} c_j(t) X_j(x) \right\|^2 \leq \max_{x \in \bar{G}} \varrho(x) \left\| \sum_{j=k}^{\infty} c_j(t) \sqrt{\varrho(x)} X_j(x) \right\|^2$$

$$= \tilde{C} \sum_{j,l=k}^{\infty} c_j(t) c_l(t) (X_j, X_l)_\varrho = \tilde{C} \sum_{j=k}^{\infty} c_j^2(t)$$

we conclude that this series converges in $\mathscr{L}_2(G)$ uniformly with respect to $t \in [0, T]$ for any $T > 0$.

We shall denote by u_k, u_{k0}, u_{k1}, and F_k the partial sums of the series (30), (33), (34), and (35), respectively, involving k terms, for example:

$$u_k(x, t) = \sum_{j=1}^{k} T_j(t) X_j(x), \qquad k = 1, 2, \ldots$$

Since

$$T_j''(t) = -\lambda_j T_j(t) + c_j(t), \qquad T_j \in C^2([0, \infty))$$

$$LX_j = \lambda_j \varrho X_j, \qquad \alpha X_j + \beta \frac{\partial X_j}{\partial \mathbf{n}} \bigg|_S = 0$$

$$X_j \in C^2(G) \cap C^1(\bar{G})$$

then the functions u_k belonging to $C^2(\mathit{Ц}_\infty) \cap C^1(\bar{\mathit{Ц}}_\infty)$ satisfy Eq. (1')

$$\varrho \frac{\partial^2 u_k}{\partial t^2} + L u_k = \sum_{j=1}^{k} (\varrho T_j'' X_j + T_j L X_j)$$

$$= \sum_{j=1}^{k} (-\lambda_j \varrho T_j X_j + \varrho c_j X_j + \lambda_j \varrho T_j X_j) = \varrho \sum_{j=1}^{k} c_j X_j = F_k(x, t)$$

§ 29. EQUATION OF HYPERBOLIC TYPE

boundary condition (3'), and initial conditions (2')

$$u_k\big|_{t=+0} = \sum_{j=1}^{k} a_j X_j(x) = u_{k0}(x)$$

$$\frac{\partial u_k}{\partial t}\bigg|_{t=+0} = \sum_{j=1}^{k} \sqrt{\lambda_j} b_j X_j(x) = u_{k1}(x)$$

In this way, we have constructed a sequence $u_k(x, t)$, $k = 1, 2, \ldots$, of classical solutions of the problem (1')-(2')-(3') such that the limiting results (22) are valid. It was proved in Sec. 29.4 that this sequence [and, consequently, the formal series (30)] converges in $\mathscr{L}_2(G)$ uniformly with respect to t on $[0, T]$ for all $T > 0$ to the generalized solution $u(x, t)$ of the problem (1)-(2)-(3). The generalized solution $u(x, t)$ which has been constructed has the properties (a)–(c) which were established in Sec. 29.4. So the following theorem is proved:

THEOREM. *If $u_0 \in \mathscr{M}_L$, $u_1 \in \mathscr{L}_2(G)$, and F is continuous in $\mathscr{L}_2(G)$ with respect to t on $[0, \infty)$, then the generalized solution of the problem (1)-(2)-(3) exists and is represented as the series (30), which is the formal solution of this problem.*

If we impose more stringent demands on the data of the problem (1)-(2)-(3), then it is possible to prove that the formal solution (30) defines a function $u(x, t)$ of the class $C^2(\Pi_\infty) \cap C^1(\bar{\Pi}_\infty)$, and in this way represents the classical solution of this problem (cf. the exercises below).

7. Exercises. (a) Prove: If the series (30) and the series obtained by single differentiation with respect to all the independent variables converges uniformly in any cylinder $\bar{\Pi}_T$, while the series obtained by double differentiation converge uniformly over any compactum $K \subset \Pi_T$, then the series (30) defines the classical solution of the problem (1)-(2)-(3).

We shall consider the mixed problem in the half-strip $\Pi_\infty = (0, l) \times (0, \infty)$

$$\varrho \frac{\partial^2 u}{\partial t^2} = \frac{\partial}{\partial x}\left(p \frac{\partial u}{\partial x}\right) - qu \tag{37}$$

$$u\big|_{t=+0} = u_0(x), \quad \frac{\partial u}{\partial t}\bigg|_{t=+0} = u_1(x) \tag{38}$$

$$u\big|_{x=0} = u\big|_{x=l} = 0 \tag{39}$$

(b) Prove: If $u_0 \in \mathscr{M}_L$ and $u_1 \in \mathscr{L}_2(0, l)$, then the series (30) for the problem (37)-(38)-(39) converges uniformly on $\bar{\Pi}_\infty$, and therefore the generalized solution $u \in C(\bar{\Pi}_\infty)$.

(c) Prove: If u_0, u_0' and u_1 belong to $\mathcal{M}_L$, then the series (30) represents the classical solution of the problem (37)-(38)-(39). [Note: Use Exercise (a) and Mercer's theorem; cf. Exercise (b) of Sec. 18.9.]

§ 30. The Mixed Problem for an Equation of Parabolic Type

In this section we shall consider a mixed problem for an equation of parabolic type (cf. Sec. 4.5)

$$\varrho \frac{\partial u}{\partial t} = \operatorname{div}(p \operatorname{grad} u) - qu + F(x, t) = -Lu + F(x, t) \tag{1}$$

$$(x, t) \in \varPi_\infty = G \times (0, \infty)$$

$$u\big|_{t=+0} = u_0(x), \qquad x \in \bar{G} \tag{2}$$

$$u\big|_S = v(x, t), \qquad (x, t) \in S \times [0, \infty) \tag{3}$$

under the conditions of Sec. 28.

1. Classical Solutions. The Maximum Principle. The function $u(x, t)$ of the class $C^2(\varPi_\infty) \cap C(\bar{\varPi}_\infty)$ which satisfies Eq. (1) in the cylinder $\varPi_\infty$, the initial condition (2), and boundary condition (3) is said to be the *classical solution* of the mixed problem (1)-(2)-(3).

The smoothness conditions

$$F \in C(\varPi_\infty), \qquad u_0 \in C(\bar{G}), \qquad v \in C(S \times [0, \infty))$$

and the consistency conditions

$$u_0\big|_S = v(x, 0)$$

are necessary conditions for the existence of the classical solution of the problem (1)-(2)-(3).

In studying boundary value problems for an equation of parabolic type the following maximum principle is extremely useful.

MAXIMUM PRINCIPLE. *Let the function $u(x, t)$ of the class $C^2(\varPi_\infty)$ $\cap C(\bar{\varPi}_\infty)$ satisfy Eq. (1) in $\varPi_\infty$. Let T be an arbitrary positive number. Then, if $F(x, t) \leq 0$ in the cylinder $\varPi_T$, either $u \leq 0$ on $\bar{\varPi}_T$ or the function*

§ 30. EQUATION OF PARABOLIC TYPE

$u(x, t)$ assumes its (positive) maximum on the cylinder $\bar{\Pi}_T$ on the lower base $\bar{G} \times \{0\}$ or on the side surface $S \times [0, T]$; that is,

$$u(x, t) \leq \max[0, \max_{x \in \bar{G}, t=0} u(x, t), \max_{x \in S, 0 \leq t \leq T} u(x, t)], \quad (x, t) \in \bar{\Pi}_T \quad (4)$$

Proof. Let us suppose the opposite, that is, that the function $u(x, t)$ assumes positive values at certain points of the cylinder $\bar{\Pi}_T$ but does not attain its (positive) maximum either on the lower base $\bar{G} \times \{0\}$ or on the side surface $S \times [0, T]$ of the cylinder. This means that there is a point (x_0, t_0), $x_0 \in G$, $0 < t_0 \leq T$, such that

$$u(x_0, t_0) > \max[0, \max_{x \in \bar{G}, t=0} u(x, t), \max_{x \in S, 0 \leq t \leq T} u(x, t)] = M \geq 0 \quad (5)$$

Writing

$$\varepsilon = u(x_0, t_0) - M > 0 \quad (6)$$

we shall construct the function

$$v(x, t) = u(x, t) + \frac{\varepsilon}{2} \frac{T - t}{T}$$

Then

$$v(x, t) \leq u(x, t) + \varepsilon/2, \quad (x, t) \in \bar{\Pi}_T$$

and, by virtue of (6), for all (x, t) belonging to $\bar{G} \times \{0\}$ or $S \times [0, T]$ we have

$$v(x_0, t_0) \geq u(x_0, t_0) = \varepsilon + M \geq \varepsilon + u(x, t)$$
$$\geq \varepsilon + v(x, t) - \varepsilon/2 = \varepsilon/2 + v(x, t)$$

From this it follows that the function v also assumes its (positive) maximum value on $\bar{\Pi}_T$ at a certain point (x', t'), $x' \in G$, $0 < t' \leq T$, and

$$v(x', t') \geq v(x_0, t_0) \geq \varepsilon + M \geq \varepsilon \quad (7)$$

We shall write out the necessary conditions for a maximum of the function v at the point (x', t'):

$$\frac{\partial v}{\partial t} \geq 0, \quad \operatorname{grad} v = 0, \quad \Delta v \leq 0$$

From these conditions and also from inequality (7) it follows that at this point

$$\varrho \frac{\partial u}{\partial t} - \text{div}(p \text{ grad } u) + qu - F$$

$$= \varrho \frac{\partial v}{\partial t} - p \Delta v - (\text{grad } p, \text{grad } v) + qv - F + \frac{\varepsilon}{2}\left(\frac{\varrho}{T} - q\frac{T-t}{T}\right)$$

$$\geq qv + \frac{\varepsilon}{2}\left(\frac{\varrho}{T} - q\frac{T-t}{T}\right) = q\varepsilon\left(1 - \frac{T-t}{2T}\right) + \frac{\varepsilon\varrho}{2T} > 0$$

which contradicts Eq. (1). This means that inequality (5) is untrue and therefore the opposite inequality (4) is true, as was to be shown.

Replacing u by $-u$, we obtain the following minimum principle from the maximum principle.

MINIMUM PRINCIPLE. *If the function $u(x, t)$ of the class*

$$C^2(Ц_\infty) \cap C(\bar{Ц}_\infty)$$

satisfies Eq. (1) in $Ц_\infty$ and $F \geq 0$ in $Ц_\infty$, then the inequality

$$u(x, t) \geq \min[0, \min_{x \in \bar{G}, t=0} u(x, t), \min_{x \in S, 0 \leq t \leq T} u(x, t)], \quad (x, t) \in \bar{Ц}_T \quad (4')$$

is true for all $T > 0$.

2. Uniqueness and Continuous Dependence of the Classical Solution. We shall apply the maximum and minimum principles to establish the uniqueness and the continuous dependence of the classical solution of the mixed problem (1)-(2)-(3).

Let $u(x, t)$ be the classical solution of the problem (1)-(2)-(3) and $F \in C(\bar{Ц}_\infty)$. We shall fix $T > 0$ and write

$$M = \|F\|_{C(\bar{Ц}_T)}, \quad M_1 = \|v\|_{C(S \times [0,T])}, \quad M_0 = \|u_0\|_{C(\bar{G})}$$

We construct the function

$$\chi(x, t) = u(x, t) - \frac{M}{\varrho_0}t, \quad \varrho_0 = \min_{x \in \bar{G}} \varrho(x) > 0 \quad (8)$$

The function χ is the classical solution of the mixed problem (1)-(2)-(3) with a change of F and v to $F - (\varrho/\varrho_0)M - (q/\varrho_0)Mt$ and $v - (M/\varrho_0)t$, respectively. Allowing that

$$F - \frac{\varrho}{\varrho_0} M - \frac{q}{\varrho_0} Mt \leq 0, \qquad (x, t) \in \bar{\Pi}_T$$

$$v - \frac{M}{\varrho_0} t \leq M_1, \qquad (x, t) \in S \times [0, T]$$

and using inequality (4), we obtain the result

$$\chi \leq \max(M_0, M_1)$$

that is, by virtue of (8),

$$u(x, t) \leq \frac{M}{\varrho_0} T + \max(M_0, M_1), \qquad (x, t) \in \bar{\Pi}_T$$

Analogously, introducing the function

$$\chi_1(x, t) = u(x, t) + \frac{M}{\varrho_0} t$$

and using inequality (4'), we obtain the opposite result

$$u(x, t) \geq -\frac{M}{\varrho_0} T - \max(M_0, M_1), \qquad (x, t) \in \bar{\Pi}_T$$

So, if $u(x, t)$ is the classical solution of the problem (1)-(2)-(3) and $F \in C(\bar{\Pi}_\infty)$, then for any $T > 0$ result

$$\| u \|_{C(\bar{\Pi}_T)} \leq \max[\| u_0 \|_{C(\bar{G})}, \| v \|_{C(S\times[0,T])}] + \frac{T}{\varrho_0} \| F \|_{C(\bar{\Pi}_T)} \qquad (9)$$

is true.

Using this result we shall prove the following theorem.

THEOREM. *The classical solution of the problem (1)-(2)-(3) is unique and depends continuously on u_0, v, and F in the sense that if*

$$\| F - \tilde{F} \|_{C(\bar{\Pi}_T)} \leq \varepsilon, \quad \| u_0 - \tilde{u}_0 \|_{C(\bar{G})} \leq \varepsilon_0, \quad \| v - \tilde{v} \|_{C(S\times[0,T])} \leq \varepsilon_1 \qquad (10)$$

then the respective (classical) solutions $u(x, t)$ and $\tilde{u}(x, t)$ satisfy the inequality

$$\| u - \tilde{u} \|_{C(\bar{\Pi}_T)} \leq \max(\varepsilon_0, \varepsilon_1) + \frac{T}{\varrho_0} \varepsilon \qquad (11)$$

Proof. The uniqueness of the solution follows from the fact that, by virtue of result (9), the homogeneous problem (1)-(2)-(3) (for $u_0 = 0$, $v = 0$, and $F = 0$) has only a zero classical solution (cf. Sec. 1.9).

To prove continuous dependence we shall construct the difference $\eta = u - \tilde{u}$. The function η is the classical solution of the problem (1)-(2)-(3) with a change of F, u_0, and v to $F - \tilde{F}$, $u_0 - \tilde{u}_0$, and $v - \tilde{v}$, respectively. Applying inequality (9) to the function η and using results (10), we obtain result (11). The theorem is proved.

3. The Generalized Solution. As with the equation of hyperbolic type, we shall introduce the concept of a generalized solution.

Let there be sequences of functions $F_k \in C(\bar{\Pi}_\infty)$, $v_k \in C(S \times [0, \infty))$, and $u_{k0} \in C(\bar{G})$, $k = 1, 2, \ldots$, such that for $k \to \infty$

$$F_k \to F \text{ in } C(\bar{\Pi}_T), \qquad v_k \to v \text{ in } C(S \times [0, T]) \qquad (12)$$
$$\text{for any } T > 0; \qquad u_{k0} \to u_0 \text{ in } C(\bar{G})$$

and for each $k = 1, 2, \ldots$, there is a classical solution of the mixed problem

$$\varrho \frac{\partial u_k}{\partial t} = -Lu_k + F_k(x, t) \tag{1'}$$

$$u_k |_{t=+0} = u_{k0}(x) \tag{2'}$$

$$u_k |_S = v_k(x, t) \tag{3'}$$

We shall prove that *there is a function $u(x, t)$ continuous in the cylinder $\bar{\Pi}_\infty$ and such that for any $T > 0$*

$$u_k \to u, \qquad k \to \infty \text{ in } C(\bar{\Pi}_T) \tag{13}$$

The function $u(x, t)$ is said to be the *generalized solution* of the problem (1)-(2)-(3).

In fact, if we apply inequality (11) of the theorem of Sec. 30.2 to the difference $u_k - u_p$, for all $T > 0$ we obtain

$$\| u_k - u_p \|_{C(\bar{\Pi}_T)} \leq \max[\| u_{k0} - u_{p0} \|_{C(\bar{G})}, \| v_k - v_p \|_{C(S \times [0,T])}]$$
$$+ \frac{T}{\varrho_0} \| F_k - F_p \|_{C(\bar{\Pi}_T)}$$

from which, by virtue of (12), it follows that the sequence $\{u_k\}$ converges

§ 30. EQUATION OF PARABOLIC TYPE

in itself in $C(\bar{\Pi}_T)$. Therefore there is a function $u(x, t)$ continuous in $\bar{\Pi}_\infty$ and such that the sequence $\{u_k\}$ converges to u in $C(\bar{\Pi}_T)$ for any $T > 0$ (cf. Sec. 1.3).

From the definition of the generalized solution of the problem (1)-(2)-(3) it follows that (cf. Sec. 29.4): Each classical solution of this problem is its generalized solution; for the existence of the generalized solution the following conditions must be satisfied:

$$F \in C(\bar{\Pi}_\infty), \qquad v \in C(S \times [0, \infty)), \qquad u_0 \in C(\bar{G}), \qquad v(x, 0) = u_0|_S$$

the generalized solution satisfies initial condition (2) and boundary condition (3) at each point; the generalized solution satisfies Eq. (1) in a generalized sense, that is, for any $\varphi \in \mathscr{D}(\Pi_\infty)$ the integral equality

$$\int u(x, t)\left(-\varrho \frac{\partial \varphi}{\partial t} + L\varphi\right) dx\, dt = \int F(x, t)\varphi\, dx\, dt \qquad (14)$$

is satisfied.

4. Uniqueness and Continuous Dependence of the Generalized Solution. We shall prove that result (9) remains true for the generalized solution $u(x, t)$ of the problem (1)-(2)-(3).

In fact, let $u_k(x, t)$, $k = 1, 2, \ldots$, be the sequence of classical solutions of the problem (1)-(2)-(3) converging uniformly to the generalized solution $u(x, t)$ on any cylinder $\bar{\Pi}_T$. Applying result (9) to the solutions u_k, for all $T > 0$ we obtain

$$\| u_k \|_{C(\bar{\Pi}_T)} \leq \max[\| u_{k0} \|_{C(\bar{G})}, \| v_k \|_{C(S\times[0,T])}] + \frac{T}{\varrho_0} \| F_k \|_{C(\bar{\Pi}_T)} \qquad (15)$$

Allowing the limiting results (12) and (13) and proceeding to the limit in inequality (15) as $k \to \infty$, we see that result (9) is valid.

The uniqueness of the generalized solution of the problem (1)-(2)-(3) and its continuous dependence on u_0, v, and F in the sense of the theorem of Sec. 30.2 follow from Eq. (9).

5. Existence of the Generalized Solution. We shall prove the existence of the generalized solution of the mixed problem (1)-(2)-(3) for $F = 0$ and $v = 0$:

$$\varrho \frac{\partial u}{\partial t} = -Lu, \qquad u|_{t=+0} = u_0(x), \qquad u|_S = 0 \qquad (16)$$

In Sec. 28.3 we constructed the formal solution of problem (16) in the form of a Fourier series involving the eigenfunctions $\{X_j\}$ of the operator L,

$$u(x, t) = \sum_{j=1}^{\infty} a_j e^{-\lambda_j t} X_j(x), \qquad a_j = (u_0, X_j)_\varrho \qquad (17)$$

Let the boundary S of the region G and the coefficients ϱ, p, and q be such that Theorem 1 of Sec. 19.4 is valid. We shall suppose moreover that $u_0 \in \mathcal{M}_L$. We shall prove that in these conditions the series (17), representing the formal solution of problem (16), converges uniformly on any cylinder $\bar{\Pi}_T$ and defines the generalized solution $u(x, t)$ of this problem.

In fact, if we use the expansion theorem 1 of Sec. 19.4 (cf. note), we may represent the function u_0 in the form of a Fourier series uniformly converging on $\bar{G}$ involving the eigenfunctions of the operator L,

$$u_0(x) = \sum_{j=0}^{\infty} a_j X_j(x) \qquad (18)$$

We shall denote by u_k and u_{k0} the partial sums of the series (17) and (18), respectively, over k terms. The functions u_k, $k = 1, 2, \ldots$, are the classical solutions of problem (16) with a change of u_0 to u_{k0}, and $u_{k0} \to u_0$ as $k \to \infty$ in $C(\bar{G})$. In Sec. 30.3 it was proved that the sequence $\{u_k\}$ [and, consequently, the formal series (17)] converges uniformly on each cylinder $\bar{\Pi}_T$ to the generalized solution $u(x, t)$ of problem (16). So the following theorem has been established.

THEOREM. *If $u_0 \in \mathcal{M}_L$, then the generalized solution of problem* (16) *exists and is represented by series* (17), *which is the formal solution of this problem.*

Note. Using the fact that the fundamental solution $\mathscr{E}(x, t)$ of the heat conduction equation is infinitely differentiable for $(x, t) \neq (0, 0)$ (cf. Sec. 14.1), and proceeding as for the case of harmonic functions (cf. Sec. 21.7), it may be shown that each continuous function which satisfies the heat conduction equation in the generalized sense in a certain region is infinitely differentiable in this region (cf. also S. L. Sobolev (*I*, Chap. XXII); cf. with Note 1 of Sec. 14.4]. It follows from this that in the conditions of the theorem of this subsection, the generalized solution of the mixed problem for the heat conduction equation belongs to the class $C^\infty(\Pi_\infty)$ and is, therefore, the classical solution of this problem.

Bibliography

Arsenin, V. Ya.
[1] *Basic Equations and Special Functions of Mathematical Physics*, Iliffe, 1968.

Bogolyubov, N. N. and Shirkov, D. V.
[1] *Introduction to the Theory of Quantized Fields*, Goztekizdat, 1957; Wiley, New York, 1959.

Budak, B. M., Samarsky, A. A. and Tikhonov, A. N.
[1] *Collection of Problems in Mathematical Physics*, Goztekizdat, 1956.

Courant, R.
[1] *Methods of Mathematical Physics*, Vol. II, Wiley-Interscience, New York, 1962.

Evgrafov, M. A.
[1] *Analytic Functions*, Nauka, 1965.

Fichtengolz, G. M.
[1] *Course of Differential and Integral Calculus*, Vols. I–III, Pergamon, New York, 1965.

Gelfand, I. M. and Shilov, G. E.
[1] *Generalized Functions*, 2nd ed., Academic Press, New York, 1964.

Gradshteyn, I. S. and Ryzhik, I. M.
[1] *Tables of Integrals, Series, and Products*, Academic Press, New York, 1965.

Hörmander, L.
[1] *Linear Partial Differential Operators*, Academic Press, New York, 1963.
[2] "On the Division of Distributions by Polynomials," *Arkiv. matem.* **3** (1958), 555-568.

Kolmogorov, A. N. and Fomin, S. V.
[1] *Elements of the Theory of Functions and Functional Analysis*, Vols. I, II, Graylock, 1957, 1961.

Koshliakov, N. S., Gliner, E. B. and Smirnov, M. M.
[1] *Basic Differential Equations of Mathematical Physics*, Fizmatgiz, 1962.

Kudriavtsev, L. D.
[1] *Basis of Mathematical Analysis*, Vysshaya Shkola, 1970.

Landau, L. D. and Lifschitz, E. M.
[1] *Quantum Mechanics. Nonrelativistic Theory*, Vol. III, Pergamon, New York, 1965.

Lavrentiev, M. A. and Shabat, B. V.
[1] *Methods of the Theory of Functions of the Complex Variable*, Fizmatgiz, 1958.

Lavrentiev, M. M.
[1] *Some Improperly Posed Problems of Mathematical Physics*, Springer, Berlin, 1967.

Maltsev, A. I.
[1] *Basis of Linear Algebra*, Goztekizdat, 1956.

Marchuk, G. I.
[1] *Numerical Methods for Nuclear Reactor Calculations*, Consultants Bureau, New York, 1959.

Miranda, K.
[1] *Equazioni alle derivate parziali di tipo ellittico*, Springer, Berlin, 1955.

Petrovsky, I. G.
[1] *Lectures on Partial Differential Equations*, Wiley-Interscience, New York, 1954.
[2] *Lectures on the Theory of Integral Equations*, Graylock, 1963.

Polozhy, G. N.
[1] *Equations of Mathematical Physics*, Vysshaya Shkola, 1964.

Pontryagin, L. S.
[1] *Ordinary Differential Equations*, Prentice-Hall, Englewood Cliffs, N. J., 1966.

Reisz, F. and Nagy, B. Sz.
[1] *Lectures on Functional Analysis*, Ungar, New York, 1955.

Schwartz, L.
[1] *Mathematics for the Physical Sciences*, Addison-Wesley, Reading, Mass., 1967.
[2] *Théorie des distributions*, Vols. I–II, Paris, 1950–1951.

Shilov, G. E.
[1] *Mathematical Analysis*, Pergamon, New York, 1965.
[2] *Mathematical Analysis, Second Special Course*, Nauka, 1965.

Smirnov, M. M.
[1] *Problems on the Equations of Mathematical Physics*, Noordhoff, Groningen, The Netherlands, 1966.

Smirnov, V. I.
[1] *Course of Higher Mathematics*, Vol. II, Pergamon, New York, 1965.
[2] *Course of Higher Mathematics*, Vol. IV, Pergamon, New York, 1965.
[3] *Course of Higher Mathematics*, Vol. III, Pt. 2, Pergamon, New York, 1965.

Sobolev, S. L.
[1] *Partial Differential Equations of Mathematical Physics*, Pergamon, New York, 1964.
[2] "Méthode nouvelle à résoudre le problème de Cauchy pour les équations linéaires hyperboliques normales," *Math. Collection (Math. Sb.)* **1** (43) (1936), 39–72.

Tikhonov, A. N. and Samarsky, A. A.
[1] *Equations of Mathematical Physics*, Pergamon, New York, 1963.

Tychonoff, A.
[1] "Theorèmès d'unicité pour l'équation de la chaleur," *Math. Collection (Math. Sb.)* **42** (1935), 199–216.

Vekua, I. N.
[1] *Generalized Analytic Functions*, Pergamon, New York, 1962.
[2] "Ultraharmonic Functions," *Proc. of the Tbilisi Math. Inst.*, **XII** (1943), 105–174

Vladimirov, V. I.
[1] "Mathematical Problems of the One-Velocity Theory of Transfer of Particles," *Proc. of the Math. Inst. of the Acad. of Sci. of the USSR,* **61** (1961).

Index

A

Arzela's lemma, 233
Associated Legendre functions, 339

B

Bessel's inequality, 17
Bilinear expansion of an Hermitian continuous kernel, 242
Bilinear expansion of iterated kernels, 241
Bilinear form, 25
Bolzano-Weierstrass theorem, 2
Boundary value problem, 53
Boundary value problem for equations of elliptic type, 56

C

Cauchy-Buniakowski inequality, 12
Cauchy principle of convergence, 2
Cauchy's problem, 53
Cauchy's theorem, 5
Characteristic cone, 43
Characteristic equation, 43
Characteristic function of a set, 2
Characteristic numbers of a kernel, 202
Characteristic surface, 43
Closure equation, 17

Closure of a set, 2
Compactum, 2
Complete system of functions, 17
Continuation of a linear functional, 21
Continuity, equation of, 35
Convolution of generalized functions, 102, 104, 105
Convolution of generalized functions of slow growth, 121

D

D'Alembert's formula, 176
Derivative, correct normal, 281
Derivative of a generalized function, 80
Descent, method of, 144
Differential equations of characteristics, 47
Diffusion equation, 30
Diffusion of heat in a bounded rod, 382
Dini's lemma, 5
Dirac's equation, 38
Dirac δ-function, 64
Direct product of generalized functions, 96
Direct product of generalized functions of slow growth, 96
Dirichlet, external problem, 307
Dirichlet, internal problem, 307
Distributions, 75
Du Bois Reymond's lemma, 72

E

Eigenfunctions of a kernel, 202
Eigenfunctions of an operator, 24
Eigenvalues of a kernel, 202
Eigenvalues of an operator, 24
Elliptic equation, 380
Elliptic type, equation of, 40, 49
Energy integral, 259
Euler's equation of motion, 35

F

Fourier coefficients, 16
Fourier method, 263, 373
Fourier series, 16
Fourier transform of convolution, 129
Fourier transform of functions belonging to $\mathscr{S}$, 121
Fourier transform of generalized functions belonging to $\mathscr{S}'$, 123
Fredholm's alternative, 225
Fredholm integral equations, 201
  adjoint, 201
  homogeneous, 201
Fredholm theorems, 233, 225, 226, 227, 229
Fubini's theorem, 11
Fundamental solution, 141, 147
  of the Cauchy-Riemann operator, 154
  of the heat conduction operator, 148
  of the Helmholtz operator, 153
  of the Laplace operator, 151
  of the transport operator, 154
  of the wave operator, 149
Function with compact support, 4
Functional, linear, 21

G

Generating function for Legendre polynomials, 337
Generalized Cauchy problem for heat conduction equation, 197
Generalized Cauchy problem for wave equation, 172

Generalized function, 69
  regular, 72
  singular, 72
  of slow growth, 116
Generalized harmonic function, 287
Generalized solution of linear differential equation, 140
Green's first formula, 256
Green's formula, 279
Green's second formula, 257
Green's function, of a boundary value problem, 274
  of a Dirichlet problem, 318, 358
  construction, 322
  of an operator, 273

H

Hadamard's example, 60
Harmonic function, 278
Heat conduction equation, 31, 44
Heine-Borel lemma, 3
Helmholtz equation, 33, 362
Helmholtz equation, homogeneous, 363
Hilbert-Schmidt theorem, 237, 240, 247, 249
Hölder continuous function, 5
Huygens principle, 179
Hydrodynamics, equations of, 35
Hyperbolic homogeneous equation, 374
Hyperbolic inhomogeneous equation, 376
Hyperbolic type equations, 40, 46

I

Influence, function of, 141
Interior point of a set, 2
Integral equation, 201
  with degenerate kernel, 217
  Fredholm, 201
  with Hermitian kernel, 202
  with polar kernel, 211
  Volterra, 209
Iterates of a function, 204

Q

Quadratic form, 25

R

Reflections, method of, 322
   finite string, 191
   semi-infinite string, 198
Region, 3
Regularization of generalized functions, 108
Riesz-Fischer theorem, 14
Resolvent, 207

S

Schmidt's formula, 244
Schmidt's orthogonalization process, 16
Schrodinger's equation, 37, 379
Schwartz's theorem, 115
Sequence, converging, 2
   in itself, 2
Sequence of functions, converging in the mean, 13
Set, bounded, 2
   closed, 2
   connected, 2
   equicontinuous functions, 5
   of functions continuous in $\mathscr{L}_2(G)$, 13
   measurable, 6
   of measure zero, 6
   open, 2
Small longitudinal vibrations of a rod, 29
Small transverse vibrations of a membrane, 30
Small transverse vibrations of a string, 29
Sokhotski's formula, 75
Sommerfeld radiation conditions, 362
Space, C^p, $p \geq 1$, 3
   $C^p(G)$, 3
   $C^p(\bar{G})$, 3

$C(T)$, T closed, 4
$\mathscr{D}$, 66
$\mathscr{D}'$, 69
$\mathscr{D}(G)$, 4
$\mathscr{L}$, 113
$\mathscr{L}'$, 114
$\mathscr{L}_2(G)$, 12
Spherical function, 333, 345
State, equation of, 35
Steklov's theorem, 277
Sturm-Liouville problem, 270
Sufficiently-smooth surface, 310
Summable function, 6, 8
Superposition of waves, principle of, 179
Support of a generalized function, 70
Support of a piecewise continuous function, 4

T

Telegrapher's equation, 36
Test functions, 63, 65, 113
Transformation, Fourier, of convolution, 129
   of the function belonging to $\mathscr{L}$, 121
   of the generalized function belonging to $\mathscr{L}'$, 123
Transport equation, 33
Tricomi's equation, 41, 51
Types of equations, 40

V

Vibration equations, 27
Vibration of a fixed membrane, 383
Vibration of a spherical volume, 384
Vibration of a string, 381

W

Wave equation, 29
Weak convergence, 70
Weierstrass theorem, 5

INDEX

J
Jentsch's theorem, 249

K
Kellogg's method, 251
Kelvin's transformation, 309
Kernel, 21
  continuous, 202
  degenerate, 217
  Hermitian, 231
  of integral equation, 201
  of positive type, 246
  polar, 211
  Volterra, 209
  weakly polar, 211
Kernels, Hermitian adjoint, 202
  iterated, 206
  symmetric, 249
Kirchhoff's formula, 176
Klein-Gordon equation, 38
Kowalewski's theorem, 59

L
Laplace equation, 33
Laplace's formula, 343
Lebesgue integrable function, 5, 6
Lebesgue integral, 5, 6
Lebesgue theorem, 8
Legendre polynomials, 335
Liapunov surface, 297
Linear normed space, 4
Lipschitz continuous function, 5

M
Maximum principle, 283, 285, 405
Maxwell's equations, 36
Mean value theorem, 283
Measurable function, 6
Minimum principle, 406
Mixed problem, 54, 57
Mixed type, equation of, 41
Multiplicity of an eigenvalue, 24

N
Neumann, external problem, 307
  internal problem, 307
Neumann series, 204
Normalized function, 15
Normally hyperbolic type equation, 40
Normally parabolic type equation, 40

O
Operator, bounded from $\mathcal{M}$ to $\mathcal{N}$, 20
  Cauchy-Riemann, 94
  continuous from $\mathcal{M}$ to $\mathcal{N}$, 20
  Hermitian (Lagrange self-adjoint), 25
  Laplace Δ, 30
  linear, 20
    differential, or order m, 22
    integral, 21
    integrodifferential, 22
  positive, 26
  wave (D'Alembert $\square_a$), 30
Orthogonal functions, 15
Orthonormal system of functions, 15

P
Parabolic type equations, 40, 49, 378
Parseval's equality, 17
Poisson's equation, 33, 44
Poisson's formula, 186, 325, 326
Potential, double layer, 111, 293, 302, 303
  heat with density f, 194
  logarithmic, 110, 350
  Newtonian, 110, 291
  Newtonian, physical sense, 296
  retarded, 163
  Robin's, 316
  Robin's, physical sense, 317
  simple layer, 111, 293, 300, 304
  surface retarded, 167
Potentials, 365
Principle, of limiting absorption, 368
  of amplitude, 369
Propagation of waves, 178
Property fulfilled almost everywhere, 5